COMPETITION AND COOPERATION IN SOCIAL AND POLITICAL SCIENCES

PROCEEDINGS OF THE ASIA PACIFIC RESEARCH IN SOCIAL AND HUMANITIES, DEPOK, INDONESIA, 7–9 NOVEMBER 2016: TOPICS IN SOCIAL AND POLITICAL SCIENCES

Competition and Cooperation in Social and Political Sciences

Editors

Isbandi Rukminto Adi & Rochman Achwan

Faculty of Social and Political Sciences, Universitas Indonesia, Indonesia

Routledge
Taylor & Francis Group

LONDON AND NEW YORK

First published 2018 by CRC Press/Balkema

2 Park Square, Milton Park, Abingdon, Oxon, OX14 4RN
605 Third Avenue, New York, NY 10017

Routledge is an imprint of the Taylor & Francis Group, an informa business

First issued in paperback 2020

Typeset by V Publishing Solutions Pvt Ltd., Chennai, India

ISBN: 978-1-138-62676-8 (hbk)
ISBN: 978-0-367-73549-4 (pbk)

Table of contents

Competition and Cooperation in Social and Political Sciences – Adi & Achwan (Eds)
© 2018 Taylor & Francis Group, London, ISBN 978-1-138-62676-8

Preface

This Conference Proceedings volume contains the written versions of most of the contributions presented during the 1st Asia-Pacific Research on Social Sciences and Humanities (APRISH). The Conference took place at the Margo Hotel in Depok, 7–9 November 2016. The main theme of the APRISH Conference was "Competition and Collaboration in Globalized World". This book contains articles related to the Social and Political Issues and has been reviewed by peer reviewers.

There are 56 articles in this book which can be divided into six themes. The first theme is the issue of competition and collaboration related to international relationship and foreign affairs. The second theme is the competition and collaboration related to the issue of national development. The third theme of this book is the competition and collaboration related to the issue of local development and community empowerment. The fourth theme of this book is the competition and collaboration related to gender issues. The fifth theme of this book is the competition and collaboration related to spirituality, development and political movement. And the last theme is the competition and collaboration related to environmental issues and sustainable development in local areas.

This proceeding provide the permanent record of what was presented. It will be invaluable to all people who interested with the topics. Finally, it is appropriate that we record our thanks to our fellow members of the Organising and Steering Committee, and the financial support from Universitas Indonesia. We are also indebted to those who served as chairmen. Without their support, the conference could not have been the success as it was. We would like to express our appreciation and gratitude to the scientific committee and the reviewers who have selected and reviewed the papers, and also to the technical editor's team who helped carry out the page layout and check the consistency of the papers with the publisher's template. It is an honor to publish selected papers in this volume by CRC Press/Balkema (Taylor & Francis Group).

The Editorial Board of the 1st APRISH Proceedings for Topics in Social and Political Sciences

Prof. Isbandi Rukminto Adi & Prof. Rochman Achwan
Faculty of Social and Political Sciences, Universitas Indonesia, Indonesia

Competition and Cooperation in Social and Political Sciences – Adi & Achwan (Eds)
© 2018 Taylor & Francis Group, London, ISBN 978-1-138-62676-8

Organizing committee

STEERING COMMITTEE

Rosari Saleh, *Vice Rector of Research and Innovation, Universitas Indonesia*
Topo Santoso, *Dean Faculty of Law, Universitas Indonesia*
Ari Kuncoro, *Dean Faculty of Economics and Business, Universitas Indonesia*
Adrianus L.G. Waworuntu, *Dean Faculty of Humanities, Universitas Indonesia*
Arie Setiabudi Soesilo, *Dean Faculty of Social and Political Sciences, Universitas Indonesia*

INTERNATIONAL ADVISORY BOARD

Peter Newcombe, *University of Queensland, Australia*
Fred Piercy, *Virginia Tech University, Australia*
Frieda Mangunsong Siahaan, *Universitas Indonesia, Indonesia*
James Bartle, *University of New South Wales, Australia*
Elvia Sunityo Shauki, *University of South Australia, Australia*

SCIENTIFIC COMMITTEE

Manneke Budiman
Isbandi Rukminto Adi
Beta Yulianita Gitaharie
Surastini Fitriasih
Sri Hartati R. Suradijono
Elizabeth Kristi Poerwandari

CONFERENCE DIRECTOR

Tjut Rifameutia Umar Ali

CONFERENCE VICE-DIRECTOR

Turro Wongkaren

ORGANIZING COMMITTEE

Dewi Maulina, *Faculty of Psychology, Universitas Indonesia*
Intan Wardhani, *Faculty of Psychology, Universitas Indonesia*
Elok D. Malay, *Faculty of Psychology, Universitas Indonesia*
Josephine Rosa Marieta, *Faculty of Psychology, Universitas Indonesia*
Teraya Paramehta, *Faculty of Humanities, Universitas Indonesia*

Keynote speech

Competition and Cooperation in Social and Political Sciences – Adi & Achwan (Eds)
© 2018 Taylor & Francis Group, London, ISBN 978-1-138-62676-8

Reconciliation after recognition? Indigenous-settler relations in Australia

A. Little
School of Social and Political Sciences, University of Melbourne, Victoria, Australia

ABSTRACT: Since 2010 the political agenda on addressing Indigenous-settler relations in Australia has been dominated by the debate on constitutional recognition. On many levels this debate has been unsatisfactory but in this paper we focus on three particular issues that we argue should be the focus of continuing analysis of indegenous-settler relations regardless of the outcome of the constitutional recognition process. First, we contend that efforts to reconcile Australia need to recognize the conflictual nature of Indigenous-settler relations and that, as many conflicts, this requires management rather than resolution. Second, we suggest that this debate needs to pay greater attention to ongoing conflictual relations rather than a mere accounting for the wrongdoing of the past. Third, we unpack the constitutional recognition debates to demonstrate that—regardless of the core issue of Indigenous-settler relations—the ongoing process of reconciliation needs to address the conflicts within non-indigenous people over the appropriate course of action. Therefore, rather than solely relying on indigenous peoples to drive the process (and bear responsibility if it fails), there needs to be a future-oriented engagement involving non-indigenous people across their political divisions if the aspiration towards an ongoing process of reconciliation is to be achievable. Until the internal conflicts within non-indigenous people are identified and ventilated, Australia falls a long way short of being in a position to address and managed internally the conflict between indigenous peoples and the state.

1 INTRODUCTION

For the last four years a team of political, legal and criminological scholars based across four continents has been engaged in a research project examining 'Reconciliation, Recognition and Resistance in South Africa and Northern Ireland—lessons for Australia'.[1] The aim of the project has been to examine the differing dynamics of the transitional processes in South Africa and Northern Ireland since the 1990s to develop comparative knowledge that can inform ongoing debates on reconciliation in Australia, and, in particular, initiatives focused on constitutional recognition of Aboriginal and Torres Strait Islander peoples.

The transitional processes out of widespread political violence in South Africa and Northern Ireland since the 1990s have been quite divergent reflecting the differing dynamics of the two conflicts.[2] However, in this paper I concentrate on three common characteristics of the South Africa and Northern Ireland processes that are not as yet widely accepted in Australia. I contend that it is the failure to recognize these three conditions that is inhibiting the current debate in Australia and is at least partially responsible for preventing the development of a more mature debate in Australia. The conditions are, first, that the country is dealing with

---

1. The research project is funded by the Australian Research Council (DP130101399). The investigators on the team are Adrian Little (Melbourne), Mark McMillan (Melbourne), Paul Muldoon (Monash), Juliet Rogers (Melbourne), Erik Doxtader (South Carolina) and Andrew Schaap (Exeter).
2. See Little 2017.

a conflict situation, second, that it needs to establish a social and political framework for the future rather than concentrating solely on the past, and third, that this involves political engagement both within and between the various parties to the conflict.

While I will not dwell for long on the theoretical and methodological background to the project, it is worth briefly identifying that the team has worked with a broadly agonistic framework based on their previous research outputs (see, for example, Schaap 2005, Muldoon 2008, Muldoon and Schaap 2009, and Little 2002, 2004, 2008 and 2014, and Little and Lloyd 2009) and other recent scholarly works (Maddison 2015). Agonistic theories focus on the inevitable enduring nature of conflict and the ways in which that inflects political life. While concerned with the politics of conflict and the mechanisms and processes through which conflict can be managed, negotiated and transformed, agonistic theory sees the existence of such conflict as immutable. Rather than wishing away conflict or seeking to eradicate or resolve it, agonistic theories are concerned with the way in which it is expressed in political life (Mouffe 2000). Instead of trying to resolve conflicts, agonists are concerned with ways in which it can exist and indeed characterize political life without necessitating forms of overt political violence.

The methodology of the project has involved a series of workshops, interviews and documentary analysis across the three comparator countries in the study. The named researchers on the project attended all of the workshops with a range of stakeholders in each of the societies and conducted the interviews in pairs. While much of the initial work was in South Africa and Northern Ireland, the latter part of the project has almost exclusively been focused on the Australian case. As we reach the final stages of the project (with one final workshop in 2017), this paper provides some positioning and reflection on where the team has ended up as well as indicators of where this program of research needs to go next.[3]

2 CONTEXTUALIZING THE AUSTRALIAN DEBATE

Since 2010, the Australian debate has been dominated by discussion over the rights and wrongs of constitutional recognition of Aboriginal and Torres Strait Islander peoples as the original inhabitants of Australia and for whom there is a specific relationship with the land and waters of the country as well as a diverse range of cultural traditions, languages and practices that need to be reflected in the primary legal document underpinning the Australian state. While there is not sufficient space here to reflect on the historical backdrop to the constitutional recognition debate, it is part of a broader process going back to the referendum in Australia in 1967 and then, from the early 1990s a series of events that have been part of a process dubbed 'reconciliation'. These events have included the formation of the Council for Aboriginal Reconciliation, Prime Minister Paul Keating's Redfern Park speech, and the Mabo case in the early 1990s through to Prime Minister Kevin Rudd's apology to the Stolen Generations in 2008.

However, since the narrow election of the Gillard government in 2010, there has been a bi-partisan commitment across Australia's political divide to pursue constitutional recognition of Indigenous peoples. An Expert Panel that had been established by the Gillard government reported on a proposed wording for constitutional change in 2012, a proposal that was subsequently discussed by a parliamentary Joint Select Committee that reported in 2015. Meanwhile, the subsequent Liberal Prime Minister, Tony Abbott had been discussing these issues with a select group of Indigenous leaders after his election in 2013, before Abbott was deposed by the current Prime Minister, Malcolm Turnbull, in 2015. While Turnbull's approach has lacked the alacrity of Abbott's pursuit of a referendum on constitutional change in 2017, he did establish a Referendum Council to pave the way towards a proposal for constitutional change. At the time of writing, and while consultations take place, the proposal appears to be in abeyance. Indeed, in one such consultation exercise with Aboriginal

---

3. When I use the term 'we' in what follows, I am particularly referring to the research conducted with Mark McMillan in the course of the project.

people in the state of Victoria, the gathering in Melbourne voted unanimously to reject the idea of constitutional recognition with a much stronger call for treaty.[4]

3   THE SHORTCOMINGS OF THE AUSTRALIAN APPROACH TO CONSTITUTIONAL RECOGNITION

Given the project's focus on the mechanisms and processes of managing and governing the politics of conflict, the researchers have become increasingly concerned by the ways in which the Australian debate on constitutional recognition has been conducted. Indeed, one of the most obvious observations has been the unwillingness in Australia to construe Indigenous-settler relations in conflictual terms. This is reflected in the mainstream media coverage and the conduct of the debate in the main political parties, but is most certainly not the message that is being delivered by many Aboriginal and Torres Strait Islander participants in this project.[5] This is not to say that there is overt advocacy for violence, but there is a desire to see the relationships established since the initial colonization of Australia recognized as grounded in violent upheaval that has permeated Indigenous-settler relations ever since.

Here, and in other publications (Little and Macmillan 2017), we contend that his has been due to the invisibility of at least one of the parties to the conflict. For decades this was, of course, Aboriginal and Torres Strait Islander peoples who had to live with the legal fiction of *terra nullius* and were therefore invisible as participants in a conflictual relationship. Since the 1990s, however, with much greater acceptance of the need for a process of reconciliation, contemporary non-Indigenous Australians have become invisible. The debate is conducted as one involving the (very important) relationship between Indigenous people and the Australian state rather than one that should be partly focused on the relationship between Indigenous and non-Indigenous *people*. This exonerates non-Indigenous people from responsibility for their role as parties to a conflict and places all of the burden on Aboriginal and Torres Strait Islander peoples to reconcile with the Australian *state*.

Therefore, we contend that while it is vital not to underplay the significance of the legal relationship with the state, the conflict in Australia is also about the political relationship between people. This much is apparent to most Indigenous activists but it is far less clear for non-Indigenous Australians who have been exonerated from engagement. This is important because the management and/or governance of these relationships is an ongoing process rather than something which can simply be enacted through legal change or changes to the practices of political institutions.

The process of putting the Australian debate on a more substantive footing requires a number of significant developments. Not the least of these is the recognition that the focus on past misdemeanours and violence is a necessary but not sufficient part of the discussion of Indigenous-settler relations. The focus on the past that has predominated in the debate thus far pays insufficient attention to the need for changes that are directed towards the politics of these relationships in the future. Considerations (and apologies) for the failings of the past are not substantive enough to reset the structure of contemporary and future relationships. Therefore, whatever the outcome of moves towards constitutional recognition of Aboriginal and Torres Strait Islander peoples, there remains a set of relationships between people that need to be recalibrated and transformed on an ongoing basis.

4   A STRONGER FOOTING FOR AUSTRALIAN RECONCILIATION DEBATES

Based on both the various engagements in Australia during the research process and the comparative analysis of the processes in South Africa and Northern Ireland, we contend that

---

4. Weblink.
5. Reference to JCR Network conference at Melbourne in October.

5

there are three basic background conditions that are impeding the progress of the Australian reconciliation debate. First, in both South Africa and Northern Ireland, it was clear and widely accepted—at no obvious cost to any of the participants—that the situation being transformed was conflictual and involved conflict between people (as well as or not just between states). Second, the transitional processes that were established were focused just as much on creating the conditions for co-existence in the future as they were about accounting for the past. Third, and most significantly, much of the effort expended on building new conditions and institutions was as much about conducting debates within groups that were on one side of the divide, as it was about conversations across political divides. While we find Australia's debate deficient on all three counts, we are particularly concerned that the final point about the need for greater engagement *within* non-Indigenous Australia is not currently taking place. As a result, the onus for delivering a reconciled Australia is directed towards Aboriginal and Torres Strait Islander peoples.

The first issue is a very basic one but it seems to be a formidable challenge for Australia to recognise the conflictual relationships within its boundaries. This is significant because it changes the dynamics of the conversation from 'how do we right this historical wrong?' to 'how do we manage this relationship between people now and in the future?'. This is not just a semantic issue because the focus on 'acts' and 'wrongs'—necessary as that process is—can deflect attention away from thinking through the social and political dynamics of a recalibrated relationship between Indigenous and non-Indigenous people. To be clear, this in no way should detract from the need to account for the experiences of the Stolen Generations or the reason behind deaths in custody, but we also need to recognize that a debate about the future of relations between different people in Australia needs to grapple with the question of what it means to 'share this place'. A process directed toward understanding the conflictual nature of the relationships that exist within Australia between Aboriginal and Torres Strait Islander peoples and non-Indigenous citizens demands a much clearer public discourse about the recognition of difference. That is, the pursuit of legal equality between all inhabitants of Australia is not a process that does enough to reflect that there is an inherent difference in the standing of Indigenous people that will not (and indeed should not) go away. This difference does not need to be resolved through legalistic egalitarianism—it needs to be recognized and accentuated.

This first point leads inexorably to the second issue around managing relationships that is objectively simple but politically difficult. That is, the basic understanding of reconciliation needs to be recalibrated so that it is better understood as the ongoing management of relations rather than a mere accounting for the past. While apologies and recompense may indeed be necessary, they cannot provide a firm footing for future political relationships. This is partly because apologies and the like tend to focus on particular individuals or groups of individuals who are categorized as victims. While the apology may be on the behalf of 'the state', it is only to those perceived to be directly impacted by specific wrongful actions. As such, apologies—be they to the Stolen Generations, or the families of victims of Bloody Sunday in Northern Ireland, or those mistreated in Indian Schools in Canada—are never really focused on the impact of structural injustices on minority groups. This does not belittle their potential importance to individuals or groups who want remorse on the part of the state for their experiences, but they are rarely events that implicate the state in as any way guilty for the perpetration of ongoing structural injustice against particular sections of society. Moreover, there is often no recognition of the ways in which the behaviours of the past have created the conditions in which contemporary structural injustices have been established. In short, there is a need to ensure that processes of reconciliation in Australia are understood as an ongoing process of managing Indigenous-settler relations and the development of a future-oriented perspective. While constitutional recognition may have a role in that process, it can only be as a precursor to further initiatives including (potentially) a treaty or, to put it more precisely, a series of local agreements between different Aboriginal nations and the constituent states of the Australian Federation.

Third, and most importantly, a more specific focus on relations between people in general rather than the relationship between Aboriginal people and the Australian state is vital. While relationships *between* Indigenous and non-Indigenous people are vital in the discussion of a more reconciled society, a pre-cursor to this debate is that there is a more open debate *within* these groups especially non-Indigenous Australians. Just as there is no uniformity within Aboriginal Australia—this will come as no news to Indigenous political activists! – so there is no agreement within non-Indigenous Australians about the issue of reconciliation. And yet this is a debate that is rarely entered into between non-Indigenous people. Those who oppose reconciliation or forms of recognition direct their points towards political parties or Indigenous interlocutors. Aboriginal people are deeply engaged within their own forums as to the rights and wrongs of processes of recognition and reconciliation.

Yet, there appears to be a complete lack of engagement within non-Indigenous Australia between those who support and oppose these processes. More often than not, the supporters of reconciliation defer to Aboriginal people and effectively say 'tell us what you want and we will support it' rather than engaging on their own terms with other non-indigenous people and, in particular, opposing those who differ. The task of opposing constitutional conservatives and those on the right who oppose some of the more fundamental forms of change such as treaty is largely left to Aboriginal people with the well-meaning non-Indigenous people lined up behind. Once again, this speaks to a lack of engagement and an absence of responsibility of non-Indigenous Australians to lead in this debate. In short, the disagreements between non-Indigenous Australians are rarely ventilated and there is little onus to engage in that discussion. Therefore, not only is the burden to reconcile placed on the shoulders of Indigenous people, the job of resistance to insubstantial forms of reconciliation is also located there. Put simply, there is a need for much greater engagement within non-Indigenous Australia to fulfill the responsibility that the role of changing relationships requires.

5 CONCLUSION

The debate over constitutional recognition of Aboriginal and Torres Strait Islander peoples in Australia is mistakenly being conducted in terms of equality. That is, it is being imagined as a process that leads to the accordance of the same status as other Australian citizens through recognition that Indigenous people were the initial inhabitants of Australia and have special relationships with land and sea as well as specific languages and cultural traditions. However, the primary division in the debate—which receives very little attention—is the fact that most opponents of a thoroughgoing reform of the Australian constitution do so on the basis that all people—Indigenous and non-Indigenous—are effectively the same. This is opposed by most Indigenous activists and commentators who reject this form of procedural equality and favour instead something that appropriately recognizes the differences between Aboriginal and Torres Strait Islander peoples and other Australian citizens. Moreover, our contention in this paper is that this difference is both grounded in conflict and likely to generate further political contention.

To summarise, Aboriginal and Torres Strait Islanders are already recognized as legally different in the Australian constitution; the change that most Indigenous activists that favour recognition seek is a rectification of the terms under which that difference is understood. Of course, many others do not want constitutional recognition at all, favouring more substantive changes without the need for this intermediary step. Regardless of whether the Australian constitution is amended through the contemporary debate or not, our argument suggest that there needs to be a much more substantive, ongoing conversation about the nature of difference for Aboriginal and Torres Strait Islander peoples that goes beyond mere legal status. We suggest that this more developed understanding of difference is more likely to bring political conflict into the foreground of debate. And, finally, in turn, this places much greater emphasis on non-Indigenous people to engage with each other about these issues rather than the debate being conducted solely in terms of relations between Aboriginal and Torres Strait Islander peoples and the Australian state.

REFERENCES

Little, A. (2002) *The Politics of Community: Theory and Practice*. Edinburgh: Edinburgh University Press.

Little, A. (2004) *Democracy and Northern Ireland: Beyond the Liberal Paradigm?* London: Palgrave.

Little, A. (2008) *Democratic Piety: Complexity, Conflict and Violence*. Edinburgh: Edinburgh University Press.

Little, A. (2014) *Enduring Conflict: Challenging the Signature of Peace and Democracy*. New York: Bloomsbury.

Little, A. (2017) 'Fear, Hope and Disappointment: The Politics of Reconciliation and the Dynamics of Conflict Transformation'. *International Political Science Review*, forthcoming.

Little, A. and Juliet, R. (2017) 'The Politics of "Whataboutery": The Problem of Trauma Trumping the Political in Conflictual Societies', *British Journal of Politics and International Relations*, forthcoming.

Little, A. and Mark, Mc. (2017) 'Invisibility and the Politics of Reconciliation in Australia: Keeping Conflict in View', *Ethnopolitics*, DOI: 10.1080/17449057.2016.1219473, forthcoming.

Little, A. and Moya, L. Eds. (2009) *The Politics of Radical Democracy*. Edinburgh: Edinburgh University Press.

Little, A. and Sarah, M. (2017) 'Reconciliation, Transformation, Struggle: An Introduction', *International Political Science Review*, forthcoming, March 2017.

Maddison, S. (2015) *Conflict Transformation and Reconciliation: Multi-level Challenges in Deeply Divided Societies*. London: Routledge.

Mouffe, C. (2000) *The Democratic Paradox*. London: Verso.

Muldoon, P. (2008) 'The sovereign exceptions: Colonization and the foundation of society'. *Social & Legal Studies*, 17 (1), 59–74.

Muldoon, P. and Andrew, S. (2009) 'Aboriginal Sovereignty and the Democratic Paradox'. In Little and Lloyd. Eds. *The Politics of Radical Democracy*, Edinburgh: Edinburgh University Press, 52–72.

Schaap, A. (2005) *Political Reconciliation*. London: Routledge.

Comprehending Indonesian transformation

Bambang Shergi Laksmono
Faculty of Social and Political Sciences, Universitas Indonesia, Depok, Indonesia

ABSTRACT: This paper is intended to illustrate the challenges that Indonesia faced in formulating the agenda of development in the country. One will need to grasp the complex agendas after the Indonesian *Reformasi* (democratic reform) in 1998. In line with all the complexities, democracy will be held accountable for the progress in the living standards of the people. Within the framework of government decentralization, the paper analyses on the performance of local provincial government on three governance indicators, namely (i) investment, (ii) ecological management, and (iii) human investment. The score variation among the 34 provinces will shed light on the prospect of the long term improvements in addressing poverty and inequality in Indonesia.

1 INTRODUCTION

Comprehending Indonesian transformation seems to be a very difficult thing to do. It is not impossible, but it is also not easy. Indonesia in transition has to dwell on the political and technocratic means of boosting development, within the spheres of globalisation, decentralisation and popular democracy. Concurrently, a new public agenda has emerged, namely the upholding of human rights principles and the risks of ecological sustainability, apart from the agenda of public accountability. Does democracy, overall, improve living standards? Indonesia is basically a resource-rich country. This has shaped the history of the country. In this complex picture, how does Indonesia comprehend fully how to continue its modernity and to achieve social justice and social welfare as stipulated in the national Constitution? How does one configure national and global development agendas and sustainability within the narrow corridor of national development of this archipelagic nation?

It is not impossible, but it is also not easy. This discussion is intended to underline the importance of being able to see the 'whole'. It is even more important to manage the broad challenges of development. One of the central concepts to the whole issue of people's progress would be the term 'development'. Here, we need to remind ourselves that we should take development for granted without being critical of the direction and the context of its implementation. This involves fundamental changes around us, by social forces and political directions accepted and favourably termed as development. The subject of development has been a constant preoccupation for a long time. There seems to be a significant shift in the development orientation set by the current Indonesian government. President Joko Widodo, our 7th President of the Republic, provided a fresh perspective (and meaning) to development, promising a significant leverage for the (once) overlooked segment of the population. Still, the country is facing great challenges. We would like to discover some important dimensions to this.

A blunt and interesting question was raised during the International Conference on Social and Political Science (ICSPI) in Bali earlier this month. The question was directed to an Indonesian presenter, who was presenting a public health issue. The concern was about the persistent high national rate of maternal mortality in the country. The question came from an Afghanistan academic who was curious as to the persistence of the high rate of maternal mortality in Indonesia. The figure is particularly striking in the eastern provinces of the country.

The Afghanistan gentlemen argued that it is obvious that his country currently has a high maternal and infant mortality rate. Clearly, Afghanistan is in a state of continuous hostilities. Institutional breakdown characterise the country as a result of years of violent power feuds and civil war. But Indonesia should be different. Given Indonesia's comparatively stable situation, childbirths are less exposed to risks and therefore high incidence of maternal mortality should not occur. In addition, Indonesia has also undergone decades of intensive development which should have resulted in better indicators, including maternal mortality rate.

Development in the broadest sense is the obvious background to the relative low achievement in human development in the country. Basically, Indonesian human development rests on the country's development effectiveness. Development outcomes overall the product of various developmental inputs. Local potentials, both natural and human resources, are the subject for improvements. Technology, capital and education are central to the process of modernisation and productivity. This has been the whole idea of development as a major thrust for change in Indonesia for a number of decades. Maternal mortality is a good indication of pockets of underdevelopment, discernible through social indicator sets. Tackling the issue will require broad analysis into the development capacity of the government at the central, provincial and the district level.

It is true that maternal and infant mortality not only depend on the performance of the health system. Any social outcomes will be determined by many other factors, such as the conditions of the infrastructure, the quality of human resources, the state of the local economy, sanitation, clean water supply, etc. Maternal mortality will be specific and will relate closely to the conditions in the village and the household. It is important to realise at this point that development is a process and outcome of the dynamic conditions of several key elements, namely the capacity of the local government. Institutional performance does not happen in a void. Outcomes result from a process of different contributions of development actors. In this light, local governments will be tested on their capability to mobilise the local resources, the business sector, the civil society and the family towards social ends.

Development effectiveness, therefore, is relevant as the framework for this discussion. The broad term denotes a planned change and could be termed as development, social transformation and other terms that denote systematic planned changes. The way the term is used in practice usually involves particular domains, sectors or issues.

2 THREE SIMULTANEOUS TRANSFORMATIONS

Development involves great ideas and strategies. New development concepts and approaches will continue to emerge and will become a beacon for programme strategies in developing countries. Indonesia's situation is no exception. The current conditions are particularly critical for Indonesia, which is undergoing three major transformations, all of which are occurring at the same time. It is unique that Indonesia is undergoing three major transformations; namely *democratisation*, *decentralisation* and a further process of integration in the world market through the *globalisation* process. The two fundamental shifts in the government system have been fairly recent. The liberalisation of the economy began significantly earlier. This has made Indonesia an open economy for trade and capital, which utilizes the advantages of its labour, market and natural resources. The 1998 Indonesian *Reformasi* gave way to another two fundamental changes: the shift from a centralised to a decentralised government at the district level, and the shift from an authoritarian government to a directly elected President, elected leaders of the Provincial, District and Village Government and a continuous shift towards an open market.

Understandably, great transformations face great challenges too. The following outlines the general ramifications of the turn of events:

- **Decentralisation:** Continued divisions of local governments, 80 new Regencies and 6 Provinces waiting for the approval of Parliament, weak and divided bureucracy, development discontinuity.

- **Democracy:** Expensive, transactional, weak technocracy, primordial division, weak political party, election related fraud.
- **Globalisation:** Consequences of global integration, shared economy and the dominant extractive industry.

There are many examples of the 'local paradoxes' related to the above. There are various forms of anomalies or distortions associated with the implementation of local democracy and development management. Democracy involves the masses, while politics needs funding from the business sector. Representation of interest would then become intertwined in political collusions. This is inevitable. The complex adversities and tensions created is making the news headlines. While Indonesia is making headway in its Democracy Index and Governance, the reality of living conditions remains pitiful. Poverty generally remains unchanged. This is a paradox. While the grand policy of decentralisation and democracy is to redistribute power and authority, local reality is entangled by power struggles, rivalry and transactions. Again, this is the paradox. Realities lead us to question what systems, in what ways and within what context can these waves of transformation bring about broad benefits, equity and inclusivity.

3 DEMOCRACY AS THE BASIS OF WELFARE

Democracy is a fundamental transformation that is occurring in Indonesia right now. The system of democracy basically works on the basis of the broad participation of citizens in the political system. Indonesia has entered full democracy, which involved complex political undertakings at both national and regional level to establish legitimacy as the foremost mission, and effective public policies though a participative process. Dahl (1971) underlined the value of *contestation* and *inclusiveness* in this complex process. Democracy is the platform of governance, which is to produce responsive policies to meet the needs of the public.

In democracy, 'people's voices' are the core reference. Vote results determine the delegation that will represent them in the legislative body and also elect the rightful leaders of the government. Legitimacy, legitimate government and legitimate policies are the foremost fundamental value and action system. This process ended authoritarianism that had been in place and had characterised national leadership for a long time. For Indonesia, this implies the adoption of a new election system and leadership organisation. More than this, the shift from a closed political system to an open and deliberative one requires a new political culture among the political actors and the voting masses. The voters are the constituents, the subject for accountability of those designated in power.

It is important to underline that democracy requires the essentials of contestation, an element that is rather new to Indonesian political culture. The terms contest and contestation imply a broad understanding of the competition of good ideas and effective public policies. For our purposes, democracy is to be relevant and understood as the vehicle for progress in living conditions. The question we wish to raise is, does democracy promote welfare? There are several answers to this question. It is suffice to say, at this point, that we need to study the policy choices and decisions that relate to macro fiscal instruments (subsidies, social spending) and local basic provisions in health and education. It is important to remember that some major investments/industries may also create forms of externalities that, in the end, harm the interests of the locals.

- Challenge the idea of the 'race to the bottom' of welfare spending and social policy due to the pressure of globalisation. Overall the retrenchment of social spending has been limited in Latin America, Asia and Eastern Europe. Democratic rulers have to consider the interests of wider and competing social interest groups—most specifically the 'interests of the poor'. The opposite is true for the dictator governments (Haggard & Kaufman, 2008).
- 'Nations fail when they have extractive economic institutions, supported by extractive political institutions that impede and even block economic growth' (Acemoglu & Robinson, 2012, p. 83).

Local development and the politics of development implementation create different forms of development policies. The following figure describes the variations of indicator scores drawn from the Indonesian Governance Index Report. The index composites, although complex in their items measurement, provide us with a picture of the performance of local (Provincial) governments in Indonesia with reference to grouped governance indicators.

There are some interesting combinations of item rankings among the Provincial Government in the figure. The figure shows three broad indicators that represent policies and context connections. The three selected development indicators, that is represented by the following: investment climate, the environment and education. The three indicators were selected to gain road understanding on how inter-sector policies may produce positive or negative long-term consequences in the quality of life indicators. We can see that the provinces can be categorised under three general conditions: first, the provinces that perform well in the investment climate but fail to secure grounds for the environment. These provinces may diverge into two directions: those who have good education spending and those who have low education spending. Second, some provinces score low in the economy, while keeping good scores in the environment. Third, provinces that consistently have low achievements in every dimension. They score low in the economy, environment and education. The categories presented above may provide us with the background into conditions of welfare and the human development index. In broad terms, living conditions will be determined by the overall conditions of the economy, the ecological sustainability and human investment.

Attention needs to be given to the Indonesian provincial government that consistently scores low in the three governance components. At this point it is still difficult to conclude which score combination of sector policies that guarantee long term achievements in sustainable and inclusive development. The difficulty in making development judgement rests on the fact that some Indonesian provinces have demonstrated good score in education but have relative low score in the environment management. The facts can even be more puzzling with the fact that some these provinces that have low score in environment management have high scores in investment.

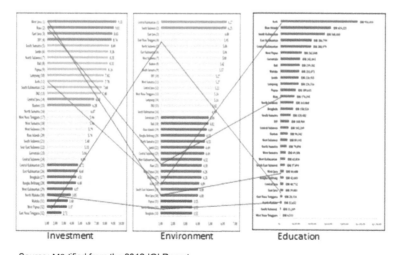

Source: MOdified from the 2012 IGI Report

Figure 1.   Variation in inter-province governance indicators.

5 EPILOGUE

Indonesian political transformation will be tested by the progress made in development. However, development outcome rest on the performance of politics, both at the national and local level. The question about the development outcome of a political regime is a long term discussion subject. Understanding Indonesian transformation provides good platform to predict how the country is able to address the issue of poverty and inequality. Some good reference can be taken from experiences of Indonesia neighbour countries. Last month I was giving a lecture on the broad issue of development in Indonesia. It touched upon national development strategies and how Indonesia should best foresee future challenges. In the lecture, I specifically focused on two points, namely the success stories of Thailand's community development projects in the agricultural sector and the success of Singapore under the long and consistent leadership of Lee Kuan Yew. The question time was interesting. Not only one but most of the questions directed to me were criticising the reference I made about Thailand and Singapore being the ideal references. I was a little surprised, as the audience was adamant that an authoritarian regime in Singapore and the Thailand government in turmoil should not be considered best practice. It seems that the audience was missing the point I made about the two fundamental things concerning the cases of these two countries. First, about the consistency of 60 years of development and nation building in Singapore. Second, the implementation of the robust roadmap of Thailand in agricultural development. Notably with the strong support of the late Thai King Bhumibol Adulyadej, in developing their global oriented agricultural sector.

Obviously, leadership and robust development plans are required in any country. Indonesia has its own records on the regimes and national leaders. Going through decades of development, Indonesia has undergone different forms of social tensions through its decades of global political and economic spheres.

REFERENCES

Acemoglu, D. & Robinson, J. A. (2012). *Why nations fail: The origins of power, prosperity and poverty*. London: Crown Publishing.

Chomsky, N. (2007). *Failed states: The abuse of power and the assault on democracy*. New York: Owl Books.

Curchin, K. (2016). From the moral limits of markets to the moral limits of welfare. *Journal of Social Policy*, *45*(1), 101–118. https://doi.org/10.1017/S0047279415000501.

Dagdeviren, H., Donoghue, M. & Promberger, M. (2016). Resilience, hardship and social conditions. *Journal of Social Policy*, *45*(1), 1–20. https://doi.org/10.1017/S004727941500032X.

Dahl, R. (1971). *Polyarchy: participation and opposition*. New Haven: Yale University Press.

Deming, C. & Smyth, P. (2015). Social investment after neoliberalism: Policy paradigm and political platforms. *Journal of Social Policy*, *44*(2), 297–318. https://www.ncbi.nlm.nih.gov/pmc/articles/PMC4648188/.

Drake, R. F. (2001). *The Principles of social policy*. New York: Pelgrave Publishing.

Haggard, S. & Kaufman, R. (2008). *Development, democracy and welfare states: Latin America, East Asia and Eastern Europe*. Princeton: University Press.

Halperin, M. H., Siegle, J. T. & Weinstein, M. M. (2005). *The democracy advantage: How democracies promote prosperity and peace*. New York: Routledge.

Jamrozik, Adam. (2001). *Social policy in the post-welfare state: Australian society in a changing world*. Sydney: Pearson Education.

Kaletsky, A. (2010). *Capitalism 4.0: The birth of a new economy in the aftermath of crisis*. London: Bloomsbury Publishing Plc.

O'Neill, J. (2012). *The growth map: Economic opportunity in the BRICs and beyond*. New York: The Penguin Press.

Ow, R. (2010). Negotiating challenges: Social development in Asia. *Asia Pacific Journal of Social Work and Development, 20*(1), 82–94.

Ricard, M. (2015). *Altruism: The power of compassion to change yourself and the world*. New York: Little, Brown and Company.

Sachs, J. D. (2008). *Common wealth: Economics for a crowded planet*. New York: The Penguin Press.

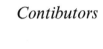

Contibutors

Contibutors

The policy of community empowerment in implementing Corporate Social Responsibility (CSR) of a coal mining company (a case study of CSR implementation by PT BARA as a state-owned company)

R. Resnawaty & I.R. Adi
Department of Social Welfare, Faculty of Social and Political Sciences, Universitas Padjadjaran, Bandung, Indonesia

ABSTRACT: This paper discusses the discrepancy between a coal mining company and society in Indonesia. One of the efforts to decrease that discrepancy is through CSR activities, which aim to empower and develop society. PT BARA, as a state-owned enterprise, has both internal and external policies for the implementation of CSR. As a government company, PT BARA links CSR with various policies (laws, regulation ministers, ministerial decrees, etc.). PT BARA aims to contribute to the escalation of society's quality of life through their CSR programmes. To achieve that, PT BARA has a policy of implementing CSR through the Partnership and Environmental Development Programme, in accordance with the provisions of the regulations of the Minister of State Enterprises. As a state-owned enterprise, PT BARA then becomes a mainstay of the local government in the implementation of regional development. Therefore, PT BARA's policies on CSR are followed by various dynamics on achieving their goals in order to increase the quality of life of the community.

1 BUSINESS SECTOR IN COAL MINING IN INDONESIA

The complexities in the mining industry cover several concerns, which are not only related to the production capacity of the company, technology, capital sufficiency and mineral supply availability, but also include the relationship between the mining company and the surrounding community. Most of the time, the existence of mining companies brings up several issues regarding the surrounding area, especially in relation to the disparity of wealth between the two. Mining employers are supposedly concerned and are expected to be responsible for the well-being of the surrounding area, both for the people and the natural environment. The company's concern towards the surrounding community is reflected through various CSR programme initiatives.

Every company has certain goals when conducting CSR activities. At mining companies, CSR is aimed at securing the mining operational activities and achieving targets from mining exploration. To achieve these goals, mining companies have both internal and external policies. The internal policy is a range of strategies that the company uses to encourage the achievement of the vision, mission and goals of CSR. Meanwhile, the external policy towards the community is a foundation or a guideline, which rules how the entire company's components should be conducted and should act towards the implementation of CSR.

PT BARA (not the real name), located in Muara Enim Regency, South Sumatera Province, is a state-owned mining enterprise (*BUMN*) that certainly has both internal and external policies for conducting their CSR programme. As a state-owned company, PT BARA's CSR implementation is also regulated under various central government regulations, which makes it more complex compared to that of private mining companies. Therefore, besides aiming to support the mining operational activities, it also needs to empower the surrounding community.

A company's motives in administrating CSR will affect the form of both the company's internal and external policies in conducting CSR. An internal policy is a policy made by the company that is related to those efforts in achieving CSR goals that have been applied by the company. Bowie (1990) states that a company is the reflection of capitalist behaviour, which collects as much profit as it can. The steps taken in the implementation of CSR are the reflection of moral concerns towards society. Therefore, most often, CSR is conducted through charity or philanthropic activities towards clients, consumers and society.

Appropriate to what Bowie said, PT BARA is reflecting of capitalist behaviour also. It still constitutes a profit-oriented. Thus, the economic motive becomes the main motive when formulating CSR policies. Therefore, in the end, the implementation of CSR is an effort to reach a previously arranged goal, namely achieving maximum profits. However, besides the economic motive for conducting CSR, PT BARA also has the goal of making a contribution towards the improvement of the quality of life of the community. By increasing people's welfare, this will have positive effects towards the company's exploration activities, with the hope of being free from any strike protests that ask for welfare, job vacancies, and land compensation to the company. From investigating these matters, it seems the internal policies made by PT BARA are based on the company's interests. This fact is related to the statement by Prayogo (2011), that in order to reduce external pressure and to consider business safety, community development (*comdev*) programmes are undertaken to maintain a good relationship with the local community, to reduce business risks and to smooth the mining activity. The implementation of the external policy of PT BARA on CSR activities is manifested through "*Program Kemitraan dan Bina Lingkungan*/PKBL" (Partnership and Environment Development Programme). This programme supports the manifestation of the vision and missions of PT BARA. In this regard, PT BARA has a vision to achieve a prosperous, self-reliant and environmentally friendly community. Meanwhile the company's missions are (1) to support government programmes to improve the economic, social and public education of the community and to preserve the environment; (2) to empower local potential and to expand the market in order to provide more job opportunities for the company's surrounding community; and (3) to encourage public participation to support the company's long-term plans and post-mining development.

The company's internal policy is related to budgeting, staff organisation and programme monitoring and evaluation. In terms of budgeting, PT BARA has allocated 4% of its profits for conducting community development and empowerment programmes. Such a large fund has been issued by the company in order to achieve the company's vision and missions to increase the community's welfare. The amount of 4% issued by the company is twice as big as the percentage mandated in the Regulation of the Indonesian Minister of State-Owned Enterprises Number. Per-05/2007 on the amount of CSR funding taken from net income.

The chairman of PT BARA has a commitment to keep increasing the quality of life of the community. The allocated fund has been increasing each year, along with the increase of PT BARA's profit. However, it was found that the large budget has not yet been able to bring more empowerment to the society. This issue could be explained by observing how the allocated CSR budget has been utilised.

Until 2012, the budget for the Partnership Programme had been around 2%; meanwhile for the Environment Development Programme it was around 2% of the total company budget. However, in 2013, there was a policy change resulting in a change to the budget for the Partnership Programme to only 1%, while a 3% proportion was allocated to the Environment Development Programme. This policy shows that PT BARA focuses on programmes related to the processes of developing, improving and constructing infrastructures. In relation to the policy change on budget allocation, change might occur when the company gives big support to certain development plans in the regency. On one hand, this shows that the company has a commitment to conducting CSR based on the development goals of the surrounding region, such as constructing a sports centre or assisting with the construction of school buildings, as well as constructing infrastructures that require a large budget. However, on the other hand, the budget allocation is less likely to directly show the company's alignment towards society.

Table 1. PT BARA CSR fund allocation from 2010 to 2012.

Programme	(in million Rupiah)			
	2010	2011	2012	2013
Partnership	67.730	98.940	125.780	41.700
Community development	26.110	74.090	90.090	113.970

The development programmes that become the focus of the CSR programme are those that have not yet been covered entirely by the government, which are usually related to the construction of physical infrastructures that only have a small impact on society. In other words, most of the programmes are still top-down in nature.

The budgeting policy adopted by PT BARA falls into the category of *corporate philanthropy*, which is a traditional form of CSR. The *corporate philanthropy* activity conducted by the company is a direct contribution by the company, as a form of charity, towards the resolution of issues faced by society (Saidi & Abidin, 2004; Kottler, 2005). This is one of the reasons why the CSR programmes have not yet effectively empowered the community.

3 THE POLICY OF EMPOWERING THE MICRO, SMALL, MEDIUM ENTERPRISES (SMES) AND DEVELOPING THE LOCAL ECONOMY THROUGH THE PARTNERSHIP PROGRAMME

The Partnership Programme was initiated in the middle of 1989 by the Indonesian Ministry of Finance through Ministerial Decision No. 1232/KMK.013/1989. This regulation stipulates that a Partnership Programme is conducted between one big company and small enterprises under the principle of assisting, supporting and protecting the small enterprises. Ideally, there should be an equal standing in partnership relations, so there will be neither intimidation by the bigger enterprises nor dependency from the smaller enterprises. However, in reality, the Partnership Programme conducted by PT BARA is simply a programme of providing capital assistance, so that the external policy made by the company is merely a regulation on the mechanism of giving aid to small enterprises. This results in a superior-subordinate relationship pattern between PT BARA and its smaller enterprise partners.

The Partnership Programme conducted by PT BARA has been focusing on the efforts to increase the capacity of SMEs so that they will become strong and independent enterprises. The activities of the programme constitute the provision of capital assistance through soft loans to small enterprises and grants in the form of training and empowerment to improve the local economy. Basically, the Partnership Programme is an effort by PT BARA to enhance the welfare of society; it is expected that the programme will create new job opportunities for the community. From the perspective of *community development*, the efforts to improve the welfare of society constitutes the application of economic development in responding to the crisis or the economic issues faced by society. Such effort is carried out by finding out alternative new enterprises for them, revitalising and increasing their quality of life (Ife, 2012, p. 221).

The Partnership Programme is aimed at the local SMEs in order to increase their capacity through capital assistance. Meanwhile, grants, which are assistance that does not need to be returned, constitute a form of conservative economic development. PT BARA realises that if CSR is not conducted, there will be many negative impacts that can disturb the relationship between the company and the community. Therefore, the CSR programme is an integral part of the efforts to achieve the company's goals and, thus, various regulations to control the implementation of the Partnership Programme are made.

Basically, the Partnership Programme can be classified into two categories, i.e. the capital assistance programme and the grants programme. The capital assistance programme is a revolving fund programme that can be utilised by small enterprises, especially those that cannot access loan capital services from banks. However, at the beginning of the programme, small enterprises get capital assistance that is bound by a special loan agreement. Therefore, they consider the funding to be pure assistance that does not need to be returned. This results in a number of cases of non-performing loans in the Partnership Programme, which is still running now.

In 2012, PT BARA co-operated with a bank for fund distribution, especially for the revolving fund programme. A certain prerequisite for partner candidates was added, namely the obligation to have 'collateral', to separate the small enterprises that have the capacity to return the loan and the enterprises that do not have the capacity to return the loan (non-performing loans). The collateral is one of the factors determining whether or not the small enterprises are eligible to receive a loan. This policy is actually a step towards making the programme sustainable and keeping it revolving. The use of collateral will encourage small enterprises to run their businesses effectively so that they can pay the instalments every month.

However, since PT BARA issued the new regulations regarding collateral for the capital assistance eligibility for small enterprises, this programme has become more difficult to access for micro-scaled enterprises. The Partnership Programme shows no difference from the conventional bank system in terms of having a low interest rate.

The partnership funds, especially the capital assistance programme, have experienced some problems. Based on the data, the roots of the problems are: (1) the Partnership Programme, especially the capital assistance, has been conducted since 1989, when the programme was formerly named PUKK. In the past, the management of the revolving fund did not have a good system, so it was impossible for the partners to return the revolving fund to PT BARA. The instalment payment system was done manually in the past, by visiting each house to deliver an invoice, without involving any bank. This made the community see the current Partnership Programme as the same programme under the same old policy, in which the return of the revolving funds was lenient. (2) The limited human resources in PT BARA have made it difficult for the partners' to be given intensive guidance on performance. The guidance was 'so simple'; in the respect that it did not provide detailed explanations regarding the partner's rights and obligations. This caused the partners to ignore the guidance programme and, as a result, they did not experience any progress. Even worse, almost 50% of the partners could not return the revolving funds to PT BARA. (3) The name of the SMEs enhancement programme is identical to the so-called 'capital assistance programme', which is often identified as an aid programme that does not need to be returned.

On the other hand, Grant Programmes focus on the improvement of ability, skills and knowledge in the community, incorporating various types of training on how to run a business. This programme aims to give skills to unemployed people living around the company. However, the training was conducted separately to the capital assistance programmes. Unfortunately, these are only incidental programmes and have not made any significant impact on the increase of the quality of life of the community.

4 THE POLICY OF INCREASING THE QUALITY OF LIFE OF THE COMMUNITY THROUGH THE ENVIRONMENT DEVELOPMENT PROGRAMME

Law Number 4 of 2009 on Mineral and Coal Mining states that mining companies are obliged to empower and develop the community. Although PT BARA has conducted social responsibility programmes through the Partnership and Environment Development Programme since 2003, the community development programme was then manifested in the construction of public and social infrastructures. Thus, the co-ordinating staff of PT BARA still consider that the community development programme is similar to the Environment Development Programme. Meanwhile, the local community understands the community development programme to be a programme for constructing public infrastructures, such as the District Head Office, school buildings, pathways, trenches and others.

The Environment Development Programme is written in the Mechanisms of Environment Development Conduct. Several things are pointed out concerning the type of programme, society and third party involvement, and the prohibition of giving direct aid in the form of cash. The company builds infrastructures and also strengthens the local economy by providing aid to micro businesses. Basically, the policy was an effort to support continuous development. Based on the mission of **PT BARA**, the policy was carried out to encourage the community to be independent or less dependent on the company.

The improvement to the quality of life of the community through infrastructure development is the goal of the Environment Development Programme, for the development of the community's economy is supported by sufficient infrastructure. In the future, it is expected that the society's access to the health and education system will be increased, so that the society will be more empowered.

5 CSR POLICY AS A 'RED-PLATE' COMPANY

As a state-owned enterprise, **PT BARA** is usually called a *red-plate* company. So far, the implementation of CSR by **PT BARA** has been very dependent on the results of the development plans at the regency level. It is assumed that it is much easier for **PT BARA** to co-operate with the government compared to private companies. **PT BARA** is known as a company that helps the local government to execute development plans that cannot be achieved by the government alone because of lack of budget, especially in public infrastructure development that requires a large amount of money.

So far, **PT BARA** has yet to have both long-term and short-term plans in the CSR 'frame' that are integrated with their business conduct. This is because the company still gives the biggest focus to supporting the government's development programme. Therefore, CSR activities conducted through Partnership and Environment Development Programmes are still essentially considered as charity. Although it is aimed at empowering society, the CSR activities of **PT BARA** are still putting more emphasis on the provisions of aids and grants. Therefore, the community views the company as a generous party.

As a state-owned mining company, both the internal and external policies of **PT BARA** are influenced by the central government's policy. In conducting CSR activity, so far, **PT BARA**'s policy has been heavily influenced by the law. Generally, the CSR policy of **PT BARA** concerns developing and empowering society, a policy about Partnership and Environment Development Programmes, as well as the obligation to conduct a social and environmental responsibility programme. The policies concerning the development and empowerment of society have been at the heart of the administration of an infrastructure development programme. Hence, the CSR programme conducted by **PT BARA** is dominated by public infrastructure development. This matter is strengthened by a relationship between the company's CSR programme and development planning at the regency level. The company's status as a state-owned enterprise, however, has positive inputs, with an assumption that state-owned enterprises will have a greater alignment with society compared to privately-owned enterprises. Therefore, the local government feels fortunate to have the existence of its companies in their neighbourhood. Companies can be categorised as 'good' companies, since their CSR programme can help the government workload. Things to be analysed include the fact that, by having no long-term plans in CSR administration, as well as a determination to make the commitment, they increase the budget amount of CSR for the public infrastructure construction.

External policy is a policy made by **PT BARA** in relation to the implementation of CSR to the society in Ring-1. This policy becomes one reference of CSR administration, which is then taken down to several written points, which are called the Mechanisms of Work in conducting the Partnership Programme and the Environment Development Programme.

PT BARA limits the area that receives aid to the scope of Ring-1; therefore, people who live in Ring-1 become the priority as the main recipients of the CSR programme. The regulation requires the company's commitment to improving the society that is directly affected by its mining activities. Blowfield & Frynas (2005) state that companies have a responsibility for the impact of their operational activities towards society and nature, which is sometimes more than just meeting the legal aspects and individual responsibilities.

In relation to the ideas of Blowfield & Frynas (2005), it can be understood that, so far, the mining exploitation by **PT BARA** has had a significant impact towards natural degradation. Therefore, it is necessary to prioritise the provisions of the distribution of the CSR programme towards the local community. Friction between the community and **PT BARA** sometimes heats up and needs to be given primary attention by the company. This is done in order to reach the production targets of **PT BARA**.

As a state-owned enterprise running in the mining sector in certain areas, **PT BARA** has become one of the source systems that are accessed and utilised by the local government to help the administration of local development, especially on funding matters. Sometimes, the local government's expectations of **PT BARA**'s contribution to development are very high. As an example, the government requested that the Ring-1 areas that receive aid had to be expanded to include the entire Muara Enim Regency. This was intended to widen the target area receiving aid from the CSR. In this case, **PT BARA** refused the request, since the CSR activity through the Partnership and Environment Development Programmes is just dedicated to those affected directly by the coal mining activity. This regulation has been incorporated into the external policy of **PT BARA**.

As an old and established mining company, **PT BARA** is a well-respected company, both in the market and by the government. Especially with the label of a 'red-plate' company, the local government cannot interfere in **PT BARA**'s mining and CSR policies. **PT BARA** is seen as a 'good' company, since the company's contribution towards local development, especially towards infrastructure development, has been extensive and real.

6 CONCLUSION

The Partnership Programme has become important in the implementation of CSR for **PT BARA**, as it is mainly aimed at the community and economic development. As a state-owned enterprise, which since 1989 Indonesian government regulation sets each of the Public Companies should provide assistance to small businesses, **PT BARA** interpreted the government's policy into a systematic series of events called the Partnership Programme to support the economy of the local community, by, among other things, providing revolving capital assistance to small businesses, training grants and grants for the promotion of the partners and the public. The term 'foster father', a commitment by big businesses to small businesses, has been achieved by **PT BARA**.

In addition to the policy regarding the Partnership Programme, the BUMN Decree on the implementation of CSR has been itemised in the Minister of State Regulation of BUMN, PER05/MBU/2007, Article 11 on the scope of the environmental development assistance programme. This is divided in more detail into six (6) major programmes, which collectively have the first word **'aid'**: (1) the aid for victims of natural disasters, (2) the aid for education and/or training, (3) the aid for health improvement (4) the aid for the development of infrastructure and/or public facilities, (5) the aid for the development of worship facilities, (6) the aid for nature conservation.

Basically, the policy regarding the Partnership Programme and the community development programmes that exist make it easy for **PT BARA** to draw up an annual programme. The policy can be developed into empowerment programmes that can improve the quality of life. However, the external policy of **PT BARA**, in fact, adopts the Decree of the Minister of BUMN as a whole, without a different interpretation; in other words, it brings every sentence of the legislation into a naming programme. Thus, the feel of 'aid' becomes very strong in **PT BARA**'s CSR programme. The word 'aid' in the law has an influence on the implementation of CSR activities at **PT BARA**, for 'aid' has the same connotation as a good deed and 'charity' or a form of corporate philanthropy. It is then reflected in the implementation of CSR conducted by **PT BARA**. The conducted programmes are the types of aid that cannot empower the community as a whole. The aid has the connotation of giving, which has the soul of 'philanthropy' but has not yet reached the spirit of empowerment. Hence, the word 'aid' diminishes the spirit of empowerment.

Actually, there is no mistake in the interpretation of the legislation, which is directly used as the programme name in the CSR of **PT BARA**. Nevertheless, there are some interesting things that have been found regarding the external policy of **PT BARA** that are influenced by the existing central government policies, by which **PT BARA** strives to be obedient to the government regulations, thus adopting all of the policies without any exceptions.

Another issue of interest regarding the discussion of CSR policies by a state-owned coal mining company, such as PT. BARA, is that despite a series of policies on CSR leading to the efforts of community development, the enormous costs spent on CSR have not had any significant impact on the gap between the company and the communities. In other words, people who live in the areas of mining operations are still living in poverty. Therefore, the challenge faced by the whole field of programmes that have become the focus of CSR by PT BARA is to manage the budget allocation effectively. The local government demands that PT BARA makes huge efforts on local development. On one hand, PT BARA did a great job in developing infrastructure in a local regency, but, on the other hand, PT BARA has not significantly empowered the community. The CSR programmes are still dominated by charity and aid programmes, which can create the dependence of the local government and community.

The remaining challenge need to address is to ensure whether the implementation of CSR activities undertaken by mining companies, especially state-owned companies, has been in accordance with the existing policy on community empowerment.

REFERENCES

Adamson, D. (2010). Community empowerment: Identifying the barriers to "purposeful" citizen participation. *International Journal of Sociology*, *30*(3/4), 114–126.

Adi, I.R. (2003). *Empowerment, community development and community intervention*. Jakarta: FE-Universitas Indonesia.

Blowfield, M. and Jedrzej George Frynas. (2005). Editorial: Setting new agendas—critical perspectives on corporate social responsibility in developing world. International Affairs 81(3): 499–513.

Bowie, N.E. (1990). Money, morality, and motor cars. In W.F. Hoffman, R. Frederick & E.S. Petry (Eds.), *Business ethics and the environment* (pp. 89–97). New York: Quorum Book.

Elkington, J. (1997). *Cannibals with forks: The triple bottom line of 21st century*. Philadelphia: New Society.

Eweje, G. (2006). The role of Mne's in community development initiatives in developing countries. *Sage Publications, 45*, 93–129.

Fallon, G.C. (2010). Giving with one hand: On the mining sector's treatment of indigenous stakeholders in the name of CSR. *International Journal of Sociology and Social Policy, 30*, 666–682.

Fatah, L. (2008). The impacts of coal mining on the economy and environment of South Kalimantan Province, Indonesia. *Asean Economic Bulletin, 25*, 85–98.

Frank den Hond, F.G. (2007). *Managing CSR in action; Talking, doing, and measuring.* England: Ashgate Publishing Limited.

Frynas, J.G. (2009). *Beyond corporate social responsibility.* New York: Cambridge University Press.

Hopkins, M. (2007). *Corporate social responsibility and international development: Is business the solution?* United Kingdom: Earthscan.

Idemudia, U. (2009). Oil extraction and poverty reduction in Niger delta: A critical examination of partnership initiatives. *Business Ethics*, vol. 90, issue 1, p. 91–116.

Ife, Jim, (2012). *Community development in an uncertain world.* Australia: Cambridge University.

Jenkins, R. (2005). Globalization, corporate social responsibility and poverty. *International Affairs, 80*(03), 524–540.

Kapelus, P. (2002). Mining, corporate social responsibility and the "community": The case of Rio Tinto, Richard Bay Minerals and the Mbonambi. *Journal of Business Ethics*, vol. 39, issue 3, p. 275–296.

Kapelus, R.H. (2004). Corporate social responsibility in mining in Southern Africa: Fair accountability or just greenwash? *Development, 47*(3), 85–92.

Kemp, D., Owen, J.R. (2012). Assets, capitals, and resources: Frameworks for corporate community development in mining. *International Association for Business and Society*, vol. 51, issue 3, p. 382–408.

Kotler, P., Lee, N. (2005). Corporate social responsibility: doing the most good for your company. United State of America: Wiley.

Ogula, D.C. (2008). *Stakeholder involvement in corporate social strategy: An ethnographic study of the Niger Delta, Nigeria.* Phoenix: University of Phoenix.

Prayogo, D. (2011). *Socially responsible corporation: Peta Masalah, Tanggung Jawab Sosial dan Pembangunan Komunitas pada Industri Tambang dan Migas.* Jakarta: UI Press.

PSE-UI (Pusat Studi Energy-Universitas Indonesia). (2002). *Indonesia energy outlook & statistics 2000.* Jakarta: UI-Press.

Saidi dan Abidin, 2004. *Corporate social responsibility: Alternatif bagi Pembangunan Indonesia.* Jakarta: ICSD.

Interpretation of social policy in Jatinangor, Sumedang

M. Fedryansyah
Department of Social Welfare, Faculty of Social and Political Sciences, Universitas Padjajaran, Bandung, Indonesia

I.R. Adi
Department of Social Welfare, Faculty of Social and Political Sciences, Universitas Indonesia, Depok, Indonesia

ABSTRACT: This research was conducted to describe the interpretation of social policy in Jatinangor. Social policy can be defined as a holistic process that includes three spheres: political, administrative, and operational spheres. However, the interpretation process is an activity at administrative level, which will develop a framework for the operational level. There are several factors used as the foundation for interpreting the policy, such as an understanding of the social problems as the policy target, the policy purpose, and related regulations. In addition, there are some foundations of values and criteria that underlie the choices in policy interpretation. The choice aspects are bases of allocation, social provision, delivery system, and finance. This research shows that the values in policy interpretation comprise equality and equity. These values are reflected in choice aspects, such as bases of allocation and social provision. Meanwhile, the criteria that appear in policy interpretation comprise effectiveness and efficiency. These criteria are mentioned in choice aspects, such delivery system and finance.

1 INTRODUCTION

Related to social policy, there are studies that have been done to evaluate the programs which have already been implemented by the government. In general, the studies aim at knowing whether the objects or the clients experienced any changes or improvement. In other words, research related to social policy focuses more on the implementation and evaluation of social policies. The studies on evaluation of a policy are held to examine government programs (Mujiyadi & Sumarno, 2013; Suradi, 2012; Habibullah & Noviana, 2013; Widodo *et al.*, 2010; Sumarno & Roebiyantho, 2013; Anasiru, 2011; Supeno, 2006). From these studies, it can be concluded that there is an imbalance between the policy and its implementation.

Moreover, we can conclude that some supporting regulations are needed to optimize a social policy.

Besides evaluation, social policy research also focuses on its implementation (Astuti, 2013; Padmi, *et al.*, 2013; Purwanto & Syawie, 2013; Fatony, 2011; Suradi, *et al.*, 2013). The results show that there is a problem in the coordination of the implementation process. The lack of synergy between institutions may cause some programs to be done separately.

In conclusion, there are still many problems and obstacles that prevent many policies and service programs to achieve their initial goals. Although social policy, as a guidance in making programs, is already in a good shape, it does not automatically assure that the implementation will run as expected. We can say that the effectiveness of a social policy is derived from how it is interpreted by any related parties, especially how the policy is interpreted to be more applicable as guidance in the implementation.

To understand more about the interpretation of the policy, Jamrozik (2001) shares his thoughts about spheres in public policy. He believes that social policy is a holistic process that includes three spheres: the political, administrative, and operational spheres. The political sphere is a process of policy planning and formulation. Then, at administrative level, a policy will be interpreted in a series of activities to be more operational. Furthermore, it is a framework for the operational level where actual social services or service delivery is carried out directly to the public (service-receiving public).

One of the development policies is the policy that was issued by the government of Jatinangor in Sumedang. The purpose was to improve the quality of life and to give response to social problems in Jatinangor. The policy is the Regulation of Sumedang Regent Number 12/2013 on Building and Environment Plans in Provincial Strategic Area Education Jatinangor. The government expects to solve problems in Jatinangor, so that there will be improvement in people's welfare.

The problems that exist in Jatinangor include the construction of unorganized houses/buildings, narrow streets and traffic jams on the roads, and the pile of garbage. In addition, the population growth in Jatinangor also continues to increase at a rate that has reached 2.6. The number is considered to be the highest in Sumedang, and the population density in Jatinangor reaches 3,504 people per square kilometer (the Regency Profile of Sumedang, 2013).

According to the social policy sphere, the regulation shall be interpreted within administrative sphere. Therefore, this research will be centralized in the social policy interpretation of the development in Jatinangor, Sumedang. The question of this research is how regulations and base values are interpreted by the local government. The research was conducted using a qualitative approach. The informants of this research were up to 11 administrators whose task is directly related to the interpretation of the policy regulations of the Regulation of Sumedang Regent Number 12/2013.

2 FRAMEWORK

Social policy was formulated to allocate human resources in order to meet the desired goals by all means that correspond to the dominant values. Hence, the success of social policy is not only measured by the achievement of development goals, but also the strengthening of dominant values. Meanwhile, it can be considered as a success if the policy can be well interpreted by all parties, especially in the administrative sphere.

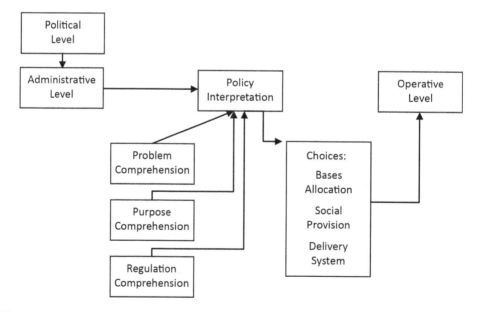

Figure 1. Framework scheme.

Activities at the administrative sphere or the interpretation phase of this policy will be perceived according to the choices in interpreting a policy: bases of allocation, social provision, delivery system, and finance. Those choices will provide values (equality, equity, and adequacy) and criteria (efficiency, effectiveness, cost-effectiveness, or cost benefit) set by the government. These values are mentioned in the government's choice of bases of allocation and social provision aspects. Meanwhile, the criteria are mentioned in the government's choice of the delivery system and finance. The following chart presents the framework of policy interpretation.

As we can see from the chart, social policy has three spheres: the political sphere (policy formulation), administrative sphere (policy interpretation), and operative sphere (policy implementation). In this case, the policy was observed in the administrative sphere or policy interpretation. Then the focus of the research was the activities carried out at the administrative level, or policy interpretation. Several factors are used as the foundation for interpreting the policy, including the understanding of social problems as the policy target, the policy purpose, and related regulations. In addition, there are some foundations of values and criteria that underlie the choices in policy interpretation. The choice aspects comprise bases of allocation, social provision, delivery system, and finance.

3   POLICY INTERPRETATION AT THE ADMINISTRATIVE SPHERE

At the administrative sphere, the Regulation of Sumedang Regent Number 12/2013 was interpreted through the administrators' understanding, the choices taken, and also the obstacles in its implementation. Interpretation at administrative sphere is described from the understanding of the administrators about the social problems, policy objectives, and related regulations. Meanwhile, the choices that were taken are related to the determination of program receivers, the program itself, program delivery, and funding.

Based on the findings, the administrators interpret the Sumedang Regent's Regulation as a policy with the purpose of coping with social problems in Jatinangor as the object of the policy. This finding is in line with Ellis' (2008) opinion on the main process in interpreting a policy. Ellis states that the main processes in interpreting a policy consist of identification, definition, and problem legitimation.

In addition to coping with the social problems in Jatinangor, the purpose, or the objective, of the policy is also understood as an effort to improve the social welfare of the society. This understanding is in sync with Midgley's views about social welfare. Midgley (1995) asserts that a good management of social issues will be able to create social welfare.

The understanding about the policy objectives also shows the government's effort in transforming the tone of voice at the administrative sphere. Jamrozik (2011) explains about how

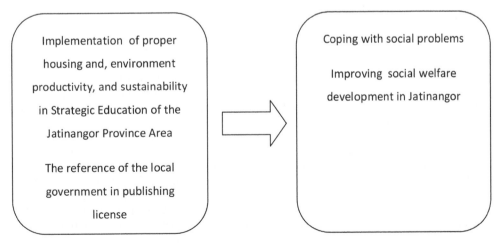

Figure 2.   The movement of "Tone of Activity". Source: Field finding, 2014.

the manifest tone becomes the assumed tone at the administrative sphere. Manifest tone is also an aim enclosed in the Regulation of Sumedang Regent Number 12/2013 as the foundation for the realization of proper housing and environment, productivity, and sustainability in Province Strategic Area Education in Jatinangor. Meanwhile, the assumed tone raised by the administrators is the regulation aimed at coping with social problems as well as improving social welfare in Jatinangor.

In interpreting the regulation, administrators did not only consider the understanding of the said regulation's goals, but also referred to other related regulations, such the Local Strategic Planning of Sumedang and the tasks of each local government agency. This is in line with Di Nitto (2003) who elaborates that interpreting policy requires rules, regulations, and guidelines.

Other findings also show that the understanding of these main tasks has been accomplished by forming work groups. The formation of work groups can be described as the preparation of human resources. Ellis (2008) proposed that one of the main processes in the translation of a policy is the preparation of human resources (staff or practitioners). This is in line with Di Nitto (2003) who assumes that forming, organizing, and employing a new institution have been conducted to develop a new policy.

Therefore, policy interpretation by administrators is based on how well their understanding of social problems, objectives, and related regulations is. Those will explain the tone of activity in the administrative sphere. The following chart shows the movement of tone of activity in the Regulation of Sumedang Regent Number 12/2013.

From the chart, we can see that the tone of voice in the Regulation of Sumedang Regent Number 12/2013 is the manifest tone. The manifest tone has a normative purpose, and it requires to be translated into more operational service programs. In the administrative sphere, the tone of activity has been changed into the assumed tone by the administrators. The administrators assume that the manifest tone of the policy is as the government's effort to overcome social problems and to increase social welfare development in Jatinangor. The administrators' assumptions about the manifest tone cannot be separated from their comprehension on the social issues in Jatinangor targeted by the policy and the understanding of how certain regulations are related with the Regents' Regulation. However, the assumed tone of the administrators has been affected by related regulations, such as the local strategic plan, and the task of each local government agency.

Besides the understanding of administrators in policy interpretation of the Regulation of Sumedang Regent Number 12/2013, the administrators also determined the implementation in terms of a program receiver category, an offered program, and a funding activity. It is known that administrators interpret the beneficiaries or the target of the Regulation of Sumedang Regent Number 12/2013. Determining the beneficiaries was categorized by Gilbert and Terrell (1993) into the bases of allocation aspect. According to them (1993), the choices in the bases of allocation are either selective or universal. From these choices, the administrators interpret the policy target or the beneficiaries as universal nature.

A further implementation category in the policy interpretation of the regulation is the determination of the program given. Gilbert and Terrell (1993) categorized the program that is given into the social provision aspect. It is a form of service or program established for the policy targets or beneficiaries. They argue that the choice in social provision is intangible and limited or concrete and diversified. The findings note that administrators' choice in the social provision aspect in the regulation is concrete and diversified service.

From the government choice in both categories, some values can be identified. Gilbert and Terrell (1993) argue that values rising from choice aspects are equality, equity, and adequacy. It is presumed that the intrinsic values of the administrators' choice in both aspects are equality and equity. It shows that the administrators consider equal treatment for all citizens. In addition, the equity value also arises from the policy, which can be seen from the recognition of citizens' rights.

The next aspect is program delivery. Gilbert and Terrell (1993) state that choices in the delivery system are maintenance linked to income or public, private, and free-standing. From the field findings, it is known that the administrators interpreted the delivery strategy of program or service from Regents' Regulation as public, private, and free-standing. The delivery strategy program chosen by government officials involved public and private sectors.

Lastly, the chosen aspects taken by administrators were funding or finance determination. Gilbert and Terrell (1993) assert that the choice in finance is an open-ended categorical grant or fixed-amount block grants. Current conditions show that financial resources opted by administrators to run the program are fixed-amount block grants. Sources of funding may come from the government's budget, both central and local, and also from the private sector through corporate social responsibility programs.

From the last two aspects (service delivery and funding), there are several criteria utilized by the administrators. Jansson (2009) argues that the criteria used in the policy are based on the aspects like efficiency, effectiveness, cost-effectiveness, and cost-benefit. From the four criteria, only two were used by the administrators to make the choices concerning the interpretation of the Regulation of Sumedang Regent Number 12/2013, namely effectiveness and efficiency. Effectiveness is perceived from the strategy of program delivery which involves the society. Likewise, efficiency appears when government officials involve private sectors as program funding resources. The following chart is made to give comprehensive understanding of the policy implementation.

From the chart above, it can be seen that there are implementation choice aspects in policy interpretation by administrators. The choices in bases of allocation and social provision include equality and efficiency, while the choices in the delivery system and financial aspects include effectiveness and efficiency. In bases of the allocation aspect, the beneficiaries are interpreted as universal. Thus, from the social provision aspect, the program is interpreted as concrete and diversified. The delivery system aspect is interpreted as public, private, and free-standing. Lastly, administrators have interpreted the financial aspect as a fixed amount block grant.

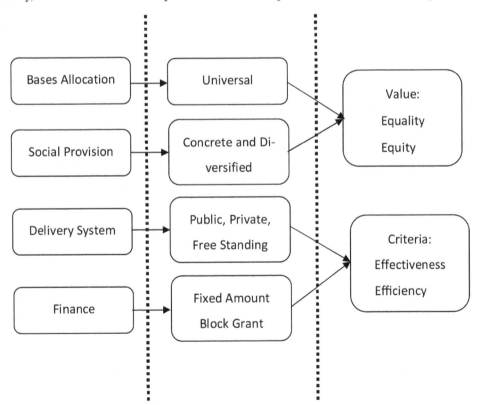

Figure 3. Implementation choice aspects in interpreting the policy of the Regulation of Sumedang Regent Number 12/2013. *Source: Research Result 2014.*

4 CONCLUSION

Administrators' understanding about the Regulation of Sumedang Regent Number 12/2013 can be seen from how the administrators interpret the purpose of the policy. It is also related to their understanding of social issues and other regulations. A comprehension of the policy's purpose also reveals a shift in the tone of voice at the administrative sphere, from a manifest tone to an assumed tone. This shows that the shift in tone of voice occurs not only from the political sphere to the administrative sphere, but also at the administrative sphere itself.

The value in policy interpretation based on choices in the four main aspects namely bases of allocation, social provision, the delivery system, and finance. The bases of the allocation aspect show that administrators chose the target universally. This means that the beneficiaries are the people in Jatinangor. Then, in terms of the social provision aspect, the administrators chose the programs and services that were concrete and diversified. From those two aspects (bases of allocation and social provision), the values that appeared are equality and equity. Equality can be seen from the equal treatment of all Jatinangor citizens. Meanwhile, equity can be seen from the principles of social justice that emphasize the importance of respecting the rights of every member of the society.

In terms of the delivery system aspect, the choices that appear are public, private, and free-standing, where administrators selected community involvement in the program delivery. Then in the finance aspect, administrators chose fixed-amount block grants from the government's funds and private sectors. From the delivery system and the financial aspect, criteria that appear are effectiveness and efficiency. Effectiveness is perceived from the administrators' choice to involve the people in the program delivery. However, efficiency can be seen from the budget provision chosen by the government which involves the corporate social responsibility program of the private sector to support the program implementation.

REFERENCES

Alcock, P., Erskine, A. & May, M. (2003) *The Students' Companion to Social Policy*. Oxford, Blackwell.

Alston, M. & Wendy, B. (1998) *Research for Social Workers an Introduction to Methods*. Sydney, NSW.

Anasiru, R. (2011) Implementasi model-model penanggulangan anak jalanan di kota Makassar. *Sosiokonsepsia*. 16 (02).

Astuti, M. *et al.* (2013) *Kebijakan Kesejahteraan dan Perlindungan Anak: Studi Kasus Evaluasi Program Kesejahteraan Sosial Anak di Provinsi DKI Jakarta, DI Yogyakarta, dan Provinsi Aceh*. Jakarta, P3KS.

Babbie, A.R. (2005) *Research Methods for Social Work*. Belmont, Thomson Brooks/Cole.

Bappeda Kabupaten Sumedang. (2009) Laporan Akhir Studi Kelayakan Kawasan Jatinangor Sebagai Kawasan Perkotaan.

Bappeda Kabupaten Sumedang. (2013) Profil Daerah Kabupaten Sumedang Tahun 2013.

Blackmore, K. & Edwin, G. (2007) Social Policy an Introduction. New York, McGraw-Hill.

Chambers, D.E. & Kenneth, R.W. (2005) Social Policy and Social Programs: a Method for the Practical Public Policy Analyst. Boston, Pearson Education Inc.

Chambers, D.E., Kenneth, R.W. & Rodwell, M. (1992) *Evaluating Social Programs*. Boston, Allyn and Bacon.

Colby, I.C. (2008) *Comprehensive Handbook of Social Work and Social Welfare (Social Policy and Policy Practice)*. John Wiley & Sons, Inc.

Creswell, J.W. (2002) *Desain Penelitian Kualitatif*. Jakarta, KIK Press.

Deacon, A. (2002) *Perspectives on Welfare*. Buckingham, Open University Press.

Di Nitto, D.M. (2003) *Social Welfare Politics and Public Policy*. USA, Allyn and Bacon.

Dobelstein, A.W. (1991) *Social Welfare Policy and Analysis*. Chicago, Nelson-Hall Publishers.

Dubois, B. & Karla, K.M. (2010) Social Work: An Empowering Profession. Boston, Pearson Education Inc.

Ellis, R.A. (2008) Policy Practice. In: Colby, I.C. (eds.) *Comprehensive Handbook of Social Work and Social Welfare (Social Policy and Policy Practice)*. John Wiley & Sons, Inc.

Fatony, A. (2011) Kebijakan pengentasan kemiskinan berbasis *participatory poverty assessment*: Kasus Yogyakarta. *Sosiokonsepsia*. 16 (02).

Friedlander, W. (1980) *Introduction to Social Welfare*. Englewood Cliffs, NJ, Prentice Hall.

Gilbert, N. & Paul, T. (1993) *Dimensions of Social Welfare Policy*. Massachusetts, Allyn and Bacon.

Gough, I. & Geof, W. "Welfare Regimes: Linking Social Policy to Social Development". http://www.staff.bath.ac.uk.

Habibullah & Ivo, N. (2013) *Kebijakan Pendamping Program Keluarga Harapan*. Jakarta, P3KS.

Hall, A. & James, M. (2004) *Social Policy for Development*. London, Sage Publications.

Jamrozik, A. & Nocella, L. (1998) *The Sociology of Social Problems: Theoretical Perspectives and Methods of Intervention*. Cambridge, Cambridge University Press.

Jamrozik, A. (2001) *Social Policy in the Post-Welfare State: Australians on the Threshold of the 21st Century*. Australia, Longman.

Jansson, B. (2008) *Becoming an Effective Policy Advocate: Form Policy Practice to Social Justice*. Pacific Groove, CA, Brooks/Cole.

Kirst-Ashman, K. (2010) *Introduction to Social Work and Social Welfare: Critical Thinking Perspective*. Canada, Brooks/Cole, Cengage Learning.

Kwok, J.K.F. (2008) Social Justice for Marginalized and Disadvantaged Groups. In Colby, I.C. (eds.) *Comprehensive Handbook of Social Work and Social Welfare (Social Policy and Policy Practice)*. John Wiley & Sons, Inc.

Mendoza, T.L. (1981) *Social Welfare and Social Work, an Introduction*. Cebu City, E.Q. Cornejo & Sons.

Midgley, J. & Michael, S. (2009) The Social Development Perspectives in Social Policy. In: Midgley, J., Tracy, M.B. & Livermore, M. (eds.) *The Handbook of Social Policy*. London, Sage.

Midgley, J. & Michelle, L. (2009) *The Handbook of Social Policy*. California, SAGE Publications Ltd.

Midgley, J. (1995) *Social Development: The Developmental Perspective in Social Welfare*. London, SAGE Publications Ltd.

Minichiello, V.R.A. (1995) In-Depth Interviewing. Melbourne Australia, Longman.

Mkandawire, T. (2005) *Social Policy in a Development Context*. Geneva, UNRISD.

Mujiyadi, B. & Setyo, S. (2013) *Evaluasi Program Bedah Kampung*. Jakarta, P3KS.

Neuman, L.W. (2006) *Social Research Methods*. Boston, Allyn and Bacon.

Padmi, T.A., *et al.* (2013) *Studi Kebijakan Penanggulangan Bencana Alam Berbasis Masyarakat: Studi Kasus Kampung Siaga Bencana dalam Mengurangi Resiko Bencana Alam, di Kota Padang, Provinsi Sumatera Barat dan Kabupaten Sleman, Provinsi D.I. Yogyakarta*. Jakarta, P3KS.

Purwanto, A.B. & Moch. S. (2013) *Kebijakan Pembangunan Kesejahteraan Sosial di Daerah Perbatasan Antar Negara: Studi di Kabupaten Sambas, Kalimantan Barat*. Jakarta, P3KS.

Rein, M. (1983) *From Policy to Practice*. London, Macmillan.

Rubin, A. & Earl, R.B. (2008) *Research Methods for Social Work*. Belmont, Thomson Brooks/Cole.

Spicker, P. (1995) *Social Policy: Themes and Approaches*. London, Prentice Hall.

Stein, T.J. (2001) *Social Policy and Policy Making: The Branches of Government and the Public-at-Large*. New York, Columbia University Press.

Suharto, E. (2005) *Analisis Kebijakan Publik*. Bandung, Alfabeta.

Sumarno, S. & Haryati, R. (2013) *Evaluasi Program Keserasian Sosial dalam Penanganan Konflik Sosial*. Jakarta, P3KS.

Supeno, E. (2006) Implementasi kebijakan jaring pengaman sosial-operasi pasar khusus beras (JPS-OPKB) keluarga pra sejahtera. *Masyarakat, Kebudayaan, dan Politik*. 19 (1).

Suradi, *et al.* (2013) *Kebijakan Pemberdayaan Sosial Komunitas Adat Terpencil*. Jakarta, P3KS.

Suradi. (2012) Studi evaluasi dampak kebijakan sosial: Rehabilitasi sosial rumah tidak layak huni bagi keluarga miskin di kota Banjarmasin. *Sosiokonsepsia*. 17 (02).

Thompson, N. (2005) *Understanding Social Work: Preparing for Practice*. New York, Palgrave Macmillan.

Weimer, D. & Vining, A. (1992) *Policy Analysis: Concepts and Practice*. Englewood Cliffs, NJ, Prentice Hall.

Widodo, N., Ruaida, M., Anwar, S. & Togiaratua, N. (2010) *Studi kebijakan pengembangan kegiatan satuan bakti pekerja sosial di panti sosial masyarakat*. Jakarta, P3KS.

Zastrow, C. (2010) *Social Work and Social Welfare*. Canada, Brooks/Cole, Cengage Learning.

Indonesia's subnational competitiveness and preparedness for ASEAN economic community 2015: Mapping literatures

A. Virgianita & S.A. Choiruzzad
Department of International Relations, Faculty of Social and Political Sciences, Universitas Indonesia, Depok, Indonesia

ABSTRACT: This article scrutinizes and develops categories on available literatures concerning "the competitiveness and preparedness of Indonesian subnational actors in the context of AEC 2015." It then classifies the literatures into four typologies, namely subjective nature (perception), objective measures, multidimensional, and specific ones. We present these typologies into quadrants. From this, we understand that most of the literatures are written in Indonesian because of the orientation of the research (i.e. policy recommendation). We also find that most of the literatures look at a particular actor/region and prefer to look at the objective indicators rather than the subjective ones. Based on these findings, we argue that there is a 'double-gap' in the subnational competitiveness and preparedness in Indonesia. The first is a regional gap of competitiveness and preparedness level, especially between Java and non-Java areas. However, a more comprehensive study is needed. Secondly, there is also a gap in the studies themselves. Most of the studies were looking at more developed areas in Indonesia and only few studies look at the less-developed areas. The fact that most studies highlight more on developed areas implies the disparity of research networks between Java and non-Java universities.

1 INTRODUCTION

The willingness of ASEAN leaders to move forward with the schedule for ASEAN Community from 2020 to 2015 in the 12th ASEAN Summit in Cebu, the Philippines, in 2007, illustrated the optimism that economic integration will bring benefit to the people in the region. This optimism mirrors the argument pointed out by many economists, most of whom are staunch supporters of the regional integration project that the ongoing process of economic liberalization and regional integration have brought tremendous benefits for all ASEAN member countries. In the 1990s, when the economic liberalization and the regional integration project started to gain stronger support in the region, the poverty rate in ASEAN was 45%.

The number declined to 14% in 2010, or 15.6% with the inclusion of Myanmar. The number of middle class people in ASEAN countries also rose from 15% in 1990 to 37% in 2010 (Intal, 2014). This optimism is also apparent in the ASEAN Economic Community Blueprint. The Blueprint envisions ASEAN Economic Community that will establish Southeast Asia as *"a highly competitive economic region, a region of equitable economic development, and a region fully integrated into the global economy."*

Amidst this optimism, the plan to establish the ASEAN Economic Community 2015 has attracted a lot of attention from policymakers and academics alike. While many literatures on the response of ASEAN member countries are plenty, there is a deficit of literature on the responses towards the ASEAN Economic Community at the subnational level of ASEAN member countries, especially those written in English for international audience. In the case of Indonesia, many of the literatures on the responses of subnational actors towards AEC, ranging from perception to policies to competitiveness, are written in Indonesian and thus are not readily available for international audience. Many of these literatures are focusing on "how prepared is a particular actor/province/municipality to face the AEC." Due to the limitation of literatures on this issue in English, this article aims to provide a better understanding on the responses towards the ASEAN Economic Community at the subnational level in Indonesia by surveying available literatures on "the competitiveness/preparedness level of Indonesian subnational actors (including but not limited to local governments) in the context of AEC 2015" (including academic articles, research reports, and policy papers) written by scholars and government officials. Then this article classifies them to help understand the trend of the literatures discussing this topic. A survey on literatures is also important to understand the characteristics of the research in Indonesia over ASEAN, as an important aspect of the preparedness itself.

2 METHOD

This study collects available literatures focusing on "the competitiveness/preparedness of Indonesian subnational actors (including but not limited to local governments) in the context of AEC 2015." This study found 42 available literatures regarding this topic. After gathering the literatures, our team identifies, compares, classifies, and maps those literatures to identify the patterns and trends on the studies on this issue.

Our literature survey found that the literatures are very diverse. To help find patterns amidst this diversity, our study classifies the literatures based on two indicators:

1. What is the aspect studied? Is it subjective or objective aspect?
2. How specific is the study? Is it multidimensional/multifactor or specific?

The aspect indicator differentiates literatures based on what aspect the literatures are focusing on. Are they focusing on subjective aspects, such as perception and level of confidence, or on objective aspects that are quantifiable, such as infrastructure, macroeconomic indicators, governance, or human resources indicators? The second indicator looks at the scope of the literatures. Are they looking at various sectors and actors simultaneously, or are they focusing on a specific sector or actor in a specific area?

These two indicators for classification are based on the general diversity of the literatures looking at different actors, ranging from general people, government officials, students, businesspersons, and others. The literatures also look at different scopes of sectors. Some of them look only at one specific sector, while some others look at various sectors simultaneously, such as sectors related to the *Priority Integration Sector* (PIS) and the *Mutual Recognition Arrangement* (MRA) of ASEAN.

Based on the abovementioned indicators, literatures are then classified into four quadrants. The first quadrant is "subjective-multidimensional." The literatures fall in this category are those looking at subjective aspects such as perception and level of confidence of multiple types of actors spread in multiple sectors. The second quadrant is the "objective-multidimensional" box, reserved for literatures that look at objective aspects of competitiveness or preparedness of multiple actors spread in multiple sectors. The third quadrant is "objective-specific." The literatures that look at objective aspects of a specific actor in a specific sector fall into this category. Lastly, the fourth quadrant, "subjective-specific", is the category for the literatures that look into subjective aspects of a specific actor in a specific sector.

3 MAPPING THE LITERATURES

3.1 *Subjective-multidimensional*

Among 42 literatures, only two look at the subjective aspect of the responses of actors in various sectors. Ariani, *et al.* (2014) from Gadjah Mada University surveyed the opinion of youths in seven different sectors in Yogyakarta to understand their perception towards the

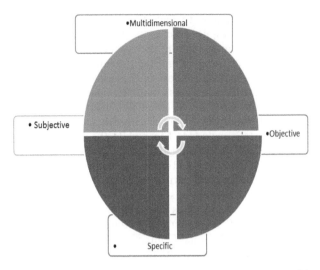

Figure 1. Aspect and scope quadrant for literatures on subnational competitiveness/preparedness towards MEA.

ASEAN Economic Community. The respondents were 400 youths spread among 7 backgrounds: high school students, university students, young entrepreneurs, academicians, artists, media workers, and activists. The 7 categories were selected because these are the youth segments with the most exposure to media and information technology and will face the biggest impacts from the implementation of AEC. Our initial assumption was that since the city of Yogyakarta is a center for education with reputable educational institutions, the youths of Yogyakarta must have strong knowledge and awareness about AEC. However, the study found that only young academicians and activists have sufficient knowledge about ASEAN. Among entrepreneurs, only 36% knows ASEAN and ASEAN Community. The perception about ASEAN is also mixed, with 32.8% respondents having positive impression towards the establishment of ASEAN Community as an opportunity and 31.5% respondents seeing ASEAN Community as a threat.

A wider study by Pudjiastuti, *et al.* (2015) on 16 cities and municipalities in Indonesia also found a similar pattern. While this study did not focus on youths (the respondents were divided into 'general public' and 'business sector' with a wider range of age), the study also found that the knowledge about the ASEAN Economic Community is relatively low. Only 27.80% respondents in the business sector that were surveyed knew the ASEAN Economic Community, and the percentage was slightly better than that of general public members that only reached 25.9%. However, it is interesting to note that despite the lack of knowledge, 41.5% of the total respondents believe that they were ready to face the ASEAN Economic Community (32.5% mentioned that they were not ready and 26% did not know). Another interesting finding from this study is that Malaysia, Singapore, and Thailand are perceived as the main competitors in the ASEAN Economic Community.

3.2 *Objective-multidimensional*

The literatures in the objective-multidimensional category look at the competitiveness and preparedness of different regions in Indonesia. Some of them compare the competitiveness level of Indonesian provinces, such as "*Membandingkan Provinsi: Kajian Kebijakan Penguatan Daya Saing Daerah dalam Rangka Peningkatan Kesejahteraan Rakyat*" (Comparing Provinces: the Study on the Regional Competitiveness Policy towards People's Welfare Development) (Kementerian Keuangan RI, 2014), "Subnational Competitiveness and National Performance: Analysis and Simulation for Indonesia" (Tan & Amri, 2013), and "*Peningkatan Daya Saing Daerah dalam Menghadapi* ASEAN Economic Community 2015"

(Increasing Regional Competitiveness to Face the ASEAN Economic Community 2015) (Joedo & Widyasanti, 2013). Despite differences in methodologies, the three literatures similarly argue that there is a regional gap in the competitiveness level among provinces in Indonesia. The provinces situated in Java and the provinces with rich natural resources are more competitive than others, especially those in the eastern part of Indonesia. In addition to comparing provinces, some literatures also compare major cities in Indonesia. The paper "*Daya Saing Kota-Kota Besar di Indonesia*" (Big Cities' Competitiveness Level in Indonesia) (Santoso, 2009) maps and elaborates 24 major cities in Indonesia whose populations are above 500,000 people (Jakarta is excluded). Meanwhile, the article "*Peningkatan Daya Saing Daerah dalam Era Globalisasi*" (Increasing Regional Competitiveness Level in the Era of Globalization) (Simorangkir, 2013) maps all cities and municipalities. Both studies found that cities in Java tend to have a higher competitiveness level.

There is also one literature that looks at some selected provinces. Compared to the literatures looking at all provinces or many cities across all provinces that tend to make rankings, the article "*Prakarsa Strategis Peningkatan Daya Saing Daerah dalam Menghadapi MEA*" (Strategic Initiatives to Increase Regional Competitiveness Level to Face the AEC) (Hadi, *et al.*, 2013) elaborates the challenges affecting the competitiveness level in some selected cities in more detailed. The selected cities are Surabaya, Medan, Makassar, and Manado. Looking at the challenges, the literature also maps recommendations to raise the competitiveness level, including identifying the leading sectors in each city.

In observing a similar issue in the national level and across provinces, the paper "*Tenaga Kerja Terampil Indonesia dan Liberalisasi Jasa ASEAN*" (Indonesian Skilled Workers and Liberalization of Services in ASEAN) (Keliat, *et al.*, 2014) looks at a more specific aspect. The study looks at the competitiveness and preparedness levels of professionals in 8 areas (architecture, engineering, nursing, medical practice, dentistry, land survey, accounting, and tourism) included in ASEAN Mutual Recognition Arrangements (MRAs).

In this category, there are also several literatures that compare cities and municipalities within a single province, such as in Southeast Sulawesi (Irawati, *et al.*, 2012) and East Java (Huda & Santoso, 2014). While the two literatures look at the competitiveness level of the cities, the article "*Kesiapan Industri Tekstil dalam Mendukung Poros Maritim dan Peningkatan Daya Saing*" (Textile Industry Preparedness to Support Maritime Fulcrum and to Increase Competitiveness) (Prasetyo, 2015) looks at the competitiveness level of cities and municipalities in textile sector. In this objective-multidimensional category, there is also a literature that compares the competitiveness of two sectors in one municipality, namely the Rattan and Batik industries in Cirebon, West Java (Rusnidah, *et al.*, 2013).

3.3 *Objective-specific*

The objective-specific category is the category with most literatures. From 42 literatures collected for this research, more than half (22 literatures) fall in this category. It is understandable since many government institutions and universities seek to provide policy recommendations for local governments.

It is interesting to note that the competitiveness level of small and medium enterprises (SMEs), tourism, and creative industries are the most frequent topic discussed by the literatures. In terms of locality, most look at cities or municipalities, mostly those located in Java and Sumatera. The detailed list of the literatures will be provided in the next section.

3.4 *Subjective-specific*

In the subjective-specific category, two literatures look at perceptions and attitudes towards ASEAN. While the paper "*Indonesia Perceptions and Attitudes towards the ASEAN Community*" (Benny & Abdullah, 2011) provides survey findings on the perceptions and attitudes of general public in five major cities (Jakarta, Surabaya, Medan, Pontianak, and Makassar), the paper "*Persepsi Publik Indonesia terhadap ASEAN* (Public Perceptions in Indonesia towards ASEAN)" (Takwin *et al.*, 2015) looks at the public perception in Jakarta. It is interesting to note that the literatures found no correlation between the respondents' knowledge, perception, and attitude towards ASEAN Community. Benny & Abdullah (2011) argue that this indicates that ASEAN remains separated from the general public.

In addition to the literatures that look at the perceptions of general public, there are also literatures that look at the perception of local governments in some specific cities and provinces (Virgianita, 2013) and people involved in particular sectors, such as the sectors of Small and Medium Enterprises (Sulistyawati & Permana, 2015) and tourism (Chandra & Munthe, 2012). Virgianita *et al.* (2013) look at the perceptions of government officials in Jakarta, Surabaya, and Makassar towards the ASEAN Economic Community. The researchers found that local government officials in Makassar have less knowledge about ASEAN compared to their counterparts in Jakarta and Surabaya. However, ASEAN is similarly not discussed frequently among local government officials in Jakarta, Surabaya, and Makassar.

Sulistyawati & Permana (2015) from ASEAN Studies Center, Gadjah Mada University look at the perceptions of small and medium business owners in Yogyakarta. Generally, the small and medium business owners in Yogyakarta have lack of knowledge about ASEAN Economic Community but still see the establishment of AEC as an opportunity. Chandra and Munthe (2012) conducted a research on the perceptions of tourism professionals in West Java towards the ASEAN Economic Community. They find that, generally, the tourism professionals in West Java are optimistic towards AEC.

4 CONCLUDING NOTES: DOUBLE GAPS

From this literature mapping, we understand that most of the literatures are written in Indonesian because of the orientation of the research (i.e. policy recommendation). We also identify that most of the literatures look at particular actors or specific areas and prefer to look at objective indicators (e.g. macroeconomic indicators and infrastructure) rather than subjective ones (e.g. perception of actors). Based on the four quadrants, most of the studies fall into the third category, which is the "objective-specific" category. Infrastructures, government policies, natural and human resources are the most popular objective indicators that the researchers looked at. In terms of sector, the most discussed topics in these literatures are the sectors of Small and Medium Enterprises, tourism, and creative industry. There are very few literatures looking at the sectors of manufacture or agriculture. In terms of provinces studied in these literatures, we found that there is no study looking comprehensively at the whole Indonesia's 34 provinces simultaneously. Most of the studies only look at one or a number of provinces (with the largest number being 13 provinces).

Based on these findings, we argue that there is a double-gap in a subnational competitiveness/preparedness level in Indonesia. The first is a regional gap of the competitiveness and preparedness levels. While a more comprehensive study is needed, existing literatures provide us with the understanding that there has been a great disparity on the competitiveness and preparedness levels among different provinces, especially between the Java and non-Java provinces.

The second gap is the gap in the studies themselves. Most of the studies only look at more developed areas in Indonesia (i.e. provinces in Java, Bali, and some other rich provinces), and only a few look at the less-developed areas (i.e. provinces in the eastern part of Indonesia). This pattern is consistent with and complementary to the previous studies on the wider issue of research in Indonesia, showing the dominance of applied research and the bureaucratization of research activities, as well as the gap of the scale of research networks between Java and non-Java Universities (as shown in Rakhmani and Siregar, 2016).

REFERENCES

Abdullah., K. & Benny, G. (2011) Indonesia Perceptions and Attitudes towards the ASEAN Community. *Journal of Current Southeast Asian Affairs*, 39–67.

Amurwanti, D.N. (2014) Dissipating Disparity: The Case for ASEAN Economic Community. *ASEAN Insights.* 1, June 2014.

Andriyoso, A. (2015) *Analisis Kesiapan Industri Kreatif Menghadapi AEC 2015 Menggunakan Metode Regresi Linier Berganda: Studi Kasus Kecamatan Pasar Kliwon, Kota Surakarta* [*The Analysis of the Preparedness of Creative Industry towards AEC 2015 Using Double Linear Regression Method: A Case of Pasar Kliwon, Surakarta City*]. Surakarta: Universitas Muhammadiyah Surakarta.

Benny, G. & Abdullah, K. (2011) Indonesia perceptions and attitudes toward the ASEAN community. *Journal of Current Southeast Asian Affairs.* 30 (1), 39–67.

Boediman, S.F. & Agie, P.R.W. (2013) *Studi Tingkat Daya Saing Destinasi Pariwisata Budaya di Indonesia: Kasus Kota Yogyakarta* [*A Study of the Competitiveness of Cultural Tourism Destination: A Case of Yogyakarta City*]. Jakarta, Sekolah Tinggi Pariwisata Trisakti.

Chandra, A. & Munthe, A.G. (2012) *Profil Kesiapan Daerah dalam Memasuki Masyarakat Ekonomi ASEAN Studi Kasus: Sektor Kepariwisataan Jawa Barat 2012* [*The Profile of the Preparedness of Subnational in Facing AEC: A Case of West Java Tourism*]. Bandung, Lembaga Penelitian dan Pengabdian Masyarakat Universitas Parahyangan.

Hadi, S. (2013) *Prakarsa Strategis Peningkatan Daya Saing Daerah dalam Menghadapi Masyarakat Ekonomi ASEAN* [*Strategic Initiatives to Increase of Regional Competitiveness to Face AEC*]. Jakarta, Badan Perencanaan dan Pembangunan Nasional.

Huda, M. & Santoso, E.B. (2014) Pengembangan Daya Saing Daerah Kabupaten/Kota di Provinsi Jawa Timur berdasarkan Potensi Daerahnya [The Development of Subnational Competitiveness at the City and Province Level in East Java]. *Jurnal Teknik ITS.* 3 (2), C81–C86.

Joedo, P. & Widyasanti, A. (2013) Peningkatan Daya Saing Daerah dalam Menghadapi ASEAN Economic Community 2015 ["Increasing Regional Competitiveness to Face the ASEAN Economic Community 2015"]. *Konferensi Seminar Nasional dan Sidang Pleno ISEI XVI.* Jambi, Ikatan Sarjana Ekonomi Indonesia. pp. 25–45.

Keliat, M. *et al.* (2014) *Tenaga Kerja Terampil Indonesia dan Liberalisasi Jasa ASEAN (Indonesian Skilled Workers and the Liberalization of Services in ASEAN).* Insist Press, Yogyakarta.

Kementerian Keuangan R.I. (2014) Perbandingan antar Provinsi: Studi tentang Kebijakan Daya Saing Wilayah terhadap Pembangunan Kesejahteraan Masyarakat (Comparing Provinces: The Study on the Regional Competitiveness Policy towards People's Welfare Development).

Mahatir M.R.E. & Juanda, B. (2013) Peranan Komoditi Kelapa Sawit dalam Perekonomian Daerah Provinsi Jambi: Analisis Input-Output Tahun 2000, Tahun 2010 dan perbandingannya [The Role of Palm Oil in Local Economy of Jambi Province: Input—Output Analysis for 2000, 2010 and Its Comparison]. *Prosiding Konferensi Seminar Nasional dan Sidang Pleno ISEI XVI.* Jambi, Ikatan Sarjana Ekonomi Indonesia. pp. 242–253.

Meliala, A.S., Matondang, N. & Sari, R.M. (2014) *Strategi Peningkatan Daya Saing Usaha Kecil dan Menengah (UKM) berbasis Kaizen* [*A Strategy to Increase the Competitiveness of SMEs based on Kaizen*]. Medan, Universitas Sumatera Utara.

Nurhadi. (2015) Pemberdayaan Manusia Produktif dalam Menghadapi Masyarakat Ekonomi ASEAN (MEA) di Provinsi Banten. *The Empowerment of Productive Human towards AEC. Prosiding Konferensi Nasional Ilmu Pengetahuan dan Teknologi.* Bekasi: STMIK STBA Nusa Mandiri.

Pasaribu, S.M. (2014) Kajian Kesiapan Sektor Pertanian Menghadapi Pasar Tunggal ASEAN 2015. *Study of the Readiness of Agriculture Sector towards ASEAN Single Market 2015.* Jakarta, Pusat Sosial Ekonomi dan Kebijakan Pertanian, Badan Penelitian, dan Pengembangan Pertanian.

Permatasari, A. (2015) Kesiapan Kabupaten Cianjur dalam Menghadapi Komunitas ASEAN 2015. *The Readiness Cianjur District towards AEC 2015.* Bandung, Universitas Padjajaran.

Pranadji, T. (2003) Otonomi Daerah dan Daya Saing Agribisnis: Pelajaran dari Provinsi Lampung. *Analisis Kebijakan Pertanian.* 1 (2), 1–15.

Prasetyo, P.E. (2015) Kesiapan Industri Tekstil dalam mendukung Poros Maritim dan Peningkatan Daya Saing [Textile Industri Preparedness in Supporting Maritime Fulcrum and Competitiveness]. *Seminar Nasional Multi Disiplin Ilmu Unisbank SENDI_U.* Semarang, Universitas Stikubank.

Pudjiastuti, T.N. (2015) Strategi Peningkatan Pemahaman Masyarakat Tentang Masyarakat Ekonomi ASEAN. Pusat Penelitian Politik. Jakarta, Lembaga Ilmu Pengetahuan Indonesia (P2Politik-LIPI).

Putri, E.K., Rifin, A., Daryanto, H.K. & Istiqomah, A. (2014) Tangible value biodiversitas herbal dan meningkatkan daya saing produk herbal Indonesia dalam menghadapi masyarakat ekonomi ASEAN 2015. *Jurnal Ilmu Pertanian Indonesia.* 19 (2), 118–124.

Rahman, M.A. (2015) Daya saing tenaga kerja Indonesia dalam menghadapi masyarakat ekonomi ASEAN. *eJournal Ilmu Hubungan Internasional.* 3(1), 117–130.

Rakhmani, I. & Siregar, F. (2016) Reforming research in Indonesia: Policies and practice. Global Development Network, *Working Paper* Series, No. 92, February 2016.

Ridwansyah, D. & Lubis, D.A. (2013) Penguatan Produk Unggulan Sektor Industri di Provinsi Jambi dalam Menghadapi Era Masyarakat Ekonomi ASEAN (MEA) 2015. *Proceeding. National Conference ISEI XVI*. Jambi, Ikatan Sarjana Ekonomi Indonesia. pp. 128–137.

Rusnidah, I. (2013) Pengembangan dan Daya Saing Produksi Rotan dan Batik dalam Menghadapi Tantangan Masyarakat Ekonomi ASEAN (MEA) 2015: Studi Kasus Pada Industri Rotan dan Batik di Kabupaten Cirebon-Jabar. *Proceeding. National Conference ISEI XVI*. Jambi, Ikatan Sarjana Ekonomi Indonesia. pp. 197–211.

Santoso, E.B. (2009) Daya Saing Kota-Kota Besar di Indonesia. National Seminar, Institute Teknologi Surabaya.

Sari, D.F. (2015) Persiapan pemerintah Provinsi Kalimantan Timur menuju masyarakat ekonomi ASEAN 2015: Kebijakan Pembebasan Penanaman Modal. *Jurnal Ilmiah Mahasiswa FEB Universitas Brawijaya 2015*. 3 (2).

Schwab, K. (2009) *The Global Competitiveness Report 2009–2010*. World Economic Forum.

Simorangkir, I. (2013) Peningkatan daya saing daerah dalam era globalisasi. *Proceeding National Conference ISEI XVI*. Jambi, Ikatan Sarjana Ekonomi Indonesia. pp. 46–65.

Soekarno, G. (2013) Pertumbuhan industri kreatif di Surabaya melalui upaya triple helix dan keunggulan bersaing. *Proceeding. National Conference ISEI XVI*. Jambi, Ikatan Sarjana Ekonomi Indonesia. pp. 212–224.

Soeroso, A. (2013) Pemetaan dan pemeringkatan acara budaya kajian daya saing produk pariwisata DIY. *Proceeding. National Conference ISEI XVI*. Jambi, Ikatan Sarjana Ekonomi Indonesia. pp. 225–241.

Sulityawati, D.R. & Permana, M. (2015) *Memperkuat Sektor UKM dalam Merespon Masyarakat Ekonomi ASEAN 2015*. Yogyakarta, Pusat Studi ASEAN, Universitas Gajah Mada.

Sumardjoko, I. (2014) *Strategi Percepatan Transformasi Ekonomi Jawa Timur Melalui Penguatan Instrumen Fiskal dan Moneter Menghadapi Masyarakat Ekonomi ASEAN*. Malang, Ikatan Akuntansi Indonesia Wilayah Jawa Timur.

Susamto, A.A., Musthofa, M. & Amiadji. (2014) *Peta Sektor Unggulan 497 Kabupaten/Kota di Indonesia*. Yogyakarta, Universitas Gajah Mada.

Susilo, Y.S. (2010) Strategi Meningkatkan Daya Saing UMKM dalam Menghadapi Implementasi CAFTA dan Masyarakat Ekonomi ASEAN. *Buletin Ekonomi*. 8 (2), 70–170.

Syukriah, A. & Hamdani, I. (2013) Peningkatan Eksistensi UMKM melalui Comparative Advantage dalam rangka Menghadapi MEA 2015 di Temanggung. *Economics Development Analysis Journal*. 2 (2), 110–119.

Takwin, B. (2015) *Persepsi publik terhadap ASEAN*. Jakarta, INFID.

Tan, K.G. & Amri, M. (2013) Subnational Competitiveness and National Performance: Analysis and Simulation for Indonesia. *Journal of CENTRUM Cathedra the Business and Economics Research Journal*. 6 (2), 173–192.

Tarigan, A. (2009) Kerjasama Antar Daerah (KAD) untuk Peningkatan Penyelenggaraan Pelayanan Publik dan Daya Saing Wilayah. *Bulletin Online Tata Ruang Maret 2009 Badan Perencanaan dan Pembangunan Nasional*.

Virgianita, A., et al. (2013) *Persepsi dan Kebijakan Pemerintah Daerah dalam Menyongsong MEA 2015. Research Report*. Depok, Center for International Relations Studies Fakultas Ilmu Sosial dan Ilmu Politik Universitas Indonesia.

Virgianita, A. (2014) Dari Lokal ke Regional: Kesiapan Pemerintah Lokal Menyongsong Komunitas Ekonomi ASEAN 2015 (Studi Kasus Kota Makassar). *Bunga Rampai, FISIP UI*.

Wahyuni, S. (2008) A Review of Indonesia's Competitiveness and its Investment Climate. *Asia Competitiveness Institute Working Paper/Case Study Series*.

Victimisation in tax collection in Indonesia

P. Harahap & M. Mustofa
Department of Criminology, Faculty of Social and Political Sciences, Universitas Indonesia, Depok, Indonesia

ABSTRACT: Non-compliant taxpayers generate tax arrears and have placed a stressful burden on the country for many years. Considering the benefits of tax, there have been a number of efforts carried out by the Indonesian government to realise such liabilities. The state tax bailiff is one of the tax officials who have the authority to collect taxes and directly interact with non-compliant taxpayers. This article explains the victimisation process that has occurred and been experienced by state tax bailiffs and other tax officials when executing tax regulations, which, apparently in one case, led to criminal action (murder). By using the literature and document studies (statute and also a conceptual approach), further enhanced by the secondary and content analysis, in this article I contend that state tax bailiffs and other tax officials have experienced victimisation when performing their duties. Taking account of victimology, the implementation of tax collection from the public is related to three components that further lead to victimisation, such as the victimisers, the victimhood and the victims. The findings suggest that these victimisation components are experienced by state tax bailiffs and other tax officials when executing their responsibilities.

1 INTRODUCTION

At the beginning of the second quarter of 2016, Indonesia was shocked by the news of the murder of a state tax bailiff, one of the officers of the Indonesian Tax Authority. People were astonished because the perpetrator was a taxpayer who had IDR 14 billion tax arrears.

'A state tax bailiff named Parada Toga Fransriano and a security Soza Nolo Lase who worked in Sibolga Tax Office were killed while performing their duties in tax collection, stabbed by the unscrupulous taxpayer' (Press Release 12 April 2016 DGT).

Why does it happen? Is a tax that important that Toga Parada Fransriano had to lose his life over it? This article will explain the task implementation process of tax officials, especially state tax bailiffs, and link it with the victimisation process. Is there any victimiser? Who is the victim? And how does the victimhood happen? Nevertheless, I suggest that state tax bailiffs and other tax officials experience victimisation when carrying out their duties. What is more interesting is that in the tax collection process, there are indeed three components that further lead to victimisation, which are the victimiser, the victimhood and the victim. Due to the absence of this kind of study, it is essential to get a proper understanding about taxes, the method of tax collection employed by the government, and its relationship to the victimisation process and to criminology in general.

In 2016, Indonesia had nearly 38,000 tax officials who were given the task and the authority to collect taxes, either from individuals, companies or other legal entities. This figure will continue to grow along with the recruitment process. Considering the importance of taxes for Indonesia, the state devotes all its energy to generating more revenue from the tax sector. One strategy that has been employed is to improve the tax collection process or, in other words, to collect the accrued tax debt from the taxpayers. Tax collection efforts consist of persuasive or administrative efforts, as well as the ultimate step of using coercive efforts.

2 RESEARCH METHODOLOGY

The study was conducted by analysing the documents of the taxation regulations of Indonesia, or, in other words, using the statute approach. Furthermore, to understand the theory and opinion of victimology and taxation and to assess the theoretical framework, I conducted the literature review, which is also called a conceptual approach, to support the purpose of this research. Moreover, in order to complete the discussion, this article conducted a secondary analysis and a content analysis utilising the results of the research and other secondary data. The murder case of a state tax bailiff is the first case in Indonesia, but it will be used in this article to delineate one of the victimisations experienced by tax officials.

3 CRIMINOLOGY, VICTIMISATION, AND STRUCTURAL VICTIMISATION

Criminology is a science that learns about everything associated with crime. Three main features concerning the object of criminology are the law-making process or the sociology of law, criminal aetiology, which elaborates the theories that lead to crime, and the social reactions towards the offence committed. To some extent, crimes in the economic and tax sectors fall into the category of white-collar crime. Hence, they are regarded as an unconventional crime. In conjunction with victimology, victims arising from unconventional crime will be analysed further through critical victimology.

Fundamentally, who can be called a victim? In the discussion of a phenomenon or case that has occurred previously, the conflicting or disputing parties will compete for this important position: the position of a victim. In general, the victim is perceived as the party in a position of weakness, who is helpless, oppressed and marginalised. Therefore, the party in this position would be less likely to commit the crime.

Z.P. Zeparovic, in Waluyo, suggests that the victim is "the persons who are threatened, injured, or destroyed by an actor or omission of another (mean, structure, organization, or institution) and consequently, a victim would be anyone who has suffered from or been threatened by a punishable act (not only criminal act but other punishable also act as misdemeanors, economic offenses, the non-fulfillment of work duties) or an accident" (Waluyo, 2011: 9).

Slightly differently, Muladi (2005) argues that victims are people who, either individually or collectively, have suffered from any physical, mental, emotional or economic losses or experienced any substantial disruption towards their fundamental rights, through any act or commission violating the applicable criminal law in every country, which also includes the act of abusing power. Such a definition is similar to that outlined by the United Nations in 1985.

Accordingly, victims will arise through a process that makes them victims. This process is called victimisation. Karmen (2010), in his book *Crime Victims: An Introduction to Victimology,* argues that victimisation is an asymmetrical interpersonal relationship that is abusive, painful, destructive, parasitical and unfair. In other words, victimisation can be described as a process or a relationship that is destructive for a particular party.

In addition to conventional victimology, there is also the notion of critical victimology, which tries to examine the objects of victimology that cannot be elucidated by conventional victimology. It includes a discussion on structural victimisation, which is a victimisation process without any violence. This non-violent victimisation pattern has a number of characteristics that are almost always similar in each case, namely they are impersonal, random and produce a massive and collective number of victims.

A structural victimisation that is associated with the elements of particular social structures often does not directly or immediately make the victims suffer mentally and socially (Gosita, 1988). Due to the inconspicuousness of structural victimisation, the victims do not feel threatened physically and also do not feel conscious that they have been exposed to victimisation, although the damage inflicted can be either physical or non-physical.

Indonesian people often inadvertently put up with structural victimisation. These are the circumstances that force them to become victims. Whether they like it or not, they have no other choice but to deal with it. Structural victimisation leads to strains or stresses in society. Stress can be caused by the routines and the frequent exploitation of space and time by the people to secure their survival within their cultural patterns (Kleinman, 1996). The emerging lifestyle structure cannot be sidestepped because it is a consequence of the types of employment, income, residence and routine activities in urban society.

In addition to states, structural victimisation could also be committed by enterprises. It is commonly referred to as white-collar crime. For example, a company can undertake structural victimisation in the form of monopoly and cartel systems to determine the price of goods and/or services in the market.

4   TAX COLLECTION AND VICTIMISATION

As stipulated by the constitutional law, the Indonesian Directorate General of Taxes, on behalf of the state, has the authority to enforce and supervise the tax obligations of the community. Disappointingly, there are still a number of citizens who deliberately evade and embezzle their tax obligations and do not pay the taxes they are charged.

> 'During the third century, many wealthy Romans buried their jewelry or stocks of
> gold coin to evade the luxury tax, and homeowners in eighteen-century England
> temporarily bricked up their fireplaces to escape notice of the hearth tax collector'
> (Webber & Wildavsky, 1986: 141).

The lack of enthusiasm for paying taxes has incited antipathy among some communities towards anything related to tax, including the tax institution and the personal tax officials. Furthermore, a number of deviations committed by unscrupulous tax officials have made it worse. In an instant, many Indonesian people gained a bad impression, not only of the unscrupulous personnel, but also of all tax officials in Indonesia.

With regards to victimology, the implementation of tax collection from the public has led to some discussions. This article will analyse the tax collection process and its relationship with the three components that lead to victimisation, which are the victimisers, victimhood and the victims.

4.1   *Victimisers*

In its literal meaning, a victimiser constitutes a party that makes another party a victim. In other words, the victimiser is the perpetrator. In his publication, *The Criminal and His Victim* in 1948, which marks the emergence of victimology as part of criminology, Hans von Hentig (1948) refers to the victimiser as the doer, while the victim is the sufferer. Meanwhile, Robert Elias (1986) uses the term oppressor to designate the victimiser.

Related to the efforts of the state to collect taxes, there are a number of parties who consciously or unconsciously could become a victimiser, for instance:

– Community
– Taxpayer
– Tax Embezzler
– Family and/or Friends
– Other Government Institutions

Nonetheless, the form of victimisation may vary for each victimiser. The victimiser, in this study, commits several acts, namely:

– Bullying
– Threats
– Murder

Moreover, as the culmination of the hatred of the people against tax officials, especially state tax bailiffs, the murder of a state tax bailiff and a member of Security Staff at the Directorate General of Tax was committed in April 2016 by a tax defaulter. The murder was conducted by the repeated stabbing of the two officers to death.

Murder is a crime that is categorised as *mala in se*, where such action is deemed evil and unacceptable in any country. Murder is one of the conventional crimes that can be analysed using conventional victimology. Murder is also one form of aggravated assault, because the goal is to eliminate a human life. According to Barkas (1978), aggravated assault can be divided into four categories based on the relationship between the victim and the wrongdoer, such as (1) *between strangers*, (2) *between peers and relatives*, (3) *between family members* and (4) *between authority figures and the persons they regulate*.

To some extent, related to the classification proposed by Barkas, the relationship that arises between the victim and the victimiser in this case is *between authority figures and the persons they regulate*. According to the National Victimisation Survey, the most unsafe work is law enforcement apparatus jobs, with the risk of murder being the most hazardous risk in executing their duties. A similar risk is also faced by the people working in the government sector (Doener & Steven, 2011).

Different victimisers perform different actions. Even though the action might not be categorised as a criminal act, their action makes others feel threatened, embarrassed, scared and humiliated. If we refer to the notion of victims according to the United Nations and Muladi, where only actions categorised as crimes under the criminal law produce victims, there will be a myriad of aggrieved people who may not describe themselves as victims. This is due to not all disturbing conduct being overtly stipulated in the criminal law as crimes. This includes some deviant behaviours that violate the norms and values in society, but that are not included as crimes under the criminal law.

Any actions that cause suffering or losses, physically or mentally, and whether or not they are regulated by law or by the social value/norms, towards people from any class and social status, are not acceptable. The victimiser or the perpetrator should still be blamed for such actions.

4.2 *Victimhood*

Israel Drapkin and Emiliano Viano (1974) outline the English vocabulary in their book entitled *Victimology*. One of the English terms pointed out is victimhood, which is defined as a state of being a victim or a situation and condition that causes a person to be a victim.

A number of things can create a condition or situation that allows a person to be a victim. Some of these include economic conditions, living in a crime-prone neighbourhood or an area of conflict, hierarchical relationships and so forth. To some extent, even policies or regulations can create victimhood for particular parties. This is also the case with the process of victimisation towards state bailiffs and other tax officials. Constitutional law designates the institution as a competent authority to collect taxes from the public, which in turn grants such authority to state bailiffs and other tax officials. Inevitably, they have to carry out all kinds of tasks set forth in the law in accordance with standard operating procedures. State tax bailiffs and other tax officials are government officials who have been sworn in, so they are required to comply with their performance contracts and integrity pacts, which must be executed in accordance with the law.

In 2016, the Indonesian Tax Authority had been tasked to collect funds to finance 74% of the state activities listed in the state budget. If the target was not achieved, there would be a subsequent punishment. By 2016, all state tax bailiffs and tax officials had their salary cut by 20%, so that they only brought home 80% of the amount they received in 2015, which was a consequence of their inability to achieve 100% of the tax revenue target in 2015.

Moreover, under the law, the state tax bailiff Parada Toga Fransriano had to deliver the Distress Warrant directly to the tax defaulter whose tax arrears had reached billions of rupiahs. However, the tax defaulter had a misunderstanding regarding the carrying out of the duties of tax officials, not uncommon among society, who thought of them as plunderers of people's properties. This subsequently triggered certain emotions, which then further led to murder.

The acts of murder committed by a tax defaulter during the tax collection process are the culmination of feelings of hatred and resentment by the public towards tax officials. After having previously been bullied by the public and other victimisers because of his status as a tax official, Parada Toga Fransriano ultimately regrettably lost his life while performing his duty to collect the money that was the financial source of the development of this country. Will the death of Parada Toga Fransriano end this kind of victimhood? Disappointingly, the answer has been no. This can be seen from the ongoing bullying on social media and the negative comments against tax officials in the electronic mass media.

This condition is related to structural victimisation. The victimhood is caused by the regulations that require them to undertake their task. Perceived victimisation burdens state tax bailiffs and other tax officials. On one hand, they have to execute the task mandated by the law; on the other hand, they have to face conditions that can make them victims.

4.3 *Victims*

While the victimhood aspect discusses the conditions and environments that can make a person a victim, the aspect of a victim discusses all matters related to the victim himself/herself. Before examining the victim in association with tax collection, we shall firstly discuss the consequences of tax evasion, whether committed intentionally or inadvertently.

Who are the victims in the case of violations against tax regulations? Who suffers from these losses? Any violation against tax regulations will reduce the amount of revenue that should be received by the state. The amount of the missing revenue, which should have been added to the state budget, will affect the state's ability to pay its national expenses. The victims are not the Directorate General of Taxes as the Indonesian Tax Authority, nor the leaders of the country. The concrete victims are the state and the entire population of Indonesia. The more vulnerable the state's power to finance all their needs, the more debts and interest will be generated to be borne by our children and grandchildren in the future (Harahap, 2016).

The tax arrears will become a burden to the state. In addition, tax bailiffs and tax officials are also Indonesian citizens. Hence, the tax bailiffs and tax officials assume the position of victims when there are people who commit tax evasion. They are in the same position as any other Indonesians in this case. The public who should have enjoyed various kinds of public facilities provided by the state could not do so, for the construction thereof is obstructed by certain unscrupulous people avoiding their tax obligations.

> "While deterrence theory continues to be popular as a framework for understanding tax compliance, the notion of voluntary taxpaying and 'intrinsic motivation to pay taxes' (Torgler, 2003:5) has attracted considerable attention (Alm et al., 1995; Andreoni et al., 1998; Lewis, 1982). The term used for this intrinsic motivation is tax morale (Frey, 1997). Others refer to a similar phenomenon when they state that people pay taxes because they believe it is the right thing to do (McGraw & Scholz, 1991; Richardson & Sawyer, 2001; Schwartz & Orliens, 1967)". (Ahmed & Braithwaite in Understanding Small Business Taxpayers, 2005: 3).

After falling victim to tax evasion actions, state tax bailiffs and other tax officials also experience structural victimisation because of the status attached to them. Doener and Steven (2012) have confirmed their approval of the condition of victimisation towards government officials. The existence of bullying or threats made by victimisers certainly produces victims, just as any other acts that violate the rules and social norms will produce victims. Even though there is a term *victimless crime*, this is still linked to the existence of a victim. *Victimless*, in this case, refers to a situation where it is difficult to distinguish between a victimiser and a victim.

Furthermore, after experiencing structural victimisation, the state tax bailiff Parada Toga Fransriano and a member of security staff had been the victims of a conventional crime. In order to further examine victimisation, we have to consider a number of aspects that can make someone become a victim. Accordingly, related to the case of conventional crimes involving tax officials, it is important to establish the victim's precipitation.

The victim's precipitation are the trigger or triggers that can make someone become a victim. Such trigger is a form of an action which is welcomed by the reaction of victimiser. In order to ensure a Distress Warrant was received by tax defaulters, the state tax bailiff personally handed the document to them. Unfortunately, this situation triggered the perpetrator to formulate a criminal action plan. Nevertheless, if the tax defaulter had tax morale in him, this action would not have been committed.

Moreover, in this murder incident, it is also necessary to examine the contribution of the victims of the conventional crimes. The tax bailiff, in doing his duties, unwittingly allowed the victimiser to make him a victim. A *participating victim* is one of the classifications of victim categories. Whatever the reason, taking a person's life is intolerable. Nevertheless, it is important to ensure our own security and to prevent ourselves from being prone to crime. The tax bailiff's action was too confident, since he did not involve enough security forces, and thus, this made him fall into the category of a participating victim.

Conventional crimes with violence, such as murder, have an interpersonal level, which involves two or more people as the victimiser and the victim, but they are not on a massive scale. In addition to experiencing the structural victimisation that is part of the critical victimology, by executing his daily duties in collecting tax arrears, conventional victimisation is also experienced by a state bailiff.

Is a tax so important that someone has to lose their life over it? Undoubtedly, tax revenue is very important for the country and for the whole population, but a state bailiff should not have to sacrifice his life during the collecting process. Tax does not exist either for the benefit of the Indonesian Tax Authority, or for the interests of the country's leaders, but it should be used for the interests of the state and the welfare of the entire population of Indonesia. It is essential to raise the tax morale of the entire population of Indonesia and to make people willing to pay their tax in the belief that paying tax constitutes the right thing to do.

5 CONCLUSION

Victims arise through a process called victimisation. Critical victimisation is known to have several categories, one of which is structural victimisation. In the tax collection process, there are three components that further lead to forms of victimisation, which are the victimiser, victimhood and victim.

There are a number of parties who consciously or unconsciously could become a victimiser, such as the community, taxpayers, tax embezzlers, family and/or friends and other government institutions. The victimisers commit acts such as bullying, threats and the murder of tax officials as a culmination of the hatred and resentment from the public to the tax officials.

The victims of tax evasion are not the Directorate General of Taxes as the Tax Authority in Indonesia, nor the leaders of the country. The concrete victims are the state and the entire population of Indonesia. It may be represented that the argument that prevailed at the beginning of this article is confirmed. Therefore, it is important to raise the tax morale among the people of Indonesia and raise awareness concerning the importance of taxes so that people are willing to pay them because they believe it to be the right thing to do.

REFERENCES

Ahmed, E. & Braithwaite, V. (2005). Understanding small business taxpayers: Issues of deterrence, tax morale, fairness and work practice. *International Small Business Journal*, 23(5), 539–568.
Barkas, J. L. (1978). *Victims*. Bristol: Gibbons Barford Print.

Doener, William. G. & Steven, P. Lab (2012). Victimization at school and work. In William G. Doener and Steven. P. Lab (Eds.), *Victimology*. Ohio: Anderson Publishing.

Drapkin, I. & Viano, E. (1974). *Victimology*. Massachusetts: Lexington Book.

Elias, R. (1986). *The politics of victimization: Victim, victimology and human rights*. New York: Oxford University Press.

Gosita, A. (1988). Viktimisasi Struktural. *Makalah dalam Seminar Viktimologi II.*, Fakultas Hukum UNAIR, Surabaya, Indonesia.

Harahap, P. (2016). Who is tax criminal? Criminology perspective in Indonesia. In *4th Asian Academic Society International Conference Proceeding*. Thailand.

Hentig, H. V. (1948). *The criminal and his victim*. New Haven: Yale University Press.

http://www.bbc.com/indonesia/berita_indonesia/2016/04/160413_indonesia_pajak_bunuh.

http://careernews.id/tips/view/1936-Orang-Dewasa-Waspadai-Bullying, accessed on 9 June 2016.

Indonesia State Budget Year 2016.

Karmen, A. (2010). *Crime victims: An introduction to victimology* (7th ed.). Belmont: Wadsworth Cencage Learning.

Kleinman, D. (1996). *The violence of everyday life: The consequences of social dynamics*. Berkeley: University of California Press.

Law of the Republic of Indonesia Number 6 of 1983 Concerning General Provisions and Tax Procedures and the Amendments.

Law of the Republic of Indonesia Number 19 of 1997 Concerning Tax Collection.

Meliala, A. E. (2011). *Viktimologi: Bunga Rampai Kajian tentang Korban Kejahatan*. Departemen Kriminologi Universitas Indonesia, Depok.

Muladi. (2005). *Hak Asasi Manusia: Hakekat, Konsep dan Implikasinya dalam Perspektif Hukum dan Masyarakat*. Bandung: Refika Aditama.

Mustofa, M. (2010). *Kriminologi*. Jakarta: Sari Ilmu Pratama.

Press Release DGT 12 April 2016.

Press Release DGT 14 April 2016.

Sahetapy, J. E. (1987). *Viktimologi Sebuah Bunga Rampai*. Jakarta: Pustaka Sinar Harapan.

Sahetapy, J. E. (1995). *Karya Ilmiah Para Pakar Hukum, Bunga Rampai Viktimisasi*. Bandung: PT. Eresco.

The Declaration of Basic Principles of Justice for Victim of Crime and Abuse of Power 1985.

Waluyo, B. (2011). *Viktimologi Perlindungan Hukum terhadap Korban Kejahatan*. Jakarta, Sinar Grafika.

Webber, Carolyn & Aaron B. Wildavsky. (1986). *History of taxation and expenditure in the western world*. New York: Simon & Schuster.

Competition and Cooperation in Social and Political Sciences – Adi & Achwan (Eds)
© 2018 Taylor & Francis Group, London, ISBN 978-1-138-62676-8

Local action for waste bank management through an environmental communication strategy and a collaborative approach for the sustainability of villages

D. Asteria
Department of Communication Sciences, Faculty of Social and Political Sciences,
Universitas Indonesia, Depok, Indonesia

T. Santoso & Ravita Sari
Department of Law, Faculty of Law, Universitas Indonesia, Depok, Indonesia

ABSTRACT: The collaborative approach in waste management can stimulate creativity and the innovation of citizens. Support from local government, Non-Governmental Organisations (NGOs), companies and other stakeholders towards the collaborative approach can increase the participation of community-based citizens. Capacity building activities for citizens, especially for women, provide knowledge and skills to help manage the environment. The implementation of environmental communication strategies using this collaborative approach aims to address the environmental problems resulting from climate change and waste problems that lead to environmental degradation. This community empowerment can improve the well-being and independence of citizens. The education of the citizens about environmental awareness using the application of the 4R principles (reduce, reuse, recycle and replant) is done by developing a waste bank to resolve the problems of garbage. The method used in this community service activity is supported by the emancipatory participation approach through counselling, education and training. The use of waste banks can educate people to be disciplined in managing waste. The citizens can also earn economic benefits from the savings of bins and the sale of seeds, from which they can pay electricity bills and buy groceries. From this activity, the environment becomes greener and healthier, and a more optimal system of waste management with a decrease in the volume of waste in the community is realised.

1 INTRODUCTION

The readiness of citizens to deal with the environmental problems caused by environmental degradation and climate change is essential. Local action is an effort to provide a means to strengthen the ability of the citizens with regard to social resilience and to increase the citizens' knowledge about the environment and its management. Related to Law No. 18 of 2008 on Waste Management, it is necessary to change the way that society treats waste. Society's perspective on rubbish should be changed; garbage should no longer be considered as a useless waste product. Rubbish should be seen as something that has a use, a value and benefits. In order to implement Government Regulation No. 81 of 2012 on Household Waste Management, the practice of processing and utilising waste must be prioritised when managing waste. Citizens must leave behind their old ways of simply dumping their waste. Citizens can be educated to sort, select and appreciate their solid waste, while developing additional income for the community through the development of waste banks.

In connection with the above explanation, as cited in Dwipasari (2014), the handling of the garbage problem in urban areas requires integrated co-operation between the government, communities, schools, universities, industries and the private sector in multilevel mechanisms of waste management. One of the waste management models involves developing waste banks using the integration of the principles of the 4Rs (reduce, reuse, recycle and replant) in its strategy in order to achieve sustainable waste management. The application of the 4Rs principles is important because it is a development of the implementation of the principles of the 3Rs (reduce, reuse and recycle). Chowdhury et al. (2014) state that the management of waste treatment will be effective and sustainable if it starts from the main source, the houses. The recycling process results in an increase of waste efficiency up to 90%. Because of that, waste management needs to be applied with a community-based collaborative approach. The role of the government and stakeholders as partners is crucial in supporting the sustainability of waste management activities.

The effort to develop waste banks is expected to involve social engineering to achieve behavioural change. The development of a community-based waste bank is a form of empowerment to turn the community into a more independent environmental user. The roles of the government and the private sector (in Figure 1) in supporting the community include fostering the people's ability to acquire knowledge and skills by implementing the principles of the 4Rs and by understanding the social problems that occur in their communities. This support and partnership can take the form of providing information, consultation, co-oper-ation, communication, exchanging information and forming joint actions and partnerships. It is important to map out the potential of the community and the possibility of developing a network at both the local and national level. Communication becomes a fundamental ele-ment in the development of society in the process of sharing information, experience and knowledge, and understanding and trying to achieve the realisation of the collective roles.

Performing local actions can be one of the ways of empowering the community through the educational activities needed to establish public awareness. Environmental awareness comes from the way of thinking and human behaviour. Changes to the behaviour of individuals are initiated by doing the little things, which then can be 'infected' to become a habit in the family or in the society, resulting in more significant changes. The continuous active participation of the citizens becomes essential to the further development and the upgrading of the actions of waste manage-ment. There are four major problems in society with regards to waste management, namely:

– In general, the implementation of waste management is from the source (household/commu-nity), who directly dump the waste into the rubbish bin to be taken away by officer carriages organised by the citizens association *Rukun Warga/RW*). The waste is then transported to garbage dumps, and after that it is transported to landfills. If there are waste banks in the com-munity, they can reduce the amount of waste, on average, by about 4–5 carts per community.
– Socially, most people from the local area do not care about waste management. In society, waste management is still at an individual level and is not organised in an integrated manner.
– Economically, the awareness of waste management is still low, because people still think that garbage is the remnant of an undesirable process and does not have any economic value.
– With regards to the problem of waste, there are still people who do not dump their garbage in the designated places, but in the rivers/canals, or they prefer to burn their waste. These activities cause the environment to become dirty, are the sources of various diseases, cause environmental pollution and, ultimately, cause the destruction of the ecosystem.

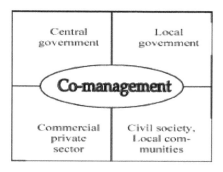

Figure 1. Stakeholder involvement in management collaborative elements (co-management). (Source: World Bank, 1999 cited in Carlsson & Berkes, 2005).

This local action programme is important because the handling of waste matter to this point has not been effective. Various waste handling and management schemes have been applied, and some are still running, but tangible results are not pronounced. Waste management is still not optimal in Indonesia, causing many ecological disasters caused by garbage piling up. In 2016, many cities and counties failed to win the Adipura Award because they could not optimise waste management, and one of the reasons for this is the failure to develop waste banks.

Women have an important role to play in waste management. In households, women generally purchase the household goods. This relates to the importance of their ability to reduce, reuse and recycle used items to reduce the volume of rubbish. Gupta (2011) also outlines that women in India are the saviours of the environment, and that women's participation in waste recycling, through the presence of the waste banks that have become a solid foundation for the recycling business, is supported by the co-operation of the private sector and the government in the protection of the environment and the improvement of the citizens' welfare through the empowerment of women by involving women in integrated waste management practices. Training women to develop waste banks leads them to become role models for other citizens. Women are also able to disseminate environmental awareness by word of mouth in the routine activities of PKK (*Pembinaan Kesejahteraan Keluarga* or Family Welfare Empowerment), prayer activities and community meetings (Asteria & Heruman, 2016). Based on previous studies, it can be seen that the involvement of women is one of the keys to accelerating the success of this waste management programme. Efforts to resolve the garbage problem in the community by using community-based programmes start with women.

The purpose of this paper is to illustrate activities in the development of waste banks as an alternative form of citizen empowerment through an environmental communication strategy and by implementing local action in waste management in the community as the basis of the collaborative management approach. In addition, this paper can propose an alternative strategy of social engineering, which can be used by policymakers to increase the citizens' participation in waste management, especially the female citizens who have the access and control to engage in environmental management. This paper also provides an input into the efforts of optimising waste management in every region in Indonesia and resolves the failures in developing waste banks. There have been three waste banks in Bengkalis, Riau; however, 25 out of the 120 waste banks in Tangerang, Banten were forced to shut down because they were not managed optimally.

2 CONCEPTUAL FRAMEWORK

To overcome the waste problems, this community service activity uses the theory of citizen empowerment within the community as the basis for enabling the community residents to undertake social control. For example, in Thailand, the efforts to change the way society thinks about waste management were achieved by encouraging a change in the behaviour of households to reduce waste at the source through the participation of citizens in conducting recycling activities and integrating this into a project of community-based waste banks (Singhirunnusorn et al., 2012). This is important, as shown in Tallei et al. (2013), because the empowerment of communities to manage waste is an important activity for people in developing countries, in order to address the problems of ecosystem damage caused by waste polluting the environment. Waste bank activity is a concept of collecting dried waste after the process of waste sorting and managing. It is similar to banking, but instead of managing money, the 'bank' manages the rubbish. Citizens who are saving (handing in garbage) are also called customers/clients. The customers/clients get passbooks and can borrow money, which they will get money with the rubbish deposit. The waste to be 'saved' is weighed and valued at a sum of money, which is paid to the customer, and then sold to a factory that has been co-operating with the waste bank. Plastic packaging can be purchased by local women's organisations (in Indonesia, called PKK) to be recycled into craft items (Pokja AMPL, 2013).

The development of waste banks by integrated co-operation between the local government and local stakeholders, including entrepreneurs, is very important to achieve sustainable waste banks. Purba et al. (2014) state that the development of waste banks actually assists the local governments in allocating funding to other management processing programmes. Educational activities regarding waste management and the presence of waste banks teach people to sort waste and raise public awareness of wise waste treatment and, in turn, will reduce the waste going into landfills. Intervention from the local government is needed to support the smooth operation of waste banks. The role of the company/local entrepreneurs is also required to increase the economic value of waste in the community. Collaborative management for environmental management through environmental communication will facilitate the process of partnership between local stakeholders and citizens in managing waste (Koda, 2012).

3 METHODS OF ACTIVITY

This community service activity focuses on a priority area, with the aim of empowering citizens in environmental management, especially female citizens and the community at large. The programme is conducted in Kampung Karangresik, Sukamanah Village, Cipedes District, Tasikmalaya, West Java. The main reasons for selecting this location are because the village generates an average of 1,627 cubic metres of waste per day, there are still piles of garbage on the roadside and many local residents throw garbage into the river. The village is located near the Citanduy River. Before the local action was carried out in this location, there had been a lot of waste dumped into the river, causing frequent floods.

The community service activities are carried out in the form of an educational programme and the development of a community-based waste bank for waste management in order to achieve environmental sustainability in village communities. The name of the waste bank is Bank Sampah Pucuk Resik (BSPR). The method uses the emancipatory participation of citizens in educational activities, through interaction and communication with citizens in the community, with dialogues to map the waste problems, socialisation activities and training activities. This programme focuses on increasing the active involvement in waste management of a multilevel collaborative partnership of various stakeholders by developing a waste bank. Some partners in this activity include the Directorate of Waste Management KLHK RI and BPLHDs West Java, as representatives of the central and provincial governments, the NGO LSPeR, as the local NGO partner, and APDUPI (Association of Entrepreneurs Recycling Bins), as the representative of industry associations. The partnership will produce a synergy of co-operation that ensures a sustainable waste bank and policy support for the marketing of those products that have a sales value (value added).

4 RESULTS AND DISCUSSION

This community service activity aims to increase community involvement by using communication strategies for local action, starting by identifying local stakeholders, as the key to collaborative management. Then, the next step is mapping the problem of waste from the experiences and needs of the citizens, by identifying the constraints that have occurred in the community. Afterwards, a working group consisting of representatives of the citizens will engage in dialogues and consultations to prepare for the development of waste banks, and educational activities for the citizens will be established. Educational activities for the citizens, using an environmental communication strategy to develop the waste bank, will consist of several activities, namely:

– Improving dissemination activity, intense publication and expanding the range of activities. The information dissemination of the waste bank programme aims to get support from the community and build the motivation and awareness of the citizens to manage their waste independently in their own homes. Publications provide continuous information on waste banks and give an environmental message to help support educational activities.
– The sustainability of entrepreneurial empowerment activities was by organising a waste bank management team [consisting of five (5) persons, the chairman/director (1), secretary/administrative staff (1), the treasurer/cashier (1) and teller/sorter (2)], previously chosen to

better direct and facilitate the implementation of the waste bank. It was expected that the management team would be better able to carry out activities at the waste bank and could better prepare the implementation of a variety of technical rules, among others: establishing procedures to save money, making the regulations for achieving savings, determining the type of waste that could be received, the exchange rate of savings, setting up the operational schedule of the waste bank, making a co-operation agreement with the waste collector agents and preparing standard equipment, such as scales, cash books and passbooks.

- Increasing recycling skills in order to increase the citizens' skills and knowledge in the application of the principle of the 4Rs, with training in recycling, the recycling of non-organic waste, such as plastic packaging, into handicraft products, and the recycling of organic waste into compost.
- There are as many as 598 bank customers/clients of Bank Sampah Pucuk Resik (BSPR), consisting of 198 people from Kampung Karangresik and 400 people at the branch of BSPR at Kampung Desa Ngenol Puspamukti, Cigalontang, District of Tasikmalaya. The branch of BSPR is in a district of Tasikmalaya located within 25 km of the village of central the BSPR. This branch was founded from the desire of some youth leaders to maintain the cleanliness of the surrounding environment, which later resulted in the idea to establish a waste bank. The idea was endorsed by the village head, who motivated all of the villagers of Puspamukti to become members.

The waste sorting process is carried out by the citizens before they deliver it to the waste bank. The citizens gained knowledge about the types of waste, especially plastic waste, from training. That kind of knowledge is essential for the development of waste banks and the production of recycled products. The delivery and garbage collection are done every Wednesday, between 8 am and noon. The activity of weighing rubbish is undertaken by unemployed youths, in an effort to involve them in the community. Now, the average amount of savings of the citizens per year has reached IDR 75,210, while the total savings in the BSPR has reached IDR 10,529,400. Plastic waste is sold directly to the collectors, with prices varying depending on the types of waste. The types of waste accepted by the collectors include paper, plastic, plastic bags, bottles and metal. Meanwhile, newspaper waste that is still good is sold to the market for traders.

Educational activities in the recycling of inorganic waste, a series of activities in the development of the waste bank, show the upgrading of women's skills in producing crafts from recycled waste. Female citizens are able to develop the skills to make five kinds of crafts from the recycled plastic waste (crackle, coffee packets and plastic bottles) and from old newspapers and rags. The BSPR in Tasikmalaya has provided benefits to the citizens, especially the immediate benefits of reducing waste generation in the community, creating a cleaner and more beautiful environment and increasing the economic independence of the citizens. This activity has also had economic benefits, including money gained from the savings produced by the waste being used to pay electricity bills and buy groceries. In addition, there are also environmental benefits, including a clean, green, comfortable and healthy environment.

To maintain the sustainability of the waste bank, based on the results of the evaluation of the community service activities, a few things need to be highlighted: (1) there is a need for data and knowledge on the development of the petroleum price associated with the selling price of plastic waste in the market. (2) In addition, a regeneration process of the cadre of citizens with the ability to independently manage the waste bank is needed, by creating a sense of belonging, a sense of responsibility and, especially, the motivation to participate in sustainable environmental management. (3) In keeping with the partnership of waste management using a collaborative approach, the development of relationships and partnerships with relevant stakeholders (especially local government) should be maintained so that local partners can facilitate the citizens in accordance with the needs of the local cultural and social conditions, in line with their levels and roles. These partnerships and relationships need to be maintained because the facilities of plastics enumeration and the supporting equipment facilities (including press machines, glue gun, sewing machine and others) are extremely necessary to improve the marketability of plastic waste in the waste bank. It also increases the income of the citizens as customers and acts to bind the citizens to the idea of continuing their savings in the waste bank.

5 CONCLUSION

The presence of waste banks has increased the capacity for citizens to be more self-reliant in waste management, with an increase of awareness, knowledge and capabilities due to the application of the 4Rs principle. The waste bank encourages community participation in environmental management. The results of these activities can be used as a recommendation for a management policy on waste management, with a community-based collaborative multilevel approach. Educational activities with an environmental communication strategy increase the creativity and innovation of the community, especially in terms of recycling waste management into finished goods with an economic value. Efforts to optimise the participation of citizens in waste management from household sources can be developed as a community-based movement with collaborative management in order to achieve the sustainability of the villages.

REFERENCES

Asteria, D. & Heruman, H. (2016). Bank sampah (Waste banks) as alternative of community-based waste management strategy in Tasikmalaya. *Human and Environmental Journal (Jurnal Manusia dan Lingkungan)*, 23(1), 136–141.

Blocker, T. J. & Eckberg, D. L. (1997). Gender and environmentalism: Result from the 1993 general social survey. *Social Science Quarterly*, 78(4), 841–858.

Carlsson, L. & Berkes, F. (2005). Co-management: Concept and methodological implications. *Journal of Environmental Management, 75*, 65–76.

Chowdhury, H., Mohammad, A. N., Ul Haque, Md. R. & Hossain, T. (2014). Developing 3Rs (reduce, reuse, and recycle) strategy for waste, management in the urban areas of Bangladesh: Socioeconomic and climate adoption mitigation option. *IOSR Journal of Environmental Science, Toxicology and Food Technology (IOSR-JESTFT)*, 8(5), 9–18.

Deputy Management B3, B3 waste & trash. (2012). *Profile of waste bank Indonesia 2012*. Jakarta: Ministry of Environment RI.

Dwipasari, L. (2014). Waste management model-based design character in junior high school as a tool of the economic empowerment. *IOSR Journal of Business and Management (IOSR-JBM)*, 16(10), 38–48.

Gupta, A. (2011). An innovative step towards women's empowerment by the successful community based garbage bank project to protect environment. *International Journal of Scientific & Engineering Research*, 2(8).

Koda, S. (2012). Theoretical approach to the collaborative environmental activities: Household waste disposal towards environmentally friendly daily life. *International Journal of Humanities and Social Science*, 2(6).

Malone, T. (2007). *The social ecology of women as transformational leaders* [Thesis]. Cultural Anthropology Vermont College of Union Institute & University.

Pokja AMPL. (2013). *Stretching action waste bank in Indonesia*. Available from http://www.ampl.or.id/read_article/geliat-gerakan-bank-sampah-di-indonesia/246.

Purba, H. D., Meidiana, C. & Adrianto, D. W. (2014). Waste management scenario through community based waste bank: A case study of Kepanjen district, Malang regency, Indonesia. *International Journal of Environmental Science and Development*, 5(2), 212–216.

Singhirunnusorn, W., Donlakorn, K. & Kaewhanin, W. (2012). Household recycling behaviors and attitudes toward the waste bank project: Mahasarakham municipality. *Journal of Asia Behavioral Studies*, 2(6), 35–47.

Tallei, T. E., Iskandar, J., Runtuwene, S. & Filho, W. L. (2013). Local community-based initiatives of waste management activities on Bunaken Island in North Sulawesi, Indonesia. *Research Journal of Environmental and Earth Sciences*, 5(12).

Utami, B. D. (2008). Community Based Waste Management (Pengelolaan sampah rumah tangga berbasis komunitas). *Soladity Journal: Journal of Transdisciplinary Sociology, Communication, and Human Ecology (Jurnal Sodality: Jurnal Transdisiplin Sosiologi, Komunikasi, dan Ekologi Manusia)*. 2.

Competition and Cooperation in Social and Political Sciences – Adi & Achwan (Eds)
© *2018 Taylor & Francis Group, London, ISBN 978-1-138-62676-8*

The relationship between political parties and women's Civil Society Organizations (CSOs) in Indonesia, South Korea, and Argentina

A. Perdana
Department of Political Science, Faculty of Social and Political Sciences, Universitas Indonesia, Depok, Indonesia

ABSTRACT: This study focuses on the factors that shape relationships between women's CSOs and political parties in Indonesia compared to those in South Korea and Argentina. Several path dependent explanations of specific issues like women and politics in policy-making and political representation can be facilitated by using these three nations as points of comparison. The main research question in this study is: why women's CSOs have distant or close relations with political parties in order to shape consolidated representation in Indonesia, South Korea, and Argentina? This study argues that the relations between women's CSOs and political parties affect the development of women's political representation in consolidated democracies. A distant and critical engagement between women's CSOs and political parties in Indonesia is evident in some progressive gender policy reforms in parliament, but the weak implementation of these laws in society still remains. Although there has been a strong, a close alignment between women's CSOs and political parties in South Korea and Argentina that has encouraged women's political representation outcomes, both countries still face some problems. It is difficult for South Korean CSOs to achieve autonomy from the state and political parties; meanwhile, in Argentina the major parties are strongly dominant in the political structure.

1 INTRODUCTION

Civil society and political society are two main arenas that should connect in democratic consolidation (J. Linz & Stepan, 1996). Civil society has several essential roles to play: to enhance the quality of democracy (Diamond, 1994; Uhlin, 2009), to act as an intermediary between democratic and anti-democratic forces (Chandhoke, 2004), to instill citizens with democratic values, and to provide external protection for democracy from any external threats that might lead to political destabilization (Merkel, 2004). Along the same lines, various scholars agree that political parties are essential for democratic regimes (La Palombara, 2007; J.J. Linz, 2002; Randall & Svåsand, 2002; Schmitter, 2010). Political parties bring a large number of citizens with similar interests into the political sphere and facilitate their political participation. Given such considerations, this study elaborates on the role of civil society and political parties in enhancing democracy. It seeks to explain the connection between women's civil society organizations (CSOs) and political parties in consolidated democracies such as Indonesia, South Korea and Argentina.

This study argues that relations between women's CSOs and political parties affect the development of women's political representation in consolidated democracies. Comparisons with other nations are pertinent to this study for two reasons. First, the understanding of the relationship between CSOs, political parties, and democracy can be enriched by considering different narratives from other democracies. The cases in South Korea and Argentina display similar evidence with the Indonesian experience, namely in terms of societies with strong patrimonial and patriarchal relationships. Second, these countries share a similar historical experience where social protests and movements have led to the emergence of pro-democratic paths. Since the democratic breakthrough, both countries have been recognized by civil society observers and experts on political parties (Wylde, 2012; Fioramonti & Fiori, 2010) as the nations where collaborative and co-operative relationships exist between CSOs, political parties, and the state. This relationship has contributed to the countries' democratization processes because during this period, both parties and CSOs played crucial parts in order to negotiate their political interests. As such, several path-dependent explanations of specific issues, like women and politics in policy-making and political representation, can be facilitated by using these two nations as points of comparison. Women's CSOs and political parties work together in order to promote gender equality in the political arena. The main research question in this study is: why women's CSOs have either distant or close relations with political parties in order to shape consolidated representations in Indonesia, South Korea, and Argentina?

This paper consists of three parts. The first section discusses theories on CSOs and political party linkages. The second section presents the role that CSOs and political parties play in establishing representation in new democracies and offers a comparison of the Indonesian, South Korean, and Argentine cases. This is followed by a section that explains authoritarian legacies in Indonesian contemporary politics that have affected the development of the relationships between women's CSO and political parties. The paper concludes with an explanation of how the collective evidence from these countries contributes to an understanding of the roles of women's CSOs and political parties in women's representation in democracies.

2 THEORETICAL FRAMEWORK

Political scientists highlight that CSOs and political parties can be defined by their connections with each other (Allern, 2010; Allern & Bale, 2012; Allern & Saglie, 2008; Lawson, 1980; Thomas, 2001). In terms of political parties, it is important to consider the aspects of party development that can build party's connections with CSOs or other groups (E.H. Allern, 2010; C.S. Thomas, 2001). Meanwhile, for CSOs, experiences in new democracies illustrate some varied and changing relationships between CSOs and political parties which are based on individual interactions (USAID, 2004). Some studies about civil society (Alagappa, 2004; Kopecký & Mudde, 2003) and political parties (van Biezen, 2003; Krouwel, 2012) indicate that the current changes in parties' organization affect the relationship between the parties and CSOs in many regions. A number of scholars believe that some relationships are distant and complex in nature and require adjustments in order to cooperate effectively (Thomas, 2001; Alagappa, 2004; Allern, 2010). In contrast, some argue that complementary and cooperative relationships can be established between these institutions in developing countries (Linz, 1996; Morlino, 1998). Nevertheless, some political party observers believe that parties are moving closer to the state, but the connections between parties and the society remain crucial for political development (Katz & Mair, 1995; Kitschelt, 2000). Therefore, these studies have shown that there is a variety of evidence on the various ways to form and maintain the CSO-party relationships.

In general, the relationships between civil society organizations and political parties have been analyzed from three specific perspectives. Firstly, some scholars believe that the relationships between them are blurred and weak (Alagappa, 2004; Allern & Balle, 2012), due to two contradicting situations (Alagappa, 2004: p. 38), namely affiliation and mobilization roles. CSOs (such as trade unions and religion groups) affiliate with some political parties; and these civil society groups mobilize support from parties to avoid co-optation by authoritarian and totalitarian regimes. Furthermore, links between weak parties and groups arise from the influence of cartel parties and the weak institutionalization of parties (Allern & Balle, 2012; Webb & White, 2007). Secondly, scholars who study democratic transitions argue that a complementary relationship is crucial in order to connect parties and civil society organizations (Linz, 1996, p. 18; Morlino, 1998, p. 26). They consider good cooperation to be essential, as parties and CSOs can help each other in terms of association, which is better than defining their relationships in sharp and dichotomous terms. Thirdly, from the perspective of social movements, the relationship between movement groups and political parties varies from close contact to oppositional (Schwartz, 2010, p. 587). There is a seemingly strong competition between social movements and political parties in the political arena. For social movements, political parties are targeted in order to push their respective political agenda. At the same time, parties consider movement groups as competitors. To establish an alliance between movement groups and political parties, a mutual agreement has to be created concerning new issues and the sources that can be used in cooperation (Schwartz, 2010, p. 588). This paper explains why relationships between CSOs and political parties in new democracies tend to be weak and distant.

3 FACTORS SHAPING THE RELATIONSHIP BETWEEN CSOS AND POLITICAL PARTIES IN THE POST-SUHARTO INDONESIA COMPARED TO SOUTH KOREA AND ARGENTINA

This study emphasizes both external and internal environmental factors in contemporary Indonesian politics as the main findings to compare with South Korean and Argentina. Furthermore, since the democratic breakthrough in 1998, socio-economic changes and international development support have also catalyzed the establishment of a relationship between CSOs and political parties. Internal factors, in terms of policy-making interactions and connectivity through political representation between CSOs and political parties, have also shaped Indonesia's democratic development. Both organizations have realized their natural limitations, which have eventually led them to work together in creating good policies for society. Therefore, parliamentary parties need technical input from CSOs in that process; meanwhile, CSOs require political support in parliament to endorse their political interests.

External factors became increasingly important when Indonesia began its transition to a democratic country. The *Reformasi* (Reform) in 1998 is recognized as a critical juncture that created new opportunities, and it has facilitated political change in Indonesia. The fall of Suharto's regime coincided with massive demands for change in Indonesia's social, economic, and political spheres. Habibie's transitional government (1998–1999), as the successor of Suharto's regime, responded to these demands. His government was committed to reviewing democratic laws and institutions. Habibie also accepted some women's groups' demands for the state to take responsibility in investigating the violence that occurred against women during the social and political riots in several major cities in May 1998.

International support, which accelerated social and political changes with numerous development programs following the 1998 student protests, is the final external factor. During the period between 1999 and 2009, many international development agencies became involved in a wide range of development sectors across the country.

If we compare the Indonesian case to that of South Korea and Argentina, it shows a similar trajectory. Democratic junctures in both countries, 1987 for South Korea and 1983 for Argentina, strongly influenced the contestation of political power among the ruling government and opposition groups led by CSOs. In South Korea, the transitional regime was successfully able to develop democratic changes after the authoritarian leader, Chun Doo Hwan, started to accept the democratic reform agreements proposed by civil society groups. The country also held transitional elections for both the legislative and executive branches at the end of 1987. Since that time, two different groups of CSOs, citizens and people's movements have worked together to monitor the implementation of democratic agreements in the public sphere. In particular, opposition leaders such as Kim Young Sam and Kim Dae Jung successfully supported the ruling party's attempts to establish cooperation for several democratic reform programs.

Argentina's democratic transition began 4 years prior to South Korea's in 1983 when the previous military junta (1976–1983) suffered a decline in trust from the army. At that time, the country lost the Falkland Islands to the United Kingdom, while civil groups successfully mobilized protests against the government. These civil groups, led by labor movements and involving major political parties (UCR and PJ), demanded the military regime to hold legislative and presidential elections in 1983. The transition to democracy in Argentina is also similar to both the South Korean and the Indonesian cases: it was triggered by civil group's far-reaching social protests. In the Argentine case, however, the cooperation between major political parties and prominent labor union groups strongly influenced the transitional period. The leader of UCR, Raul Alfonsin, did not attempt any social, economic, and political institutional reform and asked instead that the groups cooperated with each other.

Nevertheless, international development assistance, which had supported some political institutional development in Indonesia, was not evident in the South Korean and Argentine cases. Working together with major mass organizations and NGOs in the Indonesia's transitional period to develop the community's critical awareness and to design new political institutions, international agencies have successfully embedded their mission in that country[1] (Aspinall, 2010).

In terms of the internal factors that play a role in the development of relationships between women's CSOs and political parties, particularly those based on policy-making interactions and connectivity in political representation, the relationships in the post-Suharto era can be described as distant with some constructive interactions. Both groups have separated political interactions with the state and with other groups like business communities. Women's CSOs maintain a distant relationship with legislators when they advocate their interests and provide substantial input during the policy-making process. However, at the same time, women's CSOs leaders have allowed themselves and their members to become part of the state apparatus in different positions. This "critical engagement" also defines CSOs' autonomy from state and political parties, as they are free to criticize these political entities.[2] This engagement often takes two forms. In the legislation making process, there is a critical cooperation between women's CSOs and all parliamentary parties. Meanwhile, there is no clear cooperation for political representation, as individual choices are made about partisanship.

Moreover, political parties believe that distant and critical engagement with CSOs has a strong correlation with Indonesian history and is embedded as social cleavages. Historically, Indonesian political parties and civic organizations emerged at the same time, during the early national independence movement. Therefore, both organizations remained separated from each other in contrast to European countries where there has been a close relationship between labor unions and labor parties. As such, Indonesian political parties respect CSOs' independent and non-partisan positions, particularly as two large mass organizations, NU and Muhammadiyah, have a great deal of experience to share. They endorse CSO's activities that could potentially benefit the party and that they could not carry out by themselves, such as developing technical development expertise in the legislation making process.

These internal factors explain two crucial paths for the relationships between CSOs and political parties: (1) the lack of institutionalization of both organizations; and (2) the strong legacy of patronage politics, including clientelism and informal relations. These paths were also evident in the comparative cases presented above.

Current scholarly works on political parties (Ufen, 2008; Mietzner, 2008) and a few studies on CSOs (Beittinger-Lee, 2010) confirm that both parties and CSOs in Indonesia can be categorized as weak institutions. This study supports these findings based on the political involvement of CSOs and political parties with gender issues. Women's groups in Indonesia tend to be fragmented and atomized organizations. For example, although women's CSOs have endorsed gender-related legislation before the House of Representatives, they tend to be separated into particular issue-based groups, such as women's suffrage groups and women's transitional justice groups (interview with Ruth Indah Rahayu, February, 7, 2012), rather than a joint campaign to influence the Commission VIII for Religion and Social Affairs, and the Empowerment of Women. The groups tend to work individually, and thus lacking formal, institutionalized agreement.

A further example of the lack of consensus between well-structured, well-established organizations and the varied views concerning the gender quota can be found in the affirmative action discussions within the House of Representatives. In this case, however, fragmentation was easily overcome when some leaders of the women's groups came together as a united

---

1. See the perspective on democracy assistance in Carothers, 2009.
2. Some NGOs leaders, based on several interviews and informal talks with these leaders, believe that this strategy is critical for engagement with the state during the current administration.

front with similar gender perspective. It is evident that women's CSO leaders are not only involved in one single organization but are often involved in others as well. Therefore, Nur Iman Subono, a gender observer, believes that the internal dynamics of women's organizations, including conflicts and fragmentation, are part of the learning process as women's groups attempt to regroup after their period of repression under the preceding authoritarian regime (personal communication, April 2, 2012).

One of the findings of this study is that these women's organizations, along with other NGOs, are able to advocate, to campaign, and to lobby members of the House to help drive their own agenda in the political sphere. However, if we look beyond the surface of the political lobbying during the policy making process, some organizational weaknesses are evident, such as the lack of substantial arguments for supporting and rejecting specific issues and sporadic political approaches with politicians. Thus, female activists have yet to enhance their role as lobbyists by fully mapping the political mapping networks inside and outside the parliament to influence other key actors in order to gain supports for their agendas.

In this study, the weak institutionalization of political parties was evident in several phenomena, namely the limited programmatic-oriented agendas, the domination of personal and informal influences on party decisions, the loose ties between political parties and mass organizations, and the poor party discipline during the policy-making process. Parties that have limited programmatic-oriented agendas are identifiable by their unclear political agendas in prioritizing bills in the House as well as their vague platforms in terms of their political activities (Ufen, 2008). Argama *et al.* (2011) illustrate that the House was hardly productive in terms of its capacity to create laws, as it only produced 38.6 pieces of legislation per year between 2004 and 2009. At the end of 2010, only 17 out of 70 proposed bills had been passed, and only 8 of these 17 bills represented substantially new and relevant legislation. Although the party system's social cleavages are divided along religious and regional lines, most parties try to make their party accommodative to all concerns in order to attract as many voters as possible. Therefore, it is difficult to associate a political party with a particular ideology and to connect their activities in society with their political stance in the House. Only several sensitive issues, such as tolerance and religious freedom, are clearly reflected in party politics and the party's stance during the deliberation of laws pertaining to such themes.

The domination of informal influence is indicated by the role that political leaders play in deciding important party policies with some input from members of the national board. All major parties, with the exception of Golkar, rely on their party leadership structures (i.e. the chairman or the advisory board) in formulating any crucial party's policies.

The loose ties between CSOs and political parties indicate two things. First, while both organizations would rather not have political institutional agreements with each other, they would make them anyway. Second, they allow and support their members who are involved in the political sphere. Despite the fact that cultural and emotional relationships exist between major religious organizations, such as NU and Muhammadiyah, and political parties, such as PKB and PAN, organizationally NU and Muhammadiyah declare themselves to be non-partisan regardless of any political activity. Nonetheless, NU and Muhammadiyah allow their members to affiliate with any political party. A similar situation is also evident among many NGOs and other movements, such as LBH APIK and KPI, which prefer to remain non-partisan but allow their members to seek out their own political affiliations.

Furthermore, party discipline in the law-making process and other parliamentary activities is lax. During the policy making process, members of the House take on the role of members of a commission rather than representatives of a particular party in order to make decisions or monitor particular policies (Sherlock, 2012). Party functions only exist when urgent and sensitive issues emerge as controversial debates in the media and when parties need to control the members who are not following the party-political line. As such, members of the House face a political contradiction: on one hand, their voices are regulated by their respective party to avoid any distortion of the party's political stance, while on the other hand, the party recognizes the autonomy of the commission for any political decision of the House. The weakness of political parties stems from their inability to provide a firm political stance where they could take a leadership role on the commission. Therefore, at the end of policy deliberations, faction leaders prefer to arrive at a consensus to settle a deadlock, and they expect party members to follow their lead. Such contradictory dynamics reflect that the parties tend to maintain a collusive approach to make a policy rather than develop political accountability.

At the same time, clientelism and informal relations are embedded in both CSOs and political parties, which can largely be attributed to the authoritarian legacy of the New Order. To some extent, these attitudes dominate the decision-making process and weaken members' participation in both organizational structures. Political party leaders use their charismatic leadership to run the organization. Likewise, most CSOs, especially mass organizations and movement groups, share a similar pattern.

In comparison to the South Korean and Argentine cases, these endogenous factors take different paths. South Korea and Argentina have strong and powerful CSOs that are able to influence both political parties and the House. This allows both organizations to be active in multiple cooperation's; for instance, public officials from state agencies are able to work directly with development programs. During Kim Dae Jung's administration (1998–2001), the women's movement initiated and supported the establishment of the Ministry of Gender Equality in South Korea (Suh, Oh & Choi, 2011). Although there is no similar evidence to South Korea for developing gender empowerment institutions, the women's movement in Argentina also took the lead in supporting gender bills, such as the Abolition of the Patriarchal Family Registration System Act and the Anti-Domestic Abuse Act. As both of Argentina's political parties, the "Peronist" and "Menemist," have strong grassroot support from labor unions and middle class worker groups, the state has mutual cooperation with these civic groups. At different political cleavages, including under the Peronists, Menemists, and the Kirchnerists, the state attempted to accommodate and to negotiate all political/economic demands in national policies. It includes the use of different approaches to tripartite relationships (state-labor union-parties) and the institutionalization of the protest labor union (Wylde, 2012). In both countries, close ties exist between CSOs and political parties for pursuing gender policies and institutions, although these relationships have taken different paths.

Although both these countries have weakly institutionalized political parties in terms of decisions making, the alignment of political parties with major civic organizations like labor unions is highly important for the establishment of close connections. The institutionalization of political parties is affected by the party leader's strong political populism in Argentina and by the regionalism of clients in South Korea. In these countries, public awareness of certain political debates in mass media is embedded in society and such a condition can easily trigger social mobilization. Therefore, environment factors shaping the relationship between women's CSOs and political parties in the three countries can be described in Table 1.

Table 1.   Environmental factors shaping the relationship between CSOs and political parties.

	External factors	Internal factors
Indonesia (1998 until now)	Political opportunities and regime changes International aid development support	A distant and critical engagement relation between women's CSOs and political parties caused by: Weak women's CSO and party institutionalization Strong legacy patronage politics
South Korea (1987 until now)	Political opportunities and regime changes	A strong and close relationship between women's CSOs and political parties caused by: Alignment between CSOs and political parties Patronage politics Weak party institutionalization
Argentina (1983 until now)	Political opportunities and regime changes	A strong and close relationship between women's CSOs and political parties caused by: Alignment between CSOs and political parties Patronage politics Weak party institutionalization and dominant government party

Source: Author's compilation.

Since Reformasi (Reform) in 1998, Indonesian women have become increasingly engaged in the political arena, while in South Korea and Argentina women were actively involved in some movements prior to the transition to democracy. In Argentina, the women's political participation agenda was led by Eva Peron, the wife of President Peron in 1952. Although Indonesian women were organized as collective as early as the women's congress in 1928, they were unable to successfully maintain their solidarity during the Sukarno (1959–1967) and Suharto (1967–1998) regimes. Because of the high political volatility evident in the Sukarno era and the political repression of all opposition groups in the Suharto era, women's organizations were rather silent and did not engage in politics.

Therefore, once the Suharto regime fell, women's organizations used their political opportunity to support and expand gender issues. However, similar to the South Korean and Argentine cases, authoritarian legacies, such as clientelism and patriarchy, remain problematic in terms of engendering the Indonesian political sphere.

Several historical legacies remain embedded in Indonesian democracy. The first legacy is the rise of oligarchy groups defined as business-entrepreneur groups (Robison & Hadiz, 2004), and other clientelism groups[3] from the Suharto regime. Both groups continue to influence the decision-making process in the legislative and executive bodies. The second legacy is the social and political structures for instance, the combination of proportional and majoritarian electoral systems, multi-political parties, patronage, and patriarchal personal relationships. The third legacy is the fragmented and weak institutionalization (for example, flexible and personalized rules and procedures) present in most Indonesian organizations which tend to lead to low compliance with any internal organizational agreements among these institutions. The fourth legacy is related to a clear separation between civil organizations and political parties, which has led to a distant relationship between CSOs and political institutions. All these legacies are reflected in the dynamic processes of Indonesian politics and also affect the emergence of new political institutions in the young democracy of Indonesia. The influence of these legacies can be seen in recent political interactions between CSOs and political parties, when viewed from the type of issue and the group responding to the issue. However, if we examine other political issues, such as the management of electoral bodies and the electoral system, which affect the party's ability to survive, coalitions among these parties have become more fluid and dependent on what they consider to be crucial issues. Given these considerations, it is women's CSOs that are the most appropriate to support and to bring gender issues to the House, as most political parties have not articulated any strong electoral benefits or perceived community support for major gender issues such as women's equality.

4 CONCLUSION

This paper has discussed environmental factors that help explain the mode of connections between women's CSOs and political parties in Indonesia with a brief comparison to South Korea and Argentina. It has also presented some empirical evidence from these countries showing that a variety of women's CSOs and party relationships have endorsed women's political participation in democracies. A distant and critical engagement between women's CSOs and political parties in Indonesia is evident in some progressive gender policy reforms in the parliament, but the implementation of these laws in society is still poor. Although there has been a strong, close alignment between women's CSOs and political parties in South Korea and Argentina that has encouraged women's political representation outcomes, both countries are still facing some problems. It is difficult for South Korean CSOs to achieve autonomy from the state and political parties; meanwhile, Argentina's major parties are strongly dominant in the political structure. Nevertheless, the strong relationship between CSOs and political parties in these countries does facilitate the implementation of and compliance to gender policies.

---

3. Clientelism groups can be defined by ethnicity, religion, or alumni from specific universities.

Indonesian environmental factors explain that the party-society links are distant because of the weak and fragmented nature of parties and CSO institutions. Political parties that have grown too close to the state and patronage politics within the party system have also weakened the autonomy of the party. However, parties require political connections with CSOs and other groups to maintain their constituents. Although societal links and political cleavage between CSOs and political parties still exist in terms of a policy-making process, political partisanship among women's CSOs leaders is also growing. Therefore, a distant and critical engagement relationship can be found in the interaction between women's CSOs and political parties.

Some authoritarian legacies in structures and values have contributed to the relationship building between women's CSOs and political parties in different activities, such as the law-making process and electoral competitions. This paper suggests that there are two different mechanisms to promote the recruitment of women CSOs activists into political parties. Although this study has not attempted to draw generalizations from other sectoral issues, a recognition of these political structure legacies and organizational caveats and their impact on political interactions between the two organizations may be fruitful for the development of similar paths for other issues. Therefore, if CSO actors are able to critically engage with political parties and the state, women's political representation in new democracies may be strongly enhanced.

REFERENCES

Alagappa, M. (2004) Civil society and political change: An analytical framework. *Civil Society and Political Change in Asia: Expanding and Contracting Democratic Space.* 25–57.

Allern, E.H. (2010) *Political Parties and Interest Groups in Norway.* European Consortium for Political Research.

Allern, E.H. & Bale, T. (2012) Political parties and interest groups disentangling complex relationships. *Party Politics.* 18 (1), 7–25.

Allern, E.H. & Saglie, J. (2008) Between electioneering and politics as usual: The involvement of interest groups in Norwegian electoral politics. *Presented at the Non-Party Actors in Electoral Politics, Nomos Verlagsgesellschaft mbH & Co. KG,* pp. 67–102.

Beittinger-Lee, V. (2010) *(Un)civil Society and Political Change in Indonesia: A Contested Arena.* New York USA, Routledge.

Carothers, T. (2009) Democracy assistance: political vs. developmental? *Journal of Democracy.* 20 (1), 5–19.

C Suh, Oh, & Choi (2011) The institutionalization of the women's movement and gender legislation. In: *South korean social movements: from democracy to civil society.* London and New York, Routledge. pp. 151–169.

Chandhoke, N. (2004) The civil and the political in civil society: The case of India. In: Burnell, P. & Calvert, P. (eds.) *Civil society in democratization.* London, Frank Cass. pp. 144–165.

Diamond, L.J. (1994) Toward democratic consolidation. *Journal of Democracy.* 5 (3), 4–17.

Hadiz, V. & Robison, R. (2004) *Reorganising power in Indonesia: The politics of oligarchy in an age of markets.* Routledge.

Katz, R.S. & Mair, P. (1995) Changing models of party organization and party democracy the emergence of the cartel party. *Party Politics.* 1 (1), 5–28.

Kitschelt, H. (2000) Linkages between citizens and politicians in democratic politics. *Comparative Political Studies.* 33 (6–7), 845–879.

Kopecký, P. & Mudde, C. (2003) Rethinking civil society. *Democratization.* 10 (3), 1–14.

LaPalombara, J. (2007) Reflections on political parties and political development, four decades later. *Party Politics.* 13 (2), 141–154.

Lawson, K. (1980) *Political parties and linkage: A comparative perspective.* New Haven, Yale University Press.

Linz, J.J. (2002) Parties in contemporary democracies: problems and paradoxes. In: Gunther, R., Montero, J.R. & Linz, J.J. (eds.) *Political parties: old concepts and new challenges.* London, Oxford. pp. 291–317.

Linz, J. & Stepan, A. (1996) Toward consolidated democracies. *Journal of Democracy.* 7 (2), 14–33.

Merkel, W. (2004) Embedded and defective democracies. *Democratization.* 11 (5), 33–58.

Morlino, L. (1998) *Democracy between consolidation and crisis: parties, groups, and citizens in Southern Europe*. London, Oxford University Press.

Randall, V. & Svåsand, L. (2002) Party institutionalization in new democracies. *Party Politics*. 8 (1), 5–29.

Schmitter, P.C. (2010) Twenty-five years, fifteen findings. *Journal of Democracy*. 21 (1), 17–28.

Schwartz, M.A. (2010) Interactions between social movements and US political parties. *Party Politics*. New York, Sage Publications.

Sherlock, S. (2012) Made by committee and consensus: parties and policy in the Indonesian parliament. *South East Asia Research*. 20 (4), 551–568.

Thomas, C.S. (2001) *Political Parties and Interest Groups: Shaping Democratic Governance*. USA, Lynne Rienner Publishers.

Ufen, A. (2008) From alignment to dealignment: political parties in post-Suharto Indonesia. *South East Asia Research*. 16 (1), 5–41.

Uhlin, A. (2009) Which characteristics of civil society organizations support what aspects of democracy? Evidence from post-communist Latvia. *International Political Science Review*. 30 (3), 271–295.

United States Agency for International Development. (2004) *Civil society groups and political parties: supporting constructive relationships*. USA, USAID.

Van Biezen, I. (2003) *Political parties in new democracies: Party organization in Southern and East-Central Europe*. Macmillan.

Waylen, G. (2000) Gender and democratic politics: A comparative analysis of consolidation in Argentina and Chile. *Journal of Latin American Studies*. 32 (3), 765–793.

Webb, P. & White, S. (2007) *Party politics in new democracies*. London, Oxford University Press.

Wylde. (2012) *State-civil society relations in post-crisis Argentina*. London, Zed.

Fishermen's adaptation to aquatic environment changes in Jakarta Bay

T. Anugrahini & I.R. Adi
Department of Social Welfare Science, Faculty of Social and Political Sciences,
Universitas Indonesia, Depok, Indonesia

ABSTRACT: The aquatic environment in Jakarta Bay has changed since the increase of domestic and industrial waste in the waters, and culminated during the reclamation. The purpose of this study is to describe such changes and discuss how the fishermen adapt to aquatic environment changes in 3 time periods. This research was conducted in the fishermen's settlement in Kamal Muara, Penjaringan, North Jakarta using the qualitative research approach, in-depth individual interviews, and observations. The changes that occurred until 1994 (the first period) and in 1995–2007 (the second period) were caused by the increasing number of wastes that were dumped into the Jakarta Bay. Although their potential has decreased and they have minimal connectedness with their community and always surrender when pollution occurs, the fishermen still survive because, after the pollution episode, the fishery resources are still abundant. The most disturbing changes occurred in the third period (2008–2015), which was during the reclamation. The fishermen's income decreased due to the difficulty of access to the waters, the narrower fishing grounds, and the dwindling of fishery resources. Although the fishermen have adapted by establishing "*Kelompok Usaha Bersama/KUB*" (a fishermen group) and developing the cultivation of mussels, reclamation activities have made them increasingly vulnerable and threatened their livelihoods.

1   INTRODUCTION

The growth of domestic, industry, and shipping activities around the city of Jakarta have created more variety of toxins and hazardous wastes dumped into the waters of Jakarta Bay. Therefore, the aquatic environment in Jakarta Bay has become dirty and odorous. Meanwhile, population growth also requires more land for settlement and business facilities. Since the implementation of the residential Decree No. 52/1995, reclamation became a strategy of the government of DKI Jakarta to develop the municipality of North Jakarta.

As described in the Spatial Plan of DKI Jakarta 2030 and the Governor's Regulation No. 121/2012, reclamation along the North Bay will develop 17 islands to accommodate the need of settlement and business. Until 2015, C and D islands had been constructed, while the development of Island G was temporarily ceased.

The uncontrollable waste in the waters and intensive reclamation activities until 2015 had changed the aquatic environment along the Jakarta Bay. Such changes lead to a stressor for fishermen at Kamal Muara, such as the increasing difficulty of access to the waters and fishing grounds that has consequently reduced their income. Based on the monthly report of "CL" (one of the fishermen's group name in Kamal Muara) in 2015, the average trip frequency to the sea of one fisherman in that group has decreased steadily as follows: 20.75 trips/month (February), 15.58 trips/month (March), 14 trips/month (April), and a fall to 11.25 trips/month in May 2015.

The purpose of this study is to describe aquatic environment changes in Jakarta Bay and analyze how the fishermen adapt to such changes within 3 time periods. The first period extended to 1994, while the second period started from the implementation of the Presidential Decree

No. 52/1995 regarding the Reclamation of Jakarta North Bay until 2007. The third period extended since the implementation of the Presidential Regulation No. 54/2008 regarding the Spatial Plan of Jakarta, Bogor, Depok, Tangerang, Bekasi, Puncak, and Cianjur (Jabodetabek-punjur) and the Regional Regulation No. 1/2012 regarding the Spatial Plan 2030, until 2015.

This study shows that when humans ignore environmental and social factors and merely consider economic benefits in exploiting their aquatic environment, it would affect human life. The implementation of the reclamation plan of Jakarta Bay, in this case, has ignored the interest of the marginalized people, especially the traditional fishermen. The low capacity of traditional fishermen in Kamal Muara in taking an alternative action for their survival, inadequate internal relations among fishermen, and decreasing fishing resources in Jakarta Bay have made them powerless in dealing with the stressors and have made them become very vulnerable. Reclamation has led to environmental degradation as well.

Academically, this research is expected to enrich the knowledge in social welfare science, especially in creating a harmony on sustainable development. For policy makers, this study can be used as a reference in formulating social interventions for the marginalized traditional fishermen.

2 METHODOLOGY

This study was conducted in the fishermen's settlement in Kamal Muara, Penjaringan, North Jakarta. Using a qualitative approach, this study is case-study research involving field research. The selection of informants was conducted by purposive sampling. Data were collected for about 14 months (September 2014 – December 2015), using observations, in-depth interviews and literature reviews.

3 PROFILE OF INFORMANT

The main informants were 11 traditional fishermen who have lived in Kamal Muara for more than 25 years, fishing for one day, using *sero* or *bagan* (a kind of traditional fishing gear), and owning a vessel. Other informants include community leaders (4 people) and government officials (4 people) who have known about the changes of aquatic environment along Jakarta Bay, as well as the adaptation process by the traditional fishermen in Kamal Muara.

4 LITERATURE REVIEW

Principally, human welfare is inseparable from the ecological capacity, and therefore, it is important that humans live in harmony with the environment (Ife, 2013; Keraf, 2010). Some thoughts on close connectedness between humans and ecosystems have been developed. Hollingshead (1940) in Palsson (1998) states, "...nature and society were not to be seen as totally separate spheres but dialectically interlinked; each compliments and supplements others in many ways...". Berkes and Folke (1998), Holling (2001), Anderies, Janssen & Ostrom (2004) use the term 'social-ecological system' because the two are interlinked and separation of the two is arbitrary. Meanwhile, Ife (2013) uses ecological perspective to achieve a sustainable community development.

Anderies, Janssen, & Ostrom (2004) state that both social systems and ecological systems consist of interlinked and interdependent units. Each system consists of interlinked subsystems. In this case, humans have a tendency to exploit natural resources to meet their needs. On the other hand, the exploited natural resources also affect human life. However, in exploiting the resource, human actions (as resource users) have, through action and ideas, influenced public service providers. In this instance, economic consideration (made by economic actors) and political consideration (made by political actors) influence the dynamic interactions among the subsystems of humans (Scheffer, Westley, Brock, & Holmgren, 2002).

Consequently, interest competition occurs further, which leads to the emergence of stressors that endanger humans and the ecological systems.

To maintain human system stability, humans have to adapt to the stressors. According to Sofoluwe *et al.* (2011), adaptation to climate changes refers to ways and means that individuals implement to adjust to natural or human systems to offset the actual or expected impact of climate change, or its effects that may cause moderate harm or exploit beneficial opportunities (D'Silva, Shaffril, Samah, & Uli, 2012). Referring to Wilbanks *et al.* (2007), adaptation stands side by side with mitigation since they complement each other. If mitigation can keep climate change impacts at a reasonable level, adaptation would be able to deal with the impact (Saavedra & Budd, 2009). Based on such definitions, it can be stated that adaptation is the way or action of humans (individuals/groups/communities) to deal with the stressors or surprises and how they can adjust with the changes in ecological or social systems.

In relation to adaptation, human beings have noticed many different cycles in their lives. In a complex system, the cycles are not exact repetitions because the forces in a complex system are not static. Holling (2001) developed the adaptive cycle by researching the functioning of ecological system, i.e. exploitation, conservation, release, and re-organization. However, the cycle has also been used in other contexts, notably in financial and organizational systems. These cycles can only be maintained if the energy flows and attractors within the system remain dynamically stable (MacGill, 2011).

To maintain that dynamic and stable adaptation cycle, three interlinked dimensions support the running of the cycle (Holling, 2001; McGill, 2011), namely as follows:

– Potency refers to productivity or wealth of a system. It determines the number of alternative options or choices to enable the system to change.
– Connectedness or controllability reflects the level of flexibility of a system. As connectedness rises, the internal organization of system increases. A system that is too connected, particularly if it is rigidly connected, loses its flexibility to respond in new ways.
– Resilience is the ability of a system to withstand a perturbation from the outside and still maintain its function. Resilience is high where the system is free to develop in plenty of different ways and growth. In this case, resilience is contrast to vulnerability.

These are three properties that shape the responses of ecosystems, agencies, and people to crisis. Systems with low potential, connectedness, and resilience will create a poverty trap. On the other hand, the system that has high potential, connectedness, and resilience will generate a rigidity trap (Holling, 2001).

5 FINDINGS: FISHERMEN'S ADAPTATION TO THE AQUATIC ENVIRONMENT CHANGES IN JAKARTA BAY DURING 3 PERIODS OF TIME

5.1 *The first period (until 1994)*

In the first period, the fishermen in Kamal Muara were confronted with the stressors due to the increasing domestic and industrial pollutions and the increasing shipping activities in Jakarta Bay. The peak was the phenomenon of red tide at the end of 1993 resulting in the massive death of fish in Jakarta Bay. ".... in 94 or 93, my parents said (fish) all dead...nothing alive..." (Hen, fisherman).

In response to the waste pollution, the community of fishermen in Kamal Muara could not do anything. ".....when the waste comes, we just watch....do nothing...what can we do?" (Tn, fisherman). They did not know what they could do when the water is polluted. They could not go to the sea until the waste had been swept away by the waves. "When the waste comes, fish disappear ... it is better not to go to the sea...." (Bay, fisherman). They did not know to whom they could complain, since at that time the fishermen did not have any forum to channel their concerns. However, although they did not go to the sea, they were not worried at all.

At this period, Kamal Muara was a promising place for fishermen. Shallow waters in the east and west parts of Jakarta Bay were good fishing grounds with abundant fish. Although they were not able to go fishing for several days and were forced to borrow money from moneylenders when pollution occurred, they were still very optimistic. Once the pollution subsided, they went fishing again. They could repay their loan by merely fishing for one day because the fishing resources in that period were still very abundant.

5.2 The second period (1995–2007)

Entering the second period, fishermen in Kamal Muara encountered another stressor of waste pollution in Jakarta Bay. The phenomenon of plankton bloom appeared in high intensity and caused massive death of fish in 2004 (April and November), in 2005 (April, June, August, and October), and in 2007 (April and November) (Sachoemar & Wahjono, 2007). As was the case in the first period, they did not fish during the pollution period and waited until the waste had been swept away. Meanwhile, responding to the higher intensity of pollution, in 2005, fishermen in Kamal Muara established '*Kelompok Masyarakat Pengawas*' (*Pokmaswas*) or Supervisory Community Group. The group supervised uncontrolled maritime resources, fishery, and pollution. However, the report of *Pokmaswas* has never been followed up by any effective solution. Noticeably, more and more wastes were dumped into the Jakarta Bay.

Another stressor emerged since the implementation of the Presidential Decree No. 52/1995. The reclamation of Pantai Indah Kapuk (PIK) area in 2002 had decreased the area of mangrove forest (Fatahillah, 2006). At the same time, the fishing grounds in the eastern part of Jakarta Bay was removed. However, with the money they earned from the compensation, the fishermen moved their fishing gears (*bagan* or *sero*) to the west part of Jakarta Bay, close to Tangerang Regency and Kepulauan Seribu. Although the catches have decreased, the fishermen in Kamal Muara still enjoyed their activity in the shallow waters and were still optimistic, "…when PIK was developed, it was still good… we moved to the west…and the catch was good…" (Hen, fisherman). It means that although they only had fishing grounds in the west part of Jakarta Bay, the fishermen still had access to the extensive waters with good fish catch.

5.3 The third period (2008–2015)

Since the implementations of the Presidential Decree No. 54/2008 and the Spatial Plan of the government of DKI 2030, the government of DKI Jakarta has intensively developed the area of North Jakarta through reclamation along the Jakarta Bay. Aquatic environment changes in the third period became uncontrollable. It resulted in complicated stressor. Aquatic environment was degraded, and fishermen's life became endangered. They had taken the necessary efforts to deal with the stressor.

Since 2011, the Indonesian Ministry of Maritime and Fishery Affairs had facilitated the fishermen to establish '*Kelompok Usaha Bersama*' (*KUB*) or the Joint Business Group as a forum of facilitation in providing guidance, supervision, and support from the government. Until 2015, there had been 16 *KUB* fishermen in Kamal Muara. However, the institution did not perform effectively due to the high intensity of the stressor that damaged the social and economic life of the people.

The aquatic environment changes during this period were worsened by the intensive reclamation activities of the C and D islands. In 2014, the busy movement of the ships that carried and sprayed sand in the area of reclamation caused the waters to be filled with foams and turbids all day, "…sometimes the sprayed sand resulted in foam in water…." (Bay, fisherman). Besides pollution, reclamation activities also caused sedimentation. Sedimentation made the ships difficult to move. "…initially 3 *depa* of depth…now only 1 *depa*. Approximately 7 meters…now only 1.5 meters" (Ad, fisherman). Due to sedimentation, a number of fishermen did not go fishing since the access to fishing grounds were becoming more difficult. The fishermen required bamboos to move the ship to deep, adequate waters with adequate depth to ignite the engine. Due to sedimentation, fishermen's ships failed to reach the fish auction site. Therefore, the head of the neighborhood (*Rukun Warga*) took an initiative to ask the developer (KNI Company) to build a temporary wharf for landing the fish, which was quite far from the fish auction site in Kamal Muara.

The removal of fishing gears in Jakarta Bay reoccurred in 2009, 2013, and 2114. Such removals attracted capital owners to make fishing gears in the waters, only to get the compensation. "…newcomers intentionally rent local people's house …they had money and developed fish cultivation here…and expected compensation after the removal…" (SB, the head of neighborhood unit). Consequently, the fishing grounds in the west became crowded by fishing gears and green shellfish cultivation, "…currently the west part is full ……" (Abd, fishermen).

The massive removal of fishing gears resulted in suspicion among the fishermen to the 'administrators' who served as the mediator between the owners of fishing gears and the developer (KNI company). Related to such compensation money, there was no standard price of fishing gears compensation. Consequently, the compensation fund received by the fishermen was lower than they had expected. Even worse, some fishermen had not received any compensation fund at all. "…I have not received the compensation for my fishing gears…I have many fishing gears…. buried under the mud…" (Tj, fishermen).

The turbid and polluted waters, the difficult access to the waters, and the narrow fishing grounds have made the fishermen's lives worse, "…the life of fishermen was prosperous… now, we are in difficulties…even now we live very poorly…" (Ad, fisherman). Consequently, the fishermen felt that their livelihood is being endangered, "gradually fishermen would have no job…extinct…" (Bay, fisherman). The worry among the fisherman in Kamal Muara was understandable since they could not imagine how their life would be like when the reclamation plan of 17 islands has been carried out entirely.

Another problem in this period was that the 13 'bagan' (kind of fishing gears) that were located in the area of Kepulauan Seribu will be removed as well. The existence of 'bagan' breaks the rule on the Local Regulation of the Government of DKI Jakarta No. 8/2007 concerning Public Order and has hindered the shipping traffic. Consequently, there is the threat that the fishing gears of fishermen will also be removed.

The fishermen who have received compensation due to the removal of fishing gears in reclamation areas have moved their fishing gears close to Kepulauan Seribu or Tangerang Regency. Nevertheless, the fishermen who have not received any compensation would risk losing their jobs because they have no capital at all to make new fishing gears. Another problem is there are too many fishing gears and green shellfish cultivation in their fishing grounds. This has caused fewer fish to get trapped in their *bagan* or *sero*. In response to the more complicated problem in this period, the fishermen in Kamal Muara established a deliberative body, a forum to facilitate the local people of Kamal Muara, particularly the fishermen, and the developer (KNI Company) to mediate and solve the problem of reclamation. However, due to the internal conflict, the Deliberative Body was dissolved.

6 OUTCOME

In an ecological and social system, there is an interaction between social and ecological systems (Holling, 2011; Anderias *et al.*, 2004; Redman *et al.*, 2004). As Anderies *et al.* (2004) suggest the interaction between the social system and the ecological system or among the social subsystems frequently results in problems due to the conflicting interests and ideas. It leads to stressors and disruption of the social and the ecological system. To sustain it, the system will have to deal with the stressors or adapt to the changes (MacGill, 2011).

Fishermen's adaptation towards the changes of aquatic environment in each period depends on the aspects of potency, connectedness, and resilience. For the fishermen in Kamal Muara, the abundant potency of fishery resources in the Jakarta Bay and the easy access to the waters are their strongest strength to adapt to the changes. In the first and second periods, the fishermen can deal with the stressor especially when pollution occurred. Although they had limited capacity, lived in minimal internal connectedness, and did not conduct any economic activity during the pollution, the fishermen still kept their optimism. Waste pollution in Jakarta Bay did not worry them because it only happened for several days. The optimism also results from the fact that the aquatic environment provides an abundant supply of fish. Although reclamation in 2002 has reduced the extent of mangrove forests and eliminated the fishing grounds in the west part, they still remained optimistic. They adapted to the changes by moving their *bagan* or *sero* (type of fishing gears) to the east part of Jakarta Bay. During the second period, fishermen did not have any significant problem because they received capital from the compensation of removing their fishing gears. Besides that, the potency of natural resources in the west part of Jakarta Bay was abundant and access to the waters was still wide open. Therefore, in the first period and second period, they did not worry of being trapped in poverty as expected by Holling (2001).

The drastic change occurred in the third period when reclamation of C and D islands was intensively implemented. The fishermen in Kamal Muara grew powerless during the intensive reclamation in 2014. Access to waters was restricted and polluted as well. Apart from sedimentation, the waters were full of foam and turbid. Consequently, the potency of fishing resources decreased significantly. The fish moved away from their *bagan* or *sero*, so their catches decreased drastically. Meanwhile, they could not fish in the farther and deeper waters, since they lacked the skills, knowledge, and capital. Although they have formed a *KUB* and a Deliberative Body during this period, they failed to articulate the fishermen's aspiration to the relevant authorities.

The adaptation process did not run optimally. Worsening conditions of aquatic environment, decreasing potency of fish resources in the waters, non-optimal function of the local institutions in the community, and the lack of skills and adaptability of fishermen in Kamal Muara to survive have made it difficult for them to deal with stressors. If this condition persists, the fishermen, as part of the social system, will fail to adapt to the changes (mal adaption). According to Holling (2001), the system with low interconnectedness, potency, and resilience will result in a trap of poverty.

7 CONCLUSION

Adaptation of a system depends on the potency, connectedness, and resilience of such said system. The results of this study indicate that the traditional fishermen in Kamal Muara are highly dependent on aquatic resources. The potency of fishery resources in the Jakarta Bay is the most important capacity for adaptation. It is clearly described in the first and the second periods. Although the fishermen have limited internal connectedness and capacity, they chose not to do anything and only waited until the pollution subsided, and they could still survive thereafter. Meanwhile, reclamation activities in the third period have damaged the aquatic environment and have restricted access to the sea as well. Consequently, the fishermen lost their capacity to adapt (become maladaptive).

We, the researchers, expect that this article can contribute to the formulation and planning of mitigation and empowerment by the provincial government of DKI Jakarta. Fishermen's lives are inseparable from the sea. As long as they have ships and free access to the sea, the fishermen will be able to deal with the stressors. When the fishing grounds in a shallow aquatic environment are closed, these fishermen need the adequate financial capital, knowledge, and skills to fish farther away in the deeper sea, or else, they will have to take alternative jobs.

REFERENCES

Anderies, J.M., Janssen, M.A. & Ostrom, E. (2004) A framework to analyze the rebustness of social-ecological systems from an institutional perspective. *Ecology and Society*.

Berkes, F. & Folke, C. (1998) Linking social and ecological systems for resilience and sustainability. In: Berkes, F. & Folke, C. *Linking social and ecological systems: Management practice and social mechanisms for building resilience.* Cambridge, Cambridge University Press. pp. 1–25.

Berkes, F., Colding, J. & Folke, C. (2003) *Navigating Social-Ecological Systems: Building Resilience for Complexity and Change.* Cambridge, United Kingdom, Cambridge University Press.

D'Silva, J.L., Shaffril, H.M., Samah, B.A. & Uli, J. (2012) Assessment of social adaptation capacity of Malaysian Fisherman to climate change. *Jounal of Applied Science.* 876–881.

Holling, C. (2001) Understanding the complexity of economic, ecological and social system. *Ecosystems.* [Online] 390–405. Available from: http://www.jstor.org/stable/3658800 [Accessed 21st December 2011].

Ife, J. (2013) *Community Development in an Uncertain World: Vision, Analysis and Practice.* New York, Cambrige University Press.

MacGill, V. (2011) A comparison of the prochaska cycle of change and the holling adaptive cycle: Exploring their ability to complement each other and possible application to work with offenders. *Systems Research and Behavioral Science.* 526–536.

Palsson, G. (1998) Learning by fishing: Practical engagement and environmental concerns. In: Berkes, F. & Folke, C. *Linking social and ecological systems: Management practice and social mechanism for building resilience.* Cambridge, Cambridge University Press. pp. 48–66.

Saavedra, C. & Budd, W.W. (2009) Climate Change and Environmental Planning: Working to Build Community Resilience and Adaptive Capacity in Washington State, USA. *Habitat International.* 246–252.

Scheffer, M., Westley, F., Brock, W.A. & Holmgren, M. (2002) Dynamic interaction of socities and ecosystems-linking theories from ecology, economy, and sociology. In Gunderson, L.H. & Holling, C.S. *Panarchy: Understanding Transformations in Human and Natural Systems.* Washington DC: Island Press. pp. 195–239.

Competition and Cooperation in Social and Political Sciences – Adi & Achwan (Eds)
© *2018 Taylor & Francis Group, London, ISBN 978-1-138-62676-8*

Gender bias on structural job promotion of civil servants in Indonesia (a case study on job promotion to upper echelons of civil service in the provincial government of the special region of Yogyakarta)

E.P.L. Krissetyanti
Department of Public Administration, Faculty of Social and Political Sciences,
Universitas Indonesia, Depok, Indonesia

ABSTRACT: Although the number of women civil servants has increased over recent years, they, are still under-represent in upper echelons (high leader positions). This phenomenon shows that there is gender disparity in upper echelons. This paper analyses why there is a gender bias in the promotion of civil servants to the upper echelons, and identifies the factors that lead to it. Although public institutions have provided access for women civil servants to develop a career, through a process of job promotions that based on competence or merit, women civil servants still face a glass ceiling in their career achievement. This study is conducted using a qualitative method. Data are based on in dept interviews that are conducted with five women civil servants who occupy positions in the upper echelons, key people from the Local Civil Service Agency, who are associated with the selection and promotion of civil servants in the provincial government of the Special Region Yogyakarta (DIY), and key people from the National Civil Service Agency. This study's findings show that the regulations and policies concerning promotion to structural positions are based on merit or competence. The process of promotion to high leadership positions conducted by the provincial government of DIY is open. Several barriers for women in achieving the upper positions, such as societal barriers, are not proven in this case. Some organisational factors in this case really prevent women from achieving the higher positions, but other factors do not. Some individual factors also act as barriers. This paper recommends that the managers of local government institutions formulate the regulations that are gender sensitive, especially the regulations regarding promotion and selection. For example, there should be career mentoring and support for women to resolve the work-family conflict in order to achieve progress in their career.

Keywords: gender bias, promotion system, glass ceiling

1 INTRODUCTION

Women's participation in the government's workforce in Indonesia has increased significantly in recent years. The percentage of women in the bureaucracy in Indonesia was about 48.89% in 2015 (see Table 1). Nevertheless, the percentage in high leadership positions (echelons I and II) in the bureaucracy is only 16.79% of all officials in high leadership positions (see Table 2). Women's representation in high leadership positions in the bureaucracy is important because of the role of women in government decision making. According to representative bureaucracy, gender is one of the demographic groups forming the basis of representation. Theoretically, bureaucrats implement policies in ways that benefit the demographic groups they represent (Smith, 2014). Therefore, it is important for women to hold leadership positions in the bureaucracy in order to ensure that bureaucrats implement policies and regulations that are gender responsive.

Table 1. The number of civil servants in Indonesia by gender 2007–2015.

Year	Men		Women		Total
	Number	Per cent	Number	Per cent	
2011	2,403,178	52.58%	2,167,640	47.42%	4,570,818
2012	2,332,549	52.21%	2,135,433	47.79%	4,467,982
2013	2,260,608	51.82%	2,102,197	48.18%	4,362,805
2014	2,288,631	51.37%	2,166,672	48.63%	4,455,303
2015	2,319,334	51.11%	2,218,820	48.89%	4,538,154

Source: Internal data from National Civil Service Agency.

Table 2. The number of structural jobs in Indonesia's bureaucracy (2015).

Structural job	Men		Women		Total	Per cent
	Number	Per cent	Number	Per cent		
Echelon I	1,344	0.30%	497	0.11%	1,841	0.42%
Echelon II	23,913	5.39%	4,831	1.09%	28,744	6.48%
Echelon III	77,542	17.49%	20,275	4.57%	97,817	22.07%
Echelon IV	208,582	47.05%	106,297	23.98%	314,879	71.03%
Total	311,381	70.24%	131,900	29.76%	443,281	100%

Source: Internal data from National Civil Service Agency.

Table 3. The number of structural officials in the provincial government of DIY (2014).

Echelon	Men	%	Women	%	Total
I	1	100%	–	0%	1
II	28	75.7	9	24.3%	37
III	105	67.5%	52	32.5%	157
IV	241	58%	174	42%	415

Source: Internal data from Local Civil Service Agency of the provincial government of DIY.

The phenomenon of women's under-representation in the upper echelons also occurs in all of the local governments in Indonesia, including that of the Province of the Special Region of Yogyakarta (DIY). According to data from the Local Civil Service Agency of the provincial government of DIY, the percentage of women civil servants that occupy the upper echelons (high leadership positions) is only 24.3 per cent (see Table 3).

The under-representation of women in the upper echelons (high leadership positions) brings up the problem of gender inequality in the bureaucracy's workforce. This is becoming a problem, since the system of promotion and the selection of structural officials is based on a merit system. According to Law 5/2014, 'merit system is a policy and management apparatus that are based on qualification, competence, and performance in a fair and equitable manner, regardless of political background, race, color, religion, origin, gender, marital status, age, and disability or disability condition'. Therefore, the process of selection and promotion is undertaken in fair and competitive ways. Since Law No. 4 was implemented in 2015, the selection process for the high leadership positions has been conducted openly through an open system. This is intended to provide wider opportunities for the candidates, including candidates from other public institutions or even private institutions. Although the promotions and the process of selection are fair and competitive, this has not been able to drive an increase in the representation of women in high leadership positions. The fact that women are under-represented in high leadership positions needs to be analysed in order to identify the factors that lead to it. Based on this explanation, the research question in this paper is '*why has the structural position selection system that is based on the merit system not been able to encourage more women civil servants to be able to have structural positions, especially in high leadership positions?'*

2 DATA AND METHOD

This study was conducted using a qualitative method. The data were collected from in-depth interviews, through face-to-face or telephone interviews. All of the interviews were conducted individually to enable the participants to share their experiences, arguments or perceptions as openly as possible. The interviews were conducted with five women civil servants who hold high leadership positions (echelon II) in the provincial government of DIY. In-depth interviews were also conducted with key people in the Local Civil Service Agency and the National Civil Service Agency, related to the promotion system of the civil servants. The data for this study was also based on secondary data that supported the arguments of this study. The interviews and secondary data collection were conducted in about three months, from June to August 2016.

3 THEORETICAL REVIEW

3.1 *Gender bias in bureaucracy*

Gender bias is a condition favouring or harming one gender. Gender bias in leadership typically places women at a disadvantage relative to men, so that men continue to hold more powerful positions than women in both business and government (Hogue, 2016). Gender bias is usually explained as a perceived mismatch between the cognitive capabilities of women and general conceptions of leaders being masculine rather than feminine (Hogue, 2016), and/or job requirements that have historically been held by men. Gender bias in leadership has changed over time, from the first generation, which showed overt bias and discrimination, to the second generation, which showed covert bias (Ely et al., 2011; cited by Hogue, 2016). According to Ibarra et al. (2013), 'second-generation gender bias is subtle and often invisible barriers to women that arise from cultural assumptions and organizational structures, practices, and patterns of interaction that inadvertently benefit men' (p. 40). This means that second generation gender bias can be hard to detect, but when people are made aware of it, they can see possibilities for change.

Subtle and invisible barriers to women with regards to career development opportunities are evidenced by the research on the concept of the 'glass ceiling'. Glass ceiling is a term coined in the 1970s in the United States to describe the invisible artificial barriers, created by attitudinal and organisational prejudices, which block women from senior executive positions (Wirth, 2001). Glass ceiling is a metaphor that is used to characterise the situation that women encounter when they attempt to advance in managerial hierarchies (Powel and Butterfield, 2015). Bullard and Wright (1993) define the glass ceiling as the actual and perceived barriers or caps beyond which few women (or other minorities) in public or private organisational structures are able to move. For example, many women have obtained the requirements needed to be in upper or top positions, but for some reason they cannot achieve those positions, despite their qualifications and efforts.

Based on that explanation of the 'glass ceiling', it can be concluded that barriers for women in achieving upper positions are subtle, invisible and hard to perceive, but they have an actual impact. Women will be able to reach the upper or top positions if they can 'break the glass ceiling' that limits their upward mobility. As the traditional work patterns, patterns of education and training, and full-time work are based on the work patterns of men, organisations continue forcing women to conform to the stereotypes of traditional careers for male workers (Mavin, 2001). Therefore, women would reach the upper positions if they could adopt male behaviour. According to Vanderbroeck (2010), women that can 'break the glass ceiling' are seen both as strategic and as willing to take the same risks as men.

According to Jackson (2001), the literature on the glass ceiling suggests that barriers fall under the broad categories of corporate culture and corporate practices. In relation to barriers that impede women's career development, culture is further delineated into perception or stereotyping and organisational climate (Jackson, 2001). Policies should be in place to support the promotion of qualified women to management positions.

3.2 *Job promotion system*

Job promotion, or 'upward mobility' in some literature, is a term that is related to career development and career achievement. Career development can be viewed from an organisational or employment perspective. Organisations depend on promotions to fill many of their management positions. Promotions also provide direct economic and psychological reinforcement for the employees who are promoted (Sheridan et al., 1997). According to Allen (1997), the promotion system is an important human resource management mechanism. Within formal organisations, there is usually a well-established hierarchy in which advancement takes place in the form of promotions to higher level positions, in accordance with the organisational structure. According to Pergamit and Veum (1999), promotions may be used to motivate employees through increased rewards and greater responsibilities.

The system of promotion and selection, as well as other systems within the organisation, is expected to have a significant influence on the outcome of the organisation. According to the study of Sheridan et al. (1997), the job performance of employees is a key factor in determining which employees are selected for promotion. Previous research conducted by Markham et al. (1987) cited by Allen (1997) argues that, besides performance, the employees' educational attainments, on-the-job training and demographic characteristics also affect upward mobility. As job promotion or upward mobility is important for both the organisations and the employees, a person-organisational fit is necessary. According to Cole (2015), ascertaining that someone is compatible with the organisation is perhaps regarded as even more significant. The notion of 'fit' is well researched, reflecting a similarity between an organisation and the individuals concerned. The promotion process, in which there is a receipt of promotion and training, is based on an individual's revealed ability for the job. Therefore, it is necessary to determine the selection for promotion or upward mobility (Lazear and Rosen, 1990).

In this regard, Ferris et al. (1992) proposed a model of a promotion system within organisations. That model considers a range of antecedent variables and suggests that elements of the promotion system may be expected to vary according to environmental, organisational and job factors. Figure 1 describes the promotion system model that is argued by Ferris et al. (1992). This model also describes the relationship between organisational characteristics and outcomes. This model consists of three types of factors that influence the characteristics of the promotion system, namely environmental, organisational and job factors.

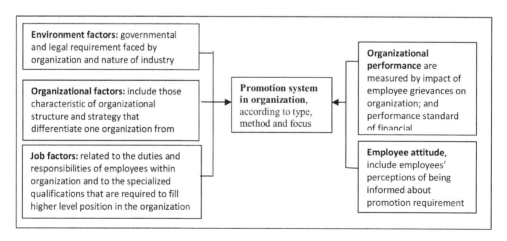

Figure 1. Model of promotion system in organization (Ferris et al., 1992, p...).

The promotion and selection system in organisations is usually prone to bias, including gender bias. Theoretical arguments for gender effects on promotions can also be grounded in human capital theory (Strober, 1990, cited by Sheridan et al., 1997). Human capital theory argues that variation in upward mobility is largely related to differences in the quantity and quality of the educational opportunities available to women (or minority groups). Gender bias is evident from the existence of discrimination in promotion decisions. Indications of this discrimination are manifested in the slowing of the promotion rates of women employees (or minority groups) or by creating 'glass ceilings' as barriers to upward mobility. Other evidence that shows that many jobs are perceived as gendered is that men and women are perceived as differentially likely to perform well in a given job (Burke & Vinnicombe, 2005). Consequently, hiring decisions are partially based on whether the job in question is considered more suitable for men or women (Hareli et al., 2008). According to Nabi and Wei (2015), several factors are related to the selection procedures that cause gender discrimination or inequality, namely transparency, male dominance on the selection committee, ambiguity in selection procedures, political issues, female stereotypes and specific networks.

As a summary of this section, the conceptual framework of this study concludes three themes concerning gender bias that are argued by Jackson (2001). They are: (1) perception about career barriers; (2) perception of workplace initiatives; and (3) perception about their chances for success in career advancement in organisations.

4 FINDING

The promotion system in public institutions, especially in local government institutions in Indonesia, is based on the regulations of the central government. According to Law 5/2014, 'the placement of the high leader positions (*madya* and *pratama* rank) is conducted openly and competitively among civil servants by taking into account the requirements of competence, qualifications, rank, education and training, office track record, and integrity as well as the requirements of other positions in accordance with the provisions of the legislation' (Law 5/2014, section 108 subsection (3)).

Promotion to high leadership positions within an open system is conducted through a selection process by a selection committee. To ensure the objectivity of the selection process, the selection committee consists of internal elements (officers who are competent, according to their task and function, and have integrity) and external elements (professionals, practitioners, academics, community leaders, NGOs, media, etc. that are competent and have integrity). For the case discussed in this study, the selection committee was made up of 45% from internal elements and 55% from external elements, namely academics, professionals or someone with the same level of competence as the job to be occupied by the candidates who were involved in the selection process. This committee was prepared to be as neutral as possible and to avoid bias, including gender bias. For example, this committee was arranged based on the aspect of gender. That means that the selection committee consist of the equal number of men and women.

The promotion system has set the standard for the selection process of candidates for higher positions. The standard is determined based on the laws and regulations on civil service management that are regulated by the central government. In addition, the local government can adapt the standards to meet the needs of the specifications of the job that will be occupied by the candidates. From this fact, it can be seen that the civil service promotion system in structural jobs is compliant with the merit system. This means that structural job promotion is based on the ability and performance of the candidates, without distinguishing between their backgrounds, including gender. Although the system of structural job promotion is based on a merit system, in practice, structural job promotions, especially in the upper echelons, are still dominated by male candidates. For the case in this study, the candidates that participated in the process of selection for a high leadership position were dominated by male civil servants. For example, during the selection process conducted at the end of 2015, only one female civil servant was promoted to a high leadership position. Regarding this fact, in relation to the issue of gender bias, one key person from the Local Civil Service Agency argues that 'women representation is not only just about a matter of the quantity, but more about the number of women civil servants that have readiness (e.g. in their competence, administrative requirement, capability) to be promoted to high leader positions'. Similarly, a key person from the National Civil Service Agency also states that 'the most important thing in the promotion system especially in high leader positions is candidates with the qualification and competency required by the job competence and specification of the high leader positions'.

However, this fact needs to be further analysed, to identify why women civil servants are less motivated to participate in the selection process for high leadership positions. The factors

Table 4. The summary of research finding.

Barriers	Issues	Detail
Societal barriers	Stereotype	Based on the participants' perception, there is no stereotype against women in the process of selection and promotion to structural positions.
	Gender differences	This argument is suggested by the fact that, although the bureaucracy of the provincial government of DIY is affected by the *Kraton* (Yogyakarta's Imperial Palace) 'royal power', there is an openness to implementing the merit system, including in the process of promotion.
	Culture	Based on the participants' perception, there is a difference between women and men in leadership styles. The leadership style has an impact on women receiving promotion to senior positions. Participants perceive that women leaders pay relatively more attention to the career development of women subordinates.
		Based on the participants' perception, the Javanese culture, which is characterised by the patriarchy system, does not hamper women in developing their careers. The Javanese culture precisely helps women in balancing their external and internal roles.
Organizational barriers	Lack of opportunities	To get a chance of being promoted to the higher positions, women often need to be able to demonstrate a greater ability and performance. They must also be able to demonstrate their commitment if they hold a higher office. This is not always the case for males.
	Mentorship	This argument is reinforced by the fact that a career development model in bureaucracy tends to be straight lined. This causes the civil servants to be provided with progressive occupational positions. That means that career progression must follow the 'pyramid structure'. Therefore, there is a smaller chance for women for upward mobility.
	Job segregation	There is no formal mentoring, but sometimes there is informal mentoring. Participants perceive that they have more support to achieve upper positions from women leaders than from men. Sometimes they get informal mentoring from their women leaders, but not always. Therefore, that lack of mentoring may hinder women to achieve the upper positions.
	Work-family balance initiatives	Participants perceive that there is no job segregation. This argument is disputable by the fact that women occupy the upper echelons (high leadership positions) in naturally supporting units, or technical units that are related to women and children, health or social welfare. Inflexible working hours. The lack of on-site day care, even though some other organisations have it.
Personal & interpersonal barriers	Chances for career advancement in organization	• Positive self perception as reach higher position • Perceived that the competence or merit based the selection to reach higher position

that inhibit women's career development can be divided into societal, organisational and personal and interpersonal factors.

The findings of this research can be summarised in the following.

5 DISCUSSION

Since the enactment of Law No. 5 of 2014, promotion and selection have been conducted openly, meaning that candidates from outside the institutions can participate in the promotion selection process. The open system is mainly used in the selection of high leadership positions. Promotion to high leadership positions using an open system is conducted through a selection process by a selection committee, whose members consist of both internal and external elements. According to Nabi and Wei (2015), the composition of the selection committee in terms of gender must be considered in order to avoid the dominance of either men or women. The findings of this study show that the selection committee has fulfilled all the criteria with regards to membership, including gender equality.

Several gender bias issues that theoretically inhibit women from achieving the high leadership positions are not perceived as inhibiting factors. Barriers that come from organisational factors, especially, are perceived to be non-existent, such as gender stereotypes against women. In addition, the women interviewees perceived the management style to be gender neutral. This gender neutrality can also be viewed in the policies that are made by the institutions, such as the policy concerning career development. The career development policy regarding gender neutrality has a different impact on male and female civil servants. Theoretically, as the model and the approach to traditional careers are 'masculine', as indicated by the career mobility that follows the organisational structure, women are often forced to leave a 'fast track' career in order to meet family responsibilities (O'Neil & Bilimoria, 2005; O'Neil, et al., 2004). That means that women are more likely to have a 'competitive disadvantage' in career advancement, as it has been structured within the organisation.

Nevertheless, women perceive that they will face the work-family conflict at some stage of their careers, especially for child care reasons. That means that women must perform multiple roles, which are public roles (work) and family roles. According to the women participants in this study, who are all Javanese women, the Javanese culture highly values women, despite putting women behind men. Javanese cultural philosophies help Javanese women to balance their domestic roles and their public roles, including their career. The work-family balance that is initiated by organisations, among others, is an informal programme that supports child care, which is the provision of on-site day care.

Interesting findings regarding the second theme are that the administration of quotas based on gender, especially for women, is actually demeaning to women because women are considered not to have the capabilities to become leaders. Therefore, for this case, affirmative action is not needed. The female participants stated that they could not be actively involved in the informal networks, which are often actually influential in the promotion decisions made by the decision makers.

The women interviewees perceive their institutions positively, as they perceive that competence or merit based selection means that the institutions give the same opportunity to both women and men to reach the higher positions. That is supported by a previous study conducted by Cole (2015). The outlook of Cole's study reflects a different approach to career success and is referred to as the 'contest-mobility model'. According to this perspective, upward mobility is more about merit, meaning that people advance in their careers by differentiating themselves from others through competition. Education and specific employment roles are the most accurate predictors of success (Cole, 2015). This finding regarding the ability of the merit system to eliminate gender bias in the promotion and selection process is not supported by a previous study conducted by Sealy (2010). According to Sealy (2010), 'Meritocracy is a principle or ideal that prescribes that only the most deserving are rewarded. As such, meritocracy can operate accurately only in an unbiased system'.

6 CONCLUSION

The purpose of this study is to analyse why gender bias exists in the promotion system of civil servants to the upper echelons and to identify the factors that lead to it. The process of promotion and selection is based on merit and is conducted openly in the provincial government of DIY. However, some factors inhibit women's careers, and these are individual factors, which are still faced by women civil servants. Organisational factors are perceived not to directly inhibit women from achieving higher positions, but some of these factors indirectly inhibit them. For example, there are a lack of programmes that support women throughout their parenting phase. Women must sometimes leave their career for a while in order to focus on the role of parenting.

The merit system that has become the basis of the promotion and selection system is used to achieve equality. Nevertheless, in the implementation of the promotion and selection systems, there are several gender biases, whether or not they are perceptually perceived by the participants, e.g. performance or merit that is a basis for promotion, using a standard that is gender neutral and free from gender stereotypes.

7 RECOMMENDATION

Basically, this study recommends that managers of local government institutions formulate a regulation regarding the promotion and selection process that is gender sensitive. The management of government institutions should increase the access enabling women to build networks and should design mentoring programmes. Public institutions should support women in order to help them balance their domestic roles with work responsibilities by providing facilities, such as day care for children under 5 years old, a pumping room for breastfeeding mothers, etc.

Theoretically, this study provides the recommendation that future research should take more participants in order to get more opinions to corroborate the conclusions. Another theoretical recommendation is that research on gender bias against women should also include cultural factors, because gender is a social and cultural construction.

REFERENCES

Allen, G. (1997). Antecedent and outcomes of the promotion system. *Human Resources Management, 36*(2), 251–260.

Bullard, A. M. & Wright, D. S. (1993). Circumventing the glass ceiling: Women executives in American State Governments. *Public Administration Review, 53*(3), 184–202.

Burke, R.J. & Vinnicombe, S. (2005). Advancing women's careers. *Career Development International, 10*(3), 165–167.

Cole, G. (2015). Up-ward career mobility. *Development and Learning in Organizations: An International Journal, 29*(4), 28–30.

Ferris, G.R., Buckley, M.R. & Allen, G.M. (1992). Promotion system in organization. *Human Resources Planning, 5*, 47–68.

Gary, P.N. & Butterfield, A.D. (2015). The glass ceiling: What have we learned 20 years on? *Journal of Organizational Effectiveness: People and Performance, 2*(4), 306–326.

Hareli, S., Klang, M. & Hess, U. (2008). The role of career history in gender based biases in job selection decision. *Career Development International, 13*(3), 252–269.

Hogue, M. (2016). Gender bias in communal leadership: Examining servant leadership. *Journal of Managerial Psychology, 31*(4), 837–849.

Ibarra, H., Robin, E. & Kolb, D. (2013, October 11). Women rising: Unseen barriers. *Harvard Business Review*, p. 40.

Jackson, J.C. (2001). Women middle managers' perception of the glass ceiling. *Women in Management Review, 16*(1), 30–41.

Mavin, S. (2001). Women's career in theory and practice: Time for change? *Women in Management Review, 16*(4), 183–192.

Nabi, G. & Wei, S. (2015). Is the selection mechanism an issue for gender equality? A study based on top three keys organizations of the state. *Journal of Leadership, Accountability and Ethics, 12*(2), 136–146.

O'Neil, D.A. & Bilimoria, D. (2005). Women's career development phases. *Career Development International, 10*(3), 168–189.

O'Neil, D.A., Bilimoria, D. & Saatcioglu, A. (2004). Women's career types: Attributions of satisfaction with career success. *Career Development International, 9*(5), 478–500.

Pergamit, M. & Veum, J. R. (1999). What is promotion? *Industrial Labor Relation Review, 52*(4), 581–601.

Sealy, R. (2010). Changing perceptions of meritocracy in senior women's careers, *Gender in Management: An International Journal, 25*(3), 184–197.

Sheridan, J.E., Slocum, Jr. J. & Buda, R. (1997). Factors influencing the probability of employee promotions: A comparison analysis of human capital, organizational screening and gender/race discrimination theories. *Journal of Business and Psychology, 11*(3), 373–380.

Smith, A.E. (2014). Getting to helm: Women in leadership in Federal regulation. *Public Organization Review, 14*, 477–496.

Vanderbroeck, P. (2010). The traps that keep women from reaching the top and how to avoid them. *Journal of Management Development, 29*(9), 764–770.

Wirth, L. (2001). *Breaking through the glass ceiling, women in management*. Geneva: International Labour Office.

Competition and Cooperation in Social and Political Sciences – Adi & Achwan (Eds)
© *2018 Taylor & Francis Group, London, ISBN 978-1-138-62676-8*

Tax amnesty policy implementation: The supporting and inhibiting factors

N. Rahayu & N. Dwiyanto
Department of Fiscal Tax Administration, Faculty of Public Administration, Universitas Indonesia, Depok, Indonesia

ABSTRACT: In the year 2015, the realization of Indonesia tax revenue only reached 81.5% of the target set by the government. There are three main causes of the weak tax collection in Indonesia: the low level of taxpayer's compliance (approximately 40%), the leakage of state revenue, and the small base of taxpayers. One alternative to overcome the problem is by implementing a tax amnesty policy published at the end of July 2016. In order to implement the tax amnesty policy successfully, the government must consider supporting and inhibiting factors. This research aims to understand the supporting and inhibiting factors in the implementation of tax amnesty by studying the experiences of other countries that have implemented the tax amnesty policy (such as Ireland, Argentina, India, Colombia, Italy, and South Africa), and Indonesia's past tax amnesty. This research uses the qualitative method with data collection through literature studies and field research by interviewing key informants from related stakeholders. The research results reveal that the supporting factors and the inhibiting factors for a tax amnesty policy which must be considered include attractive tax facility, massive dissemination to the community, legal certainty, organized data base, and post-tax amnesty law enforcement.

1 INTRODUCTION

There are some reasons why many countries in the world implement a tax amnesty policy, such as the low number of registered taxpayers, low level of tax compliance, weak taxation database, large amount of tax debts, low amount of state revenue, especially from tax revenue and low level of economic growth of the relevant countries. It is recorded that there are 37 countries in the world, both developed and developing countries, which have implemented a tax amnesty policy: the United States, Canada, Russia, Italy, France, the Netherlands, Switzerland, Germany, Austria, Hungary, Finland, Turkey, Argentina, Ireland, Spain, Belgium, Portugal, Greece, South Africa, Honduras, Uruguay, Brazil, Peru, Panama, Chile, Bolivia, Columbia, Costa Rica, Mexico, Ecuador, Australia, New Zealand, Pakistan, Sri Lanka, India, the Philippines, and Malaysia.

Similarly, in Indonesia, the government has implemented tax amnesty policy several times (1964, 1984, 2008, and 2015) for the aforementioned reasons. In July 2016, the House of Representatives issued the Law of the Republic of Indonesia Number 11 of 2016 on tax amnesty. According to the Directorate General of Taxes, the implementation of tax amnesty in 2016 was triggered by several things, among others:

1.1 *Missed tax revenue target from year to year*

Of the tax revenue target established by the government, the Directorate General of Taxes (DGT) often fails to achieve the tax revenue target. Figure 1 shows the tax revenue trends

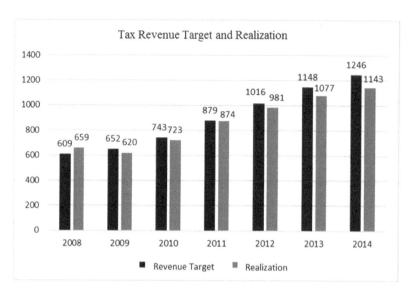

Figure 1.   Indonesia tax revenue target and realization. Source: Audited Financial Statement of the Central Government (2008–2014), reprocessed by researchers.

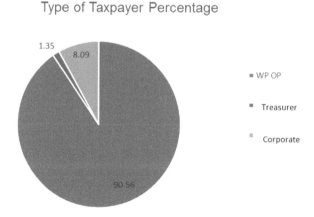

Figure 2.   The proportion of taxpayers based on taxable subject. Source: DGT, (2014, p. 3).

in the recent years indicating that the government always fails to achieve the tax revenue target.

1.2   *Low number of registered taxpayers*

Based on the Annual Report of the Directorate General of Taxes in 2014 (DGT, 2014: 3), the number of taxpayers in Indonesia until the end of 2014 was 30,574,428 inhabitants. Based on the types, the proportion of taxpayers can be seen in Figure 2 below.

1.3   *Low level of tax compliance in the implementation of tax obligations*

In comparison to the received Annual Income Tax Return in a tax year and the number of taxpayers registered at the beginning of the year, the Annual Income Tax Return for

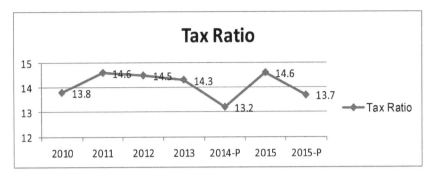

Figure 3. Tax ratio of 2010–2015 (revised state revenue and expenditure budget of 2015 p. 14).

2014 was 58.87%. One of the indicators of the low level of tax compliance in Indonesia is also reflected from the low tax ratio, especially when compared to other ASEAN countries. Based on the report data of the Revised State Revenue and Expenditure Budget of 2010–2015, the number of tax ratio in Indonesia of 2010–2015 can be seen in the following Figure 3:

1.4 Low level of economic growth

Based on the World Bank (2016: 1) data, the economic growth in 2015 weakened with the growth of only 4.8%. This figure is the lowest one compared to the percentages from previous years. Another reason why the government issued a tax amnesty policy is the Automatic Exchange of Information (AEoI) that is going to be applicable in September 2018.

2 THEORETICAL OVERVIEW

Mansury divides a fiscal policy into two definitions, namely broad and narrow definitions. Fiscal policy in the broad definition is a public policy affecting public production, job opportunities, and inflation using tax withholding and government expenditure instruments. Moreover, fiscal policy in the narrow definition is the policy related to the determination of those taxable, what the tax base will be, how to calculate the amount of taxes to be paid, and the procedure for payment of payable taxes (Mansury, 2002).

According to OECD (2010), in the formation of the tax system, policy makers should focus on the purposes to be achieved by the system. Policy makers should balance between the efficiency and the growth level. The influence in terms of tax revenues, evasion and smuggling, and the level of compliance and law enforcement should be considered. Policy makers also need to look at the tax system as a whole, rather than as separate elements.

Tax amnesty policy is one of the tax policies taken by the government to overcome existing issues. Matiello (2005) states that tax amnesty provides diverse impacts that can be classified into short and long term impacts. Within short term, tax amnesty will improve compliance and bring new taxpayers and income tax on shadow economy. However, in a long term, tax amnesty will have a negative impact in the form of a reduced level of tax compliance.

According to Andreoini, tax amnesty should be provided occasionally within a certain period so as not to raise the expectations of the taxpayers. However, tax amnesty also has negative effects. Andreoini (1991, p. 158) underlines the possibility of tax amnesty to actually have an impact on the efficiency and equity. Nar in his study states some advantages and disadvantages in a tax amnesty policy (Nar, 2015, p. 584), as shown in the following Table 1. Some of the below points should be a concern of the government in formulating policies related to tax amnesty.

Table 1. Advantages and disadvantages of tax amnesty.

Major potential advantages	Likely importance of actual effects
Recollecting tax	Important
Improving compliance in the future by lowering the compliance weight (no longer important to avoid by hiding bad behavior)	Important
Improving track records that can improve control in the future	Important
Decreasing deadweight costs of guilty feeling weight	Potentially important but not an important political argument
Politically allowing the transition towards harder compliance enforcement regime	Very important
Decreasing short term sanctions, increasing within a long term such as twisting that can increase or decrease net	Important in political approval between the soft and discipline ones
Avoiding inequity from sudden changes	Limited to lack of electors supporting tax management
Allowing encumbrance on several sanctions to those rejecting amnesty	Important
Allowing to create law enforcement against violation in the future	Very important
Making honest taxpayers angry	Important
Decreasing guilty feeling from tax evasion	Invalid if the amnesty is accompanied by a high level of law enforcement
Decreasing fear against sanctions in the future which allow to get amnesty again	Only relevant when poorly regulated

Source: Summers (Nar, 2015).

3 RESEARCH METHOD

This research is conducted using a qualitative approach since this approach provides a comprehensive overview of the phenomenon raised. Meanwhile, the data were obtained through literature studies and a field study through in-depth interviews with key informants, namely the people who are competent in understanding the issues raised in this research. The informant groups include the following: The Directorate General of Tax officials associated with the topic, members of the House of Representatives of the Republic of Indonesia, tax consultants and practitioners, taxpayers, and academics. Furthermore, the data were analyzed by applying them to appropriate theories to draw conclusions. However, the comparison of the implementation of a tax amnesty policy in different countries cannot be analyzed with the same parameters because the secondary data retrieved are different in nature.

4 RESULTS AND DISCUSSION

The implementation of a tax amnesty policy in Indonesia in the past and in other countries shows some success and some failure influenced by several supporting and inhibiting factors.

4.1 *The implementation of a tax amnesty policy in Indonesia in the past*

In the past, Indonesia implemented tax amnesty several times. From the literature and field data (interviews with key informants), the past tax amnesty application is deemed to have failed because of various factors, namely:

4.1.1 *Tax amnesty policy in 1964*

A tax amnesty policy in l964 was the first policy implemented in Indonesia. This policy was considered to have failed because the target amount was not achieved, and this tax amnesty program was designed without careful thoughts. Based on the research conducted by Lusiana (2008, p. 69), there were no structural corrective measures carried out after the implementation of tax amnesty. At that time, law enforcement still did not work well, the administration was disorganized, and the tax authorities were still closely linked to bribery. Based on interviews with Herman Butarbutar from DGT II, there were several things that made the implementation of tax amnesty that year did not succeed, and those, among others, were the insufficient administrative systems and the minimum dissemination program; therefore, there were not many taxpayers who knew about the tax amnesty program.

4.1.2 *Tax amnesty policy in 1984*

- This tax amnesty policy was the second implementation of tax amnesty in Indonesia. This policy was carried out to coincide with the tax reform in 1983, which required a stimulus to encourage taxpayers to be honest and open. Based on the publication of Sinar Harapan on 8 July 1985, the number of individual and corporate taxpayers recorded in 1985 was 1,035,989. Only 20% of the recorded taxpayers participated in this tax amnesty program. The government at that time extended the period for six months, but it did not have much influence. In the implementation of tax amnesty in 1984, Indonesia was still very much unaware of the state revenues from other sectors because the revenues from oil were still the largest state revenue, making this tax amnesty policy a failure. According to Susanto (2007, p. 85), there were several reasons why the implementation of tax amnesty that year failed, and among other things are:
- The legal foundation for the implementation of the program was not based on laws but rather limited to the presidential decrees. There was a lack of dissemination of information from the government to the public;
- The lack of openness and the increased access to public information made the DGT unable to control this program;
- The tax administration system and DGT employees were not ready in implementing the tax amnesty program;

4.1.3 *Tax amnesty policy in 2008*

The 2008 tax amnesty was the third implementation in Indonesia. This program was better known as the 'sunset policy'. This program was based on the Law Number 28 of 2007 Article 37 A stating that taxpayers reserve the right to a reduction or elimination of administrative sanctions in the form of interest. That year was also the tax reform year when some tax regulations were amended, among others:

- The Law on CTP previously based on the Law Number 16 of 2000 into the Law Number 28 of 2007.
- The Law on Income Tax previously based on the Law Number 17 of 2000 into the Law Number 36 of 2008;
- The Law on VAT previously based on the Law Number 18 of 2000 into the Law Number 42 of 2009.

The 2008 sunset policy is a tax modernization program within the period of 2001–2007. In that year, the number of successful new TIN issued was 5,365,128. The TIN growth only increased in the last period of 2008, namely November and December with the addition of 7,000–8,000 TINs per day (Santosa, 2008). The annual Tax Return also increased as much as 804,814 Tax Returns with income tax revenues increased by Rp7.46 trillion.

One of the factors behind the success of the sunset policy in 2008 was the dissemination factor. The research conducted by Rakhmindyarto in 2011 proves that there is a strong relationship between the dissemination and the amount of tax revenues gained from the sunset policy program in 2008 (2011, p. 202).

Although the implementation of tax amnesty in 2008/a sunset policy in 2008 was considered more successful than the implementation in 1964 and 1984, there were some constraints, one of which was that the DGT did not have any system that could distinguish taxpayers utilizing a sunset policy and taxpayers delivering normal annual Tax Returns. Related to data management, Herman ButarButar added that there were at least 3 things that became the obstacles in the implementation of the sunset policy in 2008, among others:

– The amnesty only covered administrative sanctions
– An unprepared tax administration system
– Too short-term implementation

4.1.4 *Tax amnesty policy in 2016*

In July 2016, the Government of Indonesia issued the Law Number 11 of 2016 on Tax Amnesty. This policy aims to attract funds deposited by Indonesian people abroad and provide opportunities for tax payers who have not reported their assets to report them at a small redeemable rate and with assurance that the tax obligations of taxpayers for 2015 and before will not be examined and considered tax crime.

The amount of the redeemable rate applied for disclosing domestic assets and repatriation of foreign assets can be divided into three periods, namely: 2% (for period 1: July to September 2016), 3% (for period 2: October to December 2016), and 5% (for period 3: January to March 2017).

This success or failure of this policy cannot be assessed because the policy remains in force until the end of March 2017.

4.2 *The implementation of tax amnesty policy in other countries*

The implementation of the tax amnesty policy in other countries shows that some countries succeed and some others fail. The success and failure of the implementation of tax amnesty in other countries are also affected by several factors.

4.2.1 *Ireland*

In the 1980s, Ireland was the most successful country implementing tax amnesty. The factors supporting the success include providing a period of 10 months to pay the underpayment tax without penalty or sanction. Taxpayers were also exempted from lawsuits. Dissemination to the public was conducted by using massive mass media and the names of taxpayers who were in arrears were included in the national newspapers of Ireland.

Along with the implementation of tax amnesty, Ireland increased the number of tax officials, named the tax sheriffs, in charge of collecting taxes. The tax amnesty conducted in 1988 was the first and final tax amnesty for Ireland. At the end of the implementation, Ireland introduced a new tax system by increasing penalties and fines for violating taxpayers and authorizing the tax authorities to confiscate assets and inventory of taxpayers in arrears and to freeze the bank accounts of the taxpayers.

4.2.2 *Argentina*

Argentina implemented the first tax amnesty in 1987. All abroad incomes for investment purposes that had not been reported were exempted from income tax. This was conducted to encourage taxpayers with assets abroad to bring their assets back to their country. This activity was also part of the debt to equity program.

Each capital converted into debts had to be added with additional fresh capital by investors. The capital used in this program would be exempted from taxes and would not be examined. Furthermore, in 1994, Argentina implemented its second tax amnesty, followed by its most recent one in 2009.

The implementation of a tax amnesty policy in Argentina was considered a failure because the policy was not supported by the improvement of a taxation system and strong law enforcement.

4.2.3 *Italy*

Italy has implemented tax amnesty several times (Malherbe *et al.*, 2010, p. 224). The first tax amnesty was conducted in 1990 with the issuance of the Royal Decree number 367 on 11 November 1990. The first tax amnesty was conducted for all taxpayers followed by tax reform as an instrument to facilitate the transition period towards the new taxation system. The reason why tax amnesty was conducted so often in Italy was because Italy did not have a strong legal system indicated by the low level of compliance and inefficiency of tax administration.

At the end of 2001 and 2002, the Government of Italy issued two regulations, namely Law-Decree 35 dated 25 September 2001, which introduced the tax shield (*Scudofiscale*) to repatriated assets from abroad and the Law 289 (27 December 2001) containing 3 amnesty standards. The goal of this tax amnesty program in Italy was to repatriate illegal assets abroad. Taxpayers would be charged a redeemable rate which varied each year for assets that would be repatriated into the country.

In 2001, the redeemable rate was 2.5%, while in 2003 it increased to 4%. However, taxpayers had to also pay a surcharge of EUR 300 corresponding to predetermined sectors based on statistical parameters. Beyond that, those using this tax amnesty facility would pay EUR 600.

In 2009 the redeemable rate payable was 5% of the amount repatriated or 1% over 5 years. To reach the underground economy sector, Italy provided a special rate for social security contributions at the rates of 8%, 10%, and 12% and tax rates of 10%, 15%, and 20%.

Interestingly, before implementing the tax amnesty, the Italian government sent their intelligence agents to track the whereabouts of people's funds parked abroad. Therefore, before the amnesty, Italy sent their economic intelligence to check where the monies of the Italians had been saved. As much as 600 billion Euros was found to have been saved in Monaco and the Vatican (Sindonews 23/04/2016).

The implementation of a tax amnesty policy in Italy was considered successful because Italy was able to attract people's funds parked abroad up to 80 billion Euros, before implementing the tax amnesty, the Italian government sent intelligence agents to track the whereabouts of people's funds parked abroad.

4.2.4 *India*

In 1981, the government of India introduced a unique form of tax amnesty. The government sold a kind of bonds that were specifically used to reach non-taxable incomes. Anyone who had an income from dark or black market sectors was allowed to buy the bonds provided by the government. The origin or source of income to obtain these bonds would not be examined and subject to the income taxes at all levels (property tax, income tax). Investing money in these special bonds would be more profitable than investing in the normal bonds of the government. However, the tax amnesty policy in 1981 was considered a failure because the policy was not followed by the improvement of the taxation system and strengthening of law.

In 1997, India again held tax amnesty. The tax amnesty was the most successful tax amnesty in comparison with the previous years. When taxpayers reported their income that had not been reported, they could use lower tax rates than previous tax rates applicable. Taxpayers could also report assets that had not been reported by using book value. This tax amnesty was the last amnesty for this kind of facility and has never been held again.

4.2.5 *Colombia*

In 1987, Colombia conducted tax amnesty which allowed individuals who had assets that had not been reported or who had tax debts to correct their reports without penalty or lawsuit. To participate in this program, individuals had to report their minimum incomes amounting to the incomes reported during the year (Alm, 2998, p. 6). Along with the implementation of the tax amnesty, the government of Colombia also changed some taxation systems, among others:

– Lowering income tax rates
– Eliminating double taxation on dividends
– Increasing *pot-put* tax rates

Besides changing the tax system, Colombia also improved the quality of its law enforcement. As a result of this tax amnesty, Colombia managed to collect nearly $100 million or 0.3% of GDP in that year.

4.2.6 *France*

France conducted tax amnesty in 1986 which aimed to cover losses due to incomes taken out illegally. To attract the interest of taxpayers, the government of France lowered the rate for repatriated assets and eliminated the estate tax. The redeemable rate used at the time was 10%. This rate was lower than the normal rate applicable. This tax amnesty was an improvement from the tax amnesty in 1982 with a lower rate. However, the tax amnesty policy in France was considered a failure because the policy was not supported by the improvement of law enforcement or severe sanctions.

4.2.7 *South Africa*

South Africa has implemented tax amnesty 4 times in ultimately aiming to repatriate the South African assets kept abroad. The tax amnesty was first conducted in 1994 and 1995. The target was taxpayers who had not paid their tax debts. The third tax amnesty was in 2003 after South Africa changed its taxation jurisdiction from a source basis to a residence basis. The last tax amnesty policy which was held in 2006 was targeted for small businesses. Taxpayers who wanted to report the violation they had committed would be treated in a special way. Taxpayers would be exempted from sanctions. However, taxpayers who were not willing to disclose it would be fined up to 200% of their debts.

The period of tax amnesty was from 1 August 2006 to 30 June 2007. The redeemable rates for those following this program were 0%-5%. Confidentiality of information was guaranteed by a special agency established to regulate the implementation of tax amnesty, and government financial institutions would not be able to access the confidential data.

The implementation of the tax amnesty policy in South Africa was considered a success, because, in addition to the very attractive facilities by the government, South Africa conducted the amnesty following a political reconciliation. The South African government, as guaranteed by President Mandela, also pledged to continuously improve their economic reforms over the following 20 years if the tax amnesty was conducted successfully (Sindonews 23/04/2016).

5 CONCLUSION

Based on the research about the implementation of tax amnesty policy in Indonesia and other aforementioned countries, it can be concluded that the supporting and hindering factors of the success and failure of the implementation of a tax amnesty policy are as follows:

– Interesting facilities (amnesty redeemable rates, a guarantee of not being examined, guarantees for the elimination of tax crime, etc.).
– Massive dissemination to public
– Good tax administration & organized data base
– Further action from the tax amnesty officials, namely the improvement of the taxation system and post-tax amnesty law enforcement

REFERENCES

Alm, J. (1998) *Tax Policy Analysis: The Introduction of Russian Tax Amnesty*. Georgia State University, Working Paper, pp. 98–6.
Andreoni, J. (1991) The desirability of a permanent tax amnesty. *Journal of Public Economics*. [Online] 45, 143–159. Available from: http://econ.ucsd.edu/~jandreon/Publications/JPubE1991.pdf [Accessed 10th February 2016].

Arnold, J. (2012) *Improving the Tax System in Indonesia*. OECD Economics Department. Working Papers, No. 998, OECD Publishing.

Bernardi, L., Fraschini, A. & Shome, P. (2012) Tax Systems and Tax Reforms in South and East Asia.

Creswell, J.W. (200) *Qualitative Inquiry and Research Design; Choosing Among Five Approaches*. 2nd edition. California, Sage Publications.

Franzoni, L.A. Punishment and grace: on the economics of tax amnesties. *Foundation Journal Public Finance*. [Online] Available from: http://www2.dse.unibo.it/franzoni/Punish.PDF [Accessed 10th February 2016].

Geurts, T. (2014) *Public Policy Making: The 21st Century Perspective*. The Netherlands.

Herman, B.L. & Richard, J.Z. (1987) Amnesty, enforcement, and tax policy. *Tax Policy and The Economy*. [Online] 1. Available from: http://www.nber.org/chapters/c10929.pdf [Accessed 4th January 2016].

Kementerian Keuangan. (2014) *Laporan Keuangan Kementerian Keuangan Tahun Anggaran 2013*. [Online] Available from: http://www.kemenkeu.go.id/Publikasi/laporan-keuangan-kementerian-keuangan-ta-2013 [Accessed 27th February 2015].

Kementerian Keuangan. (2015) *Nota Keuangan dan Rancangan Anggaran Pendapatan dan Belanja Negara Perubahan Tahun Anggaran 2015*. [Online] Available from: http://www.kemenkeu.go.id/Data/nota-keuangan-rapbn-p-tahun-2015 [Accessed 27th February 2015].

Lusiana, R.E. (2008) *Kajian atas Formulasi Sunset Policy melalui Kebijakan Pengurangan atau Penghapusan Sanksi Administrasi berupa Bunga*. [Thesis] Departemen Ilmu Administrasi, Fakultas Ilmu Sosial dan Ilmu Politik, Universitas Indonesia, Depok, Indonesia.

Malherbe, *et al.* (1998) Tax Amnesties in the 2009 Landscape. *Bulletin for International Taxation*, IBFD.

Mansury, R. (2000) *Kebijakan Perpajakan*. Jakarta, Yayasan Pengembangan dan Penyebaran Pengetahuan Perpajakan.

Mansury, R. (2002) *Pajak Penghasilan Lanjutan Pasca Reformasi 2000*. Jakarta, YP4.

Mattiello, G. (2005) *Multiple tax amnesties and tax compliance (forgiving seventy times seven)*. [Online] Available from: http://www.unive.it/media/allegato/DIP/Economia/Working_papers/Working_papers_2005/0506.pdf. [Accessed 4th January 2016].

Nar, M. (2015) The effects of behavioral economics on tax amnesty. *International Journal of Economics and Financial Issues*. [Online] 5. Available from: http://www.econjournals.com/index.php/ijefi/article/download/1044/pdf [Accessed 4th January 2016].

Rakhmindyarto. (2011) Evaluating the sunset policy in Indonesia. *International Review of Social Sciences and Humanities*. 2 (1), 198–214.

Susanto, H. (2007) Implikasi Tax Amnesty terhadap Kebutuhan Investasi dan Tax Compliance. [Thesis] Departemen Ilmu Administrasi, Fakultas Ilmu Sosial dan Ilmu Politik, Universitas Indonesia, Depok, Indonesia.

The World Bank. (2016) *Indonesia economic quarterly: Private investment is essential*. World Bank Group.

Uchielle, E. (1989) FRBNY Quarterly Review/Autumn.

Urinov, V. (2015) Tax Amnesties as a transitional Bridge to Automatic Exchange of Information. *Bulletin for International Taxation*, IBFD.

The impact of microfinance as a community development program

E. Rahayu & I.R. Adi
*Department of Social Welfare, Faculty of Social and Political Sciences, Universitas Indonesia,
Depok, Indonesia*

ABSTRACT: Program Pengembangan Ekonomi Masyarakat Kelurahan (PEMK) or the
sub-district community economic empowerment program is a community development program for low-income people who run micro and small business in DKI Jakarta. The program
provides capital access to the members of Financial Service Cooperatives or Koperasi Jasa
Keuangan (KJK) as the beneficiaries in the form of revolving fund of two to five million
rupiahs. Some members received ten million rupiahs. Besides that, the program also provides
capacity building for the beneficiaries. The study aims to identify the impacts of the program
on the beneficiaries. The qualitative research was conducted in Kelurahan Lagoa and Kelurahan Kelapa Gading Barat. Although the main purpose of the program is to empower the
local economy, it turns out this program impacts not only the economic aspect, but also the
social, personal and spiritual aspects.

1 INTRODUCTION

Beneath the rapid economic development, poverty alleviation remains a serious problem in
Jakarta. We can see that poverty in Jakarta has fluctuated but also tended to increase over
the last 5 years. In March 2010, the percentage of poor people in Jakarta amounted to 3.48%,
and in September 2014, the poverty rate in Jakarta reached 4.09% of the total population.
This is the highest poverty rate over the last 5 years (Central Bureau of Statistics 2015).

One of the attempts to alleviate poverty is empowerment. According to Ife (2006) empowerment is intended to leverage the power of the disadvantaged. Ife (2006) classified poor
community into the Primary Structural Disadvantaged Groups.

Microcredit and microfinance have been recognized as good strategies to alleviate poverty
and empowerment (Joseph, 2005; Haque 2009; Roxin *et al.,* 2010; Sultana & Hasan, 2010).
Roxin *et al.* (2010) found that microcredit has a significant effect on the economic empowerment of women in Sierra Leone and has an initial effect on social empowerment. Uddin (2011)
similarly found that microcredit plays an important role in the social and economic development in Hakaluki, particularly in the increase of the household income, the diversification of
means of livelihood, the creation of entrepreneurship, the poverty alleviation and the empowerment of women in Hakaluki Haor. Alam et al. in a study in Pakistan (2010) found that
microcredit has been proven to improve the income of the poor. In conclusion, microcredit
and microfinance are good strategies to alleviate poverty, create employment and improve the
welfare of the poor community (Chowdhurry, 2005; Hamdani, 2012; Al-Shami, 2014).

One of the poverty alleviation programs in DKI Jakarta is Program Pemberdayaan
Ekonomi Masyarakat Kelurahan (PEMK) or the Sub-district Community Economic
Empowerment Program (Wordpress, 2012). PEMK was initially named Program Pemberdayaan Masyarakat Kelurahan (PPMK), or the Sub-district Community Empowerment Program and was launched in 2001.

This program provides a revolving fund for small and micro business for communities in
the sub-district. The goal of the distribution of the revolving fund is to provide an easy access
to capital for communities of the sub-district. Meanwhile, the objectives are to:

– Improve the entrepreneurship capacity of community in sub-district.
– Improve the economic capacity of community in sub-district.
– Create new employment. (Pocket Book of Pemantapan Pelaksanaan Penyaluran Kredit Bergulir 2013)

The institution that manages this program is Unit Pengelola Dana Bergulir (UPDB) or the Revolving Fund Management Unit as the technical management unit in Dinas Koperasi, Perdagangan dan Usaha Mikro, Kecil dan Menengah, DKI Jakarta Province. In the Jakarta Governor Regulation No. 36 Year 2012 on the Technical Guidelines for the Management of PEMK Revolving Fund, it is stated that the Financial Service Cooperatives or Koperasi Jasa Keuangan (KJK) is a microfinance institution incorporated as the cooperative partner UPDB in the management of revolving funds. KJK is a business entity that is free and independent and becomes partners with Lurah (the Head of sub-district) in empowering the community's economy. Based on the interview with the Chairman of UPDB, in the early stages of each KJK, the entity gets funding that amounts to 540 million rupiahs. After that, they can apply for additional funds.

Up to August 2014, the fund that had already been disbursed by UPDB to KJK throughout Jakarta totaled to IDR 308,250,800,000,-. (Three hundred and eight billion two hundred fifty million eight hundred thousand rupiahs). Meanwhile, the fund that had been returned to UPDB amounted to IDR 179,705,648,935,-. (One hundred and seventy-nine billion, seven hundred and five million six hundred and forty-eight thousand nine hundred and thirty-five rupiahs). The rest of the fund is still in KJK.

KJK Lagoa in North Jakarta has the largest rate of the non-performing loan, namely 108 months with the total loan 67% of their loan. Meanwhile, KJK Kelapa Gading Barat, also in North Jakarta, has the largest loan and yet the smallest rate of the non-performing loan. Considering this, it is interesting to study the impacts created by the Sub-district Community Economic Empowerment Program or Program Pemberdayaan Ekonomi Masyarakat Kelurahan (PEMK) on the beneficiaries at KJK PEMK Lagoa and Kelapa Gading Barat.

This research is important considering that the funds that have already been spent by the Jakarta Provincial Government to empower the low-income people in Jakarta have been so great, and so it has to be examined whether the program has made the expected impact. If the impact is as expected, the program can continue. However, if the impact is not as expected, it may be necessary to make some improvement to the program, so that the funds are not spent in vain.

Research on the impact of the economic empowerment through microcredit has already been done, as conducted by Alam *et al.* (2010), Appah, Sophia M. John and Soreh Wisdom (2012), Annim and Alnaa (2013) and Al-Shami (2014). These studies examined the impact of the economic empowerment through microcredit on poverty. Other studies were conducted by Woller and Robert Parsons (2002), Roxin (2010) and Sultana and Hasan (2010), which examined the impact of the program on the economic empowerment. In addition, there are also studies by Shah and Huma Butt (2011) and Uddin (2011), which examined the impact of the program on the socio-economic empowerment. Within the special economic empowerment program implemented by the Jakarta Provincial Government, no one has studied the impact of the PEMK program. There are studies on the impact of the Sub-district Community Empowerment Program to the beneficiaries carried out by UKM Center FE-UI (2006) and the impact of the Sub-district Community Empowerment Program on community resilience by Atmono (2007). However, none of the studies above that uses the theoretical framework of Jim Ife's community development in the discussion.

Based on earlier research, the research question posed in this study is: what are the impacts of the Sub-district Community Economic Empowerment Program for the beneficiaries in the light of Jim Ife's conceptual framework of community development?

2 RESEARCH PURPOSE

This research aims to identify the impacts of the Sub-district Community Economic Empowerment Program for the beneficiaries in the light of Jim Ife's conceptual framework of community development.

3 METHOD OF THE STUDY

This study uses a qualitative approach and the type of research is descriptive. Descriptive research presents a picture of the specific detail of a situation, social setting, or relationship (Neuman, 2006). The study is conducted at **KJK PEMK** Lagoa with the irregular repayment of revolving funds and at **KJK PEMK** Kelapa Gading Barat with the regular repayment of revolving funds.

Informants were collected through purposive sampling or judgmental sampling, which is a non-randomized sampling method used in a particular situation. Neuman (2006) reveals that purposive sampling is suitable for unique cases. There are 36 informants who took part in this study. They are: (1) Chairman and staff of Unit Pengelola Dana Bergulir; (2) Regional coordinator/Person in charge of **KJK** Jakarta Utara; (3) Counselor of **KJK PEMK**; (4) Legal Consultant in **KJK PEMK**; (4) Administrator and Manager of **KJK PEMK** Lagoa and Kelapa Gading Barat; (5) Government officials of Kelurahan Lagoa and Kelapa Gading Barat; (6) Beneficiaries of **KJK PEMK** Lagoa and Kelapa Gading Barat.

Data were collected through literature review, documentary study, interview and observation. The data analysis technique in this study used the steps proposed by Grinnell (2005):

– Preparing your data in transcript form
– Establishing a plan for data analysis
– First-level coding
– Second-level coding
– Interpreting data and theory building

To improve the quality of this study, we used triangulation (data source and collection techniques). Lincoln and Guba in Bryman (2008) argue that the triangulation requires you to use more than one method or data source in the study of a social phenomenon.

4 FINDINGS AND ANALYSIS

The field findings of the impacts of the economic empowerment through **PEMK** on the program beneficiaries are analyzed using Jim Ife's concept of community development. According to Jim Ife (2013), community development refers to an extensive sense. Community development has 8 dimensions: (1) social development; (2) economic development; (3) political development; (4) cultural development; (5) environmental development; (6) spiritual development; (7) personal development; and (8) survival development.

From the community development perspective, our response to an economic crisis is to develop an alternative approach that seeks to relocate an economic activity within the community to improve the quality of life. Community economic development can take a variety of forms, but they all can be grouped into two categories. The first category, as the more conservative approach, seeks to develop the community's economic activity largely within the conventional parameter, namely attracting industry, initiating local industry, and tourism. The second category, as the more radical approach, seeks to develop an alternative community-based economy, namely cooperatives, community banks, credit unions, microfinance, and local currency (Ife 2013).

Furthermore, the data will be analyzed using Ledgerwood's concept (1999) concerning the impact of microfinance activities in empowerment programs. They are (1) economic; (2) social political or cultural; and (3) personal or psychological aspects. The economic impacts expected from the microfinance activities are (1) business expansion; (2) community's ability to earn livelihood from informal economic subsector; (3) asset accumulation community level or family level; and (4) availability of "protective" economic sources for reducing poverty vulnerability (Ledgerwood, 1999).

The economic empowerment through **PEMK** is paired with training to the **KJK** members. The impact of empowerment efforts through **PEMK** related to the economic aspects of perceived members of **KJK** is to develop the **KJK** business members. Before the beneficiaries get the revolving fund from **KJK**, they only sell one type of merchandise. After getting their revolving fund, they could add another merchandise. For example, there is a **KJK PEMK** member that previously borrowed money to sell sausages and now he can add his products with boiled noodles and coffee.

A business development was experienced by one **KJK** member who has a catering business. After getting a loan from **KJK**, he could accept more boxes of catering orders. Besides that, their income also increased. For example, there are members who had the income of 20 thousand rupiahs per day. After getting a revolving loan, their income has increased to 50 thousand rupiahs per day.

Another impact revealed is that beneficiaries can have savings as reserve funds for them whenever needed. There are **KJK** Lagoa members who can set aside some money for savings for about 3 thousand rupiahs per day, which can be used for unexpected things such as buying medicine if they are sick. There is also a member of **KJK** Kelapa Gading Barat who can save about 100 to 150 thousand rupiahs per day. Another member of **KJK** Kelapa Gading Barat could increase his asset in line with the increasing business and income, so he could buy a house and a car.

The economic impact as proposed by Ledgerwood is exactly what happened to the **KJK** members, both members of **KJK** Kelapa Gading Barat and Lagoa after the business development loan. Ledgerwood (1999) calls it the expansion. **KJK** members also experience an increase in income. Ledgerwood (1999) points out that the public can get income from informal economy and members can also save the money and then use it for treatment if sick. By Ledgerwood (1999), this is referred to as the source of economic "protection" (to reduce the vulnerability of the poor).

PEMK's main objective, as stated in the Pocket Book of Pemantapan Pelaksanaan Penyaluran Kredit Bergulir, is primarily related to the economic aspects, namely to improve the entrepreneurial capacity of the sub-district community, to boost the economy of the sub-district community, and to create new jobs. Increasing the entrepreneurial capacity may only be achieved by some members, namely those who are creative and can develop their business. In improving the economy of the sub-district community, recently some **KJK** members manage to increase their income after borrowing the revolving fund from **KJK**. As for creating new jobs, it may only be a small part of the members who can create new jobs, such as the members of **KJK** Kelapa Gading Barat who have a growing catering business so they can recruit employees. This is because in general the revolving funds are given to **KJK** members who have already had a business.

Improvements in the economic aspect of the **KJK** members are seen in the general incomes which are slightly above the poverty line. Meanwhile, for those whose incomes are well above the poverty line do not feel the improvement in the economic field. Improvements are only seen by a few **KJK** members whose income is at the edge of or slightly above the poverty line. Generally, **KJK** members are above the poverty line. In fact, they are actually not poor, and revolving funds are not used for business development, but for other things. It may be concluded that **KJK** has not touched those who are under the poverty line. **KJK** members should be those who have small or micro businesses. For Jakarta residents who are poor and do not have small or micro businesses, they are not covered by this **PEMK** program.

Meanwhile, the impacts of non-economic empowerment associated with the increase of members' knowledge are more appreciated by family members and friends, such as raising the awareness of responsibility of the **KJK** members, having some free time to meet spiritual needs, being able to meet the needs of food and having the fund to finance their children's and grandchildren's education. Ledgerwood (1999) calls this non-economic impact, particularly social impact. Ife (2013) calls this social development.

Being able to borrow from **KJK** has allowed the members of **KJK** to feed their family and finance the education of their children or grandchildren. By Ledgerwood, this is referred to as the political and social impact of microfinance, a change in terms of education and child nutrition (Ledgerwood, 1999).

KJK members also become more valued by other family members such as husbands or children. By Ledgerwood (1999), this is referred to as the personal or psychological impact of microfinance that has greater power both within the household or compared to before the community participates in microfinance.

When linked with the concept of community development by Jim Ife (2013), in which the development community should also develop other aspects described as balanced development, community economic development through microfinance mainly happens as an increase in economic community (economic aspect) because the main goal of the program is to boost the economy of the sub-district community. The economic development that occurs in members include the increase of business and income, the ability to have a savings, and the increase in assets.

However, in addition to improving the local economy (economic development), PEMK also improves social development, personal development, and spiritual development. Personal development here is to improve their knowledge, as this is experienced by the beneficiaries of KJK Lagoa and Kelapa Gading Barat. The spiritual development happens when members of KJK, after borrowing money from KJK, can move their business which allows time for spiritual moments. While this is not a direct effect of empowerment, spiritual development is an issue that should not be ignored.

In addition to developing their businesses that make them more respected by other family members, the social relationships with KJK family members have also enhanced. It shows the development of the social aspects (social development). With a reference to the concept of community development from Ife (2013), the impacts of the Sub-district Community Economic Empowerment Program (PEMK) on the beneficiaries are described in the following table:

The above table indicates that the impacts of the program primarily reach economic aspects. This may be because of PEMK goal which is to empower the economy of the sub-district community. Nevertheless, this program brings impact not only in the economic aspect, but also in the personal, social and spiritual aspects.

As for the beneficiaries of KJK Lagoa, their business is improved, their income increases, and they can have a savings. As for the beneficiaries of KJK Kelapa Gading Barat, in addition to the improvement of their business, the increase of their income, and the opportunity to have a savings, their assets also increase.

In social aspects, the impact of this program for the beneficiaries of KJK Kelapa Gading Barat can be seen in the better respect from family members, such as when they have the fund to finance their children's or grandchildren's education. In personal aspects, the impact of this program for the beneficiaries of KJK Kelapa Gading Barat is the increased knowledge. Besides the increased knowledge, the beneficiaries of KJK Lagoa are able to buy goods for their households.

Table 1. The impact of sub-district community economic empowerment on the beneficiaries.

Impact of PEMK	Beneficiaries of KJK Lagoa	Beneficiaries of KJK Kelapa Gading Barat
Economic	Improved business Increased income Having a savings	Improved business Increased income Having a savings Increased assets
Social	Having fund to finance children's or grandchildren's education	Having funds to finance children's or grandchildren's education Better respecting from family members
Personal	Increased knowledge Able to buy goods for household	Increased knowledge
Spiritual	Having leisure time for spiritual moments	–

Source: Study results.

5 CONCLUSION

The Sub-district Community Economic Empowerment Program or Program Pemberdayaan Ekonomi Masyarakat Kelurahan (PEMK) is a community development program in the form of microfinance. Although the main purpose of the program is to empower the local economy, it turns out that this program has an impact not only on the economic aspect, but also on the social, personal, and spiritual aspects.

REFERENCES

Agency for Cooperatives, Micro-Small-Medium Enterprises, and Commercial Affairs DKI Jakarta Province. 2013. *Buku Saku Pemantapan Pelaksanaan Penyaluran Revolving fund.* Revolving Fund Management Unit for Public Economic Empowerment in Kelurahan.

Al-Shami, S.S.A., Majid, I. Bin A., Rashid, N.A. & Hamid, Mohd.S.R.Bin.A. (2014) Conceptual Framework: The Role of Microfinance on the Wellbeing of Poor People Cases Studies from Malaysia and Yemen. *Asian Social Science.* 10 (1), 230–242. [Online] Available from: http://search.proquest.com/docview/1490613883/fulltextPDF/1433D57EB104C5BF2D4/4?accountid=17242.

Alam, T., Ellahi, N., Shahid, H. & Arif, M. (2010) Micro Credit and Poverty Eradication: A Case Study of First Micro Credit Institution of Pakistan. *Interdisciplinary Journal of Contemporary Research Business.* 2 (8), 182–192. [Online] Available from: http://search.proquest.com/docview/848430713/fulltextPDF/324F7B54DCAD4022PQ/1?accountid=17242.

Annim, S.K. & Alnaa, S.E. (2013) Access to Microfinance by Rural Women: Implications for Poverty Reduction in Rural Households in Ghana. *Research in Applied Economics.* 5 (2). [Online] Available from: http://search.proquest.com/docview/1428930364/fulltextPDF/6E1DE68C470D4F49PQ/79?accountid=17242.

Appah, E., John, M.S. & Wisdom, S. (2012) An analysis of microfinance and poverty reduction in Bayelsa state of Nigeria. *Kuwait Chapter of Arabian Journal of Business and Management Review.* 1 (7), 38–57. [Online] Available from: http://search.proquest.com/docview/1266941883/fulltextPDF/1433E035D3857001802/9?accountid=17242.

Atmono, H.D. (2007) *Pengaruh Penyaluran Dana Program Pemberdayaan Masyarakat (PPMK) terhadap Ketahanan Masyarakat (Studi Kasus Kelurahan Johar Baru).* [Thesis] Universitas Indonesia, Depok, Indonesia.

Banks, S. (2004) *Managing Community Practice: Principles, Policies, and Programs.* Great Britain, The Policy Press.

Bryman, A. (2008) *Social Research Methods.* 3rd edition. New York, Oxford University Press.

Central Bureau of Statistics. (2015) *Berita Resmi Statistik BPS Provinsi DKI Jakarta.* [Online] Available from: http://jakarta.bps.go.id/backend/brs_ind/brsInd-20150918101513.pdf.

Choudhary, A.S. (2009) *Development as social transformation: Assessing the value of social capital in microfinance and its role in the success of the Grameen Bank.* Villanova University. [Online] Available from: http://search.proquest.com/docview/305009510/CD9E485BE19149B5PQ/2?accountid=17242.

Chowdhurry, M.J.A., Ghosh, D. & Wright, R.E. (2005) The impact of micro-credit on poverty: evidence from Bangladesh. *Progress in Development Studies.* 5 (4): 298–309. [Online] Available from: http://search.proquest.com/docview/218139800/fulltextPDF/143234C51EE6548ED21/1?accountid=17242.

Grinnell, R.M. & Unrau, Y.A. (2005) *Social Work Research and Evaluation: Quantitative and Qualitative Approach.* 7th edition. New York, Oxford University Press, Inc.

Hamdani, S.M.Q. & Naeem, H. (2012) The impact of microfinance on social mobility, an empirical evidence from Pakistan. *Interdisciplinary Journal of Contemporary Research Business.* 3 (9): 81–89. [Online] Available from: http://search.proquest.com/docview/964018217/fulltextPDF/1433E035D3857001802/5?accountid=17242.

Haque, M.A. & Harbin, J.L. (2009) Micro Credit a Different Approach to Traditional Banking: Empowering the Poor. *Academy of Banking Studies Journal.* 8 (1). [Online] Available from: http://search.proquest.com/docview/215112255/fulltextPDF/14342 AE64 A1E16 A4E2/4?accountid=17242.

Ife, J. & Tesoriero, F. (2006) *Community Development: Community-Based Alternatives in an Age of Globalization.* Australia, Pearson Education, Australia.

Ife, J. (2013) *Community Development in an Uncertain World: Vision, Analysis, and Practice.* New York, Cambridge University Press.

Joseph, J.S. (2005) The relevance of involvement in micro credit self-help group and empowerment: findings from a survey of rural women in Tamilnadu. Montreal, McGill University. [Online] Available from: http://search.proquest.com/docview/304933862/fulltextPDF/1431858B95E68336FA7/5?accountid=17242.

Ledgerwood, J. (1999) *Sustainable Banking with the Poor: Microfinance Handbook an Institutional and Financial Perspective.* Washington D.C.,The International Bank for Reconstruction and Development/The World Bank.

Neuman, W.L. (2006) *Social Research Method: Qualitative and Quantitative Approaches.* 6th edition. USA, Pearson Education, Inc.

Roxin, H., Berkmuller, H., Koller, P.J., Lawonn, J., Pooya, N. & Schappert, J. (2010) Economic Empowerment of Women through Microcredit: the Case of the Microfinance Investment and Technical Assistance Facility (MITAF). *SLE Publication Series-S240.* [Online] Available from: http://edoc.hu-berlin.de/series/sle/240/PDF/240.pdf.

Shah, T.H. & Butt, H. (2011) Income generating activities through microcredit and women's socio-economic empowerment: a study of district Kasur, Pakistan. *Academic Research International.* 1 (3), 218–226. [Online] Available from: http://search.proquest.com/docview/1034969519/fulltextPDF/1431856316550C3D787/13?accountid=17242.

Sultana, S. & Hasan, S.S. (2010) Impact of Micro Credit on Economic Empowerment of Rural Women. *Bangladesh Journal Online.* 8 (2), 43–49. [Online] Available from: http://www.banglajol.info/index.php/AGRIC/article/view/7576/5715.

Uddin, M.S. (2011) *Role of Microcredit Role of Microcredit and Community-Based Organizations in a Wetland Area in Bangladesh.* The University of Manitoba. [Online] Available from: http://search.proquest.com/docview/1030793486/fulltextPDF/14342B7404BE16 A4E2/4?accountid=17242.

UKM Center FE-UI. (2006) Evaluasi program PPMK (Program Pemberdayaan Masyarakat Kelurahan). [Online] Available from: http://www.ukm-center.org/page.php?lang=id&menu=news_view&news_id=51.

Woller, G. & Parsond, R. (2002) Assessing the community economic impact of microfinance institutions. *Journal of Development Entrepreneurship.* 7 (2), 133–150. [Online] Available from: http://search.proquest.com/docview/208435331/fulltextPDF/14342814335247FAD82/15?accountid=17242.

Work-family balance: A dilemma for women workers in the manufacturing industry (case study at PT. Bintang, Tangerang City, Banten)

I.L. Fawzi
Department of Social Welfare, Faculty of Social and Political Sciences, Universitas Indonesia, Depok, Indonesia

ABSTRACT: There has not been a lot of attention paid towards women workers who work in the labour-intensive manufacturing industry and the problem of how they juggle the demands of family and work, as well as overcome their work-family problems. The importance of this study is shown due to the inability of previous research to show that the balance between work and family has a negative impact on women workers and their environment. This paper is the result of a case study conducted between 2013 and 2015 on women workers (working as operators) in the quality export sports shoe factory in Tangerang, Banten. The purpose of this study is to determine the source and impact of the problems facing women workers, the environmental support required to achieve a 'work-family balance' and how they interpret the conditions of a 'work-family balance'. Research findings reveal a number of causes at work, in the family and in individual women workers that inhibit them from achieving a 'work-family balance', as well as a number of other factors that support the achievement of a 'work-family balance'. The support of both the government and companies are required to assist women workers to overcome their problems, through policies and programmes that are sensitive to the needs and limitations of women workers in managing a work-family balance, in order to achieve 'worker well-being'.

1   INTRODUCTION

The implementation of an industrialisation policy in Indonesia since 1970 has led to a significant increase in women's participation in the labour force as factory workers in the production process (as operators). In manufacturing industries that produce shoes, clothing, textiles, food and electronics (computers) of export quality and that are labour-intensive, the majority of workers are women. Therefore, the existence of women workers is very important. However, although their roles are crucial in the production process, the attention of the government and companies towards the normative problems of women workers (especially married women), which derive from the relationships between their roles at work and in the family (work-family problems), still needs to be improved, because these problems can cause an adverse impact on the workers themselves and on their environment (family, company and government). These can include such things as physical and psychological disorders (stress), work accidents, decreased motivation and productivity, failure to achieve the production targets and the disruption of the familial relationship. The results of this study form the basis of recommendations to improve the workers' well-being and to create a working environment that is humane (to humanise the workplace).

The study aims to identify the source and impact of the problems that arise from the relationship between work and family life (work-family problems), the environmental support (family and company) needed to achieve a 'work-family balance' and the extent to which 'work-family balance' conditions are interpreted. This paper consists of the following points: methodological aspects, theoretical framework and concepts used for the analysis—especially system theory and perspective PIE (Person-in-Environment) - a literature review on the concept of 'work-family balance', findings and the analysis of the research results, recommendations and research implications.

2 METHODOLOGY

This research is descriptive qualitative research with an interpretive/phenomenological approach and aims to learn what is considered important by the research subjects and the experiences of their everyday lives from their own point of view in order to obtain their 'subjective meaning' (Neuman, 2006). The data collection uses the techniques of in-depth interviews, focus groups and observations. This study also uses the female perspective as the starting point of concern for the lives of women and the problems they face, in order to help women and find results that can be helpful to them. The main characteristic of the research from the women's perspective is their 'emic view', which in terms of epistemology is the 'most knowledgeable' (the best known) on the phenomenon being studied, as it is involves their own experiences (Reinharz, 1992).

As the case study was in the context of the manufacturing industry environment, where women work as operators, the research was conducted at PT. Bintang, Tangerang City, Banten. PT. Bintang is a labour-intensive company with a majority of women workers. Its products are quality export sports shoes for the American market and also for some European and Asian countries. The number of employees was around 9,025 people (data from 2012) and the majority (80%) were women who work as operators. Besides the work environment, an understanding of the lives of the women workers was also seen in the context of their family environment.

3 PROFILE OF INFORMANTS

The informants were chosen purposively and consisted of three women workers (key informants) sharing similar characteristics, namely, they were married, working as operators, had worked for more than five years, were members of the company's union and were known to have obtained help from the union and the company to overcome their personal problems. The complementary informants were the company (HR manager), labour union officials, family informants, the government (Disnaker's staff of Tangerang City), practitioners and a HR consultant in a labour-intensive industry. The total number of informants involved in this research was 15 people.

4 THEORETICAL FRAMEWORK

This research topic is relevant to the areas of Industrial Social Work that focus on the individuals as workers, the workplace and trades unions, the work performed and the social policies that connect the worlds of work and social welfare (Kurzman & Akabas, 1993). With respect to the focus of the study, which is to reveal, understand and analyse the lives of women workers in the manufacturing industry, this study uses the system theory, which sees humans as part of the environment (Payne, 2005; Dale et al., 2006; Healy, 2005). According to this theory, problems arise when there is an imbalance between the sub-systems and adaptability becomes an important concept for the equilibrium. Women workers are seen as a sub-system and positioned as the core of its environment. They affect and are affected by the environment (Person-in-Environment) and are also seen as biopsychosocial beings.

The results of the literature review on the relationship between work and family—which was mostly conducted in Western countries—reveal that there are several sub-studies under the topic of 'work-family relations', including 'work-family conflict', 'work-family balance', 'work-family integration', 'work-family enrichment', and so on. Several studies have been conducted on the topic of 'work-family balance' among workers of various occupations (teachers, nurses, government employees, professionals, etc.), but research on the topic of women workers (operators) was almost non-existent until today. From the methodological perspective, quantitative approaches using questionnaires or attitude scales have been more widely used than qualitative ones. This study also used a qualitative approach in the form of a case study.

The main concept used in this study is a 'work-family balance' because it shows the relationship between work and family in a positive sense, which puts the two domains in the same position and regards them as important, interrelated, complementary and requiring management to ensure the balance between the two. The balance between work and family is not easy to achieve, because each differs in the expected end result (valued ends) and differs in the way of achieving the end result (valued means). Nevertheless, the state of 'work-family balance' is a condition that needs to be achieved because of, among other things, its positive impact on the individuals and the environment, such as self-confidence, happiness, job satisfaction, increased productivity and loyalty (Greenhaus, J.H & Powel, G.N, 2006). For the company, the challenge is to create a supportive corporate culture so that employees can work in a conducive environment that increases their productivity and performance.

Balanced conditions between work and family can be seen from three aspects, namely the balance of time spent between the two domains (time balance), the balance of physical and psychological involvement between the two domains (involvement balance) and the balance between work and family in the case of obtained satisfaction (satisfaction balance) (Greenhaus, J.H, Collins, K.M and Shaw J.D, 2003). Balance is also an individual perception that is subjective in terms of individuals' interpretations of their roles and responsibilities at work and in the family (Kossek, 2005).

In order to achieve a 'work-family balance', there are two types of sources of support. The first is formal support, which comes from the company (work environment) and the government. Meanwhile, the second type is informal, from outside of the workplace, especially from the family, neighbourhood, etc. (Greenhaus, J.H., Collins, K.M., & Shaw, J.D, 2003). In order to protect the rights of workers, the Indonesian government has enacted Law No.13 of 2003 on Employment and adopted the conventions ratified by the ILO (International Labour Organisation), which aim to create prosperity among workers (Law No 13 of 2003, chapter 4). From the results of previous studies, it is known that the most important source of support from the working environment, which is the most significant in achieving a 'work-family balance' in the workers' lives, is the support of supervisors and their behaviour towards their subordinates (Ayuningtyas, L & Septarini, B.G, 2013). Moreover, other sources of support are, among others, the presence/absence of stress, family characteristics, job characteristics and the presence or absence of programmes for the workers of the company (Tremblay, 2004).

When considering the positive or negative impact towards the company that may result from 'work-family problems', the problems faced by women workers cannot be considered as mere personal problems of the workers (not the responsibility of the company), but the company should aim to design policies and programmes that support the achievement of the 'work-family balance' (Brough & Kalliath, 2009; Greenhaus J.H, Collins K.M & Shaw J.D, 2003; Kinder et al, 2008).

5 FINDINGS

The results of the interviews with the three key informants reveal that the problems that are commonly experienced by women workers (operators) who have a family are the problems that result from the demands of their roles and responsibilities both at work and in the family. According to the informants, to implement both roles in harmony and balance is not easy and often creates a dilemma where they are faced with a situation of having to choose one or the other. This is because work and family are both important to their lives and they complement each other and influence each other.

Nevertheless, the three informants had different experiences. Informant An was faced with the choice between having to work and dealing with the demands of caring for her husband and in-laws who are sick. Informant Bi was faced with the choice between the need to work to support the family and her role as a mother, including childbirth, during the breastfeeding period, when her son was a toddler and when her children reached school age. Finally, Informant Ce experienced a more complicated dilemma, between the economic pressure that forced her to continue working and the demands of her role in the family and her domestic duties, while enduring a lack of respect and support from her husband.

The three informants mentioned that the problems they faced were related to economic issues. In addition, the three informants also mentioned that some of their problems came from their work, namely: strict and long working hours (from 07.00 to 16.00 with a one-hour break for lunch and praying), a daily production target that was too high, difficulty in getting permission to go to the bathroom, a relatively small bonus for a meal allowance (Rp. 5,000, -/day), rather inhumane treatment by the supervisors and an inadequate quantity and poor conditions of the physical facilities in the workplace.

The impacts of the problems/situations experienced by these three informants are, among others: disruption of concentration at work, stress/anxiety, a decrease in motivation and productivity, sickness, errors or accidents and being laid-off from work. In addition, the impacts on the family came in the form of a disruption in the relationship with their spouses and children, and in carrying out their domestic roles. On the other hand, the impact on the company, as cited by the informant companies, are, among others: not achieving the daily production target, the declining product quality, the declining image of the company and complaints and protests from buyers, all of which mean a loss for the company.

With regards to the solutions to these problems, the three informants acknowledged the difficulty in overcoming the problems arising from the relationship between work and family (work-family problems). The main obstacle to the process of overcoming such a dilemma is the job characteristics of an operator in the production process, which are repetitive, monotonous, strict and have long working hours (7:00 a.m. to 4:00 p.m.). In addition, they also have to depend on the machinery used and they are bound by daily production targets, making it difficult to leave (by permission or on leave) the daily routine of work. If they ask permission to leave work, their daily wage will be cut. Although they experience difficult times in maintaining a balance between their work and family, the three informants had been able to adapt and survive in this job for more than 10 years, and they had been satisfied with their achievements regarding the 'work-family balance', because they felt that both their work and their family lives ran well.

According to the informants, the support received from the environment—that of the company and the family—is very important and helps to implement their roles in the two domains. Family support is very important in order to achieve a 'work-family balance', as are the co-operation and characteristics of the spouse (the husband) and the number and ages of the children. The spouse (husband) can co-operate by doing household chores and working in the same field (in this case if the husband also works as a factory worker), and this can be regarded as easing their burden. In this study, PT. Bintang had already met the workers' rights determined by the government, such as the minimum wage (regional minimum wage) and allowances, as well as health and accidents insurance, menstruation leave, maternity leave, counselling for employees and facilities in the workplace: a bathroom/toilet, a supermarket, a shuttle bus, clinics, lactation rooms, and so on.

Various programmes for the workers and their families have strengthened the relationships between the employers and the employees in the workplace, such as home visits, company lunch and picnic, workers' co-operatives, scholarships for workers' children, night school (high school and college level) for workers, and so on. However, in reality, some of the activities for workers have not been maximised in their implementation, for example: counselling for workers, lactation room facilities, co-operatives, and so on. A particular concern in this study is a counselling service for actual workers, which has become problematic and less effective because this facility cannot be used during the recess time due to time constraints and the difficulty of obtaining professional counsellors. To solve the problem, workers are also trying to find their own solutions by stating their problems and requesting help from their co-workers, requesting help from neighbours to keep an eye on their toddlers when the workers are at work, or leaving their children to stay with family in their village (parents, brothers/sister).

Finally, the meaning of 'balance' between work life and family life for the informants is subjective and can only be felt by the individual concerned. In this study, two informant workers (An and Bi) said that the condition of 'balance' for them is when they feel that what they are doing at work and for their family has met the demands. They feel satisfied with what they have achieved so far; holding down a secure job and having an harmonious family (high level of satisfaction of work and family roles). Meanwhile, the other informant (Ce) states that she is less satisfied with what she has achieved because she has been less appreciated by her partner and there are too many household chores that must be done (low level of satisfaction of work and family roles).

6 OUTCOMES

The system theory emphasises the interaction between sub-systems and also between sub-systems within a larger system to achieve a balance. The facts of working life and family life for women workers (operators) describes the two sub-systems of the individual concerned. The interaction and interplay between the two can be observed when there are problems in their family life that will affect their performance at work and vice versa. Associated with the study of 'work-family relations', the results of this research on women working in the manufacturing industry as operators show that family issues have more of an impact on employment (family to work) than the other way around (work to family) (Kossek, 2005). This is probably due to the fact that 'family' has a more emotional effect on a person, whereas 'work' has a more formal effect. Although both of these domains are important to their lives, they acknowledge that both work and family can be a source of help as well as the source of the problem of achieving a 'work-family balance'.

The characteristics of the job of an operator in a production process are known to be one factor that complicates the achievement of a balance between work and family. This is identified by comparing the time they devote to work and family, i.e. ± 10–12 hours for work, 5–7 hours for family and other activities, and approximately 5–7 hours for rest. It can be concluded that, from the aspect of 'time' alone, there is no balance. In addition, the aspect of 'involvement' also shows an imbalance, because the operator's job requires the involvement of maximum physical and psychological strength at work. For them, work is a must, so it appears that the informants fight hard to keep their job, as expressed by the following: 'The work as factory workers is our life'.

The findings of this study show that the problems experienced by workers can affect their environment, and this means that this problem can be considered neither as just a personal matter nor as only the company's responsibility. As a system, the working environment is composed of workers at various levels, which together contribute to the sustainability of the system. From the perspective of PIE (Person-in-Environment), the experiences retold by the three informants of this study demonstrate the importance of an holistic understanding regarding the female workers and their relational necessities as part of the social environment, in which they have to obtain a clear picture about the sources and types of problems (assessment) in order to provide appropriate targeted assistance (intervention) (Payne, 2005; Healy, 2005; Zastrow C & Ashman, K, 1992).

Clark's definition of 'work-family balance' is as follows: 'work-family balance is satisfying and functions well at work and at home, with a minimum of role conflict'. (Rantanen J. et al, 2011). This is put forward by the three informants of this study as their experience. They also agreed that the support and co-operation of their spouses (husbands/wives) have enabled them to manage the balance between work and family, as well as understanding the following notion: 'Work-family balance is the accomplishment of role-related expectations that are negotiated and shared between an individual and his or her role-related partners in the work and family domains' (Grzywacz & Carlson, 2007).

Another interesting thing that should be mentioned here is the similarity between the three informant female workers (operators), namely that they have a good adaptability to change and the ability to resolve problems, with or without outside assistance. They realise that it is very difficult to find another job if they get fired, so they strive with all of their abilities to quickly resolve problems while still working. The system theory also emphasises the importance of 'adaptation' in order to achieve 'equilibrium', because stability is required in a system. Work-family balance is interpreted by the three informants as satisfaction with what they have achieved, where their status as factory workers can be maintained, the family always remains harmonious and their needs are met. The meaning of satisfaction refers to their current situation after they have got through the difficulties in their life. In fact, information from practitioners and HR consultants states that women working as operators in the manufacturing industry said that their work and family lives have already achieved a 'balance'.

7   IMPLICATIONS

To improve the welfare of women workers in the manufacturing industry, it is not enough just to increase their wages (the economic aspect), but other factors must also be considered that may not be visually observable. The large number of female workers (operators), which amounts to several thousand people, makes it difficult to identify them personally and help them if they encounter a problem. They tend to be recognised by the number of the building in which they work every day, who their supervisor is, and what their job is in the shoe section. In fact, as is the case with women workers holding other jobs, they also experience the common problems faced by any other female workers who are married: the problems arising from the demands of their roles at work and in the family.

The implication of this research is the need for better awareness by governments and companies in how to assist women workers who are facing a dilemma between the demands of their roles at work and in the family. The results of this study on women workers indicate that there are differences in treatment between female and male workers, between married and single women, women who have children and those who have not, and so on. For example, by providing facilities for obtaining permission to attend an urgent family matter, additional time off for pregnant workers or those who are breastfeeding their babies, permission to leave work, care for elderly people who live at home, a sick child, etc. In addition, the counselling service is one alternative to be considered. There is a need to study and design the right kind of counselling that can be adapted to factory working conditions, job characteristics and the characteristics of women operators, because the working environment in the manufacturing industry (factories) is not the same as in other working environments, for example in banks, hospitals, offices, etc. The job characteristics of factory 'workers' are also very specific.

Providing a child day care facility in the living areas mainly inhabited by factory workers, in co-operation with the local government or housing developers, should be considered, especially to lighten the burden of women workers. Rather than bringing their children to the factory, having a child day care facility is more helpful and easier for working women.

The research looks at the importance of emphasising that labour unions should advocate for women workers who have problems by empowering staff through training in order to develop knowledge and skills on 'how-to-handle-problems', such as psychosocial problems, problems regarding the role and function of the counsellors, the sources of help, and so on. Thus, the limited number of counsellors provided by the company could be assisted by the efforts of the labour unions. Unions also need to proactively identify any vulnerable female labourers who are problematic or troubled by holding regular meetings with supervisors from each production hall and representatives of women workers.

Given the lack of attention to this specific area of study, the development of the study of 'work-family relations' should be encouraged through discussions, seminars, research, lectures and journals, considering the increasing number of working women who have a family, children and parents who are still living together. In addition, it is necessary to study this topic in other types of employment and in the context of the working environment and the living environment of workers. An understanding of the workers, the work itself and the environment with which the workers interact is required to identify the problems faced by the workers, as well as the sources and impacts of the problems, in order to develop the proper methods and programmes to assist them. Thus, the correct and precise solution between individual workers and the environment will be obtained in order to achieve 'worker well-being' (person and environment fit).

Finally, the government should consider formulating an umbrella policy regarding 'work-family policy', which would encourage the companies and trade unions to assess and formulate a programme in favour of the rights and needs of workers (male and female) to help them to balance their roles at work and in the family through the creation of a humane working environment (to humanise the workplace), as has been done in some other Asian countries, such as Malaysia, Thailand, the Philippines and India (Caparas V, 2011).

REFERENCES

Ayuningtyas, L & Saptarini B.G [2003]. Relationship between family supportive supervision behaviors and work-family-balance among working women. *Jurnal Psikologi Industri & Organisasi, 2* (1).

Brough, P. & Kalliath, T. (2009). Work-family balance: Theoretical & empirical advancements. *Journal of Organizational Behavior, 30*, 581–585. Available from: http://onlinelibrary.wiley.com/doi/10.1002/job.../pdf [Accessed 15th December 2012].

Brough, P., Holt, J., Bauld, R., Biggs, A. & Ryan, C. (2008). The ability of work-life balance policies to influence key social/organizational issues. *Asia Pacific Journal of Human Resources, 4*, 261. Available from: http://apj.sagepub.com/content/46/3/261 [Accessed 5th February 2013].

Caparas, V. (2011, June 1–3). Work family balance and family poverty in Asia: An overview of policy contexts, consequences and challenges. *Paper for United Nations Expert Group meeting on assessing family policies: Confronting family poverty and social exclusion and ensuring work-family balance.* Manila, Philippines: University of Asia and the Pacific.

Carlson, D. S., Grzywacs, J. G. & Zivnuska, S. (2009). *Is work-family balance more than conflict and enrichment?* Available from: http://www.uk.sagepub.com/journalsPermissions.nav [Accessed 5th February 2014].

Dale, O., Smith, R, Norlin J.M & Chess A.W. [2006]. Human Behavior and the Social Environment. Social System Theory [5th ed.]. New York: Pearson Education Inc.

Greenhaus, J.H & Powell, G.N [2006]. When work and family are allies: A theory of work-family enrichment. *Academy of Management Review,* 31[1], 72–92.

Greenhaus, J.H (2003). The relation between work-family balance and quality of life. *Journal Vocational Behavior, 63,* 510–531.

Greenhaus, J.H., Collins, K.M., & Shaw, J.D. [2003]. The Relation between work-family balance and quality of life. *Journal of Vocational Behavior, 63,* 510–531.

Grzywacz, J. G. & Carlson, D. S. (2007). Conceptualizing work-family balance: Implications for practice and research. *Advances in Developing Human Resources,* volume 9, 455–471. Available from: http://www.adh.sagepub.com/content/4/455.abstract [Accessed 10th October 2012].

Healy, K. (2005). *Social work theories in context: Creating frameworks for practice.* New York: Palgrave MacMillan.

Indonesian Labour Law no 13 of 2003 on Manpower.

Kinder, A., Hughes, R. & Cooper, C. L. (2008). *Employee well-being support: A workplace resource.* England: John Wiley & Sons, Ltd.

Kossek, E. E. (2005). Work-family balance. Available from: http://www.msu.edu/user/kossek/wc.b.doc [Accessed 7th December 2012].

Kurzman, P. A. & Akabas, S. H. (1993). *Work and well-being: The occupational social work advantage.* USA: NASW Press.

Neuman, W. L. (2006). *Social research methods: Qualitative and quantitative approaches* (6th ed.). USA: Pearson Education, Inc.

Payne, M. (2005). *Modern social work theory* (3rd ed.). New York: Palgrave Macmillan.

Poerwandari, E. K. (2007). *Pendekatan Kualitatif untuk Penelitian Perilaku Manusia*. Depok, LPSP3 Fakultas Psikologi UI.

Rantanen, J., Kinnunen U., Mauna S & Tillemann K. (2011). Introducing Theoretical Approaches to Work-life balance & Testing a new tipology among professionals. Available from: http://www.springer.com/...9783642161988-c.l.pdf? [Accessed 14th December 2012].

Reinharz, Shulamit (1992). *Feminist Methods in Social Research*. New York: Oxford University Press.

Saptari, R. (1997). *Perempuan, Kerja dan Perubahan Sosial*. Jakarta: PT. Pustaka Utama Grafiti.

Suziani. (1999). *Kasus Nike di Indonesia: Meneropong Kondisi Kerja Buruh Perusahaan Sepatu Olahraga*. Jakarta: Yakoma PGI.

Tremblay, D. G. (2004). *Work-family balance: What are the sources of difficulties and what could be done? University of Quebec, Canada*. Available from: www.teluq.equebec.ca/chaireecosavoir [Accessed 14th December 2012].

Warouw, N. (2008). Industrial workers in transition: Women's experience of factory work in Tangerang. In M. Ford & L. Parker (Eds.), *Women in work in Indonesia*. New York: Routledge.

Zastrow, C & Ashman, Karen K.Kirst (2010). *Understanding human behavior and the social environment* (6th ed.) USA: Thomson Brooks/Cole.

Local wisdom of landslide disaster mitigation and change in ecosocial interaction at Bojong Koneng village

R. Raharja & F.G. Wibowo
Department of Social Welfare, Faculty of Social and Political Sciences, Universitas Indonesia, Depok, Indonesia

R.V. Ningsih
Department of Psychology, Faculty of Psychology, Universitas Indonesia, Depok, Indonesia

S.V. Machdum
Department of Social Welfare, Faculty of Social and Political Sciences, Universitas Indonesia, Depok, Indonesia

ABSTRACT: Based on the hazard index in 2011, Bogor Regency ranked the 5th in the list of the most hazardous areas. In January 2014, the number of catastrophic events reached 79 cases, while 60 were the incidents of landslides. Mitigation is able to diminish the risk and avoid disaster. Mitigation in the form of local wisdom is an instrument of society in facing environmental problems. This article reflects the role of local wisdom for landslide disaster mitigation at Bojong Koneng. The study used qualitative methods through in-depth interviews with 16 participants, research findings show that local wisdom at Bojong Koneng were: (a) the use of stilt houses, (b) the environmental destruction as taboo, and (c) predictions of disaster occurrence of landslides through hereditary stories. Unfortunately, those local wisdoms were no longer used by the people at Bojong Koneng as their lifestyle. We argue that there should be reinforcing programs to strengthen local wisdom at Bojong Koneng to be used as disaster mitigation tools. Local wisdom began to vanish due to the influx of development that changed the collective vision of the community. A local wisdom strengthening program is important to change the perception of the community about safe living environment.

1 INTRODUCTION

Indonesia is an archipelagic state located at the confluence of three tectonic plates, the Euro-Asian plate in the north, the Indo-Australian plate in the South, and the plates of the Pacific Ocean in the East. Indonesia's geographical position is between two continents and two oceans, causing the Indonesian territory to be traversed by Western and Eastern monsoon winds. Its conditions potentially lead to hydro-meteorological disasters, such as floods, cyclones, and drought (ITB Disaster Mitigation Centre, 2008). During the period 1990–2000,

Indonesia was ranked the 4th country most frequently experiencing disasters among other countries in Asia. There were at least 257 recorded catastrophic events occurring in Indonesia from the overall 2,886 natural disasters in Asia during that period (ITB Disaster Mitigation Centre, 2008).

Some disasters are unavoidable, but the impact can be reduced through disaster risk management. The disaster management cycle consists of four stages: prevention/mitigation, preparedness, emergency response, and post-disaster rehabilitation and reconstruction. At the mitigation stage, the actions are taken to prevent or reduce the impact of disasters before those happen. The mitigation phase focuses on the long-term action to reduce disaster risks.

The Law No. 32 Article 1 paragraph 30 of 2009 states that local wisdom is noble values applicable in the governance of public life to, among others, protect and manage the

Figure 1. Condition of the roofs and walls of houses due to landslides. (Source: personal documentation).

environment sustainably. However, the meaning of local wisdom is the ways and practices developed by a group of people who have a deep understanding of the local environment and are formed in a hereditary manner (Fathiyah & Hiryanto, 2010). Meanwhile, according to Sartini (2004, p. 111), local knowledge is the ideas of local values that are wise, full of wisdom, embedded in good-values, and attended to by members of the community. In the sense expressed by Kongprasertamorn (2007, p. 2), local wisdom refers to knowledge that comes from the experience of a community.

Previous research conducted by Permana (2011) concerning Bedouins' local wisdom has identified the role of local knowledge in disaster mitigation. The Bedouin is a local tribe who has respected and implemented their local knowledge for generations. Therefore, local wisdom is used as an instrument of society in dealing with problems encountered in the environment. Local wisdom is all forms of knowledge, beliefs, understanding, or insights as well as customs or ethics that guide human life in an ecological community (Keraf, 2010).

According to Mercer (2007), local wisdom a traditional belief system, and indigenous knowledge evolved from past experience of geological events that are recorded through oral history and can be used as an additional source for identifying signs of geological disasters. In this case, local wisdom can play a role in the process of disaster mitigation.

2 OBJECTIVES

This study was conducted to further identify the role of indigenous communities in landslide disaster mitigation. Bojong Koneng village in Bogor Regency was selected as the study site because the village has a high level of disaster vulnerability in landslide disasters. Based on the 2011 disaster index, BNPB ranked Bogor Regency as the 5th disaster-prone area in Indonesia. In January 2014, the number of catastrophic events reached 79 events, but 60 events were landslide incidents.

In the globalized and modernization era, society has changed. Local knowledge is no longer useful and respected in integrated communities. In the meantime, this study aims to identify the role of local knowledge in disaster mitigation of landslides and community efforts in maintaining and preserving local knowledge in disaster mitigation at the Bojong Koneng village, Bogor. By utilizing local wisdoms in disaster mitigation, the expected impacts of disasters in the village can be reduced.

3 RESEARCH METHOD

This study used a qualitative approach to have inductive thinking, which is derived from facts and data in the field, assessing the approaches and theoretical thinking or being used in the formation of a new concept (Neuman, 2007). This research constitutes exploratory research that explores the findings in accordance with the facts and research purposes.

Figure 2. Views of nature in the village Bojong Koneng. (Source: personal documentation).

For data sampling, purposive sampling was used. The criteria were according to the research objectives. There were 16 informants consisting of village communities and stakeholders. The elders and villagers were set as village communities, but stakeholders were defined as the interested party at Bojong Koneng, such as the head of community associations or the hamlets and also the Indonesian National Board for Disaster Management (BNPB) and the Bogor Regency Board for Disaster Management (BPBD Bogor, Kab). Data collection techniques included in-depth interviews, observation, and literature study.

There are several stages in the analysis of data, namely data reduction, data organization, and interpretation of data (Miles & Huberman, 1992). To make it easier to categorize the data, the dimensions of local wisdom were grouped into six dimensions, namely local knowledge, local values, local skills, local resources, local-made mechanisms, and local solidarity groups (Ife, 2002), and the dimensions were separated by structural and non-structural mitigation.

4 FINDINGS AND DISCUSSION

4.1 Local wisdom at Bojong Koneng for disaster mitigation

The people in the village understand their environment well. Moreover, they know that they live in an area prone to landslides that could occur anytime. Based on the findings, there is a local wisdom gained through a deep-thinking process among the elders and the natives about their environment related to the prevention of landslides.

4.1.1 The used of a stilt house

The people of Bojong Koneng were able to build houses with a unique structure, called *Rumah Panggung* (a stilt house). The foundation of a stilt house is made of stones. It resembles beams with a length of 60–70 cm which are implanted into the ground as deep as 5–10 cm. The cantilever of a stilt house, which is made of wood, is placed just above the stone foundation until it shifts in the soil. The impact of a shift in the soil would easily place the cantilever back into an upright position as before. Thus, the damage caused by landslides would be minimized.

In Bojong Koneng, the people are aware that living or staying in stilt houses are safer than in concrete houses or any modern house. The stilt house would diminish the effects of landslides because when a movement of the ground occurs, the structure of the house is not easily shaken and cracked. Besides, the people know that their region highly vulnerable to landslides during the rainy season. The elders and the head of neighborhood stated that stone mining will also increase the risk of landslides in the village. The local knowledge becomes the basis for taking action by the public disaster mitigation agencies.

4.2 Local common law at Bojong Koneng

The people have values that have already existed in their mind, including values associated with the mitigation of landslides. The values then become taboos, which function as social

111

Figure 3.   Stilt house. (Source: personal documentation).

control to maintain the status quo and authority (Holden, 2001). There is an existence of a *pamali* (taboo) concerning environmental destruction. The taboo concerns people who cut and damage the trees in the forest and mine stone. Those who do so will be disturbed by a *jurigan* (a mythical creatures). The taboo and the myth become complex social order by which the society submit to local common law.

The people of Bojong Koneng have abundant natural resources such as bamboos, which are easy to find. The bamboos are used for making traditional avalanche anchors. On the other hand, there are two mountains of rock, namely Mount Gunung Kidul and Mount Parahu. The people believe these mountains to be a natural buffer in the Bojong Koneng village, and its raw material is used to build the foundation for stilt houses. The mountain is highly respected by the people, and this prevents the people from causing any destruction upon it.

4.2.1   *Fairytale inhabitants of the land movement*
The elders said that at some point there will be two mountains which will be fused, and in the midst of the mountains, a river will be formed. When the river is formed, it indicates that the area where these people live was once filled with water, causing the area to be prone to landslides. This prediction is used by elders to regulate people's behavior.

Then, the people are expected to be more aware of the importance of feeling safe in their environment. It creates a strong relationship among the people based on what they believe.

The people of Bojong Koneng have a spirit of mutual cooperation. It is still applied in building public facilities, such as roads, erosion embankment, and stilt houses. Moreover, the night watch system involves the participation of citizens to stand guard at night when it rains so that early detection of landslide disasters in their region is possible.

4.2.2   *Social change at Bojong Koneng*
The condition at Bojong Koneng village is currently causing the displacement of migrants and non-indigenous. In addition, the value of existing and previous embedded knowledge is beginning to influence the immigrant communities. The influx of migrant communities is causing the economies around the village Bojong Koneng to improve, and it also increases community needs. Moreover, with the existence of a waterfall that is increasingly attracting tourists, the village's tourism region has become crucial for the local resources and economics.

According to Zulkarnain & Febriamansyah (2008, p. 72), local wisdom is the principles and specific ways embraced, understood, and applied by local communities in interacting with their environment and are transformed into the value systems and customary norms. Unfortunately, community efforts to preserve local wisdom are becoming obsolete. The erosion of local wisdom changes the result of the inclusion of development in the social and cultural system of interaction between people and their environment. A majority of the communities no longer live in the stilt houses because concrete houses are generally associated with social status in the village. Concrete houses are associated with higher social status compared to the stilt houses, which are actually safer from the impact of landslides.

112

Local stories on the possibility of ground motion have not been told for generations. Local stories are just limited to legends and fairy tales. The society views these tales as just sheer nonsense. However, currently the tale is becoming a reality following the pattern of society that is more rational. Unfortunately, the warnings that are told implicitly from these tales have been ignored, and the intensity of the landslides increasingly occurs.

4.2.3 *Lifestyle versus safety*

Local wisdom at Bojong Koneng was born because of the long and deep thinking process of indigenous people who have settled and passed on the hereditary knowledge. The human's experience becomes an important point in the study of phenomena that occur in nature and the surrounding environment. Therefore, the pattern of interaction between the environment and humans in understanding local knowledge aims to achieve harmonization in the philosophy of life.

People are starting to leave the local knowledge that has built up over many generations due to the influence of development in the villages. Development naturally transforms human interaction with the environment and is influenced by the needs and demands for an improved quality of life which is measured from the economic level of the local community. It has also changed the lifestyle which has shifted the quality of life and has caused the displacement of instilled values that help prevent any potential landslides.

Development caused by modernization and globalization has changed the view of the collective perspective of life in society. The shift from the use of stilt houses into concrete houses reflects their perception regarding the choice of a comfortable home over a safe home. Modernization has changed the perspective at local community level of the trends as well as social and cultural interaction in the community over the awareness of the symptoms that may arise as the impact of development. A shared perspective to choose a safe and protected place from disaster has been dispelled by their current needs.

The sense of a place for living has encouraged the meaning of living properly. Their predecessors had the knowledge to understand their environment well. The sense of collective environmental resources was supported by the sense to protect their own environment and to be able to utilize the resources available in nature. This is conveyed by the elders to the community, even though development is considered to better improve the quality of life.

5 IMPLICATIONS OF THE STUDY

To overcome the problem of natural disasters that may occur in the future, people have to implement the values of local wisdom in a diverse society. At the same time, we argue that there should be reinforcing programs to strengthen local wisdom at Bojong Koneng as disaster mitigation tools. Through the operationalization of an existing value, efforts to understand the conditions and preconditions of the needs that exist in the community will be fulfilled. Thus, the values of local wisdom that have been built from a long and deep thinking process could be utilized, maintained, preserved, and they become a medium of learning in the study of disaster.

To accompany the development in the village of Bojong Koneng, there should be necessary efforts to strengthen the local values that are embedded in public perception. Adaptation by humans towards their environment, which includes the physical and natural processes in the environment such as disasters, shows the interrelation between humans and the environment (Marfai, 2012). The pattern of interaction between humans and their environment is a concern in this study. Local wisdom strengthening programs are important in changing the perception of community about safe living environment.

6 CONCLUSION

People in the village of Bojong Koneng have local wisdom concerning landslide disaster mitigation reflected in the use of stilt houses, making environmental destruction as their taboos, and in the predictions of landslides disaster occurrences through stories passed on from one generation to another. Community efforts to preserve local wisdom begin to be identifiable as erosion of local wisdom caused by changes as the result of development. On the other hand, it impacts on social and cultural system interaction with the environment.

The local wisdom embedded in the indigenous village of Bojong Koneng has been considered obsolete, and this results in the increasing landslides frequencies. Therefore, this study suggests the need for efforts to increase the role of local wisdom in society through community outreach. The community should reduce the negative view on the social status of the people using a stilt house and the identification of the affected area warnings based on the local wisdom of the people. This could be gained by establishing the collective perspective on the safe place to live.

In a case that has occurred in the village of Bojong Koneng, the influx of development has changed social and cultural interaction with the environment causing local wisdom to become obsolete. Therefore, this study advises the importance of strengthening the role of indigenous knowledge of the community at Bojong Koneng as a tool for landslide disaster mitigation efforts. Lifestyle and safety of living should be considered by the community in facing the structural change in modernization.

If the strengthening of the role of local knowledge in disaster mitigation landslide has been fulfilled, it will reduce the impact of landslide disaster in the village. In addition, the strengthening of the role of local wisdom will form a pattern of human interaction with the environment that is harmonious and balanced to achieve human welfare and environmental sustainability. It is an effort to prevent and abate the negative impacts of development on the environment. In turn, this will lead to the landslide disaster risk reduction as a form of local wisdom in the role of landslide disaster mitigation at Bojong Koneng.

ACKNOWLEDGEMENT

We are thankful to the Ministry of Research, Technology and Higher Education of the Republic of Indonesia which has funded this research in the Student Creativity Program (PKM) 2016.

REFERENCES

Fathiyah, K.N. & Hiryanto. (2013) Local Wisdom Identification on Understanding Natural Disaster Sign by Elders in Daerah Istimewa Yogyakarta. *Media Informasi Penelitian Kesejahteraan Sosial.* 37 (1), 453–462.

Holden, L. (2001) *Taboos Structure and Rebellion.* The Institute for Cultural Research.

Ife, J. (2002) *Community Development, Community Based Alternatives in an Age of Globalization.* 2nd edition. Australia, Pearson Education Australia Pty Limited.

Keraf, A.S. (2010). *Etika Lingkungan Hidup.* Jakarta, Penerbit Buku Kompas.

Kongprasertamorn, K. (2007) Local wisdom, environmental protection and community development: the clam farmers in Tabon Bangkhusai, Phetchaburi Province, Thailand. *Manusya: Journal of Humanities.* 10, 1–10.

Marfai, M.A. (2012) *Pengantar Etika Lingkungan dan Kearifan Lokal.* Yogyakarta, Gadjah Mada University Press.

Mercer, J. & Kelman, I. (2007) Combining Indigenous and Scientific Knowledge for PNG Disaster Risk Reduction. *ISISA Newsletter.* 7 (2), 6.

Miles, M.B. & Huberman, A.M. (1992) *Qualitative Data Analysis: A Source Book of New Methods.* Beverly Hills, Sage Publications.

Neuman, L.W. (2007) *Basic of Social Science Research, Qualitative and Quantitative Approaches, second Edition.* Boston, Pearson Education, Inc.

Permana, R.C.E., Nasution, I.P. & Gunawijaya, J. (2011) Kearifan lokal tentang mitigasi bencana pada masyarakat Baduy. *Makara, Sosial Humaniora.* 15 (1), 67–76.

Sartini. (2004) Menggali kearifan lokal nusantara: Sebuah kajian filsafat. *Jurnal Filsafat.* 37, 111–120.

The Law Number 32 Article 1 (30) of 2009 on the Protection and Environmental Management of the Republic of Indonesia.

Zulkarnain, A. & Febriamansyah, R. (2008) Kearifan lokal dan pemanfaatan dan pelestarian sumberdaya pesisir. *Jurnal Agribisnis Kerakyatan.* 1, 69–85.

Competition and Cooperation in Social and Political Sciences – Adi & Achwan (Eds)
© 2018 Taylor & Francis Group, London, ISBN 978-1-138-62676-8

Contestation of aristocratic and non-aristocratic politics in the political dynamic in North Maluku

I.R.A. Arsad
Department of Administration, Faculty of Social and Political Sciences, Universitas Indonesia, Depok, Indonesia

ABSTRACT: The local political dynamics in North Maluku is influenced by the people's political ideology and culture. In each era in Indonesia, the contestation of aristocratic and non-aristocratic group has shown the political contestation of the federation system versus the unitary system in the local government when the Republic of Indonesia was established, and the contestation on the establishment of North Maluku Province between special status and non-special status in the reformation era. The contestation was influenced by the egalitarian political culture on the political ideology of both aristocratic and non aristocratic political groups. In this article, the aristocrats still take part in both local and national political activities since they are benefited by the egalitarian political ideology in the local political culture and the patron-client relationship. This study has shown that in North Maluku, political ideology still influences political behavior, although Firman Noor (2014, p. 63) states that political ideology has been replaced by pragmatism in Indonesia, because it is influenced by the local political culture and the patron-client relationship.

1 INTRODUCTION

Political contestation between aristocratic and non-aristocratic politics has long been in line with the transformation of a political system in Indonesia, which is the process of unification of the native states into the new state of the Republic of Indonesia after the dissolution of the Federal Republic of Indonesia. Meanwhile, the variant of the association between aristocratic and non-aristocratic politics in different countries may result from the different process of unification in political transformation in those countries.

In England, the state's political transformation (unification) of the native state of monarchy to the new state proceeded gradually after the integration of the kingdoms to result in the currently known United Kingdom of Great Britain and Northern Ireland. William M. Downs (in Ishiyama & Breunings 2013, p. 284–285) describes that the process of unification of the Great Britain and Northern Ireland has proceeded for centuries. In the Republic of Italy, the unification process was resulted by the role of the totalitarian regime of Musolini through a referendum to determine the form of a republic, so that monarchic states, such as Sicily, Sardinia, and others have to be subject to the result of the referendum encoded in the state constitution (Strong 1958, p. 80).

In France, the process of unification was resulted by a social revolution that replaced the absolute monarch and bourgeois local rulers with the republic state (Skocpol, 1991, p. 21). In China, the dynasty of Manchu was defeated by a social revolution triggered by the establishment of the people's congress by Chiang Kai Sek who led the nationalist power and Mao Tse Tung who led the communist power in 1945. Later, in 1949, the communist power took control of the congress and established the People's Republic. In India and Pakistan, the states' political transformations occurred through the unification of the native states that resulted in a new province as part of the federal union states (Strong, 1958, p. 295–303).

In Indonesia, social revolution also occurred, but it was not as radical as that in France and China. Although Kahin described that Indonesian revolution driven by Indonesia's intellectuals has successfully resulted in the transformation of the nation's character during the period of 1945–1950,[1] the process of unification was partially resulted by the emergence of the new nationalism awareness among the aristocrats, as reflected in the debates concerning the form of the state in the conferences of Malino, Denpasar, and the Round Table.[2] The unification process in Indonesia reached the end after the transfer of power from the aristocratic-monarchic rulers to the government of the Republic of Indonesia.[3]

Political transformation in a state has described the role of aristocratic groups in the newly established state. However, due to the influence of the common establishment of the nation state after World War II, aristocratic politics had to face the fierce challenges from the nationalistic group with their likely orientation to the state ideology of republic, which is particularly true in Indonesia.

In North Maluku, the political role of the aristocratic group in local and national politics has been very dynamic. Earlier at the initial moment of Indonesia's independence, the aristocratic group played a significant role in local and national politics. However, at the same time they had to get involved in the political contestation with the non-aristocratic republican group. The political role of the aristocratic group ended when North Maluku had to comply with the Unitary State concept after the dissolution of the United Republic of Indonesia. The non-aristocratic group had always won the prolonged contestation of local and national politics until the New Order. When the New Order collapsed, the contestation went on in line with the political transparency and the resurgence of the traditional aristocratic rulers in North Maluku. The contestation was apparent in the establishment of the Province of North Maluku during the period of 1998–1999. The strive of the aristocratic group in making North Maluku a special region in the decentralization system in Indonesia was part of political efforts since the establishment of the Province of North Maluku up to now.

The repeated political contestation reveals the existence of strong ideology and political interest adopted by the two political groups in North Maluku. The aristocratic group has been more likely to strive for the political values focusing on the balanced position between the local and central government through the system of federalism. As a result, they required to establish Provincial-level special territory. This ideology is in line with the commonly adopted democratic culture and ideology in the local political culture of North Maluku's Moloku Kie Raha. The system comprising four kingdoms has influenced the political ideology of the aristocratic group. On the other hand, the non-aristocratic political group resisted the system of feudalism in the local politics. This idea is influenced by the democratic political culture adopted in the constitutional monarchy system in the political culture of North Maluku. After the reformation in Indonesia, the non-aristocratic political group has consistently resisted the special local governmental system for the Province of North Maluku. The contestation between the two groups is resulted from the influence of the contesting federalism and the Unitarian ideology after the independence of the Republic of Indonesia, and the democratic versus feudalistic ideology after the reform era in North Maluku.

Considering the contestation between the aristocratic and non-aristocratic groups in the local politics of North Maluku, this study would like to answer the questions of the dominant factors underlying the contestation between the aristocratic and non-aristocratic groups in local politics in North Maluku. In addition, this study would also like to identify the process of contestation between the aristocratic and non-aristocratic groups in different Indonesian political eras in North Maluku and the aristocratic political role in the local

---

1. George McTurnan Kahin, *Nasionalisme dan Revolusi Indonesia*, (Depok: Komunitas Bambu, 2013), pp. 657–658.
2. See Ida Agung Anak Agung, *Dari Negara Indonesia Timur ke Republik Indonesia Serikat*, (Yogyakarta: Gadjah Mada University Press, 1985), and Leirissa, *Kekuatan Ketiga dalam Perjuangan Kemerdekaan Indonesia*, (Jakarta: Pustaka Sejarah, 2006).
3. *Ibid.*

political dynamics in North Maluku Province, particularly in the election. Considering the questions of the study, this paper would like to argue that the political ideology in Indonesia, in spite of the abandonment as stated by Firman Noor, plays a significant role in the political behavior of the contesting aristocratic and non-aristocratic groups due to the existence of the egalitarian political culture and the patron-client relation.

1.1 *Theoretical analysis and methodology*

A number of political concepts concerning the contestation between the aristocratic and non-aristocratic politics in North Maluku can be described with the theory of political culture and political ideology concept that have accordingly encouraged the political behavior of the citizens. Political culture and local political culture are concerned with the political attitude and behavior of the citizen, which accordingly lead to the political life.

The contestation of aristocratic and non-aristocratic politics can be conceptually and theoretically explained in the political ideology that has driven the citizens' political behavior. Political culture and local political culture describe the citizens' political behavior in a particular community. It consists of a set of values or norms adopted in political life. A number of political scholars define political culture as the citizens' orientation or a set of mindset in politics and governance (Almond in Mas'oed & MacAndrews, 1997, p. 41; Ranney, 1993, p. 65). A political culture is part of culture that is generally defined as a set of knowledge that constitutes social behavior features that specifically characterize individual behaviors (Plano *et al.,* 1985, p. 53–54).

Meanwhile, an ideology is a set of beliefs influenced by political culture and later influences the establishment of mindset and behavior of a group of people, political parties, and states in the analysis of particular political phenomena, and leads the followers to an expected ideal condition.[4] Political behavior refers to a set of behavior of a number of political actors, either based on culture, ideology, norms, or political interest in a particular political system.

The political phenomena of the role of traditional rulers in the nationally acknowledged local politics are found in *neo-patrimonial* states or sultanates. This form of the authoritarian system has been defined by J.J. Linz in Anna Batta in her paper on Statism (see Batta in Ishiyama & Breuning, 2013, p. 153). Meanwhile, according to M. Bratton & N. van de Walle, in *neopatrimonial state,* the right for power is reserved for an individual instead of an institution, and benefits are exchanged to loyalty to the political parties (see Batta in Ishiyama & Breuning 2013, p. 154). In the case of local politics in North Maluku, the Sultans are the patrons and the people are the clients. Although the relationship between the patron and the clients is based on economic interest, it also derives from the obedience originating from the traditional Islamic teaching, the local knowledge of Christian community and the native ethnic of Halmahera Island. This resembles the patron-client relationship of the local political culture in Java as defined by Benedict R.O.G. Anderson (in Budiardjo, 1991, p. 51–52). The relationship of patron and client in North Maluku has beneficially favored the candidates in the general election.

2 METHOD

To examine these issues, this study employs a quantitative analysis. Methods of data collection include library research and in-depth interviews. Information and documents on local politics in North Maluku before and after the independence of the Republic of Indonesia were collected to describe the contestation between the aristocratic and non-aristocratic groups in the local political contestation in North Maluku. Meanwhile, the method of in-depth

---

4. Firman Noor, *Perilaku Politik Pragmatis dalam Kehidupan Politik Kontemporer: Kajian atas Menyurutnya Peran Ideologi Politik di Era Reformasi*, in Jurnal Masyarakat Indonesia, Vol. 40 (1), June 2014, p. 63, accessed from ejournal.lipi.go.id on 30 July 2016, at 19.10 EIT.

interview was conducted to collect information from related informants and resource persons concerning the aristocratic clan in the local political contestation in North Maluku after the reform era to explain the political role of the aristocratic group in local political contestation in accordance to the results of local legislative election. The two methods have contributed much to the description of the local political behavior in North Maluku, particularly the likelihood of contestation between the aristocratic and non-aristocratic groups in North Maluku.

3 RESULTS

3.1 *Political contestation of aristocratic and non-aristocratic political power in North Maluku until reformation era*

3.1.1 *The North Maluku statecraft*
Since 1257, the government of North Maluku has been inseparable from the existence of traditional Sultanates of Ternate, Tidore, Bacan, Jailolo, and Kingdom of Loloda. Based on the treaty of Moti in 1322, excluding the Kingdom of Loloda, the kingdoms in North Maluku established a joint government of lax confederation in North Maluku (de Clercq, 1890; Amal, 2002). The establishment of the joint state was followed by task distribution for individual kingdoms to develop the confederation of the Four North Maluku (*Moloku Kie Raha*). Each kingdom has its own task to ensure the existence of confederation, territorial security, and territorial expansion (Arsad, 2011). In the confederation model, the existence of each kingdom is reserved and equally respected.

Through the interaction with the European states of Spain, Portugal, the Netherlands and England and in attempt to develop the dignity of each kingdom, Ternate kingdom was assigned to expand the territory of the kingdoms to the North and the South, while Tidore was assigned to expand to the West and the East. The two countries developed hegemonic politics in the eastern part of the archipelago, particularly in the development of spice trade with the Netherlands and in early 1800s with the England (Willard & Alwi, 1998). The hegemonic politics of the two kingdoms was developed until the era of Dutch colonialism that conquered the sovereignty of the traditional kingdoms through the short treaties (*korte verklaring*) in 1909,[5] and through the regulation of modern bureaucracy in the Netherlands East Indies before the entrance of the Japanese in the region during the period of 1942 and 1945.[6]

3.1.2 *Contestation in the old order era*
At the initial time of Indonesian independence, North Maluku was still under the control of NICA or under the Dutch colonialism. In an attempt to establish the most suitable government form, the aristocrats and non-aristocrats were involved in the conferences of Malino and Denpasar held by the Dutch (Amal & Djafar, 2003).

The non-aristocrats, popularly known as the Republicans, consisted of a number of figures such as M.A. Kamaruddin, Chasan Busoiri, Arnold Mononutu, M.S. Djahir, M. Arsyad Hanafi, Abubakar Bachmid, and the youth figure, Abdjan Soleman. Meanwhile, the aristocratic group was represented by the Sultan of Ternate Muhammad Djabir Syah, Sultan of Tidore Zainal Abidin Syah, Salim Adjidjuddin, and Prince M. Nasir (Amal & Djafaar, 2003). The aristocratic political group attempted to develop a state federation system under the Kingdom of the Netherlands as expected by the Dutch General Governor for the Netherlands East Indies, van Mook. On the other hand, the non-aristocratic political group was

---

5. M. Adnan Amal dan Irza Arnyta Djafaar, *Maluku Utara: Perjalanan Sejarah 1800–1950*, (Ternate: Universitas Khairun, 2003), pp. 88–89. About the short treaties, see A. Dalman, *Sejarah Indonesia Abad XIX-Awal Abad XX: Sistem Politik Kolonial dan Adminstrasi Pemerintahan Hindia-Belanda* (Yogyakarta: Ombak Press, 2012), p. 83.
6. Syahril Muhammad, *Ternate: Sejarah Sosial Ekonomi dan Politik*, (Yogyakarta: Ombak Press, 2004).

very enthusiastic with the spirit of the nation state of Indonesia and had attempted to make North Maluku as a local government under the unitary state of the Republic of Indonesia. The debates over the government systems were made in the Denpasar conference on 7–24 December 1946 (Agung, 1985, p. 122). The debates resulted in the establishment of the Eastern Indonesian State, in which North Maluku is a part. The Sultans of Ternate, Tidore, and Bacan were the autonomous rulers of their respective kingdoms.

The non-aristocratic political group consistently strived to make North Maluku a part of the local government of the Republic of Indonesia. Accordingly, a demonstration led by a youth figure named Abdjan Soleman was arranged in the visit of the Prime Minister of the Eastern Indonesian State Ida Agung Gde Agung in Ternate on 23 March 1948. The demonstration, held in the form of torch-light rally and a banner, requested the integration of the Eastern Indonesian State to the Republic of Indonesia.[7] In response to the development of the Indonesian politics at that time, the Federal State of the Republic of Indonesia was established after an agreement in the Round Table Conference in 1949. North Maluku finally became a part of the Province of Maluku.

After the dissolution of the Federal State of Republic of Indonesia in 1950, the government of North Maluku was ruled by the aristocratic board of kingdoms. The resident of North Maluku was the Sultan of Ternate, Muhammad Djabir Syah. Later, there was a propaganda by the separatist group to proclaim that the Republic of South Maluku and the Sultan of Ternate, Muhammad Djabir Syah, supported the separatist movement. The government of the Republic of Indonesia then arrested the Sultan of Ternate, Muhammad Djabir Syah, for his support to the separatist movement of the South Maluku Republic. He was exiled to Jakarta (Djafaar, 2005). In Jakarta, the Sultan worked as an employee of the Ministry of Home Affairs. His position as the resident of North Maluku was replaced by the Sultan of Tidore, Zainal Abidin Syah.

The Sultan of Tidore, Zainal Abidin Syah, was later appointed as the Governor of West Irian in September 1956, and therefore the residency was assigned to the Sultan of Bacan, Dede Usman Syah. Upon the request of DPRD-GR of North Maluku, the Resident of Ternate brought the North Maluku delegation to Jakarta to reveal the dissolution of the Board of Kings in North Maluku (Muhammad *et al.,* 2015). In 1958, North Maluku covered the territories of the Sultanate of Ternate and the territory of the Sultanate of Bacan was assigned the status of regency, with M.S. Djabir (a non-aristocrat) as the caretaker of the North Maluku Regency. Since then, with the enactment of the Law of 18/1965 concerning the Local Government, the political group of the aristocrats was no longer accommodated in the formal government of North Maluku (see Marbun, 2010, p. 22–23).

In the general election of 1955, a number of national and local political parties took part. One of the local political parties, Persatuan Indonesia Raya (PIR), was affiliated to aristocratic interests. The results of the general election were described in a descending order: Masyumi ranked first (39.98%; 117,440 votes), followed by PNI (30,218 votes), Parkindo (18,920 votes), Partai Katolik (18,710 votes) and PSII (11,310 votes) (van Klinken, 2007, p. 184; Muhammad *et al.*, 2005, p. 65). In the general election of 1995 in Ternate, Partai Masyumi remained in the first rank with 51,410 votes, Parkindo 25,594 votes, PIR 17,080 votes, PNI 14,977 votes, PKI 2,729 votes and PSII 571 votes (Muhammad *et al.*, 2015, p. 64–65). The failure of PIR as an aristocrat political party was associated to the detention of the Sultan of Ternate, Muhammad Djabir Syah, by the Government of Jakarta, which might have resulted in the decrease of the number of votes in the traditional territory of the Ternate Sultanate.

3.1.3 *Contestation in the new order era*
The New Order Era was characterized by the absence of aristocratic politics in local and national politics in Indonesia. With the totalitarian system, the politics of New Order has placed the aristocratic politics as a mere auxiliary for stability of power. When Muhammad

---

7. M. Adnan Amal and Irza Arnyta Djafaar, *Op cit*, pp. 198–199.

Table 1. Composition of representation among the political parties and the armed forces in the local parliaments of the North Maluku regency in the 1982, 1987, 1992, and 1997 general elections.

Political parties and the armed forces	Representation in general election			
	1982	1987	1992	1997
Partai Persatuan Pembangunan	6	5	7	5
Golongan Karya	24	30	26	30
Partai Demokrasi Indonesia	2	1	3	1
ABRI	6	9	9	9
Total	38	45	45	45

Source: Central Bureau of Statistic, *North Maluku in Numbers 1999*.[9]

Djabir Syah passed away in 1975, his son Mudafar was appointed as the Sultan of Ternate with the full support of the government. Van Klinken revealed that the Sultan of Ternate Mudafar Syah held the position in Jakarta and was utilized by the government's political party, Golongan Karya, in each term of the general election by assigning him the role of the vote getter for Golongan Karya (Golkar) in North Maluku.[8] Golkar became the majority vote in each general election. The representation of the North Maluku Regency in the general elections during the New Order is presented in Table 1.

This era was dominated by the non-aristocratic political group, with the fusion of some political parties and screening system during the New Order to secure the development policies and political stability. The political contestation between the aristocratic and non-aristocratic groups has been apparent because the role of the aristocratic group is merely seen at the elite level. This is visible in the appointment of the Sultan of Ternate, Mudafar Syah as a member of DPRD (local parliament) of Maluku Province for the period of 1971–1977, the Chairman of Regional Board of Golkar in North Maluku for the period of 1997–1998, and a member of the local representatives of MPR (Supreme People Assembly) for the period of 1998–2002.[10]

3.1.4 *Contestation in the reformation era*

This era is characterized by the revitalization of the aristocratic politics through the establishment of Partai Demokrasi Kebangsaan. Sultans were elected to become legislative members of DPR-RI (Indonesian House of Representatives) from this political party. This party has also successfully put his two representatives at DPRD (local parliament) of North Maluku Province in the 2004 general election. Van Klinken defined this era as the revitalization of Sultans in their communitarian movement. In North Maluku in 1999, Djafar Syah was reappointed as the Sultan of Tidore after the long hiatus since 1968 after the death of Zainal Abidin Syah. Meanwhile, at the initial phase of the Reformation era, the Sultan of Bacan, Gahral Syah, was appointed as the Regent of North Maluku. The Sultan of Tidore, Djafar Syah, was elected as the representative in DPD-RI (Indonesian senate) in the 2004 general election.

The contestation between the aristocratic and non-aristocratic groups was also characterized by the contestation of the governmental form of the North Maluku Province that has been proposed since the end of the New Order era (1998–1999). The aristocratic political group expected to establish the province with a special status, while the non-aristocratic

---

8. van Klinken, Gerry, *Perang Kota Kecil: Kekerasan Komunal dan Demokratisasi di Indonesia*, (Jakarta: KITLV-Yayasan Obor Indonesia, 2007), p. 190.
9. Central Bureau of Statistics North Maluku Regency, *Maluku Utara dalam Angka 1999*, (Ternate, BPS North Maluku Regency, 1999), pp. 30–31.
10. See the article of *Inilah Kisah Panjang Gonjang Ganjing Kesultanan Ternate*, accessed from www.satumaluku.com on 14 February 2016 at 10.20 EIT.

political group expected otherwise (Muhammad *et al.*, 2015; van Klinken, 2007, p. 188). The non-aristocratic group worried that with a special status, the Sultan would become the Governor. Supported by the national government Law of 46/1999, the province was established without any special status.

This examination has shown that the contestation between the aristocratic and non-aristocratic groups derives from the ideology. At the initial phase of the government of the Republic of Indonesia, the ideological contestation between federalism and unitarianism had influenced the political dynamics in North Maluku (Amal & Djafaar, 2003). The aristocratic group, with the political culture of confederation in the government's history, strived to establish a federal state as they considered it the best form for Indonesia. Equality and autonomy at the local government have characterized the political interests of the aristocratic group in North Maluku. On the other hand, the nationalist group, with the spirit of a new nation state, preferred the unitary state. This was due to the fact that after the Independence Day and World War II, the Indonesian national politics was influenced by the emergence of the nation state, which was popular among Indonesia's new generation of intellectuals (Kahin, 2013, p. 657–658).

The political contestation between the aristocratic and non-aristocratic group was enhanced at the end of the New Order era and the initial phase of the Reformation era through the transparency in Indonesian politics, particularly in the development process of the North Maluku province. The idea of establishing the province with a special status was challenged by the non-aristocratic group. The non-aristocratic group preferred the more equal democratic system of the local government, while the aristocratic group preferred a special status in the Indonesian local governments. The dispute ended when the Province of North Maluku was established without any special status under the presidency of B.J. Habibie.

3.2 *The role of aristocratic politics in local politics in North Maluku under the 2014 general election*

The aristocratic politics in North Maluku in the reformation era has participated in the political mechanism in Indonesia. Family members of the aristocratic groups strived to become members of political parties. Alternatively, they expected to be requested by the political parties to become representatives of local parliaments. Several members of aristocratic families who dispersed in several regencies/cities in North Maluku were then elected as representatives at the regency or province level. Based on the results of the 2014 general election, the members of aristocratic families of the Sultanate of Ternate, Sultanate of Tidore and Sultanate of Bacan who were elected as the members of local parliaments are described in Table 2.

The aforementioned Table 2 reveals that the two regencies had no representatives from the aristocratic clan in the local legislative. They were the West Halmahera and the East Halmahera regencies. In the West Halmahera regency, there was such an aristocratic clan like *Kaicil*. In general, all descendants of the Sultan have the clan name *Dano* (Djoko *et al.*, 2001). However, due to the long hiatus of the Sultanate of Jailolo since the 19th century, the aristocratic clan in the West Halmahera had stopped using the aristocratic clan. This is also the case for the East Halmahera regency whose territory is far from the center of the traditional government of the Sultanate of Tidore. Therefore, the people there no longer use the aristocratic clan.

In other words, both regencies have representatives of aristocratic clan in the local parliament. It is understandable since the West Halmahera Regency is the center of the Sultanate of Jailolo which was revitalized in the reformation era with the appointment of the Sultan of Jailolo (Abdullah Syah). The representation of the aristocratic clan in local parliament shows their participations in the local development and governance through their legislative function in creating a well-organized society. After the reformation, the election of the legislative members has been based on the pragmatic behavior instead of ideology (Noor, 2014, p. 63). Mujani and Lidlle even revealed that their election was influenced by factors such as figures and close relationship to particular political parties (in Aspinall & Mietzner, 2010, p. 75–99). In the 2009 general election, Mujani and Liddle said that the election was influenced by mass

Table 2. Representation of the aristocratic clan in the local legislatives of North Maluku in the 2014 general election.

Local legislatives	Figures	Political parties	Number of representatives in DPRD
Ternate City	Azizah Fabanyo	PAN	30
	Sudin Dero	PDI Perjuangan	
Tidore Archipelago City	Ahmad Laiman	PDI Perjuangan	25
	Ridwan Moh. Yamin	P. Demokrat	
	A Karim Togubu	P. Hati Nurani Rakyat	
North Halmahera Regency	Fahmi Djuba	PPP	25
West Halmahera Regency	–	–	25
East Halmahera Regency	–	–	20
Central Halmahera Regency	Asrul Alting	PDI Perjuangan	20
	Zulkifli Alting	P. Hanura	
	Rusmini Sadaralam	PDI Perjuangan	
South Halmahera Regency	Anwar Abusama	P. Keadilan Sejahtera	25
	Muhammad Abusama	P. Golkar	
	Arsad Sadik Sangaji	P. Keadilan dan Persatuan Indonesia	
Sula Archipelago Regency	Ihsan Umaternate	P. Golkar	25
	Helman Umanahu	P. Demokrat	
	Jufri Umasugi	PDI Perjuangan	
	Julfi Umasangaji	P. Demokrat	
Morotai Island Regency	M. Rasmin Fabanyo	P. Keadilan Sejahtera	20
	M. Ali Sangaji	P. Golkar	
North Maluku Province	Rais Sahan Marsaoly	P. Nasdem	45
	Fahri K. Sangaji	P. Demokrat	
	Umar Alting	P. Hanura	
	Syahril I. Marsaoli	P. Bulan Bintang	

Source: KPU (General Election Commission) of the North Maluku Province.

media and the candidate's capability. Furthermore, Ufen said that the behavior of voters in Indonesia is also influenced by money politics and commercialization of candidates by political parties.[11]

Concerning the candidates' profiles and capabilities that have made the aristocratic clan members get elected in the 2014 general election in North Maluku, Mochtar Umamit explained that in the culture of the Sula Island, the aristocratic clans have their own figures.[12] With such charismatic figure and good capability, members of the aristocratic clan may be elected into the local legislative at another region. Therefore, the candidates' profiles from the aristocratic group in North Maluku are closely related to the local political culture. Aristocratic family members in a particular region are more likely to become the patron in the society, and therefore, they are more likely to be elected as parliament members. The people from the kingdom characterized by their patron-client relationship have a great influence on the candidates' capability and electability among the aristocratic clan members.

However, the number of the aristocratic clan members elected in the local parliament in the 2014 general election was relatively minor. This is due to the fact that the people of North

---

11. Andreas Ufen, *Electoral Campaigning in Indonesia: The Professionalization and Commercialization after 1998*, in Journal of Current Southeast Asian Affairs, 29, 4, 11–37, accessed from www.Current-SoutheastAsianAffairs.org on 30 June 2016, at 18.15 EIT.

12. Interview with Mochtar Umamit, member of aristocratic clan from Sula Islands, part of aristocratic clan of Ternate Sultanate, on 24 August 2016.

Maluku are influenced by a number of ideologies such as reformistic Islam (egalitarian) and traditional Islam (van Klinken, 2007, p. 194). In addition, ethnics and religions also influence the general election in North Maluku (Ahmad, 2012). Therefore, aristocratic politics still has adequate opportunity in the local politics of North Maluku, since both ideology and local political culture characterized by patron-client relationship facilitate the election of aristocratic clan members in each general election in North Maluku.

4 CONCLUSION

Contestation between the aristocratic and non-aristocratic groups in local politics in North Maluku is influenced by egalitarian ideology and local political culture. Egalitarian ideology has resulted in the polarization of political ideology in preference for the establishment of either the federal state as proposed by the aristocratic group, or the Unitarian state as proposed by the non-aristocratic group during the initial phase of Indonesia's independence. Democracy that reflects political equality and freedom has been adopted by the non-aristocratic group, while federal political culture in the statecraft of kingdoms in North Maluku has influenced the aristocrats' political behavior.

Contestation between the aristocratic and non-aristocratic groups after the reformation era has opened up political transparency in Indonesia, that is the establishment of the North Maluku Province, either with or without special status. The aristocratic group prefers the province with special status, while the non-aristocratic group prefers the province without special status. In the end, the province was established without special status.

The aristocratic members are likely to be elected in the local politics of North Maluku and national politics of Indonesia, because in the local political culture characterized by patron-client relationship, aristocratic figures have been put as the patrons in the society, and therefore are more likely to be elected either in local or national representatives in each general election. This study has shown that political ideology still influences the North Maluku political behavior, although Firman Noor (2014, p. 63) stated that political ideology has been replaced by pragmatism in Indonesia, because it is influenced by the local political culture and the patron-client relationship.

REFERENCES

Agung, I.A.A.G. (1985) *Dari Negara Indonesia Timur ke Republik Indonesia Serikat.* Yogyakarta, Gadjah Mada University Press.

Ahmad, S. (2012) "Politik dan Etnik: Studi Kasus Konflik Politik dalam Pemilihan Gubernur-Wakil Gubernur Provinsi Maluku Utara Tahun 2007." PhD Diss., Program Parcasarjana FISIP UI.

Amal, M.A. (2002) *Maluku Utara: Perjalanan Sejarah 1250–1800,* Jilid I. Ternate,Universitas Khairun.

Amal, M.A. & Djafaar, I.A. (2003) *Maluku Utara: Perjalanan Sejarah 1800–1950.* Ternate, Universitas Khairun.

Arsad, I.R.A. (2011) *Budaya Pemerintahan: Dari Maluku Utara untuk Kybernologi.* Ternate, UMMU Press.

Badan Pusat Statistik Kabupaten Maluku Utara. (1999) *Maluku Utara dalam Angkat 1999.* Ternate, BPS Kabupaten Maluku Utara.

Budiardjo, M. (1991) *Aneka Pemikiran Tentang Kuasa dan Wibawa.* Jakarta, Pustaka Sinar Harapan.

Daliman, A. (2012) *Sejarah Indonesia Abad XIX-Awal Abad XX: Sistem Politik Kolonial dan Adminstrasi Pemerintahan Hindia-Belanda.* Yogyakarta, Penerbit Ombak.

de Clercq, F.S.A. (1890) *Bijdragen tot de Kennis der Residentie Ternate.* Leiden, E.J. Brill.

Djafaar, I.A. (2005) *Dari Moloku Kie Raha hingga Negara Federal: Iskandar Muhammad Djabir Sjah, Biografi Politik Sultan Ternate.* Yogyakarta, Bio Pustaka.

Hanna, W. & Alwi, Des. (1998) *Ternate dan Tidore: Masa Lalu yang Penuh Gejolak.* Jakarta, Sinar Harapan.

Ishiyama, J.T. & Breuning, M. (2013) *Ilmu Politik dalam Paradigma Abad ke-21: Sebuah Referensi Panduan Tematis* (eds), (pen. Ahmad Fedyani Syaifuddin). Jakarta, Kencana.

Kahin, G.McT. (2013) *Nasionalisme dan Revolusi Indonesia.* Depok, Komunitas Bambu.

Leirissa, R.Z. (1996) *Halmahera Timur dan Raja Jailolo.* Jakarta, Balai Pustaka.

Leirissa, R.Z. (2006) *Kekuatan Ketiga dalam Perjuangan Kemerdekaan Indonesia.* Jakarta, Pustaka Sejarah.

Marbun, B.N. (2010) *Otonomi Daerah 1945–2010 Proses dan Realita: Perkembangan Otda, Sejak zaman Kolonial sampai saat ini.* Jakarta, Pustaka Sinar Harapan.

Muhammad, S. (2004) *Ternate: Sejarah Sosial Ekonomi dan Politik.* Yogyakarta, Penerbit Ombak.

Muhammad, S., Abbas, I., Hasyim, R., Barjiyah, U., Ismail, A., Mansur, M., Abdullah, T. & Ridwan. (2015) *Sejarah Pembentukan Provinsi Maluku Utara.* Ternate, Kerjasama Badan Kearsipan-Perpustakaan Daerah Provinsi Maluku Utara, Masyarakat Sejarawan Indonesia (MSI) Provinsi Maluku Utara, dan Pusat Studi Lingkungan Universitas Khairun.

Plano, J.C., Riggs, R.E. & Robin, H.S. (1985) *Kamus Analisa Politik.* Jakarta, Rajawali Press.

Skocpol, T. (1991) *Negara dan Revolusi Sosial: Suatu Analisis Komparatif tentang Perancis, Rusia dan Cina,* (pen. Kelompok MITOS). Jakarta, Penerbit Erlangga.

Strong, C.F. (1958) *Modern Political Constitutions: an Introduction to the Comparative Study of Their History and Existing Form,* 5th edition. London, Sidgwick and Jackson Limited.

Surjo, D., Nasikun, Lay, C., Falakh, F., Zaman, B.K., Mundayat, A.A. & Adelheida, M. (2001) *Agama dan Perubahan Sosial: Studi tentang Hubungan antara Islam, Masyarakat, dan Struktur Sosial-Politik Indonesia.* Yogyakarta, Universitas Gadjah Mada Press.

van Klinken, G. (2007) *Perang Kota Kecil: Kekerasan Komunal dan Demokratisasi di Indonesia.* Jakarta, KITLV-Yayasan Obor Indonesia.

Competition and Cooperation in Social and Political Sciences – Adi & Achwan (Eds)
© 2018 Taylor & Francis Group, London, ISBN 978-1-138-62676-8

Social enterprises policy approach for micro enterprises empowerment: Lessons learnt from the technology for region program in empowering micro enterprises in West and East Java, Indonesia

I.R. Maksum, A.Y.S. Rahayu & D. Kusumawardhani
Department of Public Administration, Faculty of Public Administration, Universitas Indonesia, Depok, Indonesia

ABSTRACT: Micro enterprise is one of the informal sectors in Indonesia that has a strategic position in the Indonesian economy. The enterprises absorb 99.8 percent of the country's employment and account for more than 95 percent of all the firms in Indonesia. The enterprises encounter many obstacles, such as lack of capital, skills, and technology, which result in their low competitiveness level. To solve the problem, the government has issued a number of policies, but none of these has been effective in improving the enterprises' capacity and productivity. One of the problems is the failure of the market-based approach in addressing the unique characteristics of micro enterprises in Indonesia. Regarding this, the paper highlights the implementation of Technology for the Region Program in some villages of West and East Java. The program has applied the social enterprises approach that combines the market-based and social welfare aspects of micro enterprises. The paper will use in-depth interview with the owners of micro enterprises, and show how the program benefits the micro enterprises and what lessons could be learned from the program.

1 INTRODUCTION

Since 1992, the Association of South East Asian Nations has applied continuous efforts to set up economic integration among its members. The resultant integration has been followed up by the ASEAN Economic Community (AEC) with the objective to create a single market and production base, a competitive economic region, equitable economic development and integration into the global economy. In other words, the AEC will facilitate liberalisation and trade in goods and services as well as narrowing down the development gap.

The idea of the AEC seems to have a promising future for the Micro, Small, and Medium Enterprises (MSMEs) that play a strategic role in the economies of ASEAN countries. The MSMEs' contribution to income, employment rate, and the GDP of a country has made the sector unique, and it has the potential to be a champion for the AEC. This is because free trade or open market policies may affect local market power and the ability to access foreign markets or expand improvements in technology, productivity, and efficiency (Tambunan, 2008, p. 33). Reflecting on this, the ASEAN Strategic Action Plan for SME Development 2010–2015 mission states:

> "By 2015, ASEAN SMEs shall be world-class enterprises, capable of integration into the regional and global supply chains, able to take advantage of the benefits of ASEAN economic community-building, and operation in a policy environment that is conducive to SME development, export and innovation. (ASEAN Strategic Action Plan 2010–2015 as cited in Sato 2015, p. 156)."

However, implementing the ASEAN Strategic Action Plan for SME Development has become a challenging task for Indonesia's MSMEs. This is because in Indonesia, the scope of enterprises is still dominated by micro enterprises (99.8%) which absorb the highest percentage of employment. However, they still encounter the classical problems of micro-enterprises, such as the lack of skills and capacity of the human resources, the lack of capital and technology usage, and some legal problems.

These classical problems have affected their poor export performance, resulting in their products having difficulty in competing in the global market (Tambunan, 2008, p. 55). As a consequence, the free trade that has applied the market approach will cause them to fail since they have no bargaining position at the beginning. Luna & Fajar (2010) argue that the effect of free trade that is based on the market-based paradigm will only cause suffering on the marginal sectors such as micro-enterprises. This is because at the operational level of the approach, the goal of free trade will only benefit those who have more resources and power as well as bargaining positions. Consequently, the goals of economic welfare and fairness will not be met (Luna & Fajar, 2010, p. 7).

In fact, when we review the Indonesia 1945 Constitution, it is stated that the market-based approach that undermines the role of the state is not suited to Indonesia's National Economy, since the constitution states:

– The economy shall be organized as a common endeavour based upon the principles of the family system.
– Sectors of production which are important for the country and affect the life of the people shall be under the power of the State.
– The land, waters, and natural resources within shall be under the power of the State and shall be used to the greatest benefit of the people.
– The organisation of the national economy shall be conducted on the basis of economic democracy upholding the principles of togetherness, efficiency with justice, continuity, environmental perspective, self-sufficiency, and keeping a balance in the progress and unity of the national economy.

Based upon the 1945 Constitution of the Republic of Indonesia, particularly points 1 and 4, it is implied that to meet the principle of Indonesian economic democracy, the role of cooperatives is very important. This is because both have traditional family values that focus on the principles of togetherness and self-sufficiency. Furthermore, an empirical study conducted by Idris & Hati (2011; 2013, p. 277) reveals that Indonesia's cooperative movement had been founded before its independence. Its first cooperative, called De Purwokertosche Hulp en Spaarbank der Indlandshche Hoofden by R Aria Wiria Atmaja, was established in 1895 (Abdullah, 2006; as cited in Idris & Hati, 2011, p. 3).

In other words, the national constitution of Indonesia has introduced the social enterprise system through cooperatives or MSMEs, which espouse family values in their operation. Deforuney & Nyssens (2009) and Thomas (2004) define social enterprise as an organization that works for the community's interests, stands for a significant level of economic risk, has salaried employees, has a limited distribution of profit-gain, and has an independent role in decision making which is not based on the amount of capital ownership (Deforuney & Nyssens, Thomas (2004) as cited in Hati, 2011, p. 3).

At the institutional level of policy that derives from the Constitution of 1945, the social entrepreneurship policy of the Cooperative Law Number 17 of 2012, in the Articles 3 and 4, mentions that the aim of the cooperatives is to increase its members' welfare and public welfare through trading activities such as providing products which are affordable for its members or to provide 'soft' loans to its members. For MSMEs, Article 5 of the Law Number 20 of 2008 mentions that the MSMEs are expected to increase public welfare by selling their products that generate profits and provide employment opportunities to the community.

Based upon the policies mentioned, this paper will analyze the policy of social entrepreneurship in the Technology for Region Program (Iptekda) which aims to empower micro-enterprises as the target group through a Technology Transfer Cooperatives Group called KIAT. Prasetyantoko has suggested a social enterprise approach for micro-enterprises rather than an industrial-based approach (Prasetyantoko, 2010, p. 18). This approach addresses weaknesses in the operation of both the market and the government (Leadbeater, 2007, p. 2).

The case study of the Technology for Region Program (Iptekda) has been implemented by the Indonesian Institute of Sciences (LIPI) in 33 Provinces of Indonesia since 1998. The main feature of LIPI's Iptekda is that the program does not give free technology to MSMEs as the target group. It regards technology as an investment for every micro-enterprise since they are the ultimate program beneficiaries. Universities or the LIPI research center acting as the program implementer should establish an organisation called KIAT. KIAT will function as a cooperative that provides technology (machines and equipment), materials, and trainings from the researchers from Universities or LIPI which acts as the program implementers to the micro-enterprises. All of these technology resources will be treated as 'soft' loans that should be repaid by the micro-enterprises. The repayment ideally will be done when the sales income of micro-enterprises increases because of the technology usage in their production processes.

In order to run KIAT, researchers from Universities or LIPI, as the project leaders, have the responsibility to make their technology beneficial to the recipients by improving their products' capacity and quality, so that micro-enterprises may gain a financial benefit. In addition, they should not charge any fee for the technical assistance that they give to micro-enterprises because the government has already paid their allowance. When the investment of technology from micro-enterprises has been paid off, the micro-enterprises may use their investment fund in KIAT to fulfil their business needs. For example, after using the technology they receive from Iptekda LIPI, the production capacity of their business ought to increase as well as the sales income. As a result, they need more market spending through packaging improvements, which may come at some cost to them. To cover this cost, KIAT can provide the micro-enterprises a 'soft' loan by reissuing the money that they invested in the first place.

This paper overviews the function of KIAT in the application of social enterprise policy in two West Java districts and three districts of Malang, East Java. Although Iptekda LIPI has been implemented for more than ten years, the challenges to apply social enterprises for MSMEs' empowerment remain large. This can be seen from the success rate of the program that only reached 67.5% for the years 1998–2004 (Syamsulbahri *et al.*, 2005). Addressing the issue, the research questions in this paper are as follows:

- First, what are the challenges of social enterprises through *Iptekda* LIPI implementation in West and East Java?
- Second, what lessons may we conclude from the case study of the social enterprise approach in empowering micro-enterprises?

To answer the research questions, the paper will apply a qualitative approach. The data applied come from primary and secondary coding techniques analyzed using Straus and Corbin models. The data collection method used observation and in-depth interviews.

The approach will propose a proposition of the research, which is derived from the relationship between empowerment and social enterprises concepts. Both of these concepts will be applied at the context of micro-enterprises policies. The micro-enterprises are selected for their experience in joining the *Iptekda* LIPI program during the period 2000–2011. The next section will review the literature on social enterprises and the technology for the regional (*Iptekda*) LIPI program.

2 LITERATURE REVIEW

2.1 *Policies on micro enterprises empowerment in Indonesia*

Micro enterprises have been the center of public policies in Indonesia. The history shows that since the early independence period (1950–1995), the old regime of Sukarno promulgated economic nationalism promoting State-Owned Enterprises (SOEs) and restricting import licensing. On the contrary, the New Order era under Suharto fostered the growth of a dependent capitalist class of client businessmen (Bhasin, 2010, p. 96).

Wie (2006) assert that under the Suharto regime, the government tried to assist MSMEs in the Foster Father Scheme to establish a partnership between large enterprises and cooperatives or MSMEs. However, the program was ineffective in enhancing the economic capability of the MSMEs since they were not motivated by market considerations, but by the force and fear towards the New Order (Wie, 2006, p. 15). During the Reform era, there was a shift from welfare-oriented to efficiency-based policies. The welfare-oriented policies view MSMEs as the economically weak group demanding much government's assistance, while the efficiency approach concerns market-based and demand-driven MSME programs.

The evaluation on micro-enterprises empowerment policies conducted by Widyaningrum *et al.* (2003) shows that although the development policies always prioritize MSMEs in the national development, such policies and MSME empowerment programs have never met their needs. The contributing factors of such failures are: first, the program to empower micro enterprises focuses on capital assistance; second, the corruption in the program implementation has placed more burdens on the micro-enterprises (Widyaningrum *et al.*, 2003, p. 4).

2.2 *Social enterprises*

The concept of social enterprises emerged when certain arising problems affected profit-maximization efforts which ultimately caused the Financial Crisis of 2007–2008. As a result, the social enterprise approach has become an alternative way of minimizing the negative impact of the market-based approach by regarding social aims and profit as the same side of the coin (Poon, 2011). The concept has been defined by Professor Gregory Dees, the founder of Duke University's Centre for the Advancement of Social Entrepreneurship, an institution that aims to fulfil a social mission by developing innovative problem-solving strategies and showing its accountability to the public (Dees, 2006).

Lyon and Sapulveda (2009, pp. 84–85) have focused on the concept of social enterprises based on the income they gained from trading and from their charities as well. The concept was adopted by the UK Government, which defines social enterprise as a business with primarily social objectives, whose surpluses are principally reinvested in the business for that purpose or in the community, rather than being driven by the need to maximize profits for shareholders (DTI 2002; cited in Lyon & Sapulveda, 2009, p. 85; Moizer & Tracey, 2010 a, p. 252).

From these definitions, we may conclude that a social enterprise is an organization that puts community interests ahead of profit maximization for their shareholders. The activities of the social enterprises may result in trading or charitable activities and these activities are open to public accountability. Moizer & Tracey (2010) assert that social enterprises ought to be not-for profit ventures that are established in order to meet social and commercial objectives (Moizer & Tracey, 2010, p. 252).

In the Indonesian context, based on the historical perspectives, it may be concluded that historically the approach has long been rooted in Indonesian history. This paper will apply Professor Dees' term, 'social enterprises' to explain the function of the KIAT of *Iptekda* LIPI. The next section will discuss the application of social enterprises using the case study of *Iptekda* LIPI and how it has benefited MSMEs in West Java District/Cities and Malang, East Java.

2.3 *The challenges of social enterprises of Iptekda LIPI for MSMEs' empowerment in West and East Java*

Despite the positive outcomes of the social enterprises concept in the integration of social and business needs for the community, the results of this research show that there are a lot of challenges in implementing the concept of micro-enterprise empowerment, particularly when it is related to government policies or programs. The in-depth interview with the owner of a Salak handicraft micro-enterprise revealed the following:

> "I disagree with the scheme of investment fund that KIAT of *Iptekda* LIPI requires. This is impossible for start-up business like mine. How can I invest my money in KIAT if I do not get any sales income from my handicraft? People do not buy handicrafts every day." (In-depth interview with Cineam Salak, Handicraft micro-enterprise, March 12, 2003).

As the **KIAT** project manager for a shrimp community group in Bogor District, one of the researchers explained that the social enterprises from the *Iptekda* LIPI program were successfully implemented in his activity since, at the beginning of the program intervention, but he did not mention that the technology that he provided to the local community came from the government.

> "In my community if we come as the government representatives, people would tend to think the program is charity. No need for them to pay for any of the government's aid. So, it is better to introduce ourselves as someone who has a lot of money and land and wants to invest in the shrimp business. I do not give them money, but I give them shrimp seeds to be cultivated. When they have succeeded in following my training of shrimp management farming, they will be able to produce 2.5 tons of shrimps. KIAT of *Iptekda* LIPI under my supervision will send the shrimp to restaurants in Bogor or sometimes to supermarkets in Jakarta. The sales profits will be shared between the farmers and KIAT. KIAT will keep some of the farmers' share and will invest them again to buy shrimp seeds for the farmers to cultivate. That's the cycle of production and marketing. Although it is successful at the beginning, there are conditions in the market that I could not control, such as price of the shrimp which is very uncertain; but the farmers did not understand it." (In-depth interview with a researcher from LIPI Research Center, February 12, 2013).

Based on the interview with the micro-enterprise owner and the manager of KIAT, there is a potential for the failure of social enterprise implementation in West Java. The failure is due to the fact that social enterprises that are run by KIAT have not succeeded in motivating micro-enterprises to save their investment in their organization. According to other micro-enterprises in another West Java sub-district in Cibinong, the public education of KIAT's function as an alternative funding institution that operates simply more than a bank. However, other experiences of shrimp micro-farmers in Bogor sub-districts show that KIAT has proved to be successful in transferring researchers' knowledge of shrimp management farming to local farmers as well as creating income-generating activity. The keyword 'successful' relies on the communication of the KIAT's function which is described as a non-government aid. This means that social enterprise for micro-enterprises or any other empowerment issues should be free from the term 'government aid'.

Another example of a social enterprise through *Iptekda* LIPI can be seen in some districts of Malang, East Java. The story of these micro-enterprises in implementing social enterprise by KIAT of *Iptekda* LIPI is quite different from their counterparts in West Java. In Malang Raya, most of the micro-enterprises mentioned that they needed the technology for their business growth, and they did not mind if they had to invest money for such technology in KIAT.

> "I am looking for a program in Indonesia which is suitable for my needs. I asked the Trade and Industry Agency in Batu, but they do not have the program that fulfils my business. Then, I met Mrs. E, who invited me to join *Iptekda*, on the condition that I should invest an amount of money for every machine and item of equipment that I get. I did not say yes in the first place, but then I took a risk to invest it in KIAT. I thought the procedure would be difficult, but the procedure was very simple. Now I have paid my mortgage and I can get my own machine and technology. KIAT also gives me some capital assistance when I need to expand my business without any interest. My business production and my sales have increased because of the machine and equipment that I now own." (In-depth interview with UD Dua Merpati, 25 June 2013, Batu City East Java).

The KIAT managers in Malang, East Java districts and cities admitted that the communication of social enterprises that were embedded in government aid should not be limited by time. This is because the diffusion of technology requires time to be implemented, and the outcomes of technology aid for micro-enterprises should be guaranteed by the researchers so that the micro-enterprises will want to invest their capital in KIAT.

3 DISCUSSION AND CONCLUSION

The research on the social enterprises applied in the *Iptekda* LIPI case study has some lessons to teach. First, the concept of integrating social and business aspects for empower micro-enterprise owners relies heavily on the time factor and public education. The case studies in both areas in West and East Java show that the role of the researchers as the managers in KIAT is very important, particularly in communication strategies. Second, the social enterprise can be part of an alternative solution to empower micro-enterprises' productivity; however, there are other policies that are related to trade and industrial policy which could complement the approach. The experience of researchers in both locations shows that there are some factors that they cannot control in order to guarantee income generation and that the micro-enterprise business can run sustainably in the future.

The challenges of social enterprise policy implementation through *Iptekda* LIPI has the policy implication that relates to the needs of a multidisciplinary approach particularly in anthropology and sociology. To integrate between social and business interests as the same side of the coin, any social enterprise policy implementation should be based on the local culture of each province or region in Indonesia. In addition, it would be better if social enterprise policy implementation designed to empower micro-enterprises focuses on the characteristics of micro-enterprises and the age of business, so that a model of social enterprise organization such as KIAT will not apply a single, 'one-size-fits-all' mechanism to all of its beneficiaries.

REFERENCES

Aida, I. & Hati, R.H. (2013) Social entrepreneurship in Indonesia: Lessons from the past. *Journal of Social Entrepreneurship*. 4 (3), 277–301.

Bhasin, B.B & Venkataramany, A. (2010) Globalization of Entrepreneurship: Policy Consideration for SME Development in Indonesia. *International Business and Economics Research Journal*. [Online] 9(4). Available from: http://digitalcommons.sacredheart.edu/cgi/viewcontent.cgi?article=1029&context=wcob [Accessed 28th August 2014].

Fajar, A. & Launa, P.R. (2010) UMKM dalam Pilar Kemandirian Bangsa. Jurnal Sosial Demokrasi. [Online] 9 (3). Available from: http://library.fes.de/pdf- files/bueros/indonesien/07003/2010-09.pdf July-September. Jakarta, Pergerakan Indonesia dan Komite Persiapan Yayasan Indonesia Kita. [Accessed 5th June 2010].

Hati, S.R.H. & Idris, A. (2011) The Social Movement and Social Enterprise Development in the Emergence of Indonesia. 1895–1945. [Online] Available from: https://apebh2012.files.wordpress.com/2011/05/apebh-text-of-paper-sri-rahayu-hijrah-hati.pdf. [Accessed 1st August 2016].

Leadbetter, C. (2007) *Social enterprise and social innovation: Strategies for the next ten years London*. England, Office of the Third Sector.

Lyon, F. & Sepulveda, L. (2009) Mapping social enterprises: past approaches, challenges and future directions. *Social Enterprise Journal*. 5(1) [Online] Available from: http://dx.doi.org/10.1108/17508610910956426 [Accessed 29th August 2016].

Moizer, J. & Tracey, P. (2010) Strategy making in social enterprise: The role of resource allocation and its effects on organizational sustainability. *Systems Research & Behavioral Science*. 27(3), 252–266. [Online] Available from: http://onlinelibrary.wiley.com/doi/10.1002/sres.1006/full [Accessed 18th August 2016].

Poon, D. (2011) The Emergence and Development of Social Enterprise Sectors. *Social Impact Research Experience Journal (SIRE) Wharton School*, University of Pennsylvania.

Prasentyoko, A. (2010) Pemberdayaan UMKM sebagai Perwujudan Demokrasi Ekonomi di Indonesia. *Jurnal Sosial Demokrasi*. 9(July-September). [Online] Available from: http://library.fes.de/pdf-files/bueros/indonesien/07003/2010-09.pdf [Accessed 5th August 2016].

Syamsulbahri, D. (2006) Analisis keberhasilan dan kegagalan pelaksanaan program Iptekda LIPI. Pemberdayaan UKM Melalui Program Iptekda LIPI. LIPI Press, Jakarta.

Tambunan, T. (2008) Trade liberalization effects on the development of small and medium enterprises in Indonesia: A case study. *Asia-Pacific Development Journal*. 15 (2) [Online] Available from: https://core.ac.uk/download/pdf/6278113.pdf [Accessed 28th August 2016].

Widyaningrum, N. *et al.* (2003) Pola-pola eksploitasi terhadap Usaha Kecil. Yayasan Akatiga, Bandung.

Wie, T.H. (2006) Policies for Private Sector Development in Indonesia. March. ADB Institute Discussion Paper Number 46. [Online] Available from: www.adb.org/sites/default/files/publication/.../adbi-dp46.pdf [Accessed 31st August 2016].

Competition and Cooperation in Social and Political Sciences – Adi & Achwan (Eds)
© *2018 Taylor & Francis Group, London, ISBN 978-1-138-62676-8*

This is a sample paper Confucius for non-Confucians: Understanding China from "without"

R.L. Benavides
King's College London and National University of Singapore, Singapore

ABSTRACT: Contemporary Confucian discourse may soften or alleviate tensions among nations, especially between China and other global powers. Confucianism is experiencing a revival in twenty-first century China, and its discourse is spreading throughout the world. In recent years, experts in the field have published numerous works disclosing the impact and significance of this tradition. Nevertheless, it is time to acknowledge Confucianism not only academically but, more importantly, politically. In a world where Western-dogmas are dominant and China is increasingly becoming more influential in current global affairs, political rivalry should be replaced with synergistic collaborations. The only way to harmonise with the Asian giant is by understanding Chinese culture, society, politics, and philosophy and not merely in terms of international trade and economics. The spread of Confucianism is taking place in academic circles yet reaching non-scholarly audiences. This study observes how China, through the discourse of a specific group of experts on Confucianism, is being understood from without (i.e. outside of China). The contributions of some of the major exponents of contemporary Confucianism are presented in this essay to provide the reader with a general understanding of how Confucianism is still in dialogue with the modern world.

1 INTRODUCTION

The philosophy and teachings of Confucius (551-479 BCE) have regained popularity in China in recent decades. Confucius was devoted to learning from gentlemen who inspired awe—such as the Duke of Zhou and the ancient sage kings—and embarked on a mission to transmit the ru tradition, i.e. the "doctrine of scholars." He believed "men of antiquity studied to improve themselves" and people from his generation only "study to impress others" (Analects 14, 24). Being keen of improving himself through self-cultivation and transmission of knowledge, Confucius promoted a set of principles and values borrowed from ancient Chinese sages to engender exemplary persons (junzi). The "doctrine of scholars," which evolved into what it is now known in the West as Confucianism, has been associated to religious rituals and ceremonies and learning and education. But soon after it was popularised by Confucius, his disciples, and other followers, this doctrine became the leading school of thought and official state ideology of China for over two-thousand years. It was only after the fall of the Qing dynasty (1644–1911) that Confucianism weakened as a philosophy and ceased to be the state ideology of China. In the eyes of the Chinese, Confucianism became an obstacle for progress and modernisation during the late-nineteenth and early-twentieth centuries; subsequently, the teachings of the greatest sage of China were repudiated and viewed as an impediment to remedy the turmoil China experienced at the time.

Confucianism is a humanistic tradition with philosophical creeds based on self-cultivation, ethics, loyalty, virtues, wisdom, and benevolence. Contemporary proponents of a modernised version of these doctrines may have different views on the practicality of Confucianism in the twenty-first century. Different interpretations of the tradition are reaching out to a number of people around the world through lectures and written works of scholars. With the current economic boom of China, the eyes of the global population are turning eastwards.

Not too long ago, leading social and political ideologies were established by Western nations to produce systems of governance aimed to ameliorate the livelihoods of people. Capitalism, liberal democracy, human rights, and other concepts which originated in the West became the common-standard for developed and developing nations. These and other Western ideas have been adopted by Japan, South Korea, Singapore, and Taiwan—all considered Confucian civilisations—and more recently China is taking small steps to integrate them into their existing state of affairs as they seem to bring about benefits to society. These concepts are predominant in the modern world for a good reason: they prompt prosperity, freedom, equality, equity, liberty, and so on. On the negative side, the personal attachment to consumerism and individualism has increased exponentially over time leading to an inconvenient future for humanity, the environment, and other live forms. So, can the teachings of Confucius provide better solutions to some of the modern world problems? Confucius said "the exemplary person understands what is moral. The petty person understands what is profitable" (Analects 4, 16). This proverb leads us to substitute the desire for profit and material goods with moral and admirable conducts. According to Confucius, benevolence and wisdom shall be achieved by loving and knowing your fellow men (Analects 12, 22), meaning that altruistic behaviours ultimately result in virtuous outcomes for the well-being of others.

Many teachings of Confucius remain relevant to the modern world. His philosophies are being reinterpreted in and outside China by several scholars keen on reviving the Confucian tradition, but their discourse is typically confined to academic circles and often overlooked by other audiences. Confucianism as a political ideology may bring about positive solutions to modern socio-political problems, and its contemporary discourse can help us better understand China. In the following sections, the major representatives of modern Confucian discourse are disclosed. Also, a distinction between "Confucians" and "non-Confucians" is made to differentiate those who have been influenced by the philosophies espoused by Confucianism from those who have not been exposed to them. Finally, specific works and ideas of contemporary overseas scholars of Confucianism are presented in this paper to familiarise the reader with available materials they can use to learn about Confucius, his sayings, the doctrine, and where Confucianism stands today as a political ideology. The purpose of pointing out key works of specific scholars will give the reader insights to a variety of topics and themes often discussed in Confucian studies. This work is not a detailed account of Confucius's life or his school of thought; instead, it provides a brief introduction to the development of contemporary overseas Confucian discourse. So, those unfamiliar with the tradition can have an overall idea of Confucianism and use the materials presented in this essay for future reference.

Anyone wishing to understand China or other Confucian societies is suggested to engage in a dialogue with Confucius. This paper will discuss the role of a handful of scholars who conduct research on Confucius and the school of thought named after him. It is important to recognise the significance of contemporary Confucian discourse as it provides a general understanding of Chinese history and culture which is necessary to understand the present phenomenon occurring in China today.

2   FROM CHINA TO THE WORLD

One of the first Westerners who immersed and became acquainted with the "doctrine of scholars" was the Jesuit priest Matteo Ricci (1552–1610). It has never been easy to understand a society, culture, or political system if one is not in direct contact with it. Ricci struggled to enter the mainland and was finally accepted among the literati upon his arrival. He learned Chinese and translated some of the ancient Chinese classics into Latin with the help of other missionaries; his translations reached Europe, and the teachings of Confucius began to spread outside of East Asia. The hermeneutical translations of Ricci helped Europeans better understand China and its cultural traditions which were enormously influenced by the teachings of Confucius—and other systems of thought such as Buddhism and Daoism. Similarly, in modern times, English-speakers can resort to the interpretations of contemporary scholars of Confucianism to have a profounder understanding of Chinese traditional philosophy and become familiar with the teachings of one of the greatest sages of all time.

There are a number of reasons why people, corporations, nations, and organisations world-wide would want to learn more about China. When China is studied, it is often done through a particular lens. For example, an economist may look at diverse processes involving the line of production and distribution of goods and services of a specific company or region, a political scientist will explore the ways a society is organised by authorities in terms of governance and redistribution of power, and a businessperson may study a specific market seeking to achieve profitable outcomes for the company he or she works for. But, how is China to be understood holistically? China should not be regarded solely as a modern Communist nation-state; it is a country with a complex history and rich culture with a lot of say in contemporary global affairs. Having an overall view or understanding of China is virtually impossible; however, a good place to begin is by looking at Confucianism through the lens of contemporary Confucian scholars to see how this tradition is still responsible for shaping China today.

Modern Confucian philosophy is mainly attributed to philosophers, intellectuals, or scholars of the late nineteenth-century onwards. Some of the earliest representatives include Kang Youwei (1858–1927), Tan Sitong (1865–1898), and Zhang Dongsun (1886–1962). During the transition from dynastic rule to a modern nation-state, the Chinese citizenry was exposed to science, democracy, capitalism, and communism among other Western concepts. As a result, the Confucian literati felt obliged to respond to the challenges imposed by these foreign ideas characteristic of modernisation and progress. To save the Confucian tradition from disappearing, a new intellectual-philosophical movement known as New Confucianism arose. The main figures of the New Confucian Movement have been divided into generations of scholars:

– First Generation
Ma Yifu (1883–1967), Xiong Shili (1885–1968), Zhang Junmai (1886–1969), Liang Shuming (1893–1988), Feng Youlan (1895–1990), Qian Mu (1895–1990), He Lin (1902–1992)
– Second Generation
Fang Dongmei (1899–1977),[1] Xu Fuguan (1903–1982), Mou Zongsan (1909–1995), Tang Junyi (1909–1978)
– Third Generation
Yu Ying-shih (1930–), Liu Shu-hsien (1934–2016), Cheng Chung-ying (1935–), Tu Wei-ming (1940–)

From a broad perspective, New Confucianism is concerned with keeping the teachings of Confucius alive and engaging in practical concerns faced by today's society. It has been nearly one century since this philosophical movement initiated, and it has evolved into a rich system of ideas that must be contemplated by and made available to those interested in producing positive socio-political changes. Works from the first generation such as A History of Chinese Philosophy by Feng Youlan (originally published in Mandarin 1934; second edition in English 1952, trans. Derk Bodde) and The Development of Neo-Confucian Thought by Zhang Junmai (a.k.a. Carsun Chang 1957) reached non-Mandarin speaking audiences in the 1950s. These and other works provided a broader idea of the philosophical traditions and socio-political practices that reigned in China for over two millennia. The works of second and third-generation scholars became popular in the last decades of the twentieth century. Through their works, the teachings of Confucius and his followers have been captivating Western—mostly English-speaking—audiences for more than half a century and they still remain meaningful in the East.

In the mid-1980s a group of scholars gathered in China to discuss the future of Confucianism. During this reunion, Fang Keli identified Jiang Qing, Kang Xiaoguang, and Chen Ming as "fourth-generation" Confucian scholars. Nevertheless, these generational divisions or labels do not necessarily follow a structure. One of the few things all four generations have in common is that they were born in mainland China. Many of them were educated in

---

1. Sometimes placed alongside first-generation scholars.

the West; thus, their motivations, ideas, aims, and approaches may differ extensively from one generation to another or from one individual to another. The so-called fourth generation of scholars has a slightly different approach towards Confucianism; their ideas are less concerned with a Confucianism that can adapt to the modern world but one that can be readapted in China in a more orthodox manner. For example, Jiang Qing's "Political Confucianism" is concerned with re-institutionalising the tradition and advises it should become the state religion of China. Similarly, Kang Xiaoguang advocates for the "Confucianisation" of China instead of a Westernisation. Conversely, third-generation scholars are keener to combine both Confucian and Western ideas. One particular thing distinguishes these two generations: on the one hand, the third generation is composed of Western-educated scholars who lived most of their lives outside the mainland and many of their works were written and published in English; on the other hand, the fourth generation was mainly educated in China and most of their works have been written and published in Mandarin.

An essential point needs to be stressed concerning the contemporary spread of Confucianism outside of mainland China—especially in non-Mandarin speaking countries. A number of works pertaining to New Confucians have been either published in or translated to English (and other non-Chinese languages). However, the importance of contemporary Confucian discourse should not just rely on or be confined to the works of these scholars. Much attention should be placed to a "meta" generation of Confucian scholars, or scholars of Confucianism, that has emerged in the past three to four decades, i.e. contemporary overseas scholars of Confucianism. They do not fall into the generational division of the New Confucian Intellectual Movement, but their discourse is as important and relevant to modern Confucianism as the one of those scholars associated with the movement. Consequently, anyone interested in understanding twenty-first century China should pay attention to the works of contemporary overseas scholars of Confucianism as they play an exceptional role in the development of modern Confucianism. But before examining some of their works, the next section will briefly illustrate Confucius' influence on diverse social, political, and cultural values of East and Southeast Asian civilisations.

3   WHO ARE THE CONFUCIANS?

Broadly speaking, people belonging to Confucian civilisations including Sinitic and non-Sinitic societies that have been exposed to and influenced by Confucian doctrines can be labelled as "Confucians." This tradition has shaped their existing and pre-existing conditions in terms of customs, habits, and mores for generations. In a few words, Confucianism is ingrained in the culture and values of the Chinese people and other East Asian societies. On the other hand, "non-Confucians" are individuals outside the East Asian cultural sphere with limited or no knowledge of Confucius or his teachings. Most non-Confucians are unaware of the historical significance of this tradition and its modern relevance. In recent years, Confucians have been exposed to non-Confucian ideas, especially those coming from the West. Western values have been dominant in the last couple of centuries, but with the rise of China and current economic performance of the so-called Four Little Dragons (Hong Kong, Taiwan, Singapore, and South Korea) Eastern values are gaining momentum. As a result, many non-Confucians seek to understand what China and its neighbours are doing in attempts to find new ways of dealing with obstacles to social and political progress.

During the 1980s and 1990s "Asian values" were thoroughly discussed among political leaders in Asia in response to the influence of the West in the region. Former Prime Ministers Mahathir Mohamad and Lee Kuan Yew, from Malaysia and Singapore respectively, pointed out that East and Southeast Asian societies shared a common system of principles and beliefs[2] that could be used as an alternative to the wholesale Westernisation. The system of values shared by some

---

2. These include a combination of values from the major intellectual traditions, religions, or philosophies in the region such as Buddhism, Confucianism, Hinduism, and Islam.

countries in the region is considerably inspired by the teachings of Confucius; these were and still are believed to be pertinent and applicable in Confucian societies and beyond. However, the modern notion of Asian values does not necessarily represent the views of modern Confucian philosophy. In fact, many sinologists and scholars of Confucianism prefer to dissociate the term "Asian values" from Confucianism.[3] The idea behind Asian values was that re-enforcing them in East Asian societies and employing them in other countries could produce economic stability and enhance political practices to subsequently benefit the livelihoods of people. Contemporary Confucian scholar Du Weiming (Tu Wei-ming) asserts Confucianism "has exerted profound influence on East Asian political culture as well as on East Asian spiritual life" (1989, p. 1); the teachings of Confucius have shaped East Asian civilisations creating shared values throughout the region. Du also points out that Confucianism is "a way of life in East Asia; so deeply ingrained in the fabric of society and polity, it is often taken for granted as naturally human" (p. 3). In consequence, Confucian values are often associated with Asian values. For example, filial piety (xiao) is one of the most important concepts in the ru tradition. It is an intrinsic and fundamental virtue or attitude stimulating moral behaviours and compliance towards parents and elderly family members. When Confucianism reached neighbouring territories in medieval times, it formed common values among East and Southeast Asians. Filial piety also extends from the family to the government, thus creating Confucian values of authority.

At present, it is difficult to ignore the on-going China phenomenon; people worldwide are affected by the political and economic activities of China. Because of instances of despotic rule throughout the long history of imperial China and misuses of Confucian teachings, Confucianism is often viewed as a regressive and authoritarian ideology by many—especially the non-Confucians. Those familiar with or belonging to East Asian cultures may also parallel Confucianism with authoritarianism; however, the teachings of Confucius should not be taken as valueless. In reality, many people belonging to Confucian societies are unaware that the teachings of Confucius have been a major part of their lives. Plenty of customs and conventions transmitted from father or mother to son or daughter which have been passed on and retained from one generation to another. Similarly, Confucian values of authority have remained in social and political institutions where power is exerted. There is a misconception that Confucianism promotes authoritarian rule, and it is necessary to distinguish authoritarian rule from authoritative rule. According to Sor-hoon Tan (2010), "Confucianism has often been treated as an authoritarian philosophy that exalts the absolute authority of rulers over subjects, of fathers over sons, and of husbands over wives" (p. 137). These three "bonds" have been heavily criticised for portraying hierarchical submission and less often interpreted in a constructive way. Authoritarianism is synonymous with oppressing and taking away personal freedom while authoritative conduct (ren), as epitomised in the Confucian tradition, encourages the individual to behave ethically and enact any social or political roles or responsibilities entrusted to him as a member of society or a community. Originally written as a chapter in the Book of Rites, the "Great Learning" teaches us that "the ancients who wished to illustrate illustrious virtue throughout the empire, first ordered well their own States. Wishing to order well their States, they first regulated their families. Wishing to regulate their families, they first cultivated their persons" (Legge, 1861, p. 221). This passage tells us that an authoritative conduct is required to develop personal self-cultivation that extends beyond the family leading to benevolent ruling. Hence, a ruler can practice authoritative conduct to illustrate illustrious virtue as opposed to being authoritarian.[4] Authoritative conduct is often

3. For a brief discussion on the relationship of Asian values and modern Confucianism see Jana S. Rošker (2016), Modern Confucianism and the Concept of "Asian Values". *Asian Studies* 4(1). For a more general discussion on Asian values see Chang Yau Hoon (2004), Revisiting the Asian values argument used by Asian political leaders and its validity. *The Indonesian Quarterly* 32(2).
4. For a more detailed discussion of the differences between being authoritarian and authoritative conducts see Peter D. Hershock & Roger T. Ames (eds.) (2012), *Confucian Cultures of Authority*: SUNY Press, and Sor-hoon Tan (2010), Authoritative Master Kong (Confucius) in an Authoritarian Age. *Dao* 9(2), 137–149.

misinterpreted or misunderstood as authoritarianism, and it is placed out of context from Confucius' original intent of fostering authoritative conducts among people, especially those in official positions. His teachings aimed to produce exemplary persons in and outside the ruling class to assemble a prosperous and harmonious society. The teachings of the great Master may be misunderstood by "Confucians" and "non-Confucians," yet it may be more a matter of interpretation rather than misunderstanding his teachings.

4 CONFUCIUS FOR NON-CONFUCIANS

Presently, Confucian scholars and scholars of Confucianism keen to revive this tradition are producing literary works to explicate what Confucianism signifies in today's world. Confucianism is the "DNA" of China as it aided to the development, growth, and function of its culture, so becoming acquainted with Confucianism is a starting point to better understand China. The teachings of Confucius are being made available to non-Confucians through the works of contemporary overseas scholars of Confucianism. This tradition must be acknowledged as the backbone of China as it has historically influenced in its cultural, social, and political conditions.

In 2016, Peimin Ni, Professor of Philosophy at Grand Valley State University, published Confucius: The Man and the Way of Gongfu. He presents Confucius' instrumentality in the development of the "tradition of scholars" and elucidates the role of Confucius using a unique approach called gongfu. Confucius was a pragmatist. He endeavoured to put into practice theoretical frameworks associated to social, political, and/or ethical concerns. Ni's work is a guide to the life of Confucius as an historical figure, spiritual leader, philosopher, political reformer, educator, and as a person (Ni, 2016, p. vii-viii). The overall objective of the book is to portray Confucius as a common person accessible to everyone and to help people identify themselves with the teachings of the great Master.

Confucius is known for his sayings, and these were recorded by his disciples and followers in an anthology called the Analects. Intellectual historian of China and scholar of Confucianism Philip J. Ivanhoe places particular attention to the contemporary relevance of this ancient Chinese classic in his book Confucian Reflections: Ancient Wisdom for Modern Times (2013). Ivanhoe gives the reader an insightful assessment of the weight the Analects has in contemporary times. His study reveals it is worth to exploring the content found in the Analects as it can aid to flourish an upright society in today's world (p. 86). Reading the Analects may be confusing to the modern man or irrelevant in modern times; however, Ivanhoe's reflections allow the reader to comprehend the usefulness of this canonical text by targeting genuine concerns with respect to our modern daily lives.

Rujia, the ancient doctrine or tradition of scholars which developed into what it is now known in the West as Confucianism, is a type of humanism based on the philosophical teachings of Confucius. A good resource to understand the historical evolution and creative transformations of Confucianism since its origins is Xinzhong Yao's An Introduction to Confucianism (2000). It is a fundamental reading in contemporary Confucian studies and has become a must read for people interested in learning about this religious and politico-philosophical tradition that pervaded in China for hundreds of years. Yao presents the major advocates of the tradition and discusses the main features of the three epochs of Confucianism: Classical Confucianism, Neo-Confucianism, and New Confucianism. This volume became popular among students and scholars wishing to have a general and chronological outlook of Confucianism helping to understand the modern relevance of Confucianism.

Moreover, the number of contemporary scholars concerned with Confucianism as a political ideology/philosophy is large. Inspired by the Duke of Zhou, Confucius wanted to create a model society based on the idea of the "Great Unity" (datong) supposedly experienced during the golden years of China when the legendary sage-kings Yao, Shun, and Yu lived. One of the highest ideals in Confucianism is to emulate the ancient sage-kings in attempts to produce a prosperous and harmonious society through virtuous rule. In a few words, virtue (de), harmony (he), and humanness (ren) are three key characteristics of Confucianism as a political ideology. Ideally, Confucianism as a state ideology should have a virtuous and benevolent ruler able to uphold political stability and social order. This type of rule will lead to a harmonious society where humaneness prevails among people at all levels. In ancient China, if the ruler was incapable of providing well-being to his subjects, the Mandate of Heaven (tianming) removed the incompetent ruler and replaced him with a virtuous one. Confucianism is concerned with uprightness and "promoting the virtuous and talented" (ju xiancai) to serve in the government, for humaneness shall triumph if there is virtuosity within the ruling class and harmony among society. Confucius suggested the ruler must put aside the crooked and advance decent and honest men if he wants the people to follow him (Analects 2.19). Likewise, Mencius emphasized the relationship between the people and the ruler stating that people are to be valued the most and placed above the ruler (Mencius 7B14). To quote Aristotle, human beings are and will always be "political animals," meaning that in politically organised societies there is no question about the existence of hierarchical arrangements constituted by leaders and followers. No matter how a society is structured, its political system must have a few selected—ideally virtuous—individuals leading the rest of the population. Therefore, Confucianism as a form of political ideology, which aims for ethico-political order through the indoctrination of morals to the people, coincides with some contemporary modes of governance. There are, of course, many differences in the ways politics functioned in pre-modern China, but some of the basic elements of social and political organisation espoused by Confucianism are still in dialogue with the modern world.

In recent decades, contemporary overseas Confucian discourse has generated debates on whether Confucianism and democracy are compatible. Political scientists, political theorists, and philosophers such as Bai Tongdong, Daniel A. Bell, Sungmoon Kim, Lee Ming-huei, and Sor-hoon Tan have endeavoured to revive today's political practicality of Confucianism by associating it to liberal democracy or other Western ideas. In the West, most societies practice liberal democracy; it is the dominant world-view of organising societies and doing politics. However, several East Asian societies which have already incorporated Western-democratic values into their political systems have not entirely given up their cultural values. Instead, they have created a balance by practicing democracy as their modern system of governance yet preserving their cultural traditions. So, is there room for a Confucian-style democracy in the modern world and can it replace Western liberal democracy?

Sungmoon Kim (2014) takes a pluralist approach to the question raised above. In Confucian Democracy in East Asia: Theory and Practice he proposes an alternative to modern democracy. His alternative is not a full replacement of Western liberal values with Confucian communitarianism but a subtle integration of both to form a sort of political pluralism characterised by East Asian and Western values. By integrating Confucianism, democracy, and pluralism Kim resorts to "construct a political theory of Confucian pluralist democracy by critically engaging with two dominant versions of Confucian democracy—Confucian communitarian democracy and Confucian meritocratic democracy" (p. 103). Political theorist Daniel A. Bell (2015) is also concerned with integrating political meritocracy and democracy. Although his initial encounters with meritocracy were inspired by Confucian philosophy, in recent years Bell has taken a more pragmatic approach in terms of truly spotting meritocratic behaviours in current Chinese politics (p. 12). He proposes a three-fold model for "democratic meritocracy" in China: the first at the voter level, the second at the level of central political institutions, and the third combining political meritocracy at the level of the central government and democracy at the local level (p. 9).

Similarly, other political theorists including Stephen C. Angle, Bai Tongdong, Joseph Chan, and Jiang Qing have proposed contemporary ways of organising politics in China with Confucian hints. Stephen C. Angle's (2012) "Progressive Confucianism" is concerned with four frameworks: i) political authority should not restrict the power of the people, ii) balance and harmony based on inequality (i.e. non-oppressive forms of hierarchy), iii) public participation extended to personal, social, and political domains, and iv) representation of the people ensuring all are heard and none oppressed (p. 55–56). Bai Tongdong's (2009; 2014a; 2014b) works attempt to develop a contemporary Confucian political philosophy that harmonises with Western modernity based on more orthodox approaches openly using ancient Confucian canons such as the Analects and the Mencius. Likewise, Joseph Chan (2014) seeks to refurbish ancient Confucian political values and fuse them with Western liberal democracy to create what he calls "Confucian Perfectionism," i.e. a political philosophy aiming to improve Western liberalism with Confucian values and Confucianism with Western liberal values. Jiang Qing's (2012) approach, on the other hand, is somewhat different from the aforementioned scholars. He, instead, proposes a politicised version of Confucianism which rejects Western liberal democracy (contradicting Mou Zongsan's version of New Confucianism) through the institutionalisation of Confucianism. Jiang's vision of political Confucianism is based on a tricameral legislature (p. 27–43) that materialises political legitimacy—which is lacking in contemporary Chinese politics.

There is no doubt democracy and Confucianism are compatible and that there is room for a Confucian-style democracy in the modern world. Much research has been conducted in this area by experts in the field, and they have demonstrated, theoretically speaking, how valuable and viable a Confucian-style democracy could be in the twenty-first century. Even though East Asian societies under Confucian influence (e.g. Japan, Korea, Taiwan) have been able to adopt and implement Western liberal democratic values into their political systems, Taiwanese Confucian scholar Lee Ming-huei (2010), inspired by earlier New Confucians, agrees democracy can be developed from Confucianism and employed in China. He believes "democracy cannot be directly transplanted from the West to China and can only be absorbed through the internal development and adaptation of traditional Chinese culture" (p. 246). This statement can be mirrored in the sense that Confucianism cannot be directly transplanted from China—or other Confucian civilisations—to the West and can only be absorbed through the internal development and adaptation of Confucian (political) values as an alternative to the current Western-style socio-political system based on liberal democracy Western countries have. But, why is it important to develop a Confucian democracy if liberal democracy seems to deliver proper ways of governing? Sor-hoon Tan (2009) points out that a number of Confucian enthusiasts including Western scholars "have argued that Confucian democracy offers a viable alternative, even a superior alternative, to liberal democracy" (p. 1). Her study explores how Confucianism can play a role in the development of democracy in China in future decades; a type of democratisation rooted from Chinese cultural heritage, especially Confucianism, and not necessarily drawn from Western forms of democracy (Tan, 2009, p. 7). Tan does not intend to merely combine Confucianism and democracy as this could cause one to overstep the other. Instead, she places an emphasis on the "inseparability of ethics and politics in Confucianism [which] makes the [Deweyan] conception of democracy as an ethical and social ideal the best choice for a reconstruction of Confucian democracy" (p. 12). Harmonising Confucianism and liberal democracy may not be as simple as it sounds, there are critical pragmatic concerns pointed out by Tan which should not be ignored when attempts to democratise Confucianism and Confucianise democracy are being made.

5 CONCLUDING REMARKS

The People's Republic of China is a modern nation which, at the time its establishment, endeavoured to replace its core traditional values—mostly based on the Confucian tradition—with socialist ones. As a result, China's status quo is grounded on socialism with Chinese characteristics. However, due to its long history and rich culture, modern China must not be solely understood as a socialist state based on Marxism-Leninism, Mao Zedong Thought, Deng Xiaoping Theory, and the Three Represents. Instead, much attention needs to be placed on the philosophical tradition that led the Chinese for over two-thousand years.

Confucian values are deep-seated in the Chinese culture and while China continues to develop these are to be acknowledged by the people of other nations. It is impossible to ignore the influence and impact of Confucius in China—and neighbouring regions—at this time and age. Therefore, anyone in the English-speaking world wishing to understand China will find convenient to drawn upon the works of contemporary overseas scholars of Confucianism. It may not be necessary to go as far as the ancient Chinese classics to understand modern China but recent works on Confucius or Confucianism written by these scholars should provide a basic understanding of China's current mind-set.

Contemporary overseas Confucian discourse is one of the best ways to understand China from without. Non-Confucian societies, especially the ones tied to Western dogmas, wanting to build prosperous and harmonious relations with the Asian giant would be better off by understanding China in Chinese terms. Confucian discourse has evolved, developed, and transformed for centuries embracing problems and issues encountered by society. New Confucianism originated as a response to Western ideas brought to China in the early twentieth-century and has the potential to become one of the most important intellectual movements of the modern world. The teachings of Confucius are not only pertinent in Confucian societies, they are relevant to all nations unanimously. As this intriguing humanistic socio-political philosophical tradition continues to develop in academic circles, its practicality will gradually become visible in social and political domains. Politicised forms of Confucianism can adopt many shapes and sizes, but for this tradition to be accepted more widely, it should not attempt to replace people's mores and values; it must enhance them for the betterment of humankind.

REFERENCES

Angle, S.C. (2012) *Contemporary Confucian Political Philosophy*. Cambridge, UK, Polity Press.

Bai, T. (2009) *The New Mission of an Old State: The Contemporary and Comparative Relevance of Classical Confucian Political Philosophy*. Beijing, Peking University Press.

Bai, T. (2014a) The analects and forms of governance. In: Olberding, A. (eds.) *Dao Companion to the Analects*. Dordrecht, Springer Netherlands. pp. 293–310.

Bai, T. (2014b) Early confucian political philosophy and its contemporary relevance. In: Shen, V. (eds.) *Dao Companion to Classical Confucian Philosophy*. Dordrecht, Springer Netherlands. pp. 335–361.

Bell, D.A. (2015) *The China Model: Political Meritocracy and the Limits of Democracy*. Princeton, NJ, Princeton University Press.

Chan, J. (2014) *Confucian Perfectionism: A Political Philosophy for Modern Times*. Princeton, NJ, Princeton University Press.

Ivanhoe, P.J. (2013) *Confucian Reflections: Ancient Wisdom for Modern Times*. New York, Routledge.

Jiang, Q. (2012) *A Confucian Constitutional Order: How China's Ancient Past can Shape its Political Future*. Princenton, New Jersey, Princeton University Press.

Kim, S. (2014) *Confucian Democracy in East Asia: Theory and Practice*. New York, Cambridge University Press.

Lee, M-h. (2010) Confucian traditions in modern East Asia: Their destinies and prospects. *Oriens Extremus*. 49, 237–247.

Legge, J. (1861) *The Chinese Classics: Confucian Analects, the Great Learning, and the Doctrine of the Mean*. 1st volume. London, Trübner & Co.

Ni, P. (2016) *Confucius: The Man and the Way of Gongfu*. Lanham, Maryland, Rowman & Littlefield.

Tan, S-h. (2009) Why Confucian Democracy? Paper presented at the *Varieties of Democracy, Wesleyan University, Middletown, Connecticut*.

Tan, S-h. (2010) Authoritative Master Kong (Confucius) in an Authoritarian age. *Dao*. 9 (2), 137–149.

Tu, W. (1989) *Confucianism in an Historical Perspective*. 15th volume. Singapore, Institute of East Asian Philosophies.

Yao, X. (2000) *An Introduction to Confucianism*. New York, Cambridge University Press.

Competition and Cooperation in Social and Political Sciences – Adi & Achwan (Eds)
© *2018 Taylor & Francis Group, London, ISBN 978-1-138-62676-8*

Islam and the state: Political thought contextualisation of Hamka (Haji Abdul Malik Karim Amrullah), on the state, religion and morality in Indonesia

Heri Herdiawanto
Department of Political Science, Faculty of Social and Political Sciences, Universitas Indonesia, Depok, Indonesia

ABSTRACT: This study discusses the 'political thought contextualisation of Hamka, on the state, religion and morality'. It specifically examines the phenomenon in society that politics tends to be legalising all means to achieve their political goals, it tends to be conflicting interests. The purpose of this study is to prove that Hamka's political thoughts are not the case, but that the opposite is true. This means that politics is sacred and beneficial when used as a means to achieve benefits and for the sake of the society/community. This study applies a qualitative approach and is reinforced with a literature review, including interviews with close family and a very familiar Hamka figure in the field of sociopolitical activism, namely: Masjumi, Muhammadiyah, YPI Al Azhar and the Indonesian Ulema Council (MUI). The interviews were intended to help strengthen, map, explain and analyse the political thoughts of Hamka. Subsequently, the data were analysed using a qualitative approach to the interpretation in order to derive conclusions that address existing problems. This study proves that, in the context of Indonesia, there is no separation between religion and state affairs. According to Hamka, Islam is the doctrine of revelation containing *syari'at*, worship, *muamalat* and state, rooted in *tawhid* (the supremacy of Allah). Islamic religion in this case is not narrowly conceived and executed partially as prayer alone (secular), but comprehensive in terms of society and state.

1   INTRODUCTION

Relationship problems between religion and state initially appeared in the historical development of the phenomenon in the Western world and not in the Islamic world. A combination of the power of the Church and the state took place in Europe until the medieval period. Unity was shattered due to the emergence of the recognition by the Church itself, as presented by

Pope Gelasius, that the powers of the King as the head of state and the Pope as the head of the Church were different. This doctrine is known as the doctrine of the two swords or two powers of Gelasius. In reality, it is not easy to implement because there are things that are considered to be included in their respective powers that have been disputed, some of which are more worthy of power. As a result, Europe in the middle centuries was often linked to a record of lengthy disputes, both in thought and practice, regarding the true powers of the Church and the state, which lasted until the time that two separate powers were no longer distinguishable. Thus, with the separation between the power of the Church and the power of the state, each had a symbol of power and a complete hierarchy of officials. In the Church the Pope met with his ministerial system and in the European countries there was a King with his ministers as aides (Kuntowijoyo, 1997, p. 182).

Therefore, the actual problem of religion and the state should not exist in the Islamic world. However, in fact, the debate about the relationship between Islam and the state was a classical phenomenon that continues to this day, in a local, national and international context.

The phenomenon of the debate about the relationship between religion and the country concerns the state of life in Indonesia, and, more prominently, the role played by Islam (Thaba, 1996, p. 34). Nevertheless, Indonesia recognises the existence of five other religions, namely: Christian Catholic, Protestant, Hindu, Buddhism and Confucianism, which confirms the identity of Indonesia as a country with a diverse population in terms of religion and beliefs (Jumanta, 2015, p. 36).

The relationship between Islam and the state in Indonesia has always had its ups and downs, because the relationship between the two does not stand alone but is affected by other issues, such as political, economic, social and cultural issues. Thus, addressing the relationship between Islam and the state in Indonesia remains relevant and interesting because of the objective conditions that Indonesian with majority muslim population has made Islamic image on their identity and entity for their global interaction (Hamka, 2016, p. xxiv).

The emergence of the realisation of the demands of Islamic values in the life of the state has led to a continuing debate, so there have been many opinions issued by experts regarding the position of religion in the life of the state. Most Muslims believe that the teachings and values espoused must be upheld in the life of the society, nation and state (Thayib, 1997, p. v). In almost every phase in the history of a nation, there arise problems concerning the relationship between religion and the state, and Indonesia is no exception. As an example, debates took place before and after the proclamation of independence, and they were identified into two major groups, namely between secular nationalists and religious nationalists. Among the prominent young leaders was Sukarno, a secular nationalist, and Natsir, a religious nationalist (Islam) (Iqbal, 2014, p. 32).

The general theory about the relationship between Islam and the state is classified into three types, namely (1) Secularistic, (2) Integrality and (3) Antagonistic (Suhelmi, 2012, p. 57). One of the Indonesian Islam moderate thinkers that will be discussed in this paper is Hamka, the acronym of Haji Abdul Malik Karim Amrullah. Hamka was born in Maninjau, West Sumatra on 17 February 1908 and passed away in Jakarta on 24 July 1981 (Hamka, 2012, p. 289). Hamka was well known as a scholar, writer, poet and humanist, both by the public in Indonesia and overseas, rather than as a fighter, a political activist and a national hero.

The reason underlying the discussion of Hamka in this paper is because we still rarely find viewpoints on the political thoughts of this character active in the Indonesian Islamic party Masjumi Indonesian Council of Ulama and Muhammadiyah (Noor, 2015, p. 55). Another interesting thing is that while Islamic thinkers make a reference to Al Quran, Hamka wrote a book *Tafsir Al Azhar* (Tafseer of Al Azhar) (Ridwan, 2015). Furthermore, he was also involved in the independence struggle against the Dutch and Japanese, being critical to the regime of Sukarno and becoming a political prisoner. Then when President Suharto was in power, Hamka resigned from his position as the chairman of the Indonesian Ulema Council because of a dispute with the relevant ministers about the MUI *fatwa* (edict) on Christmas in 1981 (Shobahussurur, 2008, p. xiii).

The focus of this paper is limited to the relationship between Islam and the state; the contextualisation of Hamka's political thoughts about the state and religious morality. The purpose of religious morality is the universal values of Islam based on the Qur'an, which are used as the ethical foundation in politics. Examples include the values of equality, fairness, discussion, honesty, trust and the whole universe blessing in relationships between countries (Hamka, 2015).

2 RELIGION AND STATE RELATIONS IN INDONESIA

The discussion on the relationship between Islam and the state in the context of Indonesia was started by the founders of the nation prior to independence. Before independence on 17 August 1945, the founders of the state, the religious nationalists and secular nationalists, were involved in the debate on the basis of the philosophy and ideology of how the Indonesian state was to be established. From their debate, we realise how difficult it was to formulate the basic philosophy of the Indonesian state, as the country was composed of diverse ethnic, racial, religious and political groups. The debate about the basic philosophy of the state began in the first BPUPKI Assembly, when there were three speakers, namely M. Yamin on 29 May 1945, Mr. Supomo on 31 May 1945 and Ir. Soekarno on 1 June 1945 (Latif, 2012, p. 9). As can be seen from the speeches of the three founders of the country, the issue of the basic philosophy of the state was at the centre of the debate between nationalists and Islamic groups. At first the religious groups wanted the state to be based on the Shari'a; the nationalists, however, did not agree with the proposal and wanted Pancasila as the basis (Maarif, 2003, p. 109). Then there was the signing of an agreement, namely the Jakarta Charter on 22 June 1945, which was intended as a draft Preamble of the Constitution of the Republic of Indonesia. The different formulation of Pancasila through the Jakarta Charter was the first principle, with the formulation of the Deity with the obligation to enforce Sharia Law for adherents (Kaelan, 2009, p. 11–12).

When the nation of Indonesia proclaimed its independence on 17 August 1945, which was proclaimed by Sukarno and Mohammad Hatta, on behalf of the entire nation of Indonesia, then the PPKI (Indonesian Independence Preparatory Committee), headed by Sukarno and with Hatta as the vice chairman, started their duties. Ahead of the opening of the first official office on 18 August 18 1945, Mohammad Hatta proposed changing the design of the 1945 Constitution and its contents, which was done because of complaints from the people of East Indonesia, so that the formulation of the sentence in the Jakarta Charter 'with the obligation to run sharia Islam for its adherents' should be abolished. At the historic meeting, then through an agreement, it was approved that it should become 'Belief in one God' (Kaelan, 2009, p. 112).

The founders of Indonesia made the distinctive and innovative choice about the country's position in relation to Islam. The first principle of Pancasila, 'Belief in one God', was considered as the relationship paradigm between religion and state in Indonesia. Additionally, through a very serious discussion accompanied by high moral commitment, they came to a choice that Indonesia is based on the Oneness of God. Given the diversities of the Indonesian people and a nation that consists of various ethnic groups, ethnicities, races and religions, it seemed the founding fathers found it difficult to easily decide the state principles (Hakim & Talha, 2005, p. 1).

Indonesians believe that the independence proclaimed on 17 August 1945 was not merely due to the people's struggle. Had the God Almighty not wanted it to, it would never have happened. Thus, the birth of Indonesia has been based on the values of divinity. Furthermore, the third paragraph of the 1945 Constitution Preamble also states, 'Indonesia's independence is a blessing from Allah the Almighty'. In addition, the state guarantees religious life, such as what is contained in Article 29 of the 1945 Constitution (the Secretary General of the Assembly, 2004, p. 52). In accordance with the principle stating that 'the Country is based on the Oneness of God', religion and Islam in Indonesia have become the spirit of integrity of the Unitary Republic of Indonesia. According to Adi Sulistiyono, religion is treated as one of the ideals of state-formation (*staatsidee*) (Sulistyono, 2008, p. 3).

However, this does not mean that Indonesia is a theocratic state. The relationship between the state and religion is that of symbiotic mutualism, in which one and the other give and take. In this context, religion provides 'deep spirituality', while the state guarantees religious life. Indonesia is not a theocratic state but is based on law. Law is the commander and supreme power is above the law (Syaifuddin, 2008, p. 10).

The management of the relationship between religion and state can also be constructed on the basis of checks and balances. On the one hand, the state's hegemonic tendencies, which could easily become repressive against its citizens, must be controlled and balanced by the values of the Islam-Religion that spreads mercy to all creatures in the universe, including by upholding human rights. On the other hand, the possibilities of the Islam-Religion that are misused as authoritarian practices must also be controlled and balanced by the rules and norms of democratic life, which are guaranteed and protected by the state. Therefore, both historically and legally, the Indonesian state, in its relationship with religion, has been using the paradigm of Pancasila. According to Mahfud MD, Pancasila is a prismatic concept.

A prismatic concept is one that takes the good aspects of contradictory concepts and combines them together as a concept of its own so that it can always be actualised with the reality of Indonesian society and its continuous development. The Indonesian state is not a theocratic state, as this kind of state bases itself on only one religion, yet the Pancasila state is not a secular state either, as this kind of state does not involve any religious matters, including Islam. A Pancasila state is a religious nation state that protects and facilitates the development of all the religions embraced by its people without taking into account the numbers of their followers (Busyairi, 2015).

Muslims, as the majority in Indonesia, contribute significantly to maintaining a conducive environment towards the ideology of Pancasila. The first principle of Pancasila, 'Believe in one God', is in harmony with the view that there is no separation between religion and state affairs. Religion as a political force in Indonesia has emerged since the periods of pre-independence, throughout the Old Order, the New Order and up to the present time. The New Order is a political order that was different with the Old Order. In this discourse, the political format of the New Order attempted to create a balance between conflict and consensus (Rauf, 2001, p. 127–131). If, during the Old Order, the development was emphasised in the political field, then the New Order turned it into the economic field. In politics, the ruling regime were faced with efforts to create a new political format. This effort was practically simultaneous with the growth of people's optimism as they came out of difficult times, having the optimism for a new and improved life that was more democratic and more secure. In the following sections we will discuss Buya Hamka's thoughts concerning the state and religion and the relationships among states in the context of Indonesia as part of the international community.

3   THE RELATIONSHIP BETWEEN ISLAM AND THE STATE ACCORDING TO BUYA HAMKA

Buya Hamka views that the establishment of a state is because of common beliefs. This can mean that the elements regarding the formation of a country originate from communities that have one shared view or belief, and that their centralised focus is the aim to preserve and develop their communities. Actualization of their beliefs exist in a large domain called state. Meanwhile, the establishment of a country begins once human understand deliberation and statehood (Talha, 2005, p. 94).

Based on the above, deliberation is a basic element in the formation of a state. This is due to the rational thought that it is unlikely that a country will be formed if the inhabitants of the country never consult one another. A deliberation is where members consist of various elements or different ethnicities who can reach a collective agreement if each member has the same purpose and belief. In this regard, it is the idea to establish a state (Asy-Syawi, 1997, p. 17–21).

However, the above opinion is relatively different from the opinions of other Islamic political thinkers. Al-Ghazali, Ibn Rabi and Al-Farabi, for example, argue that human beings are social beings who have a natural tendency to socialise. This is due to the fact that they are not able to meet all their needs without the assistance or co-operation of others, and this is the rationale of a country's formation. The same thing has been stated by Al Mawardi, yet he dominantly includes this element in his theory of religion (Al Mawardi, 2013).

According to Al Mawardi, Allah the Almighty creates human beings who are not able to meet their own needs with the purpose of making us aware that He is the Creator, the Provider, and that we need Him and require His help. In the Qur'an, it is not clearly explained about the state principle, and Al Mawardi more firmly argues that the Qur'an does not give a clear definition on it, be it on the purpose and the idea of an institution, the concept of sovereignty, voting principles as the conception of human rights or as the regulations of a state (Iqbal, 2014, p. 21).

As for the relationship between Islam and the state, Hamka is of the view that religion and state are inseparable, as both need one another. The reasons are that, on the one hand, the state needs religion as the fundamental basis of its efforts to establish the morality of a nation, which is essential for the survival of a state. On the other hand, religion requires the state as the main factor of its existence and development (Fuad, 2015).

The thoughts of Hamka were influenced by the thoughts of al Mawdudi and Al Mawardi, who were of the view that the political matters of a country are ruled by Islam. Sovereignty is, therefore, not in anyone but in God the Almighty. God orders that human beings do not obey anyone, except Him. This is what is meant as a straight path. As for specific cases and pragmatic realities for other cases, Al Mawardi states a symbiotic paradigm in the context of religious laws and the state in the political sense. Both are related reciprocally, and they mutually need one another. Religion requires the state in order to develop itself and the state needs religion as a guidance for ethics and morals. In this regard, Hamka argues that a state will collapse when its people neglect what is called the Islamic values and morality, and it is actually the duty of religion to establish them in society and state (Herdiawanto, 2006, p. 94).

As stated previously, Indonesia is a country that believes in the Oneness of the Almighty, and this has been made as the first principle of Pancasila of the Constitution of the Republic of Indonesia. Therefore, any policies or regulations that are adopted and enforced must be based on the spirit of this principle, either formally or substantially (Fatwa, 2010, p. 5) The principle of the Oneness of God means absolutely that in the Republic of Indonesia there is no point of contention in terms of divinity/Islam towards the attitudes and actions of anti-divinity or anti-Islam and there is no compulsion of religion (Soeprapto, 2004, p. 11).

The relationship between religion and the state has actually existed since the 1945 Constitution, particularly Article 29. The article reveals two things: the state principle based on the Oneness of God and the freedom of religion in accordance with one's religion and beliefs (Secretary of MPR, 2002, p. 51). The article, either explicitly or implicitly, states that the link between religion and state in Indonesia is inseparable. The divinity in the first principle of Pancasila is, in fact, a subject of Indonesia's socialism, more particularly religious socialism (Sahrasad, 2000, p. 5).

4 CONCLUSIONS

The author of this study focuses on the study of Buya Hamka. In Islamic heritage, his name has gained a reputation as a great scholar and writer. His thoughts are broadly welcomed by various circles, especially among Indonesian Muslims, and he is often identified as a modernist or reformer. Hamka is also a Qu'ran interpreter (*mufassir*) in Indonesian, with his phenomenal books the *Al Azhar Tafseer*.

Among the aspects of life in Islam are the principles and ethics of life in society and the state, so this is one of the indications and evidence that Islam also regulates the social system and state, known as Islamic politics, with a variety of theories to apply the basic framework of the thoughts of the Qur'an and as-Sunnah. In the view of contemporary Islamic thinkers, modern political science is not universal, and it can even be said to be rather specific.

Hamka believed that the Qur'an does not require a separation between religion (Islam) and the state. He believed that they are mutually enhancing. One example is the concept of *shura* (deliberation). Hamka viewed that, in the Qur'an, the *shura* technique is not well described, but he explained that, depending on the circumstances, *shura* is not out of religion. Then the concept of state and head of the state that declare the creation and obligations to make a welfare state are the responsibility of a leader, and his people must have the morals of the Qur'an in their daily life. On the subject of international relations, Hamka believed that Islam does not prohibit Muslims from co-operating with non-believers, as long as they do not fight and expel the Muslims from their homeland. Regarding the political morality of religion, Hamka further highlighted the consistent attitude of political actors and political behaviour. From the above interpretation, we can conclude that Hamka wanted to reconstruct the human understanding of politics from negative to positive and to interpret the Qur'anic verses related to political discourses. He hoped that Muslims could understand that politics is noble when it is connected to a moral religion.

From the above description, it can be concluded that Indonesia upholds religion and it is positioned as the basis or rationale on matters related to the affairs of state, as evidenced by the first principle concerning the God Almighty God in Pancasila. In this case, Buya Hamka concurred on the importance of the relationship between religion and the state. Both are influential and correlated, and therefore create a seamless relationship. The thoughts of Buya Hamka have a positive relevance to the practice of statehood in Indonesia with regards to the relationship between religion and state. It is also evident that the Republic of Indonesia can accommodate religious life, which was proven by the establishment of the Ministry of Religious Affairs and the Indonesia Ulema Council in 1975.

The relationship between religion and the state is reflected both in individual life and in the life of the nation. If we observe the environmental community, it continues to increase the splendour and solemnity of Islamic activities in the form of ritual and the social form of Islam. The spirit of Islam is also reflected in the life of the state, which can be found in the public documents concerning the state philosophy of Pancasila, the 1945 Constitution and the development plans that give life and colour to the speeches of statehood.

In its implementation, the national development of religious spirit becomes stronger with the enactment of basic faith and devotion to God Almighty as one of the foundations of development. This means that all enterprises and national development activities are inspired, driven and controlled by faith and devotion to God Almighty, as the noble values that form the basis of spiritual, moral and ethical development.

ACKNOWLEDGEMENT

The writer of this study acknowledges and is grateful that part of the funding for this study is subsidised by the Institute of Research and Public Service (Lembaga Penelitian dan Pengabdian Masyarakat/LP2M) of the University of Al Azhar Indonesia.

REFERENCES

Amirudin, M. H. (2000). *The concept of religion by Fazlur Rahman* (1st ed.). Yogyakarta: UII Press.

Amrullah, Haji Abdul Malik Karim. (1981). *Magazine sections from heart to heart, can a decree be withdrawn?* Jakarta: Panji Masyarakat.

Amrullah, Haji Abdul Malik Karim. (2008). *Tafsir Al-Azhar, Al Qur'an Chapter Al Maidah Juz. VII.* Jakarta: Pustaka Panjimas.

Amrullah, Haji Abdul Malik Karim.. (2016). *History of the people of religion pre prophetic until religion in the archipelago* (5th ed.). Jakarta: Gema Insani Press.

Amrullah, Haji Abdul Malik Karim. (2013). *Father: Buya Hamka story of youth, to scholars, writers, politicians, heads of household until his death.* Jakarta: Republika.

Fatwa, A. M. (2010). Pancasila as a whole work piece held nations, non group patent. Jakarta: The Fatwa Center.

Fuad, F. (2015). Pancasila: A national vision. Jakarta: UAI Press.

Hakim, A. & Talha, M. (2005). Moral politics religious, political commentary Buya Hamka. Yogyakarta: UII Press.

Herdiawanto, H. (2006). *Religion and state in a debate in the Constituent Assembly (1956–1959); Study of political thought Buya HAMKA* (Thesis). Jakarta: intro.

Herdiawanto, H. & Jumanta. (2015). *Intelligent, critical and active citizenship, civic education for college* (5th ed.). Jakarta: publisher.

Iqbal, M. (2014). *Fiqh siyasah, contextualization of Islamic political doctrine.* Jakarta: Prenada Media Group.

Kaelan. M.S. (2004) *Pancasila education: Reform process, the national constitution amendment, Pancasila as paradigm skills, nation and state.* Yogyakart: Paradigm.

Kaelan,M.S. (2009). *State and religion relations in the perspective of philosophy of Pancasila.* Yogyakarta.

Kuntowijoyo. (1997). *Identity politics Muslims* (2nd ed.). Bandung: Mizan.

Latif, Y. (2012). *National plenary, historicity, rationality and actuality of Pancasila.* Jakarta: PT. Gramedia.

Maarif, A. S. (2003). *Agama dan Pancasila as the state, the constituent assembly debates studies.* Jakarta: LP3ES.

Mawardi, Al. (2013). Al-Ahkam As-sulthaniyyah, *The laws of the state in the implementation of Islamic law.* Jakarta: Darul Fallah.

Nasution, N. (2015). *Political dynamics of foreign affairs of Indonesia.* Jakarta: UIN Press.

Noor, W. (2015). *Solidity party split & religion in Indonesia: Case PKB and PKS the first decade of reform.* Jakarta: LIPI Press.

Rauf, M. (2001). *Political consensus, a theoretical assessment.* Jakarta: Directorate General of Higher Education.

Sahrasad, H. (2000). *Islam, socialism & capitalism.* Jakarta: Madani Press.

Shaveh, A. T. (1997). *Shura not democracy.* Jakarta: Gema Insani Press.

Syamsi, Shobahussurur. (2008) *Commemorating 100 years HAMKA, personal soft smooth character figure character, modern charismatic figures.* Jakarta: YPI Al Azhar.

Suhelmi, A. (2012). *Polemics versus state religion Natsir Soekarno.* Jakarta: UI Press.

Sulistyono, A. (2008, May 8). Religious freedom in the frame law. *Seminar papers: Law Freedom of Religious Beliefs vs. Review of the Social Perspective, Religion, and Law.* Surakarta: The Faculty of Law UNS FOSMI.

Soeprapto, R. (2004). *Responding to globalization Pancasila towards world peace, peaceful and sejahtera.* Jakarta: Yayasan Pustaka Park.

Syadzali, M. (2011). *Islam and the State administration of teachings, history and thought.* Jakarta: UI Press.

Thaba, A. A. (1996). *Religion and state* (2nd ed.). Jakarta: Gema Insani Press.

Thayib, A. (1997). *Human rights and religious pluralism.* Surabaya: Center for Strategic and Policy Studies.

The MPR-RI. (2014). *1945 Results 1999–2002 amendment of article 29, paragraph 1, Jakarta, Secretary General of the MPR.* Available from: http://legal.daily-thought.info/2010/02/relasi-negara-dan-agama-jaminan-kebebasan-beragama-antara-Indonesia-dan-amerika-serikat [Accessed 15th October 2015].

Widodo, A. L. (1999). *Fiqh Siyasah in international relations* (1st ed.). Yogyakarta: Tiara Wacana.

Interview with:

H. Anwar Ranaprawira (HAMKA's private secretary) until he passed away in 1981. (Personal communication).

MUI boards, one of them is KH. Cholil Ridwan, H. Irfan, who was so close to his thought and his figure when he led MUI, 1975–1981. (Personal communication).

H. Afif, H.Irfan HAMKA, H. Yusron HAMKA. (Personal communication).

H. Badruzaman Buyairi as the one who wrote the history of YPI Al Azhar and 70 years of HAMKA. Activists Da'wa Council, the students of Masyumi's leaders, are expected to enrich the content of this paper. (Personal communication).

Competition and Cooperation in Social and Political Sciences – Adi & Achwan (Eds)
© 2018 Taylor & Francis Group, London, ISBN 978-1-138-62676-8

Halal tourism as Japan's economic and diplomatic strategy

S.A. Wibyaninggar & S. Aminah
Japanese Area Studies, School of Strategic and Global Studies, Universitas Indonesia, Jakarta, Indonesia

ABSTRACT: Based on the fact that the halal industry has now become a trend in Japan, this study attempts to investigate halal tourism in Japan as a recent tourism strategy aimed at Muslims as its niche market. This study is based on questions regarding Japan's strategy in applying halal tourism and how this halal tourism strategy is viewed from an economic and diplomatic standpoint. This study is an analytical descriptive research, which uses a qualitative approach through a data interpretation method. This study shows that the halal tourism strategy in Japan comprises restaurants, accommodation, transportation, related associations or institutions and also information services, while maintaining Japanese local wisdom. In this case, the concept of commodification has been applied implicitly to some products or facilities related to halal tourism. Hence, it should comprise the uniqueness of Japanese culture, as a non-Muslim majority country, while developing their tourism strategy in the context of halal tourism. This halal tourism strategy has implications for Japan's economic improvement and diplomatic relations.

1 INTRODUCTION

The Japanese economy has been growing painfully slowly since the asset price bubble burst in the 1990s, leading to economic stagnation for over a decade. The high exchange rate of the Yen, demographic challenges and the effects of bad investments in the 1980s have been identified as the reasons for the stagnation of the Japanese Gross Domestic Product (GDP). This gloomy economic condition was worsened by their high import levels of foodstuffs and natural resources. This condition prompted the Japanese government to create new strategies in order to develop other industrial sectors to raise their economic growth. One promising business, which is believed to have overcome the economic stagnation and can effectively increase economic growth, is tourism.

At the tourism conference held by APEC in Peru, April 2014, Susumu Kida, as the delegate from the Japan Tourism Agency, conducted a presentation entitled *Recent Progress in Tourism Policy in Japan—Policies for Economic and Social Return*. The presentation outlined the progress in Japan's tourism policy as a means of recovering their economic and social stability. Japan has set a target to attract 20 million visitors in the year 2020, when Japan is to become the host country of the Olympics. This has driven Japan to take a look at a new potential market segment: Muslim communities. This is due to the fact that the population of the world's Muslim societies is predicted to increase by up to 23% in the next 20 years.

The emergence of the halal tourism strategy in Japan in the last few years is ocular proof of Japan's effort to expand their business into the Muslim market. As a non-Muslim majority country, this poses both opportunities and challenges: how to apply halal principles that comply with Islamic principles in the context of Japanese products. This study discusses how halal tourism was created as a strategy to boost Japan's economic improvement by increasing Muslim visitors, and how this new tourism strategy is also viewed as a political strategy towards the Muslim world.

2 THEORETICAL FRAMEWORK

This research was analysed through the theory of commodification. Commodification is a process by which objects and activities come to be evaluated primarily in terms of their *exchange value* in the context of trade (Cohen, 1988, p.380;), in addition to any useful value that such commodities might have (Watson & Kopachevsky, 1994, p.283). In terms of tourism, commodification refers to using a place's culture and cultural artefacts to make profits large enough to support the economy in particular parts of the area (Fiaux, 2010).

Interpreting tourism as a commodity also discloses that it creates an asymmetrical relation. Tourism is a structure of social relationships that creates two categories of people: those who demand, and those who supply—those who serve, and those who are served. It is not hard to see how feelings of superiority and inferiority develop in tourism relationships, and why it is the locals who, more often than not, must adapt to the tourists' wishes, demands and values, and not the other way around Such relationships need not be asymmetrical, but they often are. The closer one looks at tourism as a social activity, the clearer it appears that it inherently creates dependent relationships (Krippendorf, 1987: 54, 101–103).

Regarding the definition of halal tourism, Geetanjali (2014) states that the concept of halal, meaning 'permissible' in Arabic, is not just applied to food, but includes any *shari'ah* compliant products, which range from banking, cosmetics, vaccines and, in this case, tourism. This means offering tour packages and destinations that are particularly designed to cater to Muslim considerations and needs. Furthermore, as cited by Geetanjali, another explanation of halal is given by Karim (2014); as there is no specific definition of halal tourism, it is mainly perceived as tourism products that provide hospitality services that comply with Islamic law. For example, halal hotels do not serve alcoholic beverages, offer halal certificates for food, wellness facilities for women, prayer rooms and, in general, a Muslim-friendly environment.

3 RESEARCH METHOD

This study uses a descriptive-analytical research, using qualitative methods to answer questions about halal tourism as a business strategy in Japan in recent years. Data was gained from journals or scientific research, related books, electronic newspapers, presentations and any other relevant information collected from websites or online sites. The data that were collected, both in the form of texts or numbers, were used as material to draw evidence or information. The collected data were compiled, explained and analysed through a data interpretation approach. The tourism concept in this research was analysed using a theoretical approach related to tourism, especially commodification and halal tourism.

4 DISCUSSION

Halal Tourism Strategies in Japan. Halal tourism, which means offering tour packages and destinations that are particularly designed to cater for Muslim needs, is a relatively new concept in the tourism industry. In response to the need to implement halal tourism, Japan has started to become a Muslim-friendly country, as shown by their various strategies that are closely related to aspects of halal tourism. The aspects of halal tourism can be divided into three points: restaurants offering halal food, hotel implementation and daily prayer time.

Halal Food. Nowadays, information about halal food and Muslim-friendly restaurants can easily be found in Japan. The halal restaurants are not limited to Malay, Turkey or Pakistan restaurants, but there are also many Japanese restaurants providing halal Japanese cuisines. Japan has its own diversity in traditional food, which generally contains material from pork, and/or alcohol, such as *ramen* and *gyouza*. However, in line with their purpose to provide hospitality to Muslim visitors, there are currently many Japanese cuisines or foods that are already halal certified. Several restaurants in Japan have tried to make innovations in creating halal *ramen* and *gyouza*. If this is linked to the concept of commodification in the context of tourism, in which commodification is the process by which objects and activities come to be evaluated primarily in terms of their exchange value in the context of trade, this is explicitly in line with the halal tourism phenomenon in Japan. In this case, the objects and activities of tourism come to be evaluated through combining new innovations in the form of the halal concept, in accordance with Islamic principles, in order to meet the needs of the Muslims in Japan. It is interesting to find that the application of halal tourism has not replaced existing products, but the products have been modified and adjusted to the concept of halal tourism, so that they can be sold without reducing their unique values.

Pictogram for Food and Restaurants. To help Muslim visitors have easier access to halal food in restaurants, pictograms are provided to show the identity and facilities of the restaurants. The following Table 1 is an example of food pictograms provided by Muslim Visitors Narita International Airport brochure.

Muslim-Friendly Hotels. The hotel is another important element in tourism. In order to fulfil Muslim needs, Muslim-friendly hotels should surely meet certain criteria and facilities, such as having a prayer room, a place for ablution, a sign showing the qibla direction, possessing a halal certificate for the menus, etc. There are many online sites nowadays that provide information and lists of Muslim-friendly hotels and homestays in Japan. In addition, there are traditional Japanese inns called ryokan in Japan. Ryokan contain elements such as tatami floors, futon beds, Japanese style baths (onsen) and local cuisines. Several ryokan nowadays also offer Muslim-friendly facilities.

Muslim-Friendly Transportation. Japan's railway has a women-only carriage on its metro system, which was first introduced in 2000 in order to stop sexual harassment. Although this has nothing to do with halal tourism implicitly, the policy applied on this transportation system has helped to accommodate the needs of Muslim visitors'—especially for women. It could be seen that one of the halal tourism principles has been applied by the availability of this facility, because there is a separation between men and women. Many tour guide businesses for Muslims, which provide their Muslim customers with special buses, are also popping up. Furthermore, the main airports in Japan, such as Narita International Airport, are

Table 1.   Food pictograms.

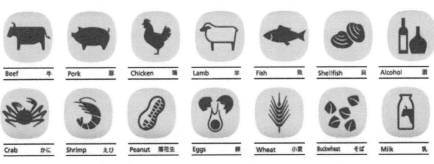

Source: Muslim Visitors Narita International Airport brochure.

already equipped with facilities for Muslim visitors, such as halal restaurants, prayer rooms and brochures containing information about Muslim-friendly bistros or restaurants.

Halal Certification. There are several associations that take part in or make a contribution to the sustainability of the halal world in Japan, including Japan Halal Association (JHA), Nippon Asia Halal Association (NAHA), Japan Halal Business Association, Japan Halal Research Institute for Product and Services (JAHARI) and Halal Development Foundation Japan Incorporated (HDFJ). They are Non-Government Organisations (NGO) and are privately managed, because there is still no organisation managed by the Japanese government dealing with halal certification management. Those NGOs are in charge of catering certification acquisition consultation, halal certificate publication, conducting various kinds of lectures and seminars regarding halal and Islam, establishing certification, developing halal-based and Muslim-friendly tourism, and also undertaking promotions through websites, social media and tourism brochures. Those NGOs are also non-profit organisations that help the establishment of halal certification in Japan. This may become another obstacle towards developing halal certification in Japan, because each organisation tends to have different standards in their screening process.

Information Services for Muslim Visitors. The Japan National Tourism Organisation (JNTO) is one of the government-owned tourism institutions that provide a special guide book for Muslim visitors. The guide book is available on their official website and can be downloaded free of charge. It is not only the JNTO who takes on the role of Japan's tourism institution and makes promotions through the media or websites. Besides the JNTO, several websites or online sites created by several cities in Japan provide information for Muslim visitors, such as Muslim-Friendly Kyoto, Hakuba Muslim-Friendly Project, etc. Other online websites can also be accessed to get further information regarding hospitality for Muslim visitors in Japan.

Halal Tourism as an Economic Strategy. The economic factor is one of the most essential points in the context of buying and selling activities, while tourism is naturally a form of business that involves demand and supply. In this case, the destination country is the host, and tourists are visitors or guests. The host offers something in accordance with the guests' demands or needs. If the rules are met, there will be many guests interested in visiting, and the national economy will grow. The international tourism industry is one of the most promising business sectors in Japan. According to Japan Macro Advisors (2015), the number of international tourists visiting Japan has significantly increased in the last decade, reaching 10 million foreign visitors for the first time in 2013.

Meanwhile, according to the Global Islamic Finance Report (2013), in *The Global Halal Industry: An Overview*, the global halal industry is estimated to make a profit of 2.3 trillion American dollars (excluding the Islamic finance industry). With the annual average growth estimated at 20%, 560 billion dollars per year, global halal tourism has a good chance of being the consumer segment with the highest growth rate in the world. This halal market is not only limited to food and everything related to food, but has already flourished beyond the food industries, to drugs, cosmetics, healthcare and the tourism products industry. In the last few years, the halal industry has developed further and is catering to a lifestyle, including a halal travel agency and service or hospitality in tourism. This development was triggered by a change of the paradigm towards Muslim consumers as a worthy market segment.

Furthermore, Muslim visitors in Japan have increased from previous years. This is proven by the increasing number of Malaysian and Indonesian visitors—the two Muslim majority countries in Asia—from 2011 to 2014. The direct contribution of the tourism industry towards Japan's GDP in 2014 was 11.900.6 billion Yen (2.4% of total GDP). This number increased to 12.287.5 billion Yen (3.3%) in 2015. From the perspective of employment, the total contribution of tourism in 2014 was 7.0% of total employment. Hence, the implementation of the halal tourism strategy is estimated to enhance the country's economy.

There are at least two main factors that have made the halal industry a global trend in recent years: the rapid growth of the Muslim population around the world and the halal rule itself, which is basically acceptable and enthused by a wider community, which is not only limited to Muslims, but also to non-Muslim consumers. In the context of halal tourism,

Japan as a non-Muslim majority country has started to expand its tourism to acquire a predicate as one of the Muslim-friendly countries.

Based on the data presented above, the remarkable contribution of the tourism industry to economic growth, and the growth of the Muslim population around the world, present a golden opportunity for Japan to create a new strategy in their tourism sector. Hence, the good prospects available in the Muslim-friendly tourism business led to Japan's new strategy to bolster their economic growth in recent years. This strategy is integrated with a number of related steps, such as the relaxation of visa requirements, the enhancement of the airlines network and a tax-free shops policy. On further interpretation, tourism is a prompt and effective tool for developing the national economy. This sector integrates many elements in social life, not only those related to certain objects in tourism, but also those involving the food and restaurant industries, accommodation, transportation, etc. Not only is it able to bring in large quantities of foreign exchange for the country, but tourism is also able to foster the creative economy.

Halal Tourism as a Diplomatic Strategy. Penn (2008) argues that Japan's attitude towards the Islamic world has a long negative history of stereotypes. In line with modernisation in the Meiji era and after, general information about Islam mostly came from the West, especially from English sources. Meanwhile, as a superpower country in the world and Japan's major ally on the international stage, the United States was considered as the centre of cultural prestige. Later on, along with US official policy, which clashed with the political movements in the Arab and Islamic world, many negative images related to Islam were published in the US media, which later found their way into the Japanese media. At the same time, when most of the Japanese people generally felt uncomfortable with everything related to Islam, at the diplomatic level the Japanese government was actually trying to create a different image. Those efforts started in January 2001 with a programme entitled Kono Initiatives, which was taken from the name of the Foreign Minister Yohei Kono, and which referred to a *Dialogue among Civilisations with the World of Islam*. The fundamental purpose of this programme was to build multilayer relations with the Persian Gulf countries through active and open dialogues among experts and academics from Japan and Islamic countries. On the other hand, the Kono Initiatives itself is closely related to an urge to secure the economic bond between Japan and those areas, in addition to increasing the intercultural understanding. This is in line with Katakura Kunio's statement (2002), which, in Japan, enhanced the dialogue with Islamic cultures and which has been given much greater priority recently as part of the country's foreign policy agenda. Many people believe that, as a country with a 'clean record' in West Asia and the Muslim countries, Japan can play a mediating role between the Islamic and the Western-Christian worlds and turn 'clash' potentials into opportunities for dialogue and co-operation. Based on Kunio's outlined argument (2002), the Japanese government and people must bear several points in mind as they develop policies towards Islam. Kunio elaborated five essential points regarding Japan's foreign policy towards Islam.

First, because Islam is not represented by any single nation or government, it is important to develop contacts and exchanges with borderless NGOs and similar networks. What is needed, in other words, is 'asymmetrical' interaction and exchange. Second, dialogue must be developed with a highly diverse and multilayered range of interlocutors, including not only states and government agencies but also private companies, academicians and research institutions, the media, women's groups and volunteer organisations. Third, Japan's strong ties with countries in the Islamic cultural sphere have so far been forged not only through trade but also through the development of energy supply systems and long-term nationally initiated oil field development projects. In light of current expectations that the Japan National Oil Corporation will be disbanded and virtually privatised in the near future, special consideration must be given to ensure that this does not weaken or reduce the exchange that has been established with Islamic oil producers in the Persian Gulf region. Fourth, as part of its internal efforts to internationalise, Japan should give due consideration to the religious beliefs of Muslim residents and travellers in Japan. This should include establishing, at airports and other public facilities, meditation/prayer rooms that are available to followers of all religious faiths, a courtesy that is gradually becoming the standard practice in other leading countries (including Japan's Asian neighbour, South Korea). Fifth, Japan must promote area studies and expertise in relevant regional languages. This effort should include the development of scholars and experts, not only in established language subjects, such as Arabic, Persian and Turkish, but also in Pashto, Uzbek, Tajiki and other regional languages of Eurasia.

Kunio (2002) emphasises that careful attention towards these five points is also crucial for developing positive interactions and dialogue with the Islamic world and Muslim people. Halal tourism, which is now blooming in Japan, and referring to the interconnection and the fraught history between Japan and the Islamic world and Muslim people in the past, could be interpreted as Japan's political and diplomatic strategy towards the Islamic world that in this phenomenon an asymmetrical relationship is occurring. Tourism provides for special types of social relationships that are a consequence of 'forced' interaction between host and guest, where the guest is a 'stranger', and the interaction is predicated on the principle of commodification. These social relations are hinted at in the interactions described in common phrases, such as 'to make tourists feel at home' and 'to provide good service'. This means that behind the 'commercialised hospitality', halal tourism cannot be separated from asymmetrical relationships between host and guest. In relation to the global Muslim market, which is growing more prospective, halal tourism, according to Kunio's suggestions, 'should give due consideration to the religious beliefs of Muslim residents and travelers in Japan'. This is one of Japan's strategies to preserve their good image in the eyes of Muslims and the Islamic world in general. This has important implications for Japan's diplomatic interests, and of course this is also closely related to Japan's economic interests.

5 CONCLUSION

Tourism is one of the industries that is able to have a direct effect on the global economy. The tourism industry is one of the largest sectors in Japan's economy. Japan has great potential tourist attractions, including nature, culture and technology. Along with the increasing global Muslim population, Japan has realised that Muslim tourists could be a large potential market that could improve Japan's tourism industry, and, with this, the halal tourism strategy becomes important. The halal tourism strategy in Japan involves food and restaurants, accommodation and transportation, halal certification through various organisations or associations, and information services. In its implementation, halal tourism in Japan has its own uniqueness, evidenced by several modified Japanese products, so that these can be accepted by the Muslim consumers, while, at the same time, retaining their attached cultural values or local wisdom.

The economic factor is one inevitable reason for this. The halal tourism strategy has had an effect on the number of Muslim visitors. The number of Muslim visitors to Japan has increased further in recent years, proven by the increasing number of visitors from Muslim majority countries. The direct contribution of the tourism industry towards the country's GDP has also increased. Hence, the implementation of the halal tourism strategy is estimated to enhance the country's economy. In terms of diplomacy and politics, Japan is actually showing their public diplomacy or soft-power to the international community, with Muslims as the main target. Therefore, this is also one of Japan's efforts to maintain a good image and good relationships with Muslim countries, which has implications for political interests and is also closely related to economic interests.

REFERENCES

Battour, M., et al. (2010). Toward a halal tourism market. *Tourism Analysis, 15.*
Cohen, E. (1984). The sociology of tourism: Approaches, issues, and findings. *Annual Review of Sociology, 10*, 373–392.
——. (1988). *Authenticity and commoditization in tourism.* In Annals of Tourism Research 15 (3) 371–386.

Directorate for Science, Technology and Industry. (2002). *National tourism policy review of Japan*. Retrieved from: https://www.oecd.org/japan/33649824.pdf

Eat Halal. (2013). *Kyoto launches new website for Muslim visitors*. Retrieved from: http://www.eat-halal.com/japan-kyoto-launches-new-website-muslim-visitors/

Eat & Stay. (n.d). *Kansai Muslim friendly guide*. Retrieved from: http://muslim-friendly.jp.net/

Fiaux, N. (2010). *Culture for sale: Commoditisation in tourism. Master study in tourism destination management*. N.p., 31 Oct. 2010. Retrieved from: http://www.tourism-master.nl/2010/10/31/culture-for-sale-commoditisation-in-tourism

Geetanjali, RameshChandra. (2014, December). Halal Tourism: A new goldmine for tourism. In International Journal of Business Management & Research (IJBMR). Vol. 4 Issue 6, Dec 2014, 45–62.

Global Islamic Finance Report (GIFR). (2013). *The global halal industry: An overview*. Retrieved from: http://www.gifr.net/gifr2013/ch_13.

Hirota, R. (2004). Air-Rail links in Japan: Present situation and future trends. *Japan Railways and Transport Review, 39*.

Japan Halal Research Institute for Products and Services. (2016). Retrieved from: http://www.ja hari.or.jp/

Japan Halal Summit 2014. (2014). Retrieved from http://www.japanhalal-summit.jp/en_hd fj.html

Japan Macro Advisors. (2015). *Foreign visitors to Japan doubled in the last decade*. Retrieved from: https://www.japanmacroadvisors.com/public/uploads/ckImages/files/Foreign%20visitors%20to%20Japan%20doubled%20 in%20the%20last%20decade.pdf

Japan National Tourism Organization. (n.d.). *Japan welcome guide for Muslim visitors*. Retrieved from: http://muslimguide.jnto.go.jp/eng/

Japan Tourism Agency. (2014). *White paper on tourism in Japan: The tourism situation in FY2013*. Retrieved from http://www.mlit.go.jp/kankocho/en/siryou/whitepaper.html

Karim, Saad. (2014). *Halal tourism: Way to muslim friendly environment and industry that goes mainstream*. Feb 13th 2014.

Kida, S. (2014, April). *Recent progress in tourism policy in Japan: Policies for economic and social return*. Presented at the Asia-Pacific Economic Cooperation in 44th Tourism Working Group Meeting. Peru.

Krippendorff K. (1987). Paradigms for communication and development with emphasis on autopoiesis. In: Kincaid D.L. (ed.) Communication theory: Eastern and western perspectives. Academic Press, San Diego CA: 189–208. Available at http://cepa.info/3042

Kunio, K. (2002, February). Japan's policy on Islam: Rethinking the dialogue approach (Nippon no Isuramu seisaku: 'Taiwa'seisaku minaoshi e no teigen). Gaiko Forum No. 163.

Mizuho Economic Impact & Analysis. (2014). *The economic impact of the 2020 Tokyo Olympic Games*. Mizuho Research Institute.

Muslim Friendly Kyoto: Kyoto Travel Guide. (n.d.). Retrieved from: http://kyoto.trave l/muslim /en

Muslim Friendly Project in Japan. (n.d.). Title?. Retrieved from: http://muslim-friendly-japan.com

New York Daily News. (2014, July 10). *Halal tourism growing in Japan as country tries to attract Muslim visitors*. Retrieved from: http://www.nydailynews.com/life-style/eats/-halal-tourism-expanding-japan-article-1.1861725

Penn, Michael. (2008, January). *Public faces and private spaces: Islam in the Japanese Context*. In Roundtable: Islam in Japan: A Cause for Concern? By Emile A. Nakhleh Keiko Sakurai Michael Penn, Asia policy, number 5 (January 2008), 61–104. http://asiapolicy.nbr.org.

Pew Research Center Forum on Religion & Public Life. (2011). *The future of global Muslim population: Projections for 2010–2030*. Washington D.C.: Pew Templeton Global Religious Future Project.

Satoyumukashibanashi Yuzanso Muslim Friendly. (n.d). Retrieved from: http://www.yuzanso.-co.jp/english/muslim/

Sendai Tourism Website. (2015). *Information for Muslim and vegetarian visitors*. Retrieved from: http://www.sentabi.jp/en/info/halal.html

Shepherd, R. (2002). *Commodification, culture and tourism*. Department of Antropology, The George Washington University.

Watson, G.L. & Kopachevsky, J.P. (1994). Interpretation of tourism as commodity. In Y. Apostolopoulos, S. Leivadi & A. Yiannakis (Eds.), (1996). *The sociology of tourism: Theoretical and empirical investigation*. London: Routledge.

World Travel & Tourism Council. (2015). *Travel and tourism: Economic impact 2015*. Japan.

Indonesian women in transnational marriages: A gender perspective on their struggles, negotiations, and transformations on ownership rights as Indonesian citizens

R. Prihatiningsih, S. Irianto & S. Adelina
Department of Gender Studies, School of Strategic and Global Studies, Universitas Indonesia, Jakarta, Indonesia

ABSTRACT: One of the impacts of globalisation is the increase of movement of people which leads to the rise of transnational marriages, in this case between Indonesian citizens and foreign nationals. Indonesian laws, which are mostly developed from a patriarchal perspective, have generally disadvantaged women, who become 'invisible' in respect to their rights as Indonesian citizens. This situation has improved, but discriminative laws against Indonesian women who are married to foreigners still exist. This paper examines the discrimination against Indonesian women married to foreigners, in their application for land ownership rights. Women have different rights, and they are not 'equal before the law' in Indonesia because of their transnational marital status. These women stand to lose their land ownership rights unless a notarised prenuptial agreement has been established prior to their marriage. With the focus on women as the subjects, this research adopts multicultural feminism and feminist legal theory perspectives, and applies qualitative methods which include in-depth interviews and participant observation. As a researcher and an Indonesian woman who is married to a foreigner, the author is also involved in the community and is experiencing the direct implication of the cause. By presenting the insights and experiences of Indonesian women as part of the women's movement, this research may contribute to changing the existing discriminative laws in Indonesia, and bring the laws in line with changes caused by the rapid mobility of people as part of the globalisation.

1 INTRODUCTION

The movement of people in and out of a country, to get a better life or a safer place to live in a new country, causes major changes such as 'demographic growth, technological change, political conflict and warfare' as pointed out by Castles and Miller (2003). This movement of people is either peaceful or forced, and fundamentally affects the economic, political and social conditions as well as the legal systems, as Castles and Miller claimed (2003). One of the impacts of today's rapid movement of people in Indonesia is that of transnational marriages between Indonesian citizens and foreigners. As stipulated in *Undang-Undang Perkawinan*, the Indonesian Marriage Law No. 1 of 1974, 'a transnational marriage is a marriage between two persons, who reside in Indonesia and are subject to different laws because of their different citizenship—one of them is Indonesian, and the other is a foreigner'.

It is stated in the United Nations research paper *Convention on the nationality of married women: Historical background and commentary* (United Nations doc. E/CN.6/389, Sales No. 62.IV.3. p. 2) that 'the nationality laws of many countries disadvantage women'. This is certainly the case in Indonesia, with its strong patriarchal culture where adult men control and dominate women and children.

Beckman and Beckman (2009) stated that the transnational family becomes an integral part of the globalising process in a globalising world thus [it] consists of a great number of chains of interactions in which legal forms are reproduced, changed and hybridised simultaneously in different contexts of interactions' (p. 11). In line with the recent global development, some Indonesian laws have been revoked or amended. However, some laws related to this research, for example the Law of the Republic of Indonesia Number 1 of the Year 1974 on Marriage, and the Law of the Republic of Indonesia Number 5 of the Year 1960 on Basic Regulations on Agrarian Principles, have not yet been changed. There are legal interpretations or understandings on the land ownership rights of Indonesian citizens who marry foreigners, which are discriminative since they have to set a prenuptial agreement prior to their marriage as an extra-legal instrument to access their ownership rights as Indonesians, although these requirements are in conflict with statements of the Indonesian government which has proclaimed adherence to 'respect, protect and preserve human rights'. The prenuptial agreement is mainly about the separation of mutual assets between the husband and the wife, intending to prevent foreigners from accessing the land and property ownership rights. However, such an extra provision or condition discriminates against Indonesians who marry non-Indonesians, in accessing their own land and property ownership rights. On the other hand, such an extra-legal instrument cannot be applied to every Indonesian woman. It may be a dilemma and cause hardship to Indonesian women, who will have economic impacts when facing separation, divorce, annulment or the spouse's death.

This paper focuses on the discrimination against Indonesian women who are married to non-Indonesians, to get Indonesian land and property ownerships rights. This research uses multicultural feminism and feminist legal theory as the conceptual framework. Multicultural feminism aims to understand that women are not a homogenous group, as highlighted by bell hook (Gloria Watkins), and feminist legal theory aims to understand how equality before the law works and is practised through the perspectives and life experiences of six women who differ in race, class, education, spouses' nationality, ethnicity, and age. The data is gathered through in-depth interview and participant observation. They discussed their agony and dilemma of a prenuptial agreement separating marital assets between the husband and the wife.

This article aims to present the voices and the struggle of Indonesian women. There are three main research questions. The first question is how their position is as Indonesian citizens and why they have different rights. Secondly, what strategies they use to cope with the discrimination in getting land ownership rights. The third question is how they struggle to change and voice the discriminative situation.

Based on these questions, the objectives of the research are: 1) to give the public awareness about the equality before the law on land and property ownership rights, which are not applied equally to every Indonesian citizen; and 2) to provide women's insights and voices through Indonesian women's experiences, as well as to provide a recommendation to the policy makers to change the discriminative laws.

In Indonesian history, the laws on marriage are divided into two divisions: before 1974 and after 1974.

2 THE LEGAL POSITION OF INDONESIAN WOMEN IN A TRANSNATIONAL MARRIAGE: BEFORE 1974

Bedner and Van Huis (2010) identified four civil groups in Indonesian society as detailed in the first Indonesian Marriage Law, effective between 1945 (the year of Indonesian independence) and 1974, when these laws were replaced. These groups, ruled by different legal regimes, were derived from the legal citizenship framework from the Dutch East Indies, or according to Bedner and Van Huis, they were 'inherited from colonial state' (p. 177).

> The first was that of the Book of the Civil Code that applied to the European part of the population (including the Japanese) and those who had been equated with the European in certain civil matters—mainly the Chinese. The second group was the Ordinance for Christian Indonesians (S. 1933 no. 74), for whom special records were kept by the colonial state. The third and most important group was subject to the regime for the Indonesians: either *adat* (customary law) or Islamic law applied

to those Indonesians and non-Chinese Foreign Orientals who were not equated with the European group. This group concerned more than 90% of the population of the Dutch East Indies. Finally, there was a special regulation on 'mixed marriages' [according to the Dutch colonial Law S.1898 no. 158 ed.] for those who were subject to different regimes but still wanted to marry.

Article 2 of S. 1898/no 158 of the Dutch Colonial Law states that '*Istri yang melakukan perkawinan campuran, selama dalam perkawinannya mengikuti kedudukan suaminya dalam hukum publik dan hukum perdata*' (*A wife in transnational marriage shall be dependent on her husband's status in the public and civil laws*). Consequently, an Indonesian woman in a transnational marriage before 1974 did not have her own legal status. She lost her own identity publicly and privately in legal matters as she became dependent on her husband's identity. She became 'invisible' as a citizen, and she was not eligible to rights and obligations as an Indonesian. To put all of these in a broader context, many countries based their legal constitution on 'patriarchy'. This has been well explained in the 'Women 2000 and beyond' (United Nations, 2003, p. 5):

> Historically, many States, at the beginning of the 20th century, adopted the patriarchal position that a woman's legal status is acquired through her relationship to a man-first her father and then her husband. .... The result of the application of this principle [dependent nationality] was that of a woman who married a foreigner automatically acquired the nationality of her husband upon marriage—and that was then usually accompanied by the loss of her own nationality. The rationale for the principle of dependent nationality was derived from two underlying assumptions: firstly, that all members of a family should have the same nationality and, secondly, that important decisions affecting the family would be made by the husband.

In summary, the principle of dependent nationality in the pre-1974 Marriage Law (article S.1898 no. 158), which was also confirmed in the Indonesian Citizenship Law No. 62 of 1958, caused Indonesian women in transnational marriage to lose their citizenship and be treated as foreigners. Indonesian women then did not have the equal rights and opportunities to the rights and obligations, as did the Indonesian men.

3 MAJOR LEGAL CHANGES IN THE POSITION OF THE INDONESIAN WOMEN IN TRANSNATIONAL MARRIAGES AFTER 1974

A variety of laws related to women and their children have been changed over the last 40 years, because of the pressure from activists of women's rights movements and their demands for equal rights for women. For the purpose of this research, the laws that are important are the new Marriage Law, decreed in 1974, and the new Citizenship Law, established in 2006. The Marriage Law No. 1 of 1974 (the Marriage Law) was a major breakthrough as it did away with (article 66) the four (colonial) regimes of the old law and established (article 31) the independent legal status and capacity of married women. Article 66 reads as follows:

> With this coming into force of this Law, in regard of marriages and any matters related thereto by virtue of this Law the provisions contained in the Civil Code, the Ordnance on Marriage of Christians Indonesians (S.1933. No. 74) and the Regulation on Mixed Marriages (S.1898 No. 158) shall cease to exist and other regulations shall be further laid down in Government Regulations.

In article 31 it is stipulated that '(1). The rights and position of the wife are equal to those of the husband both in family and in social life. (2) Either party to the marriage has legal capacity. (3) The husband is the head of the family, and the wife is a housewife the mother.' The 1974 Marriage Law also explicitly states that Indonesians who are married to foreigners would hold their Indonesian nationality. However, it did not address discrimination against transnational marriages, which are outlined in other laws such as the land ownership rights

(Agrarian Law) and the nationality status of the children from transnational marriages (Citizenship Law).

The Indonesian Citizenship Law No. 62 of 1958 was not in accordance with basic human rights and equality for women and children as stated in the Universal Declaration of Human Rights and the Convention on the Elimination of All Forms of Discrimination Against Women (CEDAW). After major pressures from local women's rights groups, particularly the transnational marriage advocacy organisations, the law was revoked in 2006 and replaced by the Indonesian Citizenship Law No. 12 of 2006. In its elucidation, this new law adheres to principles as follows:

1. '*Ius Sanguinis*': the principle that determines citizenship based on hereditary and not by place of birth;
2. Restricted '*Ius Soli*': a principle that determines citizenship based on place of birth, restricted to and only applicable for children according to the provisions set forth in this law;
3. Single Citizenship: the principle that determines only one citizenship for a person;
4. Restricted Dual Citizenship: the principle of limited dual citizenship for children as set forth by the provisions in this law.

The Law of 2006 was a major improvement for the children of transnational marriages. While previously they automatically—upon birth—acquired the nationality of their foreign fathers, they are now entitled to limited dual citizenship—limited in the sense that they can have both parents' nationality (Indonesian and foreign) until the age of 18. Thereafter, within three years, they have to choose only one nationality.

However, as of today, various laws and regulations that discriminate against Indonesian citizens married to foreigners still remain. The most conflicting ones are the laws regulating land ownership, which is the subject of this research paper.

4 LAND OWNERSHIP RIGHTS FOR INDONESIANS MARRIED TO FOREIGNERS

Land ownership is primarily affected by two major laws. They are the Marriage Law and the Agrarian Law. These laws as well as their revisions are not harmonised. These lead to conflicting and ambiguous legal interpretations towards ownership rights, particularly for Indonesian citizens married to foreigners. The key articles in the Marriage Law are articles 29 and 35, respectively outlining the 'Prenuptial Agreement' and the 'Marriage Property'. Article 29 states that:

1. At the time of or prior to the marriage performance, both parties may by mutual consent conclude a prenuptial agreement in writing, legalised by the registrar of marriage, where upon the contents shall also be binding on third parties in so far as third parties affected.
2. The contract cannot be legalised if contrary to restrictions set by the law, religion or morality.
3. The prenuptial contract takes effect as from the marriage being concluded.
4. Throughout the marriage, the contract cannot be changed except by mutual agreement between the parties, provided the changes are not prejudicial to the interest of third parties.

Article 35 states that:

1. Property acquired during marriage shall become joint property.
2. Property brought into the marriage by the husband and the wife respectively and property acquired by either of them as a gift or inheritance shall remain under their respective control, unless otherwise decided between the parties.

The relevant articles in the Agrarian Law are articles 21 (*hak milik*: right of land ownership) and article 36 (*hak guna bangunan*: right to construct and own buildings). Article 21 states that:

1. Only Indonesian citizens can have a *hak milik*.
2. The government is to determine which corporate entities can have a *hak milik* and the condition thereof.
3. A foreigner who acquires a *hak milik* through inheritance without a will or by way of joint ownership of property resulting from marriage and Indonesian citizen holding a *hak milik* who following the entry into force of this Act, loses Indonesian citizenship is obliged to relinquish that right within one year following the date the *hak milik* is acquired in the case of the latter. If following the expiry of the said time periods, the right is not relinquished, then the said right is nullified for the sake of law and the land falls to the state with the proviso that the rights of other parties which encumber the lands remain in existence.
4. A person with Indonesian citizenship, who at the same time holds foreign citizenship, cannot have land with the status of a *hak milik*, and to him/her the provision as described in paragraph 3 of this article shall apply.

Article 36 of the Agrarian Law states that:

1. Eligible for *hak guna bangunan* (the right to construct and possess buildings on land which is not one's own for a period of at most 30 years) are:
   a. Indonesian citizens (foreigners are prohibited from *hak guna bangunan*)
   b. Corporate entities incorporated under Indonesian law and domiciled in Indonesia.
2. A person or corporate entity which holds a *hak guna bangunan* but no longer fulfils the conditions which are referred to in paragraph (1) of this article is obliged to relinquish, within one year, the *hak guna bangunan* concerned, or to transfer it to another party which is eligible for such a right. This provision applies also to parties who obtain a *hak guna bangunan* in question and it is not relinquished or transferred within the said period of time, it is nullified for the sake of law—with the proviso that the rights of other parties given attention, in accordance with—provisions that are to be stipulated in a Government Regulation.

Based on the above articles of the Marriage Law and the Agrarian Law, the current legal interpretation is that Indonesian citizens married to foreigners do not have land ownership rights—*hak milik* as well as *hak guna bangunan*—UNLESS they have established a prenuptial agreement prior to marrying the foreigners. This legal interpretation is not in line with the following Indonesian laws. They, as stated in article 31 of the Marriage Law, convey the rights and positions of the wife which are equal to those of the husband. In article 23 of Citizenship Law number 12 of the year 2006, it is stated that the Indonesian women who are married to foreigners do not lose their citizenship rights. In article 9 of the Basic Agrarian Law, it is stated that Indonesian men and women equally have the right of land ownership.

The above interpretation unfortunately has recently been affirmed by the issuance of Government Regulation No. 103 of 2015 on the Ownership Rights of Residential Housing by Non-Indonesians Residing in Indonesia.

5 THE PRENUPTIAL AGREEMENT: ISSUES AND UNCERTAINTIES

The above current legal interpretation on land ownership for women in transnational marriages seems to be, in the author's opinion, legally ambiguous because Indonesian women in transnational marriages do not lose their citizenship according to the Indonesian Citizenship Law No. 12 of 2006, and as Indonesians, they would then have the right to land ownership according to article 9 of the Basic Agrarian Law as *lex spesialis*. Therefore, using the prenuptial agreement as an extra condition for the right to land ownership for women in transnational marriages should not be required.

In the author's opinion, 'the requirement of a prenuptial agreement' is inconsistent with those provisions stating 'equality before the law', stated in the Indonesian Constitution, *Undang-Undang Dasar* of 1945, as *lex generalis*. It is, therefore, deplorable that even in 2015 the Indonesian government held on to those unjust conditions in its Regulation No. 103 of 2015 that outlines the rules for Ownership Rights of Residential Housing for Non-Indonesians Residing in Indonesia. There are also legal inconsistencies in the 1974 Marriage Law between article 21 concerning the prenuptial agreement and article 35 concerning the rules for giving and inheriting property.

Additionally, a prenuptial agreement may present an Indonesian woman with a dilemma, having to choose between her Indonesian ownership rights or the joint marital property built and provided by her foreign husband. A prenuptial agreement may cause more hardship for women who do not have their own incomes and who are solely dependent on their spouses, as they cannot have access to their joint properties made and provided by their working spouses as breadwinners. In short, a prenuptial agreement cannot be applied to all women.

Finally, a major issue is that the concept of a prenuptial agreement is not very well known or culturally accepted in Indonesia. As a result, most partners in transnational marriages have no idea that a prenuptial agreement is a 'must' to retain the land ownership rights of the Indonesian partner—this phenomenon is in line with observations in the United Nations report (2003, p. 3), more specifically '… Often laws are rarely simple or comprehensive and their technical nature makes them inaccessible to many people'.

6 THE INDONESIAN WOMEN IN TRANSNATIONAL MARRIAGES—THEIR STRUGGLES AS THE 'OTHER' IN COPING WITH DISCRIMINATIONS AND THEIR TRANSFORMATIONS

Indonesia is one of the countries which was previously colonised by the Dutch and then by the Japanese. After its independence, the Indonesian people have had the strong force to establish in the name of 'nationality', referring to only Indonesian people who have access to the soil, water, and airspace within the territory of the Republic of Indonesia. Based on the nationality principle, expatriates/foreigners are not allowed to have *hak milik*; they can only have *hak pakai* (right to use for a certain period of time).

The provisions to establish a prenuptial agreement to access the right of ownership for Indonesians in transnational marriages is meant to prevent foreigners from accessing the ownership rights. As a result, the Indonesian citizens are not treated equally, because of their marital status. They are treated differently when they do not have a prenuptial agreement. The state will regard them as the 'other', as described vividly by Simone de Beauvoir (1949) where a woman is to be regarded as the 'other'. The same situation is experienced by Indonesian citizens in transnational marriages who are not regarded as members of the Indonesian community, to exercise their full status as citizens; they are required to fulfil their obligations as Indonesians. As Irianto (2008, p. 28) identified from a point of the feminist legal theory's perspectives, implications on gender from that seemingly neutral law as stated in the Government Regulations No. 103 of 2015 to acquire a prenuptial agreement for both Indonesian men and women to access the ownership rights, may cause more hardship to Indonesian women than to Indonesian men in a patriarchal society. By presenting the women's points of view and insights below, this paper may influence to change the law in accordance with 'equality before the law'.

Aside from the legal and practical issues, the Indonesian women in transnational marriages have to continue to live and cope with the situation. They need to find a way to establish a place that they can call home in Indonesia. The author conducted in-depth interviews with six Indonesian women married to foreigners, to explore their perspective on land ownerships. They are Ibu Yuyun who is married to a French man, Ibu Yessi who is the wife of a Spanish man, Ibu Nuning who is married to an American man, Ibu Inge who is married to a German man, Ibu Dias who is married to a Japanese man, and Ibu Ike Farida who is also married to a Japanese man. Their real names are not used in order to protect their identity, in line with the 'Ethics of Research', except for Ibu Nuning and Ibu Ike—these are their real names—as they are already well known publicly.

With regard to citizenship, the author agrees with Kusuma and Effendi (2002, p. vii) that citizenship comes with rights and obligations, specifically:

> Citizens as collective populations are one of the principles and basic elements of a state. Citizenship status shall create reciprocal relationship between the citizen and

his/her country. Every citizen has the rights and obligations toward their country. In return, the state has obligations to render protection towards its citizen.

In line with the above, the author's six 'research subjects' are all exemplary Indonesian women and citizens. They obey and follow the laws and regulations in Indonesia and fulfil their obligations, such as by paying taxes, and officially getting their driving licence in order to drive responsibly and safely. In contrast, Indonesia the state, has not fully provided their full rights on land ownership. All the research subjects do not have a prenuptial agreement. Consequently, they struggle in many different ways to be allowed to own land. Some have even been urged to divorce first and then remarry, after they make a prenuptial agreement. Another woman has fought for her rights up to the Constitutional Court.

Ibu Yuyun is originally from Tanjung Uban, Riau, Sumatera. She is 50 years old. Her husband is French. She got her Master's Degree in Engineering from the University of Pau and Pays de l'Adour in the city of Pau in France. When she was in France, she and her husband bought a house there, where they resided without any difficulties. However, when she was about to buy land in Indonesia, she was informed that she was required to have a prenuptial agreement. She was not aware of this, thinking that she was legally entitled to land ownership as she was still holding Indonesian citizenship. To solve this issue, Ibu Yuyun now has her sister buy land on her behalf. She does not want to make a prenuptial agreement as she thinks that it is not in her favour in the event of divorce or the death of her husband.

It is different from Ibu Yessi, a housewife who is 64 years old and originally from Bukittinggi, Sumatera. She is married to a Spanish man and is the mother of three adult children. She completed her high school and then worked as a cashier in a casino, where she met her husband in 1974. Both of them worked very hard, but at that time they could not afford to buy a house for themselves and their three children. They lived with her mother instead, occupying one of her two-bedroom houses, because they did not want to waste money by always renting a house; instead they wanted to save money to have their own house. After going through the hardship of losing and finding jobs and trying to save some money for many years, she and her husband could finally afford to buy a simple three-bedroom house, enough for the five of them. When she was about to finalise the process in acquiring the land, she was also shocked when she was asked about the prenuptial agreement by a land officer from the Agrarian Office. Her husband was the one who reminded the Agrarian officer that his wife, Ibu Yessi, is Indonesian and has no other citizenship. She was then asked by the officer of the Agrarian Office to make a statement that she is still Indonesian. After signing the statement, she was granted her land ownership rights in 1991.

Ibu Inge, who is from West Java, Indonesia, is married to a German. She was formerly a stewardess after completing her high school. She is now 45 years old, a housewife, and a mother of a 12-year-old daughter. They do not want to be bothered with potential issues regarding owning land and property in Indonesia, so they keep renting an apartment in Jakarta, and as investments, they bought apartments in Singapore, Bangkok and Germany under her own and her husband's names on those properties, which is not possible in Indonesia.

Ibu Dias is originally from Central Java. After finishing her vocational studies majoring in accountancy (under a bachelor degree), she worked in a Japanese restaurant as a cashier, where she met her Japanese husband who was as a cook there. She is now 53 years old, a business woman, and a mother of one daughter. She and her husband work hard and have built their restaurant business from scratch. Now they own three Japanese restaurants. For various reasons, she and her husband do not want a prenuptial agreement. In order to own land and property, she has made a fake ID stating that she is a single.

Ibu Nuning, 43 years old, is a PhD candidate, a professional, and an activist. She is married to an American and is the mother of three daughters. She does not have a prenuptial agreement as she was not aware of such regulation at the time of her marriage—she got married at a young age and at the time would not have had anything to put in the agreement.

Ibu Ike Farida is Sundanese and from West Java. She is a lawyer, 46 years old, and married to a Japanese citizen. She did not want to take illegal steps such as using a fake ID or getting divorced and then remarrying after establishing the prenuptial agreement. After learning that the developer backed away from his promise to give her the ownership documents, although he had been fully paid for them by her, she decided to go to court. Her case was dismissed in the East Jakarta District Court.

All these six women reside in Jakarta. They struggle, negotiate, and transform in different ways to claim their land ownership rights. They are all Indonesian women but differ from each other in terms of race, class, and ethnicity, as acknowledged by multicultural feminists. They have their own identity but like to be united in forming a sisterhood, the idea of hooks and Lorde, as pointed out by Tong (2009), *'in the sense of being political comrades'* (p. 235) in order to voice the same goal (i.e. to exercise their land ownership rights). Initiated by Ibu Ike Farida and supported by women in the transnational marriage advocacy organisations, the Indonesian women in transnational marriage continue fighting and bringing the issue to the Constitutional Court, requesting amendment of the discriminative regulations on land-ownership for Indonesian citizens married to foreigners. On 27 October 2016, based on the request of Ike Farida, the Constitutional Court decided to eliminate the requirement for a prenuptial agreement in article 29 of the 1974 Marriage Law. It is stated that the transnational marriage couple can now create a separation of marital assets and debts agreement, not only prior to their marriage but also during the marriage. However, Ike Farida other pleas, a review of Article 21 and Article 36 of the 1960 Agrarian Law, and article 35 of the 1974 Marriage Law have not been granted.

Now the transnational marriage advocacy organisations also voice these requests to the Indonesian House of Representatives.

7 CONCLUSIONS

In line with recent global developments and influenced by major pressures from local women's rights groups, some Indonesian laws have been revoked and amended with the principles to 'respect, protect, and preserve human rights'. However, various laws and regulations that discriminate against Indonesian citizens married to foreigners, in this research concerning rights on land ownership, still remain unless they have established a prenuptial agreement that may pose Indonesian women with a dilemma, as whether to choose access to marital assets or to Indonesian ownership rights. This law is not in accordance with 'equality before law' as they are still Indonesians; they do not own any other citizenship. They do and obey the rule and obligations.

Because of their marital status to non-Indonesian, the Indonesian government imposes (a prenuptial or a marriage) agreement that stipulated the separation of property ownership as an extra condition to access their ownership rights. Their rights in accessing the ownership rights are not given similarly as they are to other Indonesians who are married to Indonesians. Therefore, the Indonesian women should unite, be aware, study, advocate, and build strong networks between women within and outside the country, in order to make their voices so strong and loud that the policy makers can hear and will change the discriminative laws.

REFERENCES

Beauvoir, Simone de. (1949). Le deuxieme sexe. Paris: Editions Gallimard. *The Second Sex,* trans. Constance Borde and Sheila Malovany-Chevallier. New York: Vintage Books, 2009.
Bedner, A. & Van Huis, S. (2010). Plurality of marriage law and marriage registration for Muslims in Indonesia: A plea for pragmatism. *Utrecht Law Review, 6*(2), 175–182.
Benda-Beckmann, F von & Benda-Beckmann K von (2009). Transnationalisation of Law, Globalisation and Legal Pluralism: a Legal Anthropological Perspective. In Sulistyowati Irianto (Ed.), *Hukum Yang Bergerak: Tinjauan Antropologi Hukum* (pp. 11). Jakarta: Yayasan Obor Indonesia.
Castles, S. & Miller, S.C. (2003). *The age of migration* (3rd ed.). Hampshire: Palgrave Macmillan.

Irianto, S. (2008). Mempersoalkan 'Netralitas' dan 'Objektivitas' Hukum: Sebuah Pengalaman Perempuan. In S. Irianto & S. Irianto (Eds.), *Perempuan dan Hukum: Menuju Hukum yang Berperspektif Kesetaraan dan Keadilan*. Jakarta: Yayasan Obor Indonesia.

Kusuma, Indradi M and Wahyu Effendi. (2002). *Kewarganegaraan Indonesia: Catatan Kritis atas Hak Asasi Manusia dan Institusionalisasi Diskriminasi Warga Negara*. (Prasetyadji, Ed.) Jakarta, Forum Komunikasi Kesatuan Bangsa (FKKBI) & Gerakan Perjuangan Anti Diskriminasi (GANDI).

Tong, R. (2009). *Feminist thought: A more comprehensive introduction*. Colorado, USA: Westview Press.

United Nations. (2003). Division for the Advancement of Women, Department of Economic and Social Affairs. (2003, June) Women, nationality and citizenship. *Women 2000 and beyond* (pp. 7–17).

Yulianie, N. (2012). *Upaya Perlindungan Hukum Bagi Istri Warga Negara Indonesia Yang Melangsungkan Perkawinan Campuran*. Depok, Program Studi Magister Kenotariatan, Fakultas Hukum, Universitas Indonesia.

Laws/Regulations:

Kerajaan Belanda. (1898). *Staatsblad 1898 No. 158 Regeling op de Gemende Huwelijken (Gemende Huwelijken Regeling)*. Hindia Belanda, Keputusan Raja 29 Desember 1896 Nomor 23.

Pemerintah Indonesia. (1958). *Undang-Undang Republik Indonesia Nomor 62 Tahun 1958 tentang Kewarganegaraan Republik Indonesia*. Jakarta, Pemerintah Indonesia.

Pemerintah Indonesia. (2015). *Peraturan Pemerintah Republik Indonesia Nomor 103 Tahun 2015 Tentang Pemilikan Rumah Tempat Tinggal atau Hunian oleh Orang Asing yang Berkedudukan di Indonesia*. Jakarta: Pemerintah Indonesia.

The Republic of Indonesia. (1974). *Law of the Republic of Indonesia Number 1 of the Year 1974 on Marriage*. Jakarta, State Gazette of the Republic of Indonesia of the Year 1974 No. 1.

The Republic of Indonesia. (2006). *Law of the Republic of Indonesia Number 12 Year 2006 Regarding Citizenship of the Republic of Indonesia*. Jakarta.

Competition and Cooperation in Social and Political Sciences – Adi & Achwan (Eds)
© 2018 Taylor & Francis Group, London, ISBN 978-1-138-62676-8

The struggle of women victims of child marriage in constructing the power of their body and life (case study in Ciasihan Village, Bogor District, West Java Province)

R. Kalosa, M.U. Anshor, I.M.D. Fajriyah & S. Irianto
Department of Gender Studies, School of Strategic and Global Studies, Universitas Indonesia, Jakarta, Indonesia

ABSTRACT: Child marriage is a cruel form of gender-based violence that still happens in Indonesia. This practice has made young women victims stay voiceless and invisible as citizens in their own country. Nevertheless, no matter how oppressive their lives as being child brides, they still have the capabilities and agencies to survive. This case study in qualitative research applies in-depth interviews and observations to explore how the women victims of child marriage construct the power of their mind and body for their own life. This research reveals that the combination of poverty, religious and cultural aspects of how society values women, contributes to the practice of female child marriage. It is found that despite having big responsibilities in reproduction and production, women who are victims of child marriage have shown to have agencies in constructing the power of their body and life by taking brave decisions for their own good, personal well-being and economical independence, by sharpening their potential, and creating differences in their surroundings. This research also delivers the deepest voice and the obvious presence of the women victims of child marriage, to be heard and be visible as human beings, struggling to live the life that matters to them.

1 INTRODUCTION

> 'I prefer to be a student rather than a mother. When I was at school, everything was better' (Sari, not her real name, 18 October 2015).

Sari's voice was a bitter life portrait of a child bride that has been captured by UNICEF in the province of West Sulawesi. Sari had been married off when she was 15 because her family thought that marriage would bring a better life and pull her out of poverty. But the reality was far from the expectation. After being married, Sari immediately became pregnant and soon had a new role as a child mother. Shortly after that, Sari's husband left the house to work outside the island, without any certainty of when he would come back home or send the money to the family, leaving Sari with the multiple burdens of mothering and working, which often made her feel exhausted (Kompas, 2015).

Sari's problem is a very complicated issue that urges to be solved. The Ministry of Women Empowerment and Child Protection has revealed the data from the World Fertility Policies 2012 that shows an alarming reality. Indonesia is rated 37th in the world ranking, and runner up of the highest rank in Association of Southeast Asian Nations (ASEAN), in child marriage practices. There are 11.13% of married women aged 10–15 years and 32.10% aged 16–18 years. The data above shows how serious the problem of child marriage practice is. Even worse is how the practices found to be happening at very young ages (10–15 years) even exceeds 10%, which means girls of junior primary school are already married off. The UNICEF Annual Report 2014 finds that of the total 85 million children in Indonesia, one in six Indonesian girls would get married before the age of 18 years. This shows that there are still many other 'Saris' in Indonesia who have a voice like Sari's on a bitterness of life in child marriage, and who may remain voiceless and invisible if we do take a closer look and pay attention to this issue.

A feminist in social justice, Martha Nussbaum, in the article *Women's Capabilities and Social Justice*, published by the Journal of Human Development, revealed the 'value' of a woman which is often used as the reason behind child marriages.

> 'Women are not treated as ends in their own right, persons with a dignity that deserves respect from laws and institutions. Instead, they are treated as mere instruments of the ends of others—reproducers, caregivers, sexual outlets, agents of a family's general prosperity. Sometimes this instrumental value is strongly positive; sometimes it may actually be negative. A girl child's natal family frequently treats her as dispensable, seeing that she will leave anyhow and will not support parents in their old age.' (Nussbaum, 2000).

Child marriage has blurred girls' human rights with all the complex situations they have to face. The control over their own bodies and lives has been largely deprived by other people. Their voice is deliberately voiceless and their existence is invisible. When girls being married off, do their voices remain silent and voiceless forever? Can they have a power over their own bodies and lives for their future?

There were previous researches about child marriage in various perspectives. One of them is *The Monograph of Child Marriage*, published by Rumah KitaB, which reveals the changes of living spaces, gender relationship in families and communities, institutional roles, as well as religious values, as the causes of child marriage practices. Another study was done by Jurnal Perempuan titled *Fear of Zina, Low Education, and Poverty: Status of Girls in Child Marriage in Sukabumi, West Java*, which also elaborated on how poverty, poor access to education, religious views on taboo discussion of sexuality and fear of adultery committed by young people, as well as poor access to sexual reproductive healthcare, have caused child marriage practices. Derived from the previous researches, this study will complement the findings on the background of child marriages, provide a deep comprehension of the complex life after child marriage, and describe how those situations drove women to rise up and struggle to construct the power of their bodies and lives.

This study uses a qualitative methodology, with a case study approach providing the real situation experienced by the research subjects (Poerwandari, 2001). The data collection methods used are in-depth interviews and observations using a life story interviews approach to explore the women's lives from the point they were married off until their current age, where they have demonstrated the struggle to construct the power of the body and life and revealed significant themes in their lives. The research subjects are seven women with the characteristics including: females who were married off as children (before the age of 18 years old); had their marriages determined by others; had children when they were still children (under 18 years old), making the complexity of life as a wife and mother obvious; and came from families who could not afford the school fee, so the girl was forced to drop out of school.

2   ECONOMIC, CULTURAL, AND RELIGIOUS ASPECTS AS THE INTERLOCKING SYSTEM

The research found various reasons for families in deciding to marry off their daughters. The parents, especially the father, were the main actor to initiate marriage plans and implement them. Poor economic conditions, as well as strong religious values and family culture, became the interlocking system that triggered a child marriage practice.

Religious values and cultural stereotypes associated with the female body have put women who were already grown up as teenagers at a high moral risk. The girls who got their first period have been regarded as bringing the family honour into danger. When the girls already have boyfriends it increased their potential to break religious rules by having sexual intercourse before marriage, creating unwanted pregnancy.

Cultural values of patriarchal society in Indonesia have also attached women to domestic roles and responsibilities within the scope of their own households, such as cooking, washing, cleaning, or caring for smaller family members. This role attachment includes the concept of 'housewifisation', constructed by the state. Women are assumed to have responsibilities for domestic works, that unfortunately are not considered as paid work because they do not produce money or goods (Saptari & Holzner, 1997). As daughters in a family, all of the research subjects had the responsibility to help with the household work; some of them also did production work, helping their parents in the paddy fields.

3 THE COMPLEXITY OF WOMEN'S LIFE AS A CHILD WIFE AND MOTHER

Once the girls were married off, they immediately entered the complexities of married life that was totally new and strange to them. Sexual intercourse was the most complex situation they faced when they officially become wives. From all of the research subjects, the first sexual intercourse came at the initiative of the husbands and was faced with fear by the wives. After the first sexual intercourse passed, women had to deal with the complexity of unbalanced relationships between husband and wife, and between mother and daughter-in-law.

Production and reproduction works were part of the complex situation in child marriages. Some female research subjects had to be involved in production work to help the husband's family. However, the most challenging situation was the reproduction phase as a wife, which included pregnancy, childbirth, and mothering. Unwanted pregnancies occurred as a result of inadequate knowledge of contraception use, the realisation of the desire of the husband and in-laws, as well as being a way to reduce the loneliness of being a full-time housewife confined to domestic chores every day. Some of the pregnancy was not easy to endure, while labour became a very painful process for some women as their bodies were not ready to give birth. Motherhood is also a very tiring experience because these women had to experience this role when they themselves were still girl children. Waking up all night, feeding the baby, changing nappies and taking care of household chores were totally shocking experiences for these women when they became mothers for the first time. Nevertheless, there is little possibility for them to escape these routines of mothering because when they tried to engage in activities outside the home for themselves, such as hanging out with friends or watching movies, they faced being judged as bad mothers by their neighbourhood.

4 WOMEN'S STRUGGLE IN CONSTRUCTING THE POWER OF THE BODY AND LIFE

The complex situation of everyday life of child marriage has driven women to construct the power of their body and life. A poor relationship with the husband has influenced women to seek a way out of their unhappy marriages. Feeling bored and tired of doing reproduction work as a wife and mother, women are driven to try and escape the dull routine in everyday life. This boredom and exhaustion were closely related to the childhood dreams they had longed to realise when they grow up. In addition, economic constraints forced the women to find a way to go outside, to do productive work and fulfil their daily needs. Other driving factors departed from the way they saw marriage as a way out of distress in their childhood. When some female research subjects separated from their parents' power domination through child marriages, they tried to live a life as they wanted to.

These driving factors encouraged women to struggle with the complex life they were facing. Making a big decision in life to divorce their husband was one of the struggles faced by women who experienced unhappiness in marriage. By ending a stressful marriage at the age of 19, Teh Juju, one of the research subjects, has demonstrated her power of the body and life.

Some women showed agencies by continuing education through an informal programme and actively participating in the community activities. They became active as *Ketua Posyandu* (Head of Integrated Health Centre) and *Petugas Tabungan Keliling* (Mobile Savings Officer), which brought benefits not only for the community, but also for themselves. By doing activities outside their homes, they were increasingly exposed to new knowledge and experiences that provided a positive contribution to their lives.

Becoming economically independent was also one of the women agencies to establish the control over their body and life. Indeed, family economic constraints have encouraged them to do production works to earn money in order to fulfil their daily needs. Nevertheless, by doing the production activities, they also obtained benefits for themselves as well as increased bargaining power towards their husbands. Production works also allowed them to meet new people and give them new insights, which broadened their horizon to see the big world outside their limited surroundings.

Agencies did not have to be very obvious to be seen from the outside. Women who suffered multiple burdens during childhood used their marriages as a medium for them to live the lives that they really wanted to. By accepting the marriage, they were 'stripped off' of their obligations as a daughter to do domestic and productive works; even so, they then walked into another form of complexity of life, being a wife and a mother. However, they claimed that by getting married, they had an opportunity to redirect the life in a way they wanted to. They believed it would allow them to have the power to do things they wanted to and to enjoy the life.

5 DISCUSSION

The bitter narration on the findings have confirmed some major themes. In child marriages, women were considered as objects and **property** of people in their surroundings. They were not just property of the parents, but also property of the husbands.

When a child marriage practice occurred, parents, the prospective husband, and his families acted as if to conspire the marriage upon women. A feminist postcolonial said that the 'young man becomes incorporated into his wife's family group as he takes up residence with them and gives his services in return for food and maintenance.' (Cutrufelli in Mohanty, 2005). This quote was made to refer to matrilineal marriage; however, the same situation also occurred in patrilineal societies. Once having been married off, a girl 'provides her services' by doing domestic works as well as giving birth to one or more children, including raising them, which were categorised as 'services'. In return, the husband would give her food and fulfil her daily needs.

Girls have become both visible and at the same time invisible when people (the parents) decide to marry them off. It was closely related to women's bodies as a property. When the family's economic situation was not able to feed many family members, the daughter became visible to be 'estimated' as a property to be handed over to other people through marriage. However, she then immediately became invisible when the decision to marry was done without her consent or approval. Their existence and voice, as their reactions towards this arranged marriage, became invisible and voiceless.

After entering child marriage, women's 'ownership' has been switched to the property of the husband while they remained invisible. The husband's power was manifested in many ways: the first sexual intercourse, when his wife should be pregnant, the expected number of children, the age gap between the children, as well as kind of works that could be done by his wife. This complex life had reinforced the objectification and powerlessness of the girls, not only for their husbands but also for the larger family and neighbourhood, creating layers of power that put them in the very bottom layer. This situation has been claimed by a radical feminist who said that the root of the women's oppression is in their reproductive biology (Firestone, 2003) that places them in the weakest layer of power amongst the husband, family, and society, making women **powerless**. In the issue of child marriages, women's bodies have become the sites of problems that plunge them into the issue of child marriages, fully loaded with gender inequality as stereotyping, marginalisation, subordination, the burden compound, and gender-based violence.

The invisible female victims of child marriage were also strongly related to the position of Indonesia as a postcolonial state, which during the New Order government, still pursued the ideology of state Ibuism (*Ibu* means mother) from the Netherlands, which combined the values of the Netherlands' bourgeoisie and the Indonesian aristocracy (Suryakusuma, 1996). State Ibuism has put women in the domestic sphere and attached to the female household the primary responsibility for participating in the reproduction works, such as housework and child care as unpaid work, leaving women to become more invisible.

Women in child marriages with a labour division based on gender is illustrated by a radical feminist, Kate Millet; 'in the division of roles based on gender, women are assigned to domestic service and child care, while the interest, ambition, and the achievement of a human being handed over to men' (2000). This separation has put women at a high risk of being totally dependent on the husband (Mohanty, 2005since the reproduction work remains unpaid, and even worse, not considered as paid work. Again, women are invisible not only of their existence, but also of what they do (Saptari & Holzner, 1997).

On child marriage, family, religion, and culture have viewed women as a homogenous group, as a reproductive machine to produce and bear the children. Nonetheless, the postcolonial feminist approach regards women as a group that is **not homogenous**. The reason they 'accepted' the marriage and the way they dealt with the complex life were a diverse experience that could not be classified into a homogeneous group or situation. The state of helplessness experienced by seven research subjects at the beginning of the marriage, which has been turned into a form of struggle to construct the power of the body and life, were pursued in various ways by each of them.

The struggle of women victims of child marriage has revealed the **agency** and **capability** shown by women who tried to rise from the lowest point of their life. Nussbaum, in the capability approach developed from Amartya Sen's thought, questioned not only how satisfied was the life of a woman with what she was doing, but also what she could actually do, and in what circumstances when she did it (Nussbaum, 2000).

Agencies which were owned by everyone since they were born, especially the female victims of child marriage, grew together with their life experiences and knowledge they have. The agency became clearly obvious when being translated into concrete actions that brought changes to women's life. The concrete results of this agency then delivered the women to build brick by brick as the power over the body and life, beyond the circumstances to not only survive and follow what was in front of them day after day. Capability to achieve a satisfying life is what enables woman as a human being.

The unfulfilled capabilities occurred at the beginning of a marriage in which girls still did not have power over their body and life. Therefore, the fulfilment of a series of core capabilities was closely related to what happened in the life cycle they went through from the time they were married off until now, which has gone through a period of 16–37 years. At the beginning of marriage, girls had to focus on fulfilling the role as a wife and mother, which was full of reproduction work. However, once their children grew bigger and more independent, women had more time to do whatever they wanted to do, that was of benefit not for the kids, the husband, or the family, but for themselves as unique people who had dreams to be realised. Their agencies were slowly exposed to new experiences and knowledge they got in everyday life. Agencies which were processed together with the knowledge and life experience have created a capability to build women's power of the body and life. The struggle to fulfil this capability also happened when those capabilities were totally absent.

By tracing back the stories and experiences of the seven female research subjects, seeing their agencies being put into concrete actions such as by making important decisions, developing the potential and being active in the community, economically empowering, and enjoying life, have shown that women victims of child marriage were able to construct a meaningful life. Their struggle has proved what was said by Aristotle, that the agency which was successfully realised would make a person not only to survive like an animal, but also to flourish in human ways, as human beings (Nussbaum, 2000).

6 CONCLUSION

Child marriage is a truly harmful practice for girls. This practice has made a girl a property belonging to other people, such as her parents, the man who wants to marry her, the husband's family, and the society in which she lives. This practice also makes her existence become invisible and voiceless. Economic constraints combined with the strong religious values and culture held by the family has become the interlocking system which is strongly bonded and cannot be separated in child marriage practice.

After the marriage, the girl soon faces the complexity of the situation with regards to production and reproduction work, including sexual intercourse, pregnancy, childbirth, and parenting, as well as an imbalanced relationship with the husband and his family. Mixed reactions shown by the women in facing the complexity of marriage life had confirmed that women were not homogenous.

Seven women in this study did not look back to the past, but have focused on what was in front of them. Their struggle was shown in various actions, including: taking a big decision that changed their lives as a form of power over their own bodies; developing their self-potential and giving a meaningful contribution to the community; being economically independent and empowered; and enjoying life. They originally were the object, but now they are the subject of their own body and life. They do not live only to survive but also to lead a meaningful life by having the power over their own bodies and lives.

7 RECOMMENDATION

To prevent the practice of child marriages, the author is proposing these two following recommendations. First, the government needs to urgently revise the Marriage Law No. 1 in 1974 to raise the minimum age of marriage to 18 years, referring to the Convention on the Rights of the Child, which states that 'children' relates to every human being below the age of 18 years. Secondly, it is necessary to develop a critical awareness in society by cooperating with religious leaders and local communities to support the prevention of child marriage, and to support sex education and programmes for young people to design their future and widen their horizon to see the world. Thus, the approaches need to incorporate both top-down and bottom-up (grassroots) levels.

Associated with young women who are already married, the author proposes four suggestions. First, the government should pay special attention to the needs of children who have been married, according to the Convention of Rights of the Children. The government should publish a formal policy urging schools to remain open for those who are married and set compulsory education as an obligation, not an option. Second, the government should open up the opportunities as much as possible for the married women to develop their potential by providing alternative education such as programme Paket A, B, and C in every village; the government should also provide funds or loans for small and medium enterprises nurtured by women to encourage them to have access to the capital for their small business development. Third, the government and Non-Governmental Organisations (NGOs) need to establish an organisation for assisting and strengthening women victims of child marriages at the grassroots level. Lastly, the government, the academia, and NGOs need to do more research that focuses on the experiences of women victims of child marriages, in order to design a women's empowerment programme that can be directly targeted to the needs of women in the field.

REFERENCES

Assembly, U. G. (1979). Convention on the elimination of all forms of discrimination against women. Retrieved April, 20, 2006.

Candraningrum, D., Dhewy, A., & Pratiwi, A. M. (2016). Fear of Zina, Poor Education, and Poverty: Status of Girls in Child-Marriage in Sukabumi West Java. *Jurnal Perempuan, 21*(1), 77–95.

Firestone, S. (2003). *The dialectic of sex: The case for feminist revolution.* London: Macmillan Publishers.

http://regional.kompas.com/read/2015/10/18/04214811/Nyanyi.Sunyi.Para.Pengantin.Anak. di.Sulawesi.Barat (A Silence Singing of the Child Brides in West Sulawesi) downloaded on 18 October 2015.

Marcoes, Lies, and Nurhady Sirimorok. Seri No.1 Monografi Penelitian Perkawinan Anak: Pengantar Monografi 9 Kajian Perkawinan Anak di 5 Provinsi Indonesia—Yatim Piatu dan Kerja Kuasa Tersamar (*Introduction to 9 Monograph of Child Marriage Study in 5 Provinces of Indonesia—Orphans and Disguised Power of Work*). Rumah KitaB dan Ford Foundation, 2016.

Millet, K. (2000). Theory of sexual politics. In B.A. Crow, *Radical Feminism. A Documentary Reader* (pp. 122–153).

Mohanty, C.T. (2005). Feminism without borders: Decolonizing theory, practicing solidarity. *Hypatia, 20*(3), 221–224.

Nussbaum, M. (2000). Women's capabilities and social justice. *Journal of Human Development, 1*(2), 219–247.

Poerwandari, E.K. (2001). *Pendekatan kualitatif untuk penelitian perilaku manusia (A Qualitative Approach for Human Behavior).* Jakarta: LPSP3 Fakultas Psikologi Universitas Indonesia.

Price, J. & Shildrick, M. (1999). *Feminist theory and the body: A reader.* United Kingdom: Taylor & Francis.

Saptari, R. & Holzner, B.M. (1997). *Perempuan, kerja, dan perubahan sosial: sebuah pengantar studi perempuan* (Vol. 1) *(Women, Works, and Social Changes: An Introduction of Women Studies).* Jakarta: Pustaka Utama Grafiti.

Suryakusuma, J. (1996). The state and sexuality in New Order Indonesia. Fantasizing the feminine in Indonesia, 92–119.

Unicef. (2015). *UNICEF Annual Report 2014.* UNICEF.

The contributions of Minangkabau women, who established intercultural marriages, in creating a 'new identity' of the Minangkabau diaspora

Mina Elfira
Department of Literature Studies, Faculty of Humanities, Universitas Indonesia, Depok, Indonesia

ABSTRACT: *Pai marantau* (going voluntary, putatively temporary or permanent migration) is an activity which is usually associated with the Minangkabau of West Sumatra, Indonesia, well known as the world's largest matrilineal society and one that coexists amongst the mostly Islamic societies within Indonesia. Initially it was conducted by men only, but later on women have also participated in this activity. Some Minangkabau women do *pai marantau* through interethnic/state marriages. Based on the argument of Nira Yuval-Davis (1997b), who states the importance of home, and women as homemakers in the process of ethnicity building, as cultural rules and their practices are transmitted to the next generation at home, most strongly by women, this paper will explore the contributions of Minangkabau women, who established intercultural marriages, in modifying matrilineal *adat* within contemporary urban Minangkabau diaspora households. Using qualitative methods, it will investigate that by utilising the ambivalent roles of agents of change and defenders of *adat*, these women have created a 'new identity' of Minangkabau, which is more in lifestyle than in blood, in their daily lives. In addition, with the impact of the globalised world, people tend to look to their roots for their identity.

1 INTRODUCTION

> *Banyak malangkah* (The more steps to be performed),
> *banyak nan diliek* (the more chances to look around),
> *banyak pulo nan didapek* (the more benefits that can be gained)

The aphorism above is one of many Minangkabau aphorisms that underscore the importance of conducting a journey (*marantau*) to enrich and enlighten one's soul. It encourages a Minangkabau to *pai marantau* (go on migration) – to conduct a journey in search of wisdom and prosperity. *Rantau* originally referred to the territories outside *Luhak nan Tigo*, called *Rantau nan Tigo Jurai: Hulu* (upper end) *Batang Hari* river, *Hulu Batang Kuantan* river, *Hulu Kampar Kiri* river (Westernenk, 1981. 61). Later, *rantau* came to refer to areas outside *Alam Minangkabau* that are influenced socioculturally by Minangkabau (Azra, 2003:36). Today the term *rantau* refers to territories outside the Minangkabau homeland. Minangkabau people use the term *adat* to refer to both their oral history, pertaining to the origins of *Alam Minangkabau*, and to the proverbs and aphorisms that serve as guiding principles and rules for ceremonies, conduct, and matrilineal kinship relations (Kato, 1982:33–34).

Pai marantau (going voluntary, putatively temporary or permanent migration) is an activity which is usually associated with the Minangkabau of West Sumatra, Indonesia, well known as the world's largest matrilineal society and one that coexists amongst the mostly Islamic societies within Indonesia. Initially it was conducted by men only, while Minangkabau women stayed in the homeland. Later on, women have also participated in *marantau* activity. One of the impacts of going *marantau*, especially for those who permanently going *marantau*, is the possibility of interaction with other ethnic groups through marriages. Some

Minangkabau women, even, do *pai marantau* through interethnic/state marriages. It is interesting to know how significant is the impact of intercultural marriages to the establishing of Minangkabau identity. Matriliny and Islam are the essences of Minangkabau identity (Elfira, 2009). Moreover, it has been widely accepted within Minangkabau studies, including the work of Sanday (2002), Hadler (2009) and Elfira (2015), that Minangkabau women, symbolised as *limpapeh Minang jo Rumah Gadang* (the central pillars of Minangkabau and the Big House the *Adat* house) have played significant roles within Minangkabau society. How far is the contribution of Minangkabau women, who established intercultural marriages, in creating a 'new identity' of the Minangkabau diaspora? The research question is based on the argument of Nira Yuval-Davis, who states the importance of home, and women as homemakers in the process of ethnicity building, as cultural rules and their practices are transmitted to the next generation at home, most strongly by women (Yuval-Davis, 1997b). Woodward's argument that identity is fluid and contingent (Woodward, 1997:3) is used in analysing the identity of the Minangkabau diaspora.

The work of Elizabeth E. Graves (1971 and 1981) and Mina Elfira (2015), discussing the significant contribution of Minangkabau women in establishing Minangkabau identity in a *rantau* land, provided a foundation for my hypothesis regarding the contribution of the Minangkabau women diaspora in making a Minangkabau identity in *rantau* land. Investigating the Minangkabau response to Dutch colonial rule in the nineteenth century, shed light on the nature of Padang as the *rantau* land. Graves (1971:36) argues that the coast could never truly be Minangkabau without Minangkabau women of good family who never left their mother's house in their homeland, even after marriage. Later on, Elfira (2015) found that Padang seemingly has become a Minangkabau territory, considering the fact that these women seemingly prefer to follow their husbands migrating to Padang, and building their permanent 'home' (not just 'temporary dwelling'), rather than staying at their mothers' houses in the heartland (*darek*). Regarding the important role of Minangkabau women in establishing a Minangkabau identity, I agree with both of them. Following their arguments, I hypothesise that the Minangkabau women diaspora plays an important key in establishing Minangkabau identity, which is more in lifestyle than in blood, in *rantau* land through their homes. This research aims to contribute to the further understanding of Minangkabau identity and social relations from a Minangkabau woman's point of view, and draws on the voices of other Minangkabau women.

In order to examine the way Minangkabau women, who established intercultural marriages, modify Minangkabau identity in their daily lives outside Minangkabau mainland, this paper will rely more on qualitative data, obtained using qualitative methods. The choice is based on Nancy Lopez's argument that qualitative methods capture the contextual, real-life, everyday experiences of the individual interviewed (Lopez, 2003:7). Moreover, this method, including participant observation and in-depth interviews, is considered more effective in exploring intercultural marriages. Intercultural marriage, especially when it is also an interreligious marriage, still seems to be a sensitive issue in Indonesia. I chose my sample from women with whom I was acquainted either professionally or personally, or women whom I could contact through family members and acquaintances. The research location is Jakarta, which has become the main destination for Minangkabau migrants in Indonesia. The data was collected through a face-to-face encounter, and also from electronic media, as some of them are living abroad, including in Italy, Germany, Hong Kong, Qatar, and the USA. After being translated into English, extensive segments of the in-depth interviews, conducted in the Indonesian and Minangkabau languages, will be included in this paper. In order to protect their privacy, respondents quoted in this paper have been given pseudonyms.

2 INTERCULTURAL MARRIAGE AND ITS COMPLEXITY

As detailed by some scholars of Minangkabau, such as Krier (2000), Peletz (2005) and Elfira (2015), getting married is an important affair which is taken seriously by the kin group and the Minangkabau groom and bride. As for an individual, marriage carries an acknowledgement of being a mature individual with full rights to participate in *adat* activities within the society. A female is first expected to continue a family line. In Minangkabau, a marriage not only unites two individuals, but also two matrilineal family groups. Marriage in Minangkabau is clan exogamous, so until quite recently it was forbidden to marry someone from the same clan. Influenced by Islamic laws, nowadays it is allowed for marriage within one clan as long as the persons are not closely related by blood. The ideal marriage, according to Minangkabau *adat,* is one in which the bride is *pulang ka bako* (goes back to her paternal family). This involves a man marrying his *mamak*'s daughter—a classic case of what anthropologists term cross-cousin marriage. By applying this concept, the children will belong to the same clan as their grandfather (their father's *mamak*). This concept is also an attempt to keep *harato pusako* (ancestral property) in the same kin grouping. But, it seems that the Minangkabau younger generation, especially those who are living outside Minangkabau land such as in Jakarta, considers this practice outdated, especially as there is little contemporary interest in *harato pusako*. They also see such marriages as potentially unhappy because of strong interference from the two sides of the family, as expressed by Lenni (a 39-year-old mother with three children), whose husband is a Javanese man. In addition, in multi-ethnic Jakarta, where Minangkabau is not the biggest nor the most dominant ethnic group, there is a strong possibility to marry a person from other ethnic groups. In daily life, some Minangkabau have practised some mixed marriages, known as *kawin antar suku* (interethnic marriage), *kawin anta nagaro* (interstate marriage), and *kawin campua* (mixed marriage). In the 1970s–80 s mixed marriages were a non-preferred choice, especially within Minangkabau society in Minangkabau land. Preserving *adat* purity and the feeling of having better customs are some reasons for the rejection. In addition, the community tended to look down at a person who established it.

However, nowadays, especially in the *rantau* area, it seems that mixed marriages, meaning intercultural marriages, have been openly practised. Yerika (66-year-old), who allowed her two daughters to be married to other ethnic groups, argues that the tolerant principle is a realisation of the *adat* aphorism '*dimaa bumi dipinjak disinan langiak dijunjuang*' (where you set your foot down, there the sky is held in high esteem) as Jakarta is not the land of Minangkabau. The concept that marriage is more a personal affair than a matrilineal family affair is a common reason raised by those who established intercultural marriage, as argued by Oriza (a 25-year-old Padang man), who grew up in Jakarta and married a Javanese woman in Javanese customs. It can be said that the Minangkabau community diaspora is more tolerant on this issue, with some conditions. It seems that Minangkabau society prefers intercultural marriages between Minangkabau women and non-Minangkabau men rather than vice versa. Fear of losing Minangkabau identity may be one reason why a Minangkabau mother might tend to forbid her male children from marrying non-Minangkabau women. It can be said that this attitude is a result of the application of the matrilineal system, in which descent and inheritance go through the maternal line. Despite this acceptance, it is expected that Minangkabau *adat* is still respected in this intercultural marriage affair. For example, Rina (55-year-old) was criticised by her cousins as she let the wedding ceremony of her first son be totally in Javanese customs without any symbol of Minangkabau custom, as they said: 'it is all right we respect their customs, but our *adat* should not be ignored' (*Adaik urang diharagoi tapi adaik awak jaan dilupokan*). The criticism shows that using of Minangkabau marriage ceremony customs is a symbol of the practice of Minangkabau *adat*.

Regarding the variety of intercultural marriages, it seems that interstate marriage, especially between a female Minangkabau and a male Westerner, to some extent makes a good impression amongst the Minangkabau in Jakarta. The possibility of getting a better financial support and raising a status are two reasons for this impression. On the other hand, interreligious marriages seem still to be an exception. Influenced by Islamic laws, Minangkabau society tends to prevent interreligious marriages, especially between a Minangkabau Islamic woman and a non-Islamic man. For those who are involved in interreligious marriage, the society tends to dispense some sanctions, which can be either moral or material, or both. These people may lose their rights and responsibilities provided by the *adat*, as can be seen from Nina's case. Nina, who married a Christian Javanese man in a Christian marriage ceremony, has been excluded from the Minangkabau *adat* community. As a result, she cannot claim her rights over inheritance, especially over *harato pusako*. Because of that, many Minangkabau who established it, try to cover up their marital status. As also shown by Elfira (2009), fear of getting moral and social sanctions from the community and losing Minangkabau identity can be the main reasons for this covertness. An example of this can be seen from the case of Henny, who married an English non-Islamic man and lived in Hong Kong. Most of her kampong relatives do not yet know that her marriage is not only *kawin anta nagaro* but also *kawin campua*, which has a connotation as interreligious marriage. When I asked her through a questioner, Henny did not reply to my question. Based on the cases above, it can be said then that regarding intercultural marriage, in order to protect the Minangkabau *adat*, the Minangkabau diaspora 'plays' with their assimilation and exclusivity interests. The act has proved that identity is fluid and contingent (Woodward, 1997:3). It seems that there are some modifications of the applied *adat* in their daily life, which impacts on the creation of the Minangkabau diaspora identity.

'It's more in lifestyle than in blood': Home, women, and the creation of Minangkabau diaspora identity.

We are in Padang send our condolences ... we got the news from N Mosque through brother S, the head of our community.

The above quotation, taken from a Minangkabau family WhatsApp group (10/08/2016, 05.51), shows the closeness of overseas Minangkabau to their homeland, as can be seen from Ita's case. Although spending most of her time outside Minangkabau land, Ita (late) was still considered as a member of a Minangkabau community in the homeland. It is no wonder then that the Minangkabau are also known for their paradoxical character: their closeness to the homeland but their eagerness to *pai marantau*. These overseas Minangkabau established a viral family group through electronic media such as WhatsApp, so they can still be in contact with their family in the homeland. This media is also used to reconnect with their family members who are spreading out in the *rantau* land. In addition, it is a medium with which to educate and practice Minangkabau things, such as language, food, *adat* aphorisms, songs, and jokes. In a discussion in Jakarta (06/07/2016), one of these group members jokingly called it a kind of 'Minangisation project', a project to shape the Minangakabau character as an identity.. It seems for them the identity of Minangkabau is determined more in lifestyle than in blood; Dino's case can be taken as an example.

By blood, Dino cannot be considered as a being a Minangkabau as his mother (late) had Madurese blood. His mother was born and grew up in Padang, married a Minangkabau man, and practised Minangkabau *adat* at home. Dino, who has become accustomed to this *adat*, has practised it in his daily life in Jakarta. He is known amongst his friends, even by the Minangkabau community itself, as a Minangkabau in this multi-ethnic city. At Dino's home, who has married a Chinese Javanese woman, Minangakabu *adat* is more dominant than Javanese and Chinese customs. Dino's case shows the importance of women as the homemakers in the process of ethnicity building, as argued by Yuval-Davis (1997b). Moreover, it seems that the realisation of this project is more effective and successful at home. It can be said that most of the diaspora, whether they have full-blood or half-blood of Minangkabau, admit that they learnt these Minangkabau things from home, where they grew up as Minangkabau. They also say that their mothers have played a significant role in teaching them Minangkabau things, and in making them into Minangkabau, as expressed by Lulu (45-year-old), who was born and grew up in Jakarta, from a Sundanese father and a Minangkabau mother. Lulu, who worked overseas after finishing her graduate school, learnt Minangkabau *adat* from 'home ... mother, maternal family members ... in Indonesia I am a Minangkabau'. While in Indonesia, Lulu was more exposed to her Minangkabau roots, and when overseas she introduces herself as an Indonesian. Although she married an American man and became a USA citizen, Indonesia still means Minangkabau for Lulu, as at home she practices and educates her family members (children and husband) about Minangkabau things. It is interesting to note that the Minangkabau *adat*, which is highly respected and gives women significant roles within society, is the character that makes most *perantau* (wanderers) proud of being

Minangkabau, as expressed by Elli (43-year-old), who married an Italian man and lived in Milan (Italy). Based on her email (23/08/2016, 19.26), Elli, who included some Minangkabau *adat* practices in her marriage ceremony, said that 'A Minangkabau woman is independent and strong ...she is capable in making decisions without her brother's influence or help'.

It is interesting to note also that many Minangkabau *perantau*, even the half-blood ones, feel more Minangkabau after they live overseas, as can be seen from the case of Olla (27-year-old), who has migrated to Germany and got engaged to a German man. Olla, who grew up with Minangkabau *adat*, plans to use Minangkabau customs for her marriage ceremony at her parent's home in Jakarta, since Minangkabau is her identity:

> it is never ... how can I be lost [my root] ... it seems that when I was born I firstly heard ' oh my God my child is a daughter' than words of *adzan* [recite call to Islamic prayer over newborn child] ... it [marriage ceremony] is definitely using Minangkabau *adat*... it has been my childhood dream... in marriage ceremony there will be a Minang-kabau welcoming dance (the discussion via electronic media 16/07/2016, 17.56).

Another example can be seen from Ina's case. Since Ina's mother is a Minangkabau, she too is considered by *adat* as being a Minangkabau. But Ina, who put her Sundanese father's name as part of her full name, was better known as a Sundanese than a Minangkabau while in Jakarta. But, after moving overseas because of getting married, she became aware of her roots as a Minangkabau. This can be seen from the fact that she has been appointed as the organiser of an Indonesia-Minangkabau community in that country. It seems that for Ina, whose Minangkabau identity is getting stronger, Indonesia is Minangkabau. For many of these overseas Minangkabau, the Minangkabau *adat*, which taught a *perantau* how to behave in *rantau* land, is one of the reasons for them to go back to their roots as Minangkabau, as said by Elli, who was born in Jakarta and moved overseas after finishing her high school study:

> Minangkabau principles and modernism ... [make me able] to independently live in a foreign country, able to adjust ... [I am] rightfully proud of my own identity [as a Minangkabau] (based on her email 23/08/2016, 19.26)

Based on Ina's and Elli's cases, it can be said then that with the impact of the globalised world, people tend to look to their roots for their identity.

3 CONCLUSION

Based on the analysis above, this paper's conclusion has some main points. Firstly, in maintaining intercultural relations the Minangkabau people 'play' with their assimilation and exclusivity interests in order to protect their *adat*. The paper also argues that matrilineal principles and Islam are the defining aspects to be considered by the Minangkabau diaspora in maintaining intercultural marriage relationships. Thirdly, Minangkabau women who established intercultural marriages, modify the application of the Minangkabau *adat* in their daily lives so that the *adat* suits with the social condition of the Minangkabau diaspora. It can be said that these women act as 'defender' and the 'creator' of the *adat* in *rantau* land. These women have made significant contribution to creating a 'new' identity of the Minangkabau diaspora. It seems for them that the identity of the Minangkabau is determined more in lifestyle than in blood. Lastly, in this globalised world, the Minangkabau overseas tend to go back to their roots as Minangkabau to look for their identity.

REFERENCES

Azra, A. (2003). *Surau: Pendidikan Islam Traditional dalam Transisi dan Modernisasi (Surau:Traditional Islamic Education in Transition and Modernisation)*. Jakarta: Logos Wacana Ilmu Press.
Elfira, M. (2009). Not Muslim, not Minangkabau: Interreligious marriage and its cultural impact in Minangkabausociety.InG.W.Jones,C.H.Leng,&M.Mohamad(Eds.), *Muslim-Non-MuslimMarriage:*

Political and Cultural Contestation in Southeast Asia. Singapore: Institute of Asian Studies, National University of Singapore.

Elfira, M. (2011). Inter-ethnic relations in Padang of West Sumatra: Navigating between assimilation and exclusivity. *Wacana: Jurnal ilmu Pengetahuan Budaya, 13*(2), 293–304.

Elfira, M. (2015). *The lived experiences of Minangkabau mothers and daughters: Gender relations, adat and family in Padang, West Sumatra, Indonesia.* Germany: Scholar Press.

Graves, Elizabeth E.. (1971). *The Ever-Victorious Buffalo: How the Minangkabau of Indonesia solved their "Colonial Question"*, Ph.D. Thesis, University of Wisconsin.

Graves, E.E. (1981). *The Minangkabau response to Dutch colonial rule in the nineteenth century.* Monograph Series (Publication No. 60), Ithaca, New York: Cornell Modern Indonesia Project, Southeast Asia Program, Cornell University.

Hadler, J. (2009). *Muslims and matriarchs: Cultural resilience in Minangkabau through jihad and colonialism.* Singapore: National University of Singapore.

Kato, T. (1982). *Matriliny and migration: Evolving Minangkabau traditions in Indonesia.*, Ithaca, New York: Cornell University Press.

Krier, J. (2000). The marital project: Beyond the exchange of men in Minangkabau marriage. *American Ethnologist, 27*(4), 877–897.

Lopez, N. (2003). *Hopeful girls, troubled boys: Race and gender disparity in urban education.* New York and London: Routledge.

Peletz, M. (2005). The king is dead; long live the queen! *American Ethnologist, 32*(1), 39–41.

Sanday, P.R. (2002). *Women at the center: Life in a modern matriarchy.* Ithaca, NY, London: Cornell University Press.

Westenenk, L.C. (1981). De Minangkabausche Nagari published in 1915, (translated into Indonesian by Mahyudin Saleh, S.H.). Padang, Penerbitan dan Bursa Buku Fakultas Hukum dan Pengetahuan Masyarakat Universitas Andalas.

Woodward, Kathryn (1997). Introduction. In Kathryn Woodward (ed.), *Identity and Difference* (pp. 1–6). California: SAGE publication.

Yuval-Davis, N. (1997a). *Gender and nation.* London: Sage.

Yuval-Davis, N. (1997b). Women, citizenship and difference. *Feminist Review, 57*, 4–27.

US foreign policy towards the democratisation of Syria in 2011–2014

S. Awal & I.F.R. Agoes
Department of American Studies, School of Strategic and Global Studies, Universitas Indonesia, Jakarta, Indonesia

ABSTRACT: This article analyses the United States of America (US) foreign policy towards the democratisation process, and the FSA (Free Syirian Army) and SNC (Syirian National Council) effort to topple the Bashar al-Assad dictatorship regime that has ruled for 16 years in Syria. The paradox is that during the conflict between the opposition and Assad's regime that has been supported by the military power of Russia, US support of democracy in the world does not give an explicit reaction against the actions of the Syrian regime. A using qualitative approach is used and the theoretical framework of rational choice, rational actor models and national interest. This study finds that under Barack Obama's presidency in the fight for democracy in Syria, US foreign policy is likely to turn from being political maximalist to political minimalist. The study, suggests how this political paradigm shift happens because Syria does not have a strategic significance for US national security interests in the Middle East.

1   INTRODUCTION

From 2010 to 2012, the Middle East and North Africa regions experienced a tremendous political upheaval, the so-called "Jasmine Revolution" (Kartini, 2013). This "Revolution" urged a fundamental change over the power of the ruling regime to restore it to the people. The people demanded that development outcomes to be no longer controlled by a group of those in power, but instead to be used for public welfare.

This revolution was to spread to Egypt, Algeria, Yemen, Bahrain, Libya and the other countries in the Middle East. The "Jasmine Revolution" successfully overthrew both the Tunisian President Zine Abidin Ben Ali and the Egyptian President Hosni Mubarak. Whereas, the other heads of the state managed to survive, like the Assad regime in Syria. It which has been in power since Hafez al-Assad in 1972, and who was later succeeded by his son Bashar al-Assad since 2000 until now. That nepotism generated domestic demonstration turmoil in Syria. The Assad regime responded to Syrian's peaceful action by brutally killing the protesters (Sulaeman, 2013).

The demonstration that demanded the resignation of Bashar al-Assad caused many victims. A human rights monitoring group, *Syrian Observatory for Human Rights*, reported that the death toll has exceeded 162,000 victims, consisting of 54,000 civilians, 42,700 rebels and 65,702 Syrian military *members*. In addition, there were 62,800 victims from the pro-President Bashar al-Assad militia and foreigners who fought on the rebel side*.

Hafez al-Assad, who had been in power for 30 years, forbade his political opposition party to compete in the Syrian presidential election. One of Hafez al-Assad's actions included the slaughtering and killing of Syrian citizens who tried to drive him from the Syrian government. When his son, Bashar al-Assad, took power, positive influence remained minimal as Bashar al-Assad implemented an authoritarian government. There were many acts of discrimination committed

---

*Korban Tewas Konflik Suriah. (2016). http://www.tempo.co/read/news/2014/05/19/115578830/Korban-Tewas-Konflik-Suriah-Capai-162-Ribu-Orang, retrieved on Feb. 3, 2016.

by Bashar al-Assad, such as eliminating freedom of speech and expression. The impact led to a wide variety of demands and impeachment efforts against the Bashar al-Assad regime, which became an accumulation the Syrian people's discontent against the authoritarian Assad regime.

The United States of America (US), during President Barack Obama's administration, has taken a clear stance to not support Bashar al-Assad to remain in power. This stance has been based on the US's displeasure against Assad's leadership. The US offered a temporary reform. President Obama said that the US saw that the Syrian people wanted to achieve democratic transition safely. According to him, the Syrian people had suffered human rights violation committed by their government. (Obama, 2011; Phillips, 2011).

President Obama in August 2011 urged President Assad to resign from his post as the president and declare that the future of Syria must be determined by the people of Syria. Prior to that, in April, May and August 2011, President Obama also signed the Executive Order (EO) to freeze all the Syrian government's assets in the area of US jurisdiction and prohibit US citizens from investing in or exporting goods to Syria, or from importing petroleum products from Syria, as well as from conducting other business transactions with Syria.

Effort to overthrow the Assad regime has been part of the US foreign policy until now: "We're looking at the controlled demolition of the Assad regime," said Andrew J. Tabler, a Syria expert at the Washington Institute for Near East Policy. "But like any controlled demolition, anything can go wrong." This shows that the US rejected all forms of diplomatic relation in order to save the Assad regime in Syria and preferred to support and provide funds for the opposition.

2   RESEARCH METHOD AND THEORETICAL FRAMEWORK

This research used a qualitative approach by John W. Creswell (1994); qualitative method is a process used to understand and to dig up the behaviour of either an individual or a group who live in an environment, and to emphasise more on the process rather than the result. The research process includes asking questions and procedures, and searching data obtained directly from the background of the research object; after that, data that has been obtained is analysed inductively, so that in the end the researcher can makes an interpretation of the collected data (Creswell, 1994, p. 162).

Foreign policy can be defined as a state strategy in dealing with other countries or other international actors. Foreign policy is also a media and at the same time is controlled by the state itself in achieving national goals which are outlined specifically in terms of national interest (Plano & Olton, 1999). Meanwhile, according to Machiavelli, each state must pursue its national interests and expect neither other state nor other government (Jackson & Sorensen, 2005).

Foreign policy is an action which is various and subject to change at any time (Rosenau, 1976). Usually the change is caused by many conflicts towards the international situation and constantly changing and even sometimes inside the international environment (Mas'oed, 1997). A country's foreign policy tends to be relatively different from the other countries' foreign policies because a foreign policy generally depends on the domestic condition of the concerned country.

The US foreign policy approach is dynamic and is constantly changing in accordance with the situation and condition. Achieving world stability is the important and major goal of the US foreign policy (McCormick, 1998). In formulating a foreign policy, the US position itself in the form of unipolar is the world's political power. Furthermore, after the end of the Cold War, the US led the way as being the main power in the world's economy, politics and defence (Kagan, 2002). The US has always sought to position itself so that it becomes the important figure in the manifestation of its foreign policy. Not all international problems can be solved by war, but on the other hand there are also problems that can be solved through negotiation, and vice versa. Therefore, to respond to the international situation, politics and the US foreign policy are very flexible (Holsti, 2002).

The US foreign policy is always changing based on the international world's situation and condition. In this case, the author saw that there was a change in the determination of foreign policymaking whereby the process and outcome had become more ambiguous. Post 9/11 changed US foreign policy, which fought against terrorism, and supported democratisation and human rights in the world. This was different from what the United States did in the conflict in Syria. The US did not give a firm action in the effort to democratise Syria. Political ambiguity involved efforts to reduce the importance of the things which were conflictual (as an alternative of a specific policy) in the evaluation of politicians, so attention would be paid to a consensual appeal (peace, prosperity, fairness in democratic governance) (Page, 1976).

In the making of foreign policy, Graham T. Allison said that: "National security and national interest are the principal categories in which strategic goals are conceived. National states seeks security and a range of other objectives." (Allison, 1971, p. 33). Every process of foreign policymaking can be influenced by many factors. The process of foreign policymaking also involves domestic actors who come from various institutions. It is often the case that those involved actors in that foreign policymaking have different interests, resulting in a conflict of interest in the process. However, the result of the foreign policy is still a policy which is believed to meet national interests to the maximum, based on the consideration given to the consequences arising from that policy. Allison said that: "Governments select the action that will maximize the strategic goals and objectives" (Allison, 1971, p. 32).

The assumption that foreign policy is a value-maximising action makes out the state or the government to be a rational actor. Robert Dahl and Charles Lindblom (1994) defined a rational actor as: "an action can be called as rational if it is appropriately directed to maximize the achievement of objectives, based on consideration about that objective with the real action which is done." (Dahl & Lindblom, 1994, p. 274).

2.1   US foreign policy towards the democratisation in Syria in 2011–2014

2.1.1   Maximalism politics

In relation to the world's international politics, since the Westphalia treaty in 1648, the sovereignty of the states is a sign that there is no supreme power in the international community, if a country's national interests and security are felt threatened; an open option is the use of military force. This is a natural character of an international community. The 24th Clausewitz Dictum clearly characterises that 'war is the continuation of politics by other means' (Von Clausewitz, 1982). So, a conditional sovereignty is because of that freedom. The use of prerogative power which is patterned to defend US national interests and security is conducted in a way of non-compromise, and directly to achieve "The Ultimate Good", the American Maximalism (Muthalib, 2008).

In fact, maximalism is not a new way of thinking in the US political history. This thinking emphasises towards the achievement of demands as much as possible. Achieving demands is done either directly or revolutionary, without compromise. Political policy is to achieve a strategic breakthrough in order to transform a situation that can sustain the US influence continuously. Kagan (2007) said that the characteristics of maximalism are as follows: "They have been impatient with the status quo. They have seen America as a catalyst for change in human affairs, and they have employed the strategies and tactics of maximalism, seeking revolutionary rather than gradual solutions to problems." (Kagan, 2007, p.144). During the Cold War, this maximalism was very dominant in the US diplomacy to eliminate the Soviet Union's nuclear power. This was shown in Ronald Reagan's policy of "Zero Option" and "Strategic Defense Initiative". Negotiations were indeed conducted but those were not used as the instrument to solve a problem.

President Bill Clinton, as another example, in solving the Bosnian conflict, told the conflicting parties to stop the acts of violence or the US would use military action. Bosnia would not be stable as long as Milosevic was in power, and the appropriate policy was the Regime Change. Meanwhile, President George W. Bush who could not be separated from his unilateral policy to Afghanistan and Iraq, and was truly maximalist. That was clearly shown in his State of the Union Address. What was clearer was the view of Paul Wolfowitz towards the September 11 incident, which endorsed him to shift the US status quo in the Middle East. Wolfowitz said, "It unacceptable to just sit and live with the Middle Eastern status quo." So, the main characteristic from the maximalist tradition is that each government must be capable of taking action to save that maximalist diplomacy tradition from negligence made by its predecessors. It appeared in Stephen Sestanovich's words (2005): "The current administration in no exception. Its strategic outlook represents a particularly hard edge culmination of past practices..."

Ronald Reagan, Bill Clinton and George W. Bush applied their foreign policy based on maximalist politics. They successfully created the US shadows like "Sole Hegemon", "locomotive at the head of mankind", "indispensable nation" and *The Lonely Super Power*." The Republican party and the Democratic party play a major role in conserving maximalist diplomacy tradition through their presidential candidates elected in the general election. Fareed Zakaria (2008) implied that: "The test of a focus commander in chief is not to focus obsessively on one battlefield but to keep all of them in view and to use resources and tactics in a way that creates on overall grand strategy, one that keeps the American people safe....". What Fareed Zakaria said can be interpreted as being military force and his tactic referred to the preemptive attack.

If seen from the view of the Republican and the Democrat, especially the parties' strategic thinkers, there is an understanding that the post-Cold War has positioned the US as being the superior state or the prime state in the world. The primacy paradigm in the Republicans refers to the national liberalism (Posen, 2007). National liberalism was an ideology of George W. Bush's government. The Republican's primacy paradigm appeared in the Bush Doctrine. There were three principles that mirror the US maximalism: first, the US was free to do initiate action towards terrorists and nations that had a weapon of mass destruction; second, there was no nation in this world, or combination of those nations, that were allowed to oppose the superiority of the US military; third, one-sided action was better than the international treaties and international organisations in the prevention of the spread of nuclear weapons. The Bush Doctrine mirrored a truly approach.

The strategy of the Democrat party is Liberal Internationalism, which is basically the same with the Republican party. The difference lies in the source of legitimacy to act. According to the Democratic party, legitimacy means an action that is taken by the US, or together with its allies, that can be accepted by other nations without bothering whether they agree or not. This paradigm reflects Henry Kissinger's diplomacy concept: "The best and most prideful expressions of American purposes in the world have been those in which we acted in concert with others. Our influence in this situation has depended on achieving a reputation as a member of such concert." (Kissinger, 1969). However, in protecting the nation and the US interests, the Democratic party's vision is identical to that of the Republican party. Bill Clinton's statement showed this: "When people are (feeling) insecure, they'd rather have somebody who is strong and wrong...than somebody who is weak and right." Alli, Gordon and O'Hanlon said: "Of course, America will respond most enthusiastically to the party that promises to keep America strong and right." (Allin et al., 2003).

The statements above show that the Democrat party was also willing to finance defence and to use military power if and when needed. However, that maximalism was balanced with the US position in the international world, which was honourable besides to being feared. This means to emphasise that the war over terrorism using military power was assembled with the vision of the increase of global community welfare, in which the US had a role. The connotation from this Democratic party paradigm refers to the US military power that is capable of facing a threat, and its sustainable foreign policy in the international world. Chuck Hagel, in his article "*A Republican Foreign Policy*," clearly explained that maximalism is central to the Democratic party's paradigm: "Traditionally, a Republican foreign policy has been anchored by a commitment to a strong national defense...that a successful foreign policy must be not strong but sustainable." (Hagel, 2004). This "sustainable" demands patience, as stated by the former US Secretary of State Condoleezza Rice in her speech on the fourth anniversary of the Iraqi War: "I would ask the American people to be patient. We have invested a lot, it is worth the sacrifice." (Rice, 2007).

The matter which is important to be noted when wanting to see the process of a state's foreign policy formulation is the nation's interests. Therefore, when we want to know the process of the US foreign policy formulation towards the Middle East, then what is important for us to understand are the interests of the United States of America itself. According to Bowman

(2008, p. 78), there are three US major interests in the Middle East. First and in the long term, is its interest to secure and have unhindered oil flow from the Persian Gulf region to the United States and the other industrial countries.

To achieve that interest, therefore, the Western countries need to secure the oil reserves from terrorist nuisance and enemy states. The United States of America's second interest in the Middle East is to ensure that either the actor of the state or the non-state in that region does not develop, gain, or use weapons of mass destruction (Bowman, 2008: 79). Currently, the state in the Middle East that has become a threat to the US interest is Iran, where the country proudly shows off the nuclear technology that it has and conducts megaphone diplomacy to the Western countries that hinders their effort, so easily ignoring the resolutions and sanctions imposed upon them by the UN.

In connection with the US national interests in the Middle East, according to Robert D. Blackwill and Walter B Slocombe, the US national interests in the Middle East, among others are: preventing proliferation of weapons of mass destruction, mainly nuclear weapons; fighting against terrorism and the ideology of radical Islam from its root; promoting the transition process to the democracy and developing economic development in the region; preventing the spread of the Iranian influence and their partner as well as their proxy; ensuring the availability of oil and gas with reasonable prices; resolving Arab-Israel conflicts through a negotiation process; and protecting Israel's security (Blackwill & Slocombe, 2011).

Some of those US national interests in the Middle East mentioned above can be identified as being potentially threatened by the conflict in Syria. Those interests are: preventing proliferation of weapons of mass destruction, fighting against terrorism, preventing the spread of the Iranian and Russian influences, and protecting Israel's security that has a potential to be disrupted with the upheaval in Syria.

2.1.2 *The politics of minimalism*

In a general sense, a minimalist is the person or practitioner who proposes changes either in the government bureaucracy, politics, economics, or law on a small scale. Minimalism is an inherent minimalist principle even though they are different from one another. Cass R. Sustein stated: "There are different from minimalism, but all of them share a preference for small steps over large ones." (Sustein, 2006). A preference for small decisions that are made by political minimalism is not caused by an ideology restriction, but it is closely related to prudence in making a decision or an action. There are two concepts that refer to that restriction, which are "Narrowness" and "Shallowness". Caas R. Sustein stated:

> "Narrow rulings do not venture far beyond the problem at hand, and attempt to focus on the particulars of the dispute before the Court. When presented with a choice between narrow and wide rulings, minimalist generally opt for the former. … Shallow rulings attempt to produce rationales and outcomes on which diverse people can agree, notwithstanding their disagreement on or uncertainty about the most fundamental issue." (Sustein, 2006, p. 364).

What is interesting from the view of minimalist politics is that small changes taken as an acting paradigm basically cannot be interpreted as something normative. Even though minimalist politics respects the prevailing tradition and customs, this does not mean that it cannot be changed; a change can still be done after going through a careful analysis, and involving all existing views. This approach refers to 'civic respect' by admitting rather than refusing principles which are valid in the society. Changes to the status quo are done in the way which is accommodative, consensual, learning and adjustment, as well as understanding towards a different perspective. Ronald J. Terchek in *Political Metaphors: Markets or Oligopolies?*, clearly concluded this as follows: 'This perspective, a minimalist conception of politics must always focus on accommodation and adjustment.' (Terchek, 1999). The basic concept of minimalist politics is always to focus towards accommodative character, and adjustment to the situation and condition of the region or the state where the political crisis happens.

In this context, the author needed to explain that there was a shifting in the US people's political paradigm in dealing with the case that happened in Syria. The situation that hap-

185

pened in the Middle East region and which now has also happened in Syria, made the US issue foreign policy that showed a change; the US realism ideology was not shown in what they did to Syria. As delivered by Michael Mandelbaum:

> "The greatest immorality for realists is allowing their country to be destroyed. Humanitarian interventions that drain resources from states and make them ill-prepared to deal with challenges to their vital interests must not be undertaken. Foreign policy should not be social work." (Mandelbaum, 1996).

During the administration of President George W. Bush, the Al Qaeda terrorist attack on the World Trade Center and the Pentagon on September 11, 2001 returned to the US its credibility as the sole superpower state in the world; therefore unilateral politics was considered as the only option for the US. President Bush had to make a quick decision and take firm action, without paying attention to his allies' interests or the international community. Fighting against terrorism and supporting democratisation in the world are based on 'regime change'. This strategy was successful in destroying the Taliban regime in Afghanistan and Saddam Hussein in Iraq, which created a new balancing power that benefited the US in the South Asia and the Middle East regions. The policy that was currently applied was the foreign policy which is based on the option of "zero sum game" where "the winner takes all". President Bush's foreign policy, which was aggressive and global, was labelled as being "an angry Leviathan" by the international world (Kagan, 2008). By doing an intervention to the state which the United States considered could hinder the achievement of the US national interests.

That incident showed that the maximalist politics which were previously adopted could immediately shift to the politics of "maglicant maximalist" when the US national interests were put at stake due to those changes. Richard Holbrooke (2008, p. 4) stated: "It is a well-established historical fact that what candidates say about foreign policy is not always an exact guide to what they will do if elected". The maximalism is the United States of America's political tradition alongside the minimalism that moves from one series into another series' end.

Public speeches during presidential election political campaign and views which were expressed by Barrack Obama in his books clearly showed a combination of a visionary and minimalist leadership. His emphasis towards a change that demanded the presence of consensus, learning and being accommodative, referred to the minimalist concept. The change which Obama meant included his ambitious plans for energy independence and educational reform, as well as approving aggressive action towards climate change. Meanwhile, it was his firm rejection of the standard of social division between the "red state and blue state" liberal, conservative, the Democrats, the Republicans, and the Guantanamo Prison shutdown after being inaugurated as the president. In the 2004 Democratic convention, Barack Obama said:

> 'We worship an awesome God in the Blue States, and we don't like federal agents poking around our libraries in the Red States. We coach Little League in the Blue States, and yes we've got some gays friends in the Red States. There are patriots who supported the war on Iraq' (Sustein, 2008).

The political paradigm transition which was reported by the US in facing critical issues that touched the US national interests, such as expansion of nuclear weapons, international terrorism, democratisation and the US relationship with their allies and other countries. National interests that could be compromised or changed could be observed through President Barrack Obama's views in his article *A New Strategy for a New World* or in *Reviewing American Leadership*. It was visible that the US minimalist foreign policy referred to the "offshore balancing" strategy.

Barrack Obama, in the 44th presidential election campaign, clearly highlighted minimalist politics as being his political ideology. In his book, *The Audacity of Hope*, Obama wanted to create politics that could accept "the possibility that the other side might sometimes have a point". It was this view that made him willing to negotiate with any country, including the US's opponents. His victory speech in Iowa clearly reflected minimalist politics: 'The time has come for a president who will listen to you and learn from you, even when we disagree.' Like the other minimalist political views, President Barrack Obama's also emphasised consensus, being accommodative and learning. Judging from the domestic security condition and balance of power in the world that benefited the US, which had been established by President Bush earlier, minimalist politics was therefore the right choice for President Barrack Obama.

The US used minimalist foreign policy as an instrument to achieve national interest in Syria. To achieve it, the state as a rational actor attempted to choose each alternative option to maximise the benefit and to minimise the costs which were received. In analysing cost and benefit received by the state, this research would use 'rational choice' theory. Rational choice theory in international relations study was formed at the beginning of the 1960s. Rational choice theory is defined as being an instrument about purpose and objective, or choice from the actor's focused objective (Jackson & Sorensen, 2009). The US clearly considered cost and benefit gained when formulating their minimalist foreign policy in Syria.

The foreign policy formulation process is done by the actors who respectively have a role as players. The relationship among the actors in general is described in the process of "pulling and hauling" one another. Foreign policy is understood as "political outcomes" (Allison, 1971). According to Allison, "outcomes" are not the "completion which is" chosen by the actors, but they are the results from a compromise, coalition, and competition among actors. Capability and expertise from the actors themselves are what determines the outcomes from the process of decision making (Allison, 1971, p. 38). The US minimalist foreign policy as the rational actor towards the conflict that was happening in Syria gave an opportunity to the power existing in the region that had interests in Syria to gain benefit.

The US is the state that positions itself as the rational actor who will choose an alternative policy that has highest consequence (beneficial) in fulfilling their goals and objectives (Allison, 1971). The US minimalist foreign policy in Syria realised the consequence that would happen, by considering a policy to achieve its national interests and remaining to get benefit from the policy applied in Syria. Allison said, "The rational agent selects the alternatives whose consequences rank highest in terms of his goals and objectives". This showed the US minimalist foreign policy which was applied in Syria.

The US created minimalist foreign policy to keep its influence in the Middle East region. The US was involved in the conflict in Syria but it was seen that the attitude the US showed was to give limited assistance and also to support opposition groups to throw out the Assad regime. The minimalist political paradigm was indeed the attitude showed by the US to the Syrian democratisation and was a rational choice; there were many conflicts of interests in Syria, so the US took minimalist foreign policy as being the rational choice to still having Syria.

The US minimalist foreign policy in Syria was to weigh down everything that would happen. A cautious factor in determining foreign policy applied in Syria as the manifestation of minimalist politics, the US did the "offshore balancing" strategy and gave assistance to the opposition groups who opposed Bashar al-Assad's regime.

3 CONCLUSION

From the US minimalist foreign policy towards the democratisation in Syria, it could be concluded that the US showed an attitude which tended to be seen as ambiguous by the international world. The US minimalist foreign policy was the rational choice that considered goals and objectives, which was done by the US towards the political crisis that was happening in Syria. The US during the administration of President Barrack Obama, in determining US foreign policy took a minimalist political approach, where the approach was done in a more supportive manner and used power balance that existed in the Middle East region and political crisis that happened in Syria.

Unlike the previous US government, President Bush used maximalist politics to determine US foreign policy in order to achieve its national interests. The power owned by the US, such as military power and economic force, was used to show to the international world that the US was the only *superpower* country that had influence in the Middle East region. This was done without considering goals and objectives, nor the cost and benefit that the US would get from the foreign policy applied in the Middle East region.

The policy which was applied in the Middle East region was to promote democracy, fight against terrorism, eliminate mass destruction weapons, maintain a stable oil price, guarantee stable security and maintain human rights. The invasion of Afghanistan and Iraq showed the US as applying a policy in using all power by the Americans without calculating potential domestic sentiment arising, or the response from the international world, in determining its foreign policy.

The shifting of the US foreign policy paradigm from maximalist to this minimalist approach happened because of a rational choice that was taken by the US. The policy showed by the US saw that there was conflict of interests using the political crisis situation in Syria. The policy applied by the US of using minimalist foreign policy showed that Syria did not have a strategic meaning to the US national security interests; the US also wanted to keep its influence in the Middle East region based on what became the foundation of the US minimalist foreign policy in Syria.

REFERENCES

Abouzeid, R. (2015). *The youth of Syria: 'The rebels are on pause'*, Time World (online), http://www. time.com/time/world/article/0,8599,2057454,00.html 2012, diakses pada 1 September 2015.

Allin, D.H., Gordon, P.H. & O'Hanlon, M.E. (2003). The Democratic party and foreign policy. *World Policy Journal, 20*(1), 7–16.

Allison, G.T. (1971). *Essence of decision: Explaining the Cuban missile crisis*. Boston: Little, Brown and Company.

Creswell, J.W. (1994). *Research design: Qualitative and quantitative approaches.* California: SAGE Publications.

Dahl, R. & Lindblom, C. (1994). Politics, economic and welfare (Harper 1953). In M. Mas'oed (Eds.), *Ilmu Hubungan Internasional: Disiplin dan Metodologi* (pp. 38). Jakarta: LP3ES.

Hagel, C. (2004). Republican foreign policy. *The Article Alert.* Oct. IRC, U.S Embassy JKT.

Holbrooke, R. (2008). The next president: Mastering a daunting agenda. *Foreign Affairs*, 2–24.

Holsti, O.R. (2002). Models of international relations and foreign policy. In G.J. Ikenberry (Ed.), *American Foreign Policy: The Dynamics of Choice in the 21st Century*. Georgetown: Addison-Wesley Educational Publisher Inc.

Jemadu, A. (2008). *Politik Global Dalam Teori dan Politik.* Yogyakarta: Graha Ilmu.

Kagan, R. (2002). Power and Weakness dalam *America and the World Foreign Affairs*, New York: Council and Foreign Affairs Relations, Inc.

Kagan, R. (2007). End of dreams, return of history. *Policy Review: August/September 2007*. Research Library Core, Hoover Institution.

Kagan, R. (2008). The September 12 paradigm: America, The world, and George W. Bush. *Foreign Affairs, 87*(5), 25–39.

Kartini, I. (2013). Demokrasi dan Politik Islam di Tunisia. *Nasion, 10*(2), 70–82.

Kissinger, H.A. (1969). *American foreign policy: Three essays.* Washington: W.W. Norton & Company.

Korban Tewas Konflik Suriah. (2016). http://www.tempo.co/read/news/2014/05/19/115578830/Korban-Tewas-Konflik-Suriah-Capai-162-Ribu-Orang. diakses pada 3 Februari 2016.

Machiavelli. (2005). In R. Jackson & G. Sorensen, *Pengantar Studi Hubungan Internasional*, Translated: Dadan Suryadipura, Yogyakarta, Pustaka Pelajar.

Mandelbaum, M. (1996). Foreign policy as social work. *Foreign Affairs, 75*(1), 16–32.

Mas'oed, M. (1997). *Studi Hubungan Internasional: Tingkat Analisa dan Teorisasi*, Yogyakarta, PAU Universitas Gajah Mada.

May, R.T. (2002). *Study Strategis Dalam Transformasi Sistem Internasional Pasca Perang Dingin.* Bandung: Refika Aditama.

McCormick, J.M. (1998). *American foreign policy and process*. Illinois: F.E. Peacock Publishers, Inc.

Muthalib, A. (2008). Politik Luar Negeri Maksimalis Amerika Serikat. *Nasion, 5*(1), 111–124.

Obama, Barack. (2007). Renewing American leadership. *Foreign Affairs, 86*(5), 61–74.

Page, B.I. (1976). *The American political.* Chicago: University of Chicago.

Phillips, M. (2011). President Obama: *'The future of Syria must be determined by its people, but Presiden Bashar al-Assad is standing in their way'*. The White House Blog (online), August 18, 2011, http://www.whitehouse.gov/blog/2011/08/18/president-obama-futuresyria-must-be-determined-its-people-president-bashar-al-assad.

Plano, J.C. & Olton, R. (1999). Kamus Hubungan Internasional. Bandung: Abardin, 1999.

Posen, B.R. (2007). Stability and change in U.S. grand strategy. *Foreign Policy Research Institute, 51*(4), 561–567.

Rice, Condoleezza. (2007). April 2, Time.

Rosenau, J.N. (1976). *World politics: An introduction.* New York: The Free Press.

Sestanovich, S. (2005). American maximalism. *The National Interest, 79*, 13–23.

Sulaeman, D.Y. (2013). Prahara Suriah: *Membongkar Persekongkolan Multinasional.* Depok: Pustaka IIMaN.

Sustein, C.R. (2006). Burkean minimalist. Michigan Law Review. *105*(2), 353–362.

Terchek, R.J. (1999). Political metaphors: Markets or oligopolies? *Annual Meeting of Political Science Association,* Boston September 3–6, 1998. Published in *Associations, 3*(2), 165–183.

Von Clausewitz, K. (1982). On war. England: Penguin Books.

Zakaria, F. (2008). What Obama should say on Iraq. *Newsweek,* June 30, 2008.

Competition and Cooperation in Social and Political Sciences – Adi & Achwan (Eds)
© *2018 Taylor & Francis Group, London, ISBN 978-1-138-62676-8*

Dynamic governance in managing the urban environment— A conceptual framework of urban water governance

W. Mulyana & E. Suganda
School of Environmental Sciences, Universitas Indonesia, Jakarta, Indonesia

ABSTRACT: Rapid urbanisation creates new challenges and problems in environmental management and sustainable development. Urban environmental issues, including the degradation of the physical, social and economic environment, are complex and dynamic, requiring integrated and anticipatory problem solving. Dynamic governance is an approach that can create leverage to solve the complex problems and respond to environmental changes in a rapid, precise and innovative way. This article discusses the preliminary conceptual framework to build dynamic governance in managing the urban environment, particularly in urban water governance. It examines the attributes of dynamic governance in public sector governance and the attributes of adaptive governance in environmental governance. It contributes to interdisciplinary research in the conceptual understanding of the linkages of dynamic governance, adaptive governance and the public policy process. The expected result is that dynamic governance in urban environmental governance would be better executed if there existed an adaptive policy built on the interaction between the dynamic capabilities and the adaptive capacity of organisations and actors in all levels of government. This proposed method in applying dynamic governance in urban water governance is introduced as guidance for further research.

1 INTRODUCTION

Urban environmental conditions are under increasing pressure due to rapid urbanisation, population growth, climate change and economic development. Not only does rapid urbanisation accelerate social and economic development, but it also creates environmental problems, from the local to the global scale (Chui & Shi, 2012). Urban areas are major contributors to climate change because, although they comprise less than two per cent of the earth's surface, they consume more than 78 per cent of the world's energy and produce more than 60 per cent of carbon dioxide emissions (UNEP, 2013). At the same time, climate change is causing unavoidable impacts on urban systems and their populations (Tyler & Moench, 2012), particularly amongst vulnerable communities and the urban poor (da Silva et al., 2012). These pressures will continue to occur with increasing uncertainty, such that a transition in the approach to urban planning and management in the future is needed.

The urban area is a complex ecological system dominated by humans, where human factors greatly influence distinctive ecological patterns, process, disturbances and subtle effects (Alberti, 2008). In dealing with a complex system, urban planners need to understand the processes, mechanisms and relationship between ecological and social factors, as well as to control the dynamics and evolution of the changing process. Sustainable urban development can be achieved when the capacity of the urban ecosystem to respond and adapt to change is supported by urban environmental governance as a major prerequisite. The concept of dynamic governance emerged as an approach adopted by the government to the current development in an uncertain and fast-changing environment.

The study of the dynamic governance is carried out from the experience of Singapore in implementing its development and is emphasised more on public sector governance. Dynamic governance is the outcome of the capacity to develop adaptive pathway and policy in the face of new challenges, uncertainties and technological development (Neo & Chen, 2007). It is a condition of an ideal government that is achieved as an outcome of the development of adaptive pathways and policy, which are derived from interactions between dynamic capabilities, able people and agile processes, and organisational culture. Dynamic governance is also associated with adaptive governance, that is emerging as a new approach in resource management to respond to the complexity and uncertainty of global environmental change (Chaffin et al., 2014). It can also be defined as the ability to continuously improve and adapt to changes in the socio-economic environment (Hatfield-Dodds et al., 2007), and as the result of interaction between actors, networks, organisations and institutions that evolve to achieve desired conditions in the socio-ecological system (Chaffin et al., 2014). Adaptive governance is developed to bridge the problems of coordination amongst actors, sectors and regions, which are often found in centralised top-down approaches, as well as bottom-up approaches.

This article aims to establish a conceptual framework of dynamic governance in urban environmental management that can help decision makers to respond to various changes and uncertainties affecting the sustainability of urban development. The scope of the analysis is focused on the public policy process at the policy, organisational and operational levels. Specifically, there are two objectives in this article. First, it aims to highlight the attributes of dynamic governance and adaptive governance, and examine the link between those concepts in urban environmental management and the public policy process. It seeks to offer new perspectives and links that can be used for further research. Second, it attempts to propose methodological steps in applying the concept of dynamic governance in urban water management. These steps are important to serve as guidance for further research in urban water governance that integrates environmental governance and the public policy process.

The following section reviews the literature on the urban environment as a socio-ecological system, dynamic governance, adaptive governance and a public policy process, while presenting the attributes of each concept to be used in developing a conceptual framework. The next section explains the steps to apply the conceptual framework into future research methodology. The final section outlines the conclusion.

2 LITERATURE REVIEW

2.1 *Urban areas and socio-ecological system resilience*

The urban area is a social and ecological system where ecosystem services, biodiversity, communities and people who live in them, became a major part in the analysis of resilience and urban sustainability. The complexity and dynamics of urban areas have caused various environmental, social and economic degradations that will threaten the sustainability of future urban development. The impacts of climate change and global environmental change further exacerbate urban environmental conditions. Meanwhile, the capacity of city government to adjust and improve urban environmental conditions has always been slower than the speed of change. This condition is suggested in the increasing intensity and frequency of disasters and environmental degradation. Under such circumstances, it is important to understand the role of a city in sustainable development, as well as how the city becomes part of the solution to improve its resilience and sustainability.

Urbanisation is an important socio-economic global phenomenon in the 21st century (Allen, 2009) and carries with it social, economic and environmental challenges (Seto &

Satterthwaite, 2010; Zhang, 2016). More than 50 per cent of the world's population lived in urban areas in 2010 and the number is expected to continue to rise in the next decades (UN-Habitat, 2010). Rapid population growth provides benefits to the city's economic activities, but it also increases pressure and stress on environmental function and quality (World Bank et al., 2011). UN-Habitat (2016) identifies four challenges in the urban environment, which amongst others are: 1) equal access to resources and urban services; 2) the management of disaster risk and climate change impacts; 3) the effect of urban land expansion and its impact on biodiversity and deforestation; and 4) the changes in consumption pattern towards low-carbon development. Urban environmental issues occur at various scales and need to be addressed by involving multi-actor and multi-level governance. The socio-ecological system resilience observes the importance of human response to various environmental changes. Carpenter et al. (2001) pointed out three key aspects in a socio-ecological system, namely: 1) resilience, which in this context refers to the capacity of socio-ecological systems to continuously change and adapt within a certain threshold; 2) adaptability, which is part of resilience, that shows the capacity to respond to changes from internal processes and external factors, such that development can be implemented as intended (stability); and 3) transformability, which is the capacity to cross the threshold into a new development trajectory. These aspects are applied in building the resilience of urban areas as a complex and dynamic socio-ecological system.

2.2 *Dynamic governance*

As mentioned by Neo and Chen (2007), dynamic governance does not occur by chance, but it is the result of deliberate leadership intention and ambition to restructure social and economic interactions in order to achieve desired goals. A dynamic governance system arises when there is interaction between a supportive institutional culture and proactive organisational capabilities in producing adaptive paths and policies, including their effective execution. The main elements of dynamic governance are institutional culture, and the three dynamic capabilities of thinking ahead, thinking again, and thinking across. The main levers in developing dynamic governance capabilities are able people and agile process. These elements lead to adaptive policies. The external environment affects the governance system through future uncertainties and external practices.

Dynamic governance cannot be achieved without the understanding and interdependence between cultures and capabilities, the capabilities with the ability of people and process, the interaction between capabilities and external environment, and the capabilities with the ability to transform into adaptive paths and policies. The success of an organisation in adapting to the changing environment is reflected in knowledge building that is translated into cultural values, in the form of beliefs, rules, policies and organisational structure. Organisational capabilities refer to the knowledge, attitudes, skills and resources given to carry out the important task to achieve the desired objectives (Neo & Chen, 2007). Capability can also be perceived as a different implementation method, which is achieved through a learning process in a certain amount of time. The systemic capability is built into the people and processes, so that good ideas can be translated into realistic policies, plans, and programmes. Dynamic governance capability has three elements, namely: thinking ahead (the ability to think ahead), thinking again (the ability to rethink) and thinking across (the ability to think across).

2.3 *Adaptive governance*

Adaptive governance approaches are described in various definitions and theories and have been applied in many disciplines. The concept of adaptive governance continues to grow and is believed to be one of the new ways in environmental management. The term adaptive governance was first introduced by Dietz et al. (2003), and subsequently Folke et al. (2005) described adaptive governance as a strategy for mediating social conflicts that interfere with adaptive management of a complex ecosystem. Dietz et al. (2003) articulate the need for adaptive governance in a socio-ecological system, because of limited knowledge on a system and governance scale that keeps changing adjusting to the changes in the social and biophysical system components. Therefore, Dietz et al. (2003) underlined the necessity for an adaptive resources governance system that allows continuance of development based on feedback from the components of the human and biophysical, or combinations of those elements. Adaptive governance considers all the interactions that occur at all levels, both cross-level and cross-scale.

Adaptive capacity is the core of the new strategy in linking environmental governance to the capacity in managing resilient socio-ecological systems. Adaptive capacity is broadly defined as the ability of a socio-ecological system, or the components of that system, to be robust to disturbance and capable of responding to change (Walker & Salt, 2006). There are four main factors that help shape adaptive capacity in socio-ecological system (Folke et al. 2002), namely: 1) learning to live with change and uncertainty; 2) nurturing diversity for resilience; 3) combining different types of knowledge for learning; and 4) creating opportunity for self-organisation towards socio-ecological sustainability.

The literature discusses many of the attributes of governance associated with the capacity to increase the resilience of socio-ecological systems (Lebel et al., 2006; Armitage, 2008; Djalante et al., 2011). The identified attributes of governance include: 1) participatory and deliberative; 2) polycentric and multilayered; and 3) accountable and just (Lebel et al., 2006). Armitage formulated key attributes of the adaptive multi-level governance, including: participation, collaboration, deliberation, multiple layers, accountability, interaction, knowledge and pluralism learning. Djalante et al. (2011), in the context of adaptive governance for resistance to natural disasters, combined attributes of governance as follows: polycentric and multilayered institutions, participation and collaboration, self-organisation and networks, and learning and innovation. In the context of urban environmental governance, the attributes used are: multilayer, participation, collaboration, accountability, network, learning and innovation.

2.4   *Level of analysis: Public policy process*

Public policy is an instrument that bridges between the individual decision and the collective decision, while serving as a tool to change the resources allocation support for the community. Public policy is an institutional transaction that confirms all choices in public policy and promotes institutional change in the policy process. Institutional change has consequences for changes in the policy set, which is called institutional arrangements.

Bromley (1989) mentioned the policy process as a hierarchy, which consist of three levels: policy, organisation, and operation. The policy level is the formulation of public policy in the form of legislation and national policies, represented by the legislative and judiciary bodies. The organisational level is the second level of public policy, after the policy level, which is found in the form of institutional arrangements and technical regulations to carry out policies formulated in the form of Government Regulation, Presidential Decree, and Ministerial Decree, development plans and programmes, safeguard policy and funding support. The operational level is the implementation of the set policies, both at policy and organisation level, carried out by government agencies, firms and communities.

2.5   *Theoretical framework*

The urban area is a complex socio-ecological system that occurs as the process of interaction between humans and the environment. The challenges of future urban development lie in the degradation of the environment, the increasing risk of natural and man-made disaster, as well as the impact of climate change. We therefore need a new form of urban environmental management, that is more dynamic and adaptive to internal changes that occur and to disturbances that arise from the outside, that may threaten urban sustainability. Dynamic governance and adaptive governance are new approaches to environmental management that are more adaptive to the dynamic and changing environment. The theoretical framework for dynamic governance and adaptive governance in managing the urban environment can be seen in Figure 1.

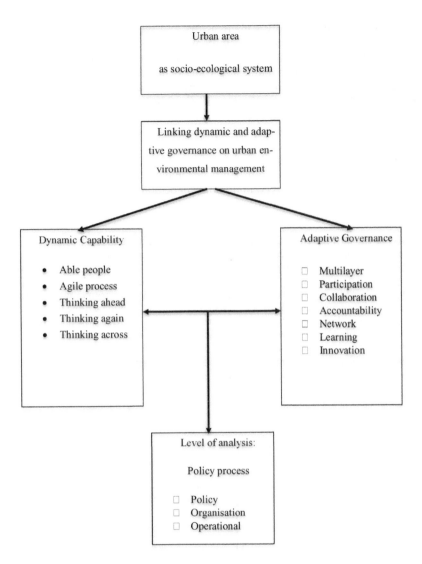

Figure 1.   Theoretical framework.

3   PROPOSED METHOD

Water is the key of sustainable urban development, and the pillar of public health and social welfare (WHO, 2012). Inadequate water and sanitation threaten human and environmental health. Water has multiple functions, in terms of society, the economy and the environment, which are needed to support the production process and contribute to economic growth and poverty reduction. Despite improved access to water and adequate sanitation in the past two decades, many countries are still facing the challenges to fulfil the needs of water and sanitation. The provision of urban basic services is under pressure from global and regional changes that affect efficient and effective environmental management. Urban water governance is closely related to a series of activities in the water cycle, or each activity in the water cycle, starting from: 1) taking water from the source (groundwater and surface water); 2) purification of water into drinking water; 3) distribution of drinking water to users; 4) collection and disposal of waste; and 5) waste processing (van Monfort et al., 2014).

Urban water governance involves many actors and sectors at different jurisdiction levels. The use of water is interconnected and may create potential conflict. Various uses of water have produced waste water, that is causing pollution if not properly managed and is threatening human and environment health. Water management is usually carried out within administrative boundaries, where various government institutions in different sectors perform limited coordination functions. In line with these jurisdictional and sectoral approaches are inter-sector, inter-regional and inter-actor policy coherence. Policy and decision-making in one sector may contradict or duplicate policies and decisions in other sectors. Taking advantage of its multi-actor and cross-sector approaches, the water issue can be used as a tool to develop an integrated solution.

Applying dynamic governance in urban water management employs different quantitative and qualitative methods. This combination will be done through a triangulation process where data collection and analysis are conducted simultaneously.

The proposed methods should be conducted using the following steps (Figure 2):

1. Identifying the current conditions of urban water governance:
   a. Stakeholder mapping is aimed to map and analyse key actors involved and their roles and relationships in public policy process at different levels of government. An actor is defined as an organisation or individual who has a significant role and is directly benefited from urban water management, be it from government institutions, non-governmental organisations, private sectors or communities. Further stakeholder analysis should be done in order to examine their interests and relationships in policy formulation and the decision-making process.
   b. The current issues of urban water governance, including regulatory framework, policy, institutional arrangement, financing, coordination and implementation.
2. Analysing key factors influencing urban water governance at policy, organisational and operational levels. This analysis is aimed at building a statistic equation model using

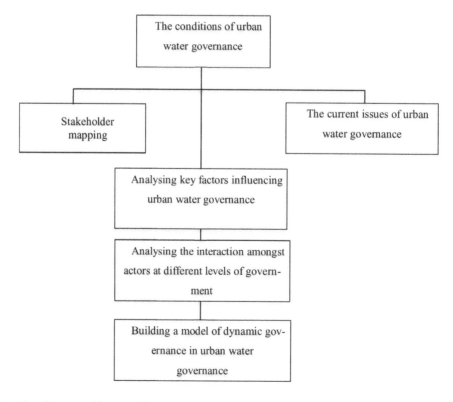

Figure 2. Conceptual framework.

Structural Equation Modelling (SEM) to determine the relationship amongst variables of dynamic capabilities and adaptive capacity in public policy and urban water governance. The SEM analysis uses primary data obtained directly from interviews with key actors as respondents. The expected output is the identification of key factors influencing dynamic urban water governance in the policy process at all levels.

3. Analysing the interaction amongst actors at different levels of government in urban water governance.

4. Building a model of dynamic governance in urban water governance using Soft Systems Methodology (SSM) and Scenario Planning. These methods are used to help in structuring the opinions and thoughts from organisations and key actors in dealing with the complexity and changes in future urban water governance. This model will be developed based on the opinions and ideas of key actors in the urban water sector at different levels of government.

4 CONCLUSION

Urban environmental conditions continue to be degraded by a variety of pressures, such as rapid urbanisation, population growth, economic development, climate change and natural disaster. Better urban environment management becomes important to deal with climate change and other global changes that would threaten the sustainability of urban development. Traditional urban environmental management is mechanistic and technocratic, and it excludes the complexity and dynamic of environmental changes, as well as the human dimension as the key actor. The critical problems of the urban environment are often unrelated to the environment itself, but are more influenced by governance failure. The water crisis in the urban area is an example that reflects such failure.

The future challenges of urban environmental management lie in developing a governance system that is able to anticipate and incorporate external and internal changes by accommodating aspects of dynamic capabilities and adaptive capacity of the organisations and the actors involved. Understanding dynamic governance and adaptive governance of the urban environment is important in implementing urban management in the context of an ever-changing socio-ecological system.

The conceptual framework discussed in this article is to build a more dynamic and adaptive urban environmental governance, especially in the public policy process at all levels of government (central, provincial, district/city), involving all actors. This concept emphasises the importance of building the dynamic capabilities and adaptive capacity of organisations or individuals involved in the public policy process, starting from policy formulation, implementation and evaluation. The conceptual framework should be applied in urban water governance as a key resource in sustainable urban development. It offers guidance for further research in urban water governance.

REFERENCES

Alberti, M. (2008). *Advances in urban ecology: Integrating humans and ecological processes in urban ecosystems*. LLC, New York: Springer Science+Business Media.

Allen, A. (2009). Sustainable cities or sustainable urbanisation? *UCL Journal of Sustainable Cities*.

Armitage, D. (2008). Governance and the commons in a multi-level world. *International Journal of Commons, 2*, 7–32.

Bromley, D.W. (1989). *Economic interest and institutions: The conceptual foundations of public policy*. New York: Basil Blackwell.

Carpenter, S.R., Walker, B.H., Anderies, J.M. & Abel, N. (2001). From metaphor to measurement: Resilience of what to what? *Ecosystems, 4*, 765–781.

Chaffin, B.C., Gosnell, H. & Cosens, B.A. (2014). A decade of adaptive governance scholarship: Synthesis and future directions. *Ecology and Society, 19*(3), 56. http://dx.doi.org/10.5751/ES-06824-190356 Synthesis.

Chui, L. & Shi, J. (2012). Urbanization and its environmental effects in Shanghai, China. *Urban Climate*, *2*, 1–15.

da Silva, J., Kernaghan, S. & Luque, A. (2012). A systems approach to meeting the challenges of urban climate change. *International Journal of Urban Sustainable Development*, *4*(2), 125–145.

Dietz, T., Ostrom, E. & Stern, P.C. (2003). The Struggle to govern the commons. *Science*, *302*, 1907–1912.

Djalante, R., Holly, C. & Thomalla, F. (2011). Adaptive governance and managing resilience to natural hazards. *International Journal Disaster Risk Science*, *2*(4), 1–4.

Folke, C., Carpenter, S., Elmqvist, T., Gunderson, L., Holling, C.S. & Walker, B. (2002). Resilience and sustainable development: Building adaptive capacity in a world of transformations. *Ambio*, *31*(5), 437–440.

Folke, C., Hahn, T., Olsson, P. & Norberg, J. (2005). Adaptive governance of social-ecological systems. *Annual Review of Environment and Resources*, *30*, 8.1–8.33. *International Edition*. Canada: Brooks/ Cole Cengage Learning.

Hatfield-Dodds, S., Nelson, R. & Cook, D. (2007). Adaptive governance: An introduction and implications for public policy. *51st Annual Conference of the Australian Agricultural and Resource Economics Sectors*.

Lebel, L., Anderies, J.M., Campbell, B., Folke, C., Hatfield-Dodds, S., Hughes, T.P. & Wilson, J. (2006). Governance and the capacity to manage resilience in regional socio-ecological systems. *Ecology Society*, *11*(1), Article 19.

Neo, B.S. & Chen, G. (2007). *Dynamic governance—Embedding culture, capabilities and change in Singapore*. Singapore: World Scientific Publishing. Co. Pte. Ltd.

Seto, K.C. & Satterthwaite, D. (2010). Interactions between urbanization and global environmental change, Editorial overview. *Current Opinion in Environmental Sustainability*, *2*, 127–128.

Tyler, S. & Moench, M. (2012). A framework for urban climate resilience. *Climate and Development*, *4*(4), 311–326.

United Nations Environment Programme. (2013). *Integrating the environment in urban planning and management key principles and approaches for cities in the 21st century*.

United Nations Human Settlements Programme (UN-Habitat). (2010). *State of the World Cities 2010/2011*.

United Nations Human Settlements Programme (UN-Habitat). (2016). *Urbanization and development: Emerging futures*. Nairobi: World Cities Report 2016.

van Monfort, C., Michels, A. & Frankowski, A. (2014). *Governance models and partnerships in the urban water sector. A framework for analysis and evaluation*. Utrecht & Tilburg, The Netherlands, The Hague.

Walker, B. & Salt, D. (2006). *Resilience thinking: Sustaining ecosystems and people in a changing world*. Washington: Island Press.World Bank, Bappenas, Swiss Economic Development Corporation, Australian AID. (2011). *Indonesia: The rise of metropolitan regions: Towards inclusive and sustainable regional development*. Jakarta, Unpublished Report 71740.

World Health Organization. (2012). *UN-Water global annual assessment of sanitation and drinking-water (GLAAS) 2012 report: The challenge of extending and sustaining services*. Switzerland.

Zhang, X.Q. (2016). The trends, promises and challenges of urbanisation in the world. *Habitat International*, *54*(3), 241–252.

Competition and Cooperation in Social and Political Sciences – Adi & Achwan (Eds)
© *2018 Taylor & Francis Group, London, ISBN 978-1-138-62676-8*

Israeli women's political participation in the peacemaking process between Israel and Palestine

M. Jamilah & Apipudin
School of Postgraduate, Department of Islamic and Middle Eastern Studies,
Universitas Indonesia, Jakarta, Indonesia

ABSTRACT: This study is conducted to explore the Israel-Palestine conflict from a postcolonial feminism perspective by enlightening the political participation of Israeli women in the peacemaking process between Israel and Palestine. Feminism generally assumes that women are inherently more peaceful than men and gender equality will naturally support policy that leads to peace. Nevertheless, Israel, which scores 70% in the Global Gender Gap Index and which has become the country with the highest level of gender equality in the Middle East region in 2015, still faces military confrontation with Palestine until recent day. The contradiction between feminist assumptions and facts which occurred in Israel will be analysed using the qualitative method and descriptive analysis approach of Miles and Huberman. The result of this study shows that there are two challenges for the Israeli feminist movement. The first is the challenge that comes from patriarchy domination in social life and the second is the challenge from colonialism values that affect Israeli feminism.

1   INTRODUCTION

Israel has the highest level of gender equality compared to all other states in the Middle East region, which is 70% based on the Global Gender Gap Index in 2015 (100% represents total equality) (World Economic Forum, 2015). This accomplishment cannot be separated from the rise of the feminist thought phenomenon that has occurred since the 1970s. There were two professors of Haifa University, Marcia Freedman and Marlyn Safir, who started a study of women in order to arouse feminism in Israel (Maor, 2000).

In the 1980s, feminist thought was articulated into feminist movement which protested against the Israeli government regarding the peace agenda issue. It started when Egypt's President Anwar Sadat visited Israel in 1978, which gave hope of peace between Israel and Egypt. Then, this spirit and hope created a peace movement named *Peace Now* (Helman, 2009). That was the first peace movement in Israel which included women in their activities.

Women's participation in conflict resolution and peace was admitted globally in 2000 in the resolution of the Security Council of United Nations. Resolution 1325 was published on 31 October 2000. The resolution relates to women, peace, and security issues. It also emphasises the importance of gender equality, and active participation of women in creating peace and maintaining international security (Finkel, 2012). The resolution had been implemented in Israel. Four years after adopting it, the Knesset Research and Information Centre published a report *Integration of Women in Peace Process*. Besides that, the Knesset approved a constitution about gender equality and the rights of women on 20 July 2005 (Finkel, 2012).

Despite the recognition of women's participation in the peacemaking process having increased both globally and domestically, conflicts and military confrontations still happen between Israel and Palestine. On 15 September, Association Press (AP) reported that violence engulfed the Temple Mount and the Muslim Quarter of Jerusalem's Old City for three consecutive days (Jerusalem Post, 2015). In addition, according to ABC News on 19 October 2015, it was reported that there were clashes between Israel and Palestine in occupied territory and this media also reported that 2,000 Palestinians had been injured (ABC News, 2015). This shows the fact that the 70 years of prolonged conflict between these two states has not ended yet.

These two facts about Israel are contradictory: whereas women's participation in the agenda of the peace process has been recognised by the Israeli government, the conflict between Israel and Palestine is still the main problem until now. Therefore, the question in this research is, how women's political participation in Israel in the peace process, and why is the participation still not affecting the peace process between Israel and Palestine. These two questions need to be studied and explained further.

2 RESEARCH METHOD

The questions of this research are answered by using qualitative research method with descriptive analysis approach. The qualitative research method aims to find a comprehensive conclusion about specific phenomena of the individual or a group of people (Lambert & Lambert, 2012). In analysing the data, this research uses Miles and Huberman's mode of analysis which consists of three steps: data reduction, data display, and the last stage is conclusion drawing or a verification stage (Miles & Huberman, 1994).

There are two basic concepts used in this article. The first is women's participation in conflict and peace by Joyce P. Kaufman and Kristen P. Williams, while the second is political participation by Barnes and Kaase. These concepts are used to describe Israeli women's political participation in the peacemaking process, by classifying whether women participate as the combatants or the peacemakers (Kaufman & Williams, 2013). These concepts study whether or not women's participation can be classified as a political participation. If it can, in which variety of political participations between conventional political participation and unconventional political participation do Israeli women participate? (Verba, 1972). Postcolonial feminism theory of Iris Marion Young (2003) is executed in the next step of analysis to answer the question of why Israeli women's political participation still cannot affect peace between Israel and Palestine. It will explain barriers and challenges to the Israeli feminist's movement (Young, 2003).

3 DISCUSSION

3.1 *Israeli women's participation in conflict and peace*

Participation of women in conflict and the peacemaking process depends on how women consider the conflict and peace themselves. There are women who support the peacemaking process, but there are also women who participate in conflict, as members of the military service. This statement is parallel with Joyce P. Kaufman and Kristen P. Williams's thought in *Women at War, Women Building Peace: Challenging Gender Norms* (2013). In the book, they used the concept of combatants and peacemakers as a dichotomy of two kinds of women's participation (Kaufman and Williams, 2013).

Israel is one of the states in the world which sets conscription for all of its citizens, both men and women, who are 18 years old or who have finished their study. This constitution is decided by the government based on Israel's circumstance that the country is surrounded by Arab countries which have rejected Israel's existence in the Middle East (Maor, 2010). The consequence of this rule can be seen from the fact that 30% of Israeli Defence Force (IDF) regular members are women (Nevo & Shur, 2003).

Before the establishment of the Israeli state, women had represented 20% of the military organisation membership. In *Palmach*, in the Israeli defence underground organisation pre-state era, women participated in various kinds of fields (Nevo & Shur, 2003). In 2000, Israeli parliament released an amendment of military service law which determined equality rights for women to choose their professions in military service (Lubin, 2002). The amendment gave more access for women to join the IDF, as data shows that 33% of soldiers who served in Israel are women, 65% of qualified women are recruited by the IDF, 24% of all regular officers are women, and 16% of carrier officers are women (Lubin, 2002).

Although some Israeli women had participated as combatants, which is opposite to women's peaceful nature, the other Israeli women have been struggling to create a better future for both Israel and Palestine by participating as peacemakers. This participation is affected by Resolution 1325 of UN Security Council in 2000. However, Israeli participation in the peacemaking process between Israel and Palestine has been actually afforded by women since years before the resolution. This article describes five of the Israeli women's peace movements and organisations; those are *Peace Now*, *Women in Black*, *Bat Shalom*, *Four Mothers*, and *Machsom Watch* (Maor, 2010).

Peace Now was established in the peace negotiation process between Israel and Egypt in 1978. While the negotiations between the two states was stuck, a group of army reservists from the IDF, which consisted of 348 people, sent an open letter to the Israeli prime minister. The letter contained their demand to ensure that the peace opportunity between Israel and Egypt would not vanish. This action then led some Israeli citizens to support them and they signed the letter. Finally, in 1979 Israel and Egypt signed a historic peace treaty. This phenomenon made *Peace Now* members and Israeli citizens realise the strength of public pressure in determining government policy, especially in the peace process between the Israelis and Arabs (*Peace Now*, 2016).

The second feminist movement is *Women in Black*. It was established in Jerusalem in January 1988 as a response to the first Palestinian Intifada. This movement had an international link that had a commitment to create peace and justice in the world. *Women in Black* actively opposed injustices, military actions, and other forms of violence. According to *Women in Black*'s official website, the main concern of this movement is to challenge militaristic policies of the government. This movement has also been considered as being an anti-occupation movement and the saying 'Stop the Occupation' is the main message of *Women in Black*. Nowadays, this movement has 30 guarded posts all over Israel and has four regular guard posts which have held demonstrations since 1988 (http://womeninblack.org/).

The next movement is called *Bat Shalom*. It is the national feminist organisation in Israel which consists of Israeli and Palestinian women who work together to create peace and conflict resolution for Israel and Palestine. As stated in Americans United for Palestinian Human Rights, this organisation respects human rights and demands equal rights for Jewish and Arab women in Israeli society (http://www.auphr.org/index.php/links/israeli-organizations).

Another movement is called *Four Mothers*. It is a protest movement that demands the withdrawal of the Israeli army from occupied territories in Southern Lebanon. Israel invaded Lebanon on 6 June 1982 as an effort to solve security issues on the northern border (Lemish and Barzel, 2000). This movement was established after a tragedy which involved two military helicopters on 5 February 1997. The accident killed 73 Israeli soldiers who were assigned to the south of Lebanon. Four mothers, who are the actual mothers of soldiers killed in this accident, are the pioneers of this protest movement. A year after the accident, this movement already had 600 activists from all around the country. Moreover, this movement also had a petition with 15,000 signatures, which declared the support of Israeli citizens to this movement.

The last movement is *Machsom Watch* which was established in January 2001 as a response to media reports about the violation of human rights of Palestinians who crossed Israeli military checkpoints in the border area. Because of exaggerated reaction of the Israeli government to Intifada El Aksa, occcupation and aggression in West Bank encouraged and motivated some activists to make an action. *Machsom Watch* is led by three women; they are Ronnee Jaeger who is a senior activist and experienced in human rights action of Guatemala and Mexico, Adi Kuntsman who is a feminist academic who emigrated from the Soviet Union in 1990, and Yehudit Keshet who is a veteran activist who is an Orthodox Jew (http://www.machsomwatch.org/).

Machsom Watch currently has 400 members in Israel. The main purposes of this movement are to supervise the behaviour of Israeli soldiers at border checkpoints, to ensure that the civil human rights of Palestinians who try to enter Israel territory are protected, and to record and report the result of their observations to all stakeholders and decision makers (http://www.machsomwatch.org/).

The Israeli women's participation as peacemakers mentioned above can be classified as a political participation. This deduction is based on two reasons. First, political participation aims to change or influence government policy. After being analysed, those five movements and organisations have the similar mission which is to create peace by demanding, protesting, and pushing the Israeli government to create peaceful policies. Second, the main characteristics of political participation are voluntary and without compulsion. These characteristics can also be seen in Israeli feminist movement action; there are no rights and duties becoming feminist movement member and it is purely based on the self-awareness of Israeli women to join and support the movement.

3.2 Israeli women's political participation in the peace process

Israeli women's participation in government or conventional political participation is lower than in other democratic states. Since the establishment of Israel in 1948, only 13 women have been involved in the ministry cabinet, including Prime Minister Golda Meir and Vice Prime Minister Tzipi Livni. Since 1992, the Israeli government has had at least one woman in the cabinet, and seven out of 33 Israeli governments did not involve women in their governments (Chazan, 2016).

Despite the lack of women in conventional political participation in Israel, the other forms of women's political participation have been growing since the 1970s until now. Israeli women's political participation in *Peace Now, Women in Black, Baht Shalom, Four Mothers,* and *Machsom Watch* can be classified as unconventional political participation due to two considerations.

First, unconventional political participation is extra-institutional actions, or in other words, the participation is done outside of the political party system. Women's participation as peacemakers by establishing peace movements and organisations are held as extra-institutional of Israel. Those movements are not embedded with any institution in any political party of Israel. Although *Peace Now* was established by reserved soldiers of IDF, they were no longer a part of the military or government institution.

Second, actions and activities of unconventional political participation are in forms such as of protests, demonstrations, petitions, and boycotts. Women's movements also participated in those forms of actions, for example the protest action of *Four Mother*s movement which complains about Israel policy regarding the occupation of Southern Lebanon, and the demonstration of *Women in Black* by delivering the message 'Stop the Occupation'.

3.3 The challenges of Israeli women's participation in the peace process

Feminism perspectives mostly conclude that women's problems come from masculine domination and discrimination of women as being the subordinate class in society. However, not all women face the same problems, because they live in different social cultural norms and they face different kinds of discriminations. This idea draws similarity with Iris Marion Young's thought that the two challenges of women are the domination of patriarchy in society, and colonialism values which construct feminism (Young, 2003).

Since the establishment of the Israeli state in 1948, Israel could not be free from military conflict with the neighbouring states. The prolonged conflict, holocaust syndrome, and years of the isolation experience of Jews, had increased Israel's awareness of its geopolitical situations. This awareness then influences the importance of military power for state survival (Klien, 1999). The centrality of military service in Israeli society constructed the gender perspective of Israel and placed women as the subordinate class in the country.

The military centrality has contributed to the lack of Israeli women's political participation. There is a conversion system of military ranking into political party ranking. In other words, people with the best military background will have the better opportunity in the political realm and military background is the main qualification to join in Israeli government (Klien, 1999). The effect of this conversion is that the Israeli political realm becomes dominated by masculinity. The data shows that Israeli generals who have retired in their 40 s then transform their professions and become politicians in the following 15 years (Klien, 1999).

Male domination does not only happen in the political aspect but also in the negotiation process between Israel and Palestine. In every peace negotiation between Israel and the Arab countries, the representatives of Israel are senior army officers, senior duty officers, and those from the top echelons. Those three sectors are only dominated by men, while women's participation is still zero (Klien, 1999).

Another challenge which constrains Israeli women's political participation is colonialist values that consider 'the West' as modern and civilised, while 'the East' as being traditional and backward. These values influence the social structure in Israel and create a dichotomy of Jews itself. The Diasporas Jews who had lived in Europe and America are called *Ashkenazi*, and the origin Jews who had lived in the Middle East area are called *Mizrahi Jews.*

The state of Israel has never been colonialised by Western countries, but Jews, who are the nation of Israel, had been living under Western power for so many years before Israel was established in 1948. The Jews who live in Europe and America indirectly have influenced their perspective and their way of life. One of the biggest impacts of Western values to Diaspora Jews is the New Jews Doctrine or the Super Modern Europe of Theodore Herzl in *Altneuland*. That doctrine also became the vision and mission of Zionism. Its aim is to transform the image of Jews, which is known as being traditional, religious, and passive. The Diaspora oriented the nation to be modern, brave, strong, and secular (Gurion, 1957).

As a consequence, this doctrine caused demands for Jews to change themselves to be more like modern Europeans. This thinking comes from Herzl who had experienced living in Germany, and it is not well-suited to the majority of Jews who had been living together with Arab countries and other Islamic countries, because they think that modernisation is not suitable with their social life. According to the *Ashkenazi* or European Jews, *Mizrahi* are lazy, backward, and primitive ethnic. The description about *Mizrahi* or original Jews is not very different from the Arab description of *Ashkenazi*'s point of view (Dahan-Kalev, 2001).

Western values in *Ashkenazi* are also affected by the development of feminist thought in Israel. The emergence of feminism in the 1970s was pioneered by some *Ashkenazi*, like Marcia Freedman who is a professor in Haifa University and also a woman activist in the United States (Maor, 2010). It can be concluded that the root of Israel feminism comes from America. This phenomenon was explained systematically by Hannah Safran in her writing entitled *Don't Want to be Nice Girls: The Struggle for Suffrage and New Feminism in Israel* (Benjamin, 2007). Israeli feminism begins with belief that there is a similar experience and problem shared by Israeli and American women, which is that of being discriminated against and dominated by patriarchy in social life (Dahan-Kalev, 2001).

This belief then became a problem for *Mizrahi* women as they face different kinds of issues from those of *Ashkenazi* women. For example, when *Ashkenazi* women deal with the mutual salary issue between women and men, the *Mizrahi* and Palestinian women who live in Israel still struggle with their non-professional jobs like low-wage labourers. It is caused by racism, poverty, and lack of education which constrains them from getting professional jobs like the *Ashkenazi* had already done. *Mizrahi* women cannot separate their lives from poverty, which most of *Ashkenazi* had never experienced (Giladi, 1990).

Ashkenazi and *Mizrahi* women also had different opinions about women's contribution in military service; the *Ashkenazi* women demanded equal rights for women to choose their career in the IDF, which was triggered by Alice Miller's case, who was not allowed to join pilot training in the Israeli Air Force (Jabareen, 2000). This case was the background to the Israeli amendment of military service law regarding women's rights (Rimalt, 2003). On the other side, this freedom excluded *Mizrahi* women (Dahan-Kalev, 2001). These feminist struggles to get equal rights in military service were only based on the interests of *Ashkenazi* women who were at the top of the social class pyramid, as they were more credible candidates to be military pilots in the Israeli Air Force. There are only a third of *Mizrahi* women who could be involved in the military, while the *Ashkenazi* women only made a 57% contribution.

It was caused by a lack of education of *Mizrahi* women which eliminates and disqualifies them from joining the IDF (Horowitz & Kimmerling, 1974).

The priority and main problem of *Ashkenazi* and *Mizrahi* are different. However, this fact tends to be ignored by the *Ashkenazi* and their discrimination of *Mizrahi* women turns into oblivion. Different from the *Ashkenazi* women who are eager to participate in the peace process between Israel and Palestine, *Mizrahi* women tend more to prioritise the elimination of discrimination and poverty than to participate in the peace process.

4 CONCLUSION

From the elaboration above, it can be concluded that Israeli women's unconventional political participation as peacemakers consists of five movements, which are *Peace Now*, *Women in Black*, *Four Mothers*, *Maschom Watch*, and *Bat Shalom*. These movements struggle for peace between Israel and Palestine by doing protests, demonstrations and petitions demanding that the Israeli state stop its occupation.

Israeli geopolitics is surrounded by Arab states, which constructed their social order of Israeli society, which is used to war and is military centric. This condition caused men as the main pillar to preserve the stability of the country. Almost all of their lives are spent as Israeli military. Thus, this condition placed women as the subordinate class compared to the men.

Many feminist's movements in Israel have been growing since the 1970s, as well as the movements which support peace with Palestine. However, the women themselves in the feminist movement experience discrimination and injustice from other feminist women. The internal conflict of Israeli feminists between *Mizrahi* and *Ashkenazi* has happened for years and latterly has caused the peacemaking effort by women in Israel to become less effective.

REFERENCES

ABC News (2015), *What behind Escalating Violence in Israel,* Retrieved from http://abcnews.go.com/International/escalating-violence-israel/story?id=34579329 [Accessed on August 14, August 2016].

AUPHR (2016) Americans United for Palestinian Human Rights). Retrieved from http://www.auphr.org/index.php/links/israeli-organizations, [Accessed on August 14, 2016].

Chazan, Naomi. (2016). *Women in Israel: In politics & public life.* Retrieved from: http://www.jewish-virtuallibrary.org/jsource/Society_&Culture_Women_in_Public_life.html [Accessed on August 14, 2016].

Dahan-Kalev, H. (1999). *Feminism between Mizrahi women and Ashkenazi women.* Hakibbutz Hameuchad, Hebrew.

Dahan-Kalev, H. (2001). Tensions in Israeli feminism. *Women's Studies International Forum, 24*(6), 669–684. doi:10.1016/s0277-5395(01)00206-0.

Discord in Zion: Conflict between Ashkenazi & Sephardi Jews in Israel. (1991). *Choice Reviews Online, 28*(08). doi:10.5860/choice.28-4462.

Dromi, U., Nevo, B. & Shur-Shmueli, Y. (2003). *Women in the Israel defense forces: A symposium held on 21 November 2002 at the Israel Democracy Institute.* Jerusalem: Army and Society Forum.

Finch, G., Verba, S. & Nie, N.H. (1974). Participation in America: Political democracy and social equality. *Political Science Quarterly, 89*(3), 674. doi:10.2307/2148474.

Finkel, L. (2012). *The Role of women in Israeli-Palestinian peace negotiations.* The Atkin Paper Series.

Giladi, G.N. (1990). *Discord in Zion: Conflict between Ashkenazi and Sephardi Jews in Israel.* Scorpion Press, Essex.

Gurion, D.B. (1957). A vision and a way (Vol. 4). D&E Tel-Aviv, Am-Oved (Hebrew).

Helman, S. (2009). Peace movements in Israel. *Jewish women: A comprehensive historical encyclopaedia.* Retrieved from http://jwa.org/encyclopedia/article/peace-movements-in-israel [Accessed on March 27, 2016].

Horowitz, D. & Kimmerling, B. (1974). Some social implications of military service and the reserves system in Israel. *Eur J Soc European Journal of Sociology, 15*(02), 262. doi:10.1017/s0003975600002939.

Hunter, M., Keneally, M. & Momtaz, R. (n.d.). *What's behind escalating violence in Israel?* Retrieved from: http://abcnews.go.com/International/escalating-violence-israel/story?id=34579329 [Accessed on August 14, 2016].

Jabareen, H. (2000). *Toward a critical Palestinian minority approach: Citizenship, nationalism and feminism in Israel law.* Peliliem.

Jerusalem Post. (2015). Riots breaks out on Jerusalem Temple Mount for third straight day. jpost.com, Retrieved September 15, 2015.

Kaufman, J.P. & Williams, K.P. (2013). Women at war, women building peace: Challenging gender norms. Boulder, Co.: Kumarian Press.

Klien, Uta. (1999). Our best boys: The making of masculinity in Israeli society. *Male Roles Masculinities and Violence*, 2(1), 47–65. UNESCO Publishing. doi:10.1177/1097184X99002001004.

Klien, Uta. (2000). Our Best Boys: the making of masculinity in Israel society, *Male Role, Masculinities and Violence: A Culture of Peace Perspective,* Hamburg: UNESCO PUBLISHING.

Lambert, V.A. & Lambert, C.E. (2012). Qualitative descriptive research: An acceptable design. *Pacific Rim International Journal of Nursing Research*, 16(4).

Lemish, D. & Barzel, I. (2000). 'Four Mothers': The womb in the public sphere. *Communication Theory and Research*, 125–138. doi:10.4135/9780857024374.d15.

Lubin, O. (2002). Gone to soldiers: Feminism and the military in Israel. *Journal of Israeli History*, 21(1–2), 164–192. doi:10.1080/13531040212331295902.

Maor, Anat (2016). *Women in Israel*. Jewish Virtual Library Publication. Retrieved from http://www.jewishvirtuallibrary.org/jsource/isdf/text/Maor.html [Accessed on April 09, 2016].

Mascom Watch. (n.d.) Retrieved from, http://www.machsomwatch.org/ [Accessed on April 09, 2016].

Miles, M.B. & Huberman, A.M. (1994). *Qualitative data analysis.* Sage Publications.

Nevo, B. & Shur, Y. (2003). *Women in the Israel defense forces.* In a symposium held on 21 November 2002 at Israel Democracy Institute. Jerusalem: IDI, Old City Press.

Orly, B. (2007). *Review of Hannah Safran, don't wanna be nice girls: The struggle for suffrage and the new feminism in Israel.* Indiana University Press. Retrieved from http://www.jstor.org/stable/10.2979/nas.2007.-.13.265 at Sept 14 2016.

Rimalt, N. (2003). When a feminist struggle becomes a symbol of the agenda as a whole: The example of women in the military. *Nashim: A Journal of Jewish Women's Studies and Gender Issues*, 6(1), 148–164. doi:10.1353/nsh.2004.0018.

Verba Sidney & Norman H. Nie (1972) *Participation in America: Political Democracy and Social Equality,* New York: Harper and Row.

Women in Black for justice against war. (n.d.). Retrieved from http://womeninblack.org/ [Accessed on August 14, 2016].

World Economic Forum. (2015). *Insight report: The Global gender gap report 2015.* World Economic Forum.

Young, R.J.C. (2003). *Post colonialism: A very short introduction.* New York: Oxford University Press.

Competition and Cooperation in Social and Political Sciences – Adi & Achwan (Eds)
© 2018 Taylor & Francis Group, London, ISBN 978-1-138-62676-8

Black lives matter in the progress of the African-American movement

A. Gracilia, T. Paramehta, N.N. Soeyono & I.F.R. Agoes
Department of American Studies, School of Postgraduate, Universitas Indonesia, Jakarta, Indonesia

ABSTRACT: This research discusses the *Black Lives Matter* movement as a progress of African-American movements in addressing racial inequality issues against the black people. The race issues that are discussed in this research relate to the emergence of the term *post-racial* after Obama's election as the United States president, which further highlights the inequality that African-Americans experience. In this research, *post-racial* refers to the racial problem's myth and framework that appear after Barack Obama was elected as the president. By applying historical and sociological approach methods, this qualitative research focuses on the development of the *Black Lives Matter* movement since it was initiated in 2013 until 2015, whereby a campaign using the 'black lives matter' hashtag spread throughout cyber-space. By contextualising the *Black Lives Matter* movement in the history of the African-American movement, this research shows that *Black Lives Matter*'s views and actions are more inclusive than those of the previous movements; in other words, this movement operates within an intersectional relationship with the other black people movements, namely the black feminist movement and the black LGBT movement. This involvement also puts *Black Lives Matter* forward as the new black people movement that lifted a boundary against other minority groups that are historically overlooked.

1 INTRODUCTION

This research focuses on the progress of minority group roles of the African-American movement. However, a movement which was specifically voiced by a group of black women often had no place in the history of movements and suffrage. The *Civil Rights Movement* in the 1960s was at first directed towards toppling racial segregation issues. Paula S. Rothenberg (1995) explained, "While minorities were suffering from white supremacy, women were suffering from male Supremacy" (Rothenberg, 1995, p. 70). White male dominated government in the US may have caused injustice towards certain groups' interests. Therefore, the *Civil Rights Movement* has its origin formed by the black people (Black Movement) and the women (Women Movement), gay people (Gay Liberation), as well as other ethnic groups (Hamby, 2005).

Since Obama was elected, the term *post-racial* began to surface, which was based in the myth believing that racial issue in America has ended. However, Obama, as the first president of black descendants, apparently still struggle to abolish racial inequalities experienced by African-Americans (i.e. in social problems such as crime, education, employment, and health that affected black people's welfare level in America). According to crime statistics of the FBI in 2013, US citizens who suffered homicide based on race are as follows: whites 3,005 inhabitants, blacks 2,491 inhabitants, other races 159 inhabitants, and unknown 68 inhabitants (Federal Bureau of Investigation, 2013). Relative to its total population, the killing of black people is the highest percentage of all inhabitants. According to the percentage, it is described by Harris and Lieberman as follows:

> African-Americans, for example, are nearly three times as likely as non-Hispanic whites to be poor, almost six times as likely to be incarcerated, and only half a likely to

graduate from college. The average wealth of white households in the United Stated is 13 times as high as that of black households (Harris & Lieberman, 2015, par. 4).

African-Americans often get stereotyped as criminals, compared with white people and other ethnic minorities. As a result, murders of innocent black people in self-defence are sometimes justified. In 2012, there was the murder case of Trayvon Martin, an unarmed black civilian, committed by George Zimmerman, a neighbourhood watch security guard who was not charged for his action because he was considered not guilty. While in 2014, there was a shooting case of Michael Brown, another unarmed black civilian, by a police officer named Darren Wilson. The handling of these cases suggests the possibility of racial prejudice as both Zimmerman and Wilson were not charged for the crimes despite evidences. This is relevant to a study by Tim Wise (2010) on the 'shoot and hold fire' simulations, that shows how unarmed blacks were quicker to be shot than were armed and dangerous whites (Wise, 2010, p. 84).

The case of Trayvon Martin's death triggered an onset of actions from the black people community that condemned the violent action committed by the police (*police brutality*). The black movement, popularly known as *Black Lives Matter*, began to campaign using terms such as 'Hands Up, Don't Shoot' and 'Black Lives Matter' in 2013. In 2015, this social movement wrote on its official website, "#blacklivesmatter is an affirmation and embrace of the resistance & resilience of black people" (blacklivesmatter.com, 2013). According to its official website, *Black Lives Matter* was a calling to act and respond towards racist issues of anti-black American sentiment that was hostile and had been deep-rooted in the American community (blacklivesmatter.com, 2013). This research argues that *Black Lives Matter* is a progressive form of the African-American people movement that becomes more inclusive than the previous movements because it includes the role of minority groups that are historically neglected. The research questions are: how do changes within *Black Lives Matter* be seen as the progress of the African-American movement and what are the contexts that have driven the movement? The study will contribute to the discourses of contemporary racial issues, which at present are part of a debate between the progressive and conservative studies of American society.

2 METHODS

I approached this study from a historical and sociological perspective to contrast the progress of *Black Lives Matter* with the previous similar movements. The findings of this research stemmed from social media analysis of *Black Lives Matter*'s official Twitter and website, as well as literature studies. This method is used because Twitter is the main domain used by *Black Lives Matter*, and therefore serves as an effective tool to monitor the progress of the issues that were raised by the movement. The literature studies employ critical race theory to investigate how differences in skin colour become significant in forming the power structure within America (Brooks, 2009). Ferris and Stein (2010) explained how social inequality in the US refers to the white male positions as being the highest compared to other groups (Ferris & Stein, 2010, p. 211). I also use the indicators within social changes factors to explain the changes in American society (Gabriel, 1991), in order to explain the changes specifically within African-American society. This research aims to explain how the movement becomes more inclusive. The intersectionality theory between race, class, and gender is used in order to examine linkages between the interests of each minority group, namely the working class, women, and LGBT (Anthias, 2012).

3 RESULTS

3.1 *Factors shaping black lives matter in the history of the African-American movement*

Black Lives Matter was a continuity form from the African-American movement that has opened boundaries and space against the involvement of other minority groups which were historically neglected, namely black women and black LGBT. Its form is a part of the African-American movement progress by contextualising it within the following factors:

3.1.1 *Ideology*

The *Black Lives Matter* movement identifies itself as the continuity of the *Black Liberation Movement*, as stated in the website: "It centers those that have been marginalized within Black Liberation Movements. It is a tactic to (re)build the Black Liberation Movement" (blacklivesmatter.com, 2013). This movement was radical but not militant, and used repressive ways in demanding its idealism, like the *Black Panther Party* movement, for example, the action of closing the public facility in *Mall of America* on 24 December 2015, in St. Paul, Minneapolis. However, this movement also used ways like campaign action and peaceful protest, similar to the *Civil Rights Movement*, namely *Freedom Ride to Ferguson*.

Black Lives Matter admits that its organisation's identity had a 'hard-line' ideology, like the *Black Panther Party Movement* that became an FBI target in the American history. This could be identified through its ideology, as an article titled 'The Persecution of Black Speech' elaborates:

> The Black Panther Party was targeted by the FBI for their work protecting people from the police and providing resources to the community. More recently, Black Lives Matter activists in Minneapolis were surveilled by the FBI during and leading up to the shutdown of the Mall of America (blacklivesmatter.com, 2016).

This movement shares a common form as the previous black movement because both aim to fight against a system in a racialised community. Winston A. Grady-Willis explained in his writings that the *Black Liberation Movement* was intended to end the stratification structure that was formed in a racialised community (Martha Biondi, 2007, para. 1). *Black Lives Matter* has the same purpose in applying the ideology to 'liberate' black individuals; as implemented by the *Black Liberation Movement*, they attempted to liberate the fate of black people as complete human beings, while *Black Lives Matter* attempts to liberate marginalised people using the rhetoric of the worth of black people's body and life, which are as worthy as that of other human life, of any group.

3.1.2 *Leadership*

The *Black Lives Matter* movement admits a new leadership model which is not only limited to heterosexual men with a single individual as the leader, as the following statement demonstrates:

> First, focusing on heterosexual, cisgender black men frequently causes us not to see the significant amount of labor and thought leadership that black women provide to movements, not only in caretaking and auxiliary roles, but on the front lines of protests and in the strategy sessions that happen behind closed doors. (blacklivesmatter.com, 2013).

This movement indeed still acknowledges the leadership of Martin Luther King Jr. as the symbol of the African-American struggle movement, but it does not require to be perceived as the continuation of King's leadership model. Even though its ideology is similar, *Black Lives Matter* is not influenced by Huey P. Newton's *Black Liberation Movement* leadership style that tends to be masculine and individual (requiring one male leader). *Black Lives Matter* possesses a leadership style that is open to feminist ideology and reaches out to a diverse spectrum of genders. This was explained on its website, "Black Lives Matter affirms the lives of black queer and trans folks, disabled folks, black-undocumented folks, folks with records, women and all black lives along the gender spectrum" (blacklivesmatter.com, 2013). On the other hand, Malcolm X's 'black rage' movement included activity with feminist ideology, as expressed by Frederick Douglass and W. E. B. Du Bois (West, 1993). However, this research observed that this 'black rage' movement did not elect the women as the leaders, as does *Black Lives Matter. Black Lives Matter* also does not acknowledge an individual leadership but a collective one. In other words, the leader figure is replaced with a collective voice delivered through the social media.

The *Black Lives Matter* movement focuses on the women's leadership. The role of heterosexual men in the black people movement leadership factor is used as a comparative variable, as West (1993) explains:

> Black male sexuality differs from black female sexuality because black men have different self-images and strategies of acquiring power in the patriarchal structures of white America and black communities. Similarly, black male heterosexuality, owing to the self-perceptions and means of gaining power in the homophobic institutions of white America and black communities (West, 1993, p. 127).

Before the emergence of the term *post-racial*, heterosexual black men were placed as the highest among the overall black people. This stood out in black movements such as the *Black Liberation Army* and the *Black Panther Party*, in which masculinity and heterosexual men dominated the leadership symbolism. Such imageries of heterosexual men and masculinity are no longer adopted by *Black Lives Matter*.

3.1.3 *The role of black youth*

Black Lives Matter is an intergenerational movement because while it involves the role of black youth, it does not shut off the role of the older generation. According to Cornel West (1993), the 'black rage' movement focused on the involvement of young people (West, 1993, p. 149). On the other hand, the *Black Lives Matter* movement focuses more towards Malcolm X's articulation of 'black rage' to the life of black people; it emphasises the challenges and difficulties of living as a black person, especially in the urban area that is prone to drug abuse or guns. *Black Lives Matter* tends to ignore formal symbolic forms that were previously considered as the leadership movement symbol dominated by intellectuals and the old generation; the reason for this is its consideration that the suffering experienced by black victims such as Trayvon Martin, Eric Garner or Michael Brown cannot be compared with the interest of the black intellectuals and elites.

The transition role of the *Black Lives Matter* movement on the involvement among generations (*intergenerational*) showed the view of Tocqueville on American society's political culture as an association in the social order that formed people's behaviour (in Goldfarb, 2013). There is a strong connectivity between the equality and freedom as the centre of modern political culture.

3.1.4 *Sexual orientation*

Black Lives Matter acknowledges the sexual orientation of LGBT and Queer (LGBTQ). This movement abolished the patriarchal system and homophobia that had been long deep-rooted in American society. *Black Lives Matter*'s amicable manner and acceptance towards black LGBT shows that there is a changing view that abolishes homophobia and the patriarchal system, which perpetuate white straight male as the default identity. As expressed by Ralph Gabriel, this explained that there is a shift of range of values that transforms into a declaration (Gabriel, 1991, p. 101). This shift of values is to no longer view heterosexuality as the only sexual orientation that exists.

The involvement of *Black Lives Matter* on issues of black LGBT, black women, and working class shows that this movement is more inclusive because it highlights its intersectional relation with the other minority groups. The *Black Lives Matter* movement can reach intersectional issues of race, class and gender as one inseparable issue. In other words, *Black Lives Matter* can be part of a LGBT movement by adopting a 'sexual orientation spectrum', while it can be part of a women's movement by adopting a 'gender spectrum', and it can also be part of a black movement by putting forward the 'skin colour spectrum'. As explained by Crenshaw, "the location of women of color at the intersection of race and gender makes their experiences structurally and qualitatively different to that of white women" (in Floya Anthias, 2012). This showed that *Black Lives Matter* could be viewed from multiple angles.

3.1.5 *The role of church and religion*

Since the commencement of the abolition until the era of the *Civil Rights Movement*, the black movement was very close with the role of religion, and had a very important religious leader, such as Martin Luther King Jr. whose background was as a priest, or Elijah Muhammad who was the founder of *Nation of Islam*. *Black Lives Matter* tends to be oriented to the Christian faith, but does not consider religion as the primary identity because it regards itself as 'The Revolutionary Jesus' and uses a radical way, namely 'Movement Pastors'. The role of church and religion is no longer considered as the movement's primary identity, that there is a shift from the church's conservative view that prioritised a peaceful way, love, prayer and submission, to become courageous to resist injustice and to demand structural change. *Black Lives Matter* shows its movement's identity as the new African-American movement that maintains revolutionary Christian values, namely remodelling the justice system that was considered unfair, somewhat portrayed as like the time Christ remodelled the rigid Jewish justice tradition.

3.1.6 *The issues that were brought up*

The *Black Lives Matter* movement prioritises criminal justice issues caused by police brutality that branches out to wider issues, such as level of prosperity, gender, and LGBT. If compared to the The National Association for the Advancement of Colored People (NAACP) organisation, the issue brought up was the injustice issue on lynching and on legal assistance for the case of *Brown v Board of Education* that annulled the decision of the Supreme Court over the 'separate but equal' doctrine. Both similarly addressed the biased justice system.

According to the research result, there are four primary issues that are brought up on its Twitter account, namely the criminal justice issue, LGBT issue, worker wage system issue, and gender issue. This Twitter account was made in 2013, and the first tweet from *Black Lives Matter*'s account was on 20 July 2013: a photo of a black boy wearing a blue *hoodie*, carrying a poster that read 'Will I be the next Trayvon Martin? Keep Fighting For Justice!!'. Throughout 2013, the criminal justice issue was the only issue that was brought up in the Twitter discussions. The discussion kept continuing throughout the year from 2014 to 2015. Meanwhile, the LGBT issue began to be campaigned on 19 December 2014; it used the hashtag #BlackLivesMatter and #BlackTransRevolution. Discussion on the LGBT issue took place until 2015. On the issue about the worker wage system issue, on 10 November 2015, the *Black Lives Matter* Twitter account posted on one of their tweets, 'When we fight for a living wage, we are fighting for the dignity of black lives. #BlackLivesMatter #FightFor15.' On the gender issue, the Twitter account wrote a tweet, 'Hands off our sister! #BlackLivesMatter' on 20 December 2014. Tweets publication on this Twitter account has shown that by the end of 2014 and entering year 2015, issues that were brought up by this movement extended to the movements of women's rights, LGBT, and working class.

The issue extension happened because there is a tendency to address masculinity as a social construction (Ferris & Stein, 2010). The criminal justice issue for men referred to shooting victims, while for the women it referred to rape cases. The criminal issue in the previous movements considered a black man's 'loss of a life' caused by relentless assaults as 'only' an individual problem and violation against someone's rights to live; on the other hand, *Black Lives Matter* put forward the women's position as a mother who lost her child, or a wife who lost the head of family figure, which therefore affects the discourse of American family well-being. In other words, when an innocent black man is killed, it not only affects the individual's rights but also affects the women in the family.

3.1.7 *Black lives matter ideology vs all lives matter*

The term of 'Black Lives Matter' raised an argument that 'All Lives Matter'. Ideologically, the *Black Lives Matter* movement supports the idea of 'All Lives Matter'. This movement requires to highlight the fact that the existing justice system is still biased against the lives of black people. This movement made the well-being of black people as the parameter for freedom as a whole. This was expressed through its statement on the webpage of 'A Her Story of the #BlackLivesMatter Movement' on its official website: 'When Black People Get Free, Everybody Gets Free' (blacklivesmatter.com, 2013).

3.2 Contextualising the birth of black lives matter in the progress of the African-American movement

This research also contextualises the beginning of *Black Lives Matter* as part of the progress of the African-American movement, which can be divided into three aspects, namely: the lack of leadership from the minority group, the use of social media technology, and the *post-racial* myth. This movement previously focused on the interest of black men, but in its progress the roles of black queer women activists voiced the black people's rights in the interests of black women, black LGBT and black working class. In the era of the *Civil Rights Movement*, the black church played a strong role and focused in its leadership on a masculine and religious role; therefore the role of women as the leaders inclined to be minimum. The rise of *Black Lives Matter* with black queer women as the driving force differentiates this movement from the other previous movements, in that it only focuses on one dimension of sex and skin colour. *Black Lives Matter* is inclusive to the rights of black women and black LGBT.

Leadership by women is a part of the development of the role of black women in America. It was influenced by the existence of gender movement in America. When *NOW* (1966) was established, women's rights still focused on the interests of white women's rights, as the needs and interests of black women were still highly overlooked. Black women were only represented in the interests of the white women's movement; in other words, they were represented in the problems that were specifically for women as a whole, but did not raise the issue of racism that women of colour experienced. This was in line with what was explained by Evelyn M. Simien, "While black women were more supportive of feminist items that tapped perceptions of social reality, they were far less supportive of those items that measured attitudes towards abortion and tactics for social change" (Simien, 2004, p. 315). This explained that the mainstream women's movement focus did not raise racism issues as questioned by black women. On the contrary, *Black Lives Matter* is less concerned in raising women's issues that focused on general women's problems, such as abortion; instead, *Black Lives Matter* focuses more on the black women's interests and experience such as losing a male family member to police brutality.

The use of information technology is one of the things that helped with the birth of the *Black Lives Matter* movement because it is able to generate communication with cyberspace users, namely through personal websites and Twitter social media accounts. Issues are delivered not in rhetorical speech nor in direct action, but through the use of hashtag #BlackLivesMatter. The use of hashtag in its Twitter account from 2013 to 2015 showed that *Black Lives Matter*'s attention was put mainly on the police brutality victims' names and on support for LGBT issues, for example #JusticeForEricGarner, #BlackTransRevolution, and #DefendBlackTransWomen. This was in line with what Ralph Gabriel said, that improving the science that supported civilisation was one of the causes of changes in society (Gabriel, 1991, p. 101). Despite utilising the cyberspace and appealing to a general netizen, the publication of victim names on Twitter was able to humanise these victims.

In the end, the *post-racial* myth was popular after Obama became the president, yet in reality the elected president met difficult challenges to make black American people feel represented. Because of this, there is therefore a stronger tendency to highlight that the problem of race sentiment is not yet over, which leads to the birth of the *Black Lives Matter* movement. One way to exemplify, based on the population census in 2014, American black people's income per capita was $35,902 per year on average. The data positioned African-American people averagely in the *working class* or *lower-middle class* groups. This shows that the election of Obama does not immediately improve the living conditions of black people, as the data gap still noted that the lives of African-Americans were not yet prosperous, and police brutality that happens to black people still exists. Such reality shows that *post-racial* America is only a myth, and such a paradigm is what drives *Black Lives Matter* to continue its movement.

4 CONCLUSION AND RECOMMENDATION

The *Black Lives Matter* movement is a historical progression of the African-American movement. Basically, this movement shares the same purpose with previous black movements that fought against the racialised system and white supremacy, but it shows its inclusivity towards other marginalised communities previously overlooked in mainstream history. *Black Lives Matter* reads the lack of the minorities' roles. By addressing issues from the viewpoint of the minorities, making use of new technology, and demanding equality for its people, the movement highlights the need of addressing their rights, even though they have a president who represented their 'skin colour'. This showed that *Black Lives Matter* is not only a movement that demanded equality, but also one that expected a continuity from the success of the Black and White desegregation in the 1960s as the movement demands equality in a more diverse and more significant aspects of black community lives and their allies.

The research shows that *Black Lives Matter* is historically notable for other racial social movements in the global sphere. Maintained by the development of cyberspace technology, *Black Lives Matter* makes human rights aspirations echo easily through social media. It has linked the interests of race, class, and gender. In the future, this research can still be developed further to interrogate the effectiveness and impacts of the *Black Lives Matter* movement, especially after the 2016 presidential election.

REFERENCES

Anthias, F. (2012). Transnational mobilities, migration research and intersectionality. *Nordic Journal of Migration Research, 2*, 102–110.
Biondi, M. (2007). Challenging U.S. apartheid: Atlanta and black struggles for human rights, 1960–1977. *The Journal of Southern History, 73*(4), 954–955. Retrieved from: http://search.proquest.com/docview/215768556/D8C401736C764B1EPQ/5?accountid=17242.
Black Lives Matter (Twitter account). Available on: https://twitter.com/Blklivesmatter/media.
Black Lives Matter (website). Available on: http://blacklivesmatter.com/.
Brooks, R.L. (2009). *Racial justice in the age of Obama.* Princeton: Princeton University Press. [ProQuest ebrary. Web.] Retrieved from: http://site.ebrary.com/lib/indonesiau/reader.action?docID=10320506&ppg=114.
Federal Bureau of Investigation. (2013). Crime in the United States 2013. *fbi.gov.* Retrieved from https://www.fbi.gov/about-us/cjis/ucr/crime-in-the-u.s/2013/crime-in-the-u.s.-2013/offenses-known-to-law-enforcement/expanded-homicide/expanded_homicide_data_table_6_murder_race_and_sex_of_vic-itm_by_race_and_sex_of_offender_2013.xls.
Ferris, K. & Stein, J. (2010). *The real world: An introduction to sociology* (2nd ed.). New York: W.W. Norton & Company, Inc.
Gabriel, R.H. (1991). *Nilai-nilai Amerika: Pelestarian dan Perubahan* (American values; continuity and change). Hargosewoyo, P.S. (trans.) Yogyakarta: Gadjah Mada University Press.
Goldfarb, J.C. (2013). *Reinventing political culture: The power of culture versus the culture of power.* New York: John Wiley & Sons. Retrieved from https://books.google.co.id/books?id=_-4M2RvRb9 AC&pg=PT21&dq=political+culture+tocqueville&hl=en&sa=X&ved=0ahUKEwi71-i5wu7MAhVLLo8 KHfGdB_wQ6 AEIGzAA#v=onepage&q=political%20culture%20 tocqueville&f=false.Hamby, A.L. (Ed.). (2005). *Garis Besar Sejarah Amerika Serikat* (Overview of the History of the United States of America). Departemen Luar Negeri AS.
Harris, F.C. & Lieberman, R.C. (2015). Racial inequality after racism: How institutions hold back African Americans. *Foreign Affairs, 94*(2), 9–20. Retrieved from http://search.proquest.com/docview/165 8668591/7166B79C61BA4698PQ/1?accountid=17242.
Rothenberg, P.S. (1995). *Race, class, and gender in the United States. An integrated study* (3rd ed.). New York: St. Martin's Press.
Simien, E.M. (2004). Gender differences in attitudes toward black feminism among African Americans. *Political Science Quarterly, 119*(2), 315–338. Retrieved from http://search.proquest.com/docview/208 288080/82CAF2D5059C412DPQ/4?accountid=17242.
The persecution of black speech. (2016). blacklivesmatter.com. Retrieved from http://blacklivesmatter.com/the-persecution-of-black-speech/.
West, C. (1993). *Race matters.* New York: Vintage Books.
Wise, T. (2010). *Color-Blind: The rise of post-racial politics and the retreat from racial equity.* San Francisco: City Lights Books.

Asset management analysis and development strategy of Muara Angke fishing port

I. Iklina, A. Inggriantara & K. Djaja
Department of Urban Studies, School of Postgraduate, Universitas Indonesia, Jakarta, Indonesia

ABSTRACT: Muara Angke Fishing Port is a vital asset in supporting industrialization of fisheries and economic growth in DKI Jakarta. The increasing activities in Muara Angke Fishing Port have caused suboptimal performance. Therefore, this study aims at identifying internal and external factors, formulating development strategy and determining the performance measures of Muara Angke Fishing Port. This quantitative research uses survey method with data based on questionnaires, interviews and observations. The methods of analysis applied are SWOT analysis, Balanced Scorecard, facility utilization rate, PESTEL and foreland-hinterland linkages. Using TOWS matrix, some strategies are formulated, such as cooperating with private sectors, synchronizing mainland and coastal spatial plan, creating a zone arrangement to support marketing of fishery and making improvement for slum areas. All the strategies are later translated into Balanced Scorecard performance to measure what must be implemented by the port management. In conclusion, the implementation of strategies and performance measures is expected to improve the performance and management of Muara Angke Fishing Port in providing the best public services.

1 INTRODUCTION

Port is an infrastructure that serves as a gateway to enter a region and as a liaison between local infrastructures (Triatmodjo, 2015). Port is very closely related to the city environment as port operations indirectly support the economic growth of the city; among others are the development of industrial plants, transportation, shopping, banking, and even the expedition offices and architecture firms. One type of ports that has special functions is the fishing port, which is a port that host fishing vessels in loading and unloading their catches at sea or loading material supplies (Lubis, 2012). One of the fishing ports in Jakarta that hosts integrated fishery activities, from unloading catches to processing and marketing is the fishing port in Muara Angke area. Muara Angke Fishery Port is one of the fishing ports that play important roles for the city, including in urban food security by providing fish for consumption of the inhabitants of Jakarta and the surrounding areas. The port also serves as the center for marketing fishery products, as shown by the data of Potensi Perikanan DKI Jakarta, which indicates that in 2014, approximately 58% supply of fish for Jakarta and the surrounding areas were delivered from the fishing port in Muara Angke. The port has also provided job opportunities, given more income for locals, and supported the development of fishery industry.

Therefore, Muara Angke Fishery Port has become a vital asset to the city that should be maintained. Effective and efficient port management is expected to provide maximum services and to support the achievement of the objectives, goals, vision and mission of the port management unit. According to Maerdiasmo (2008), maintaining and improving the level of performance and economic growth are very important to most public and private companies. The goals are often linked to the performance of the organization's assets, both tangible and intangible, and the management of the assets is labeled as asset management.

The problems that occur in the Muara Angke Fishing Port today are caused by the increasing activities that have not been supported by the availability of adequate port facilities, causing suboptimal port performance. In addition, the rapid constructions around the port in both the hinterland and foreland, such as the development of reclaimed G Island and Giant Sea Wall, may affect fishery activities in the area.

Accordingly, the research question which forms the basic issue of this paper is what the appropriate development strategy for the Muara Angke Fishing Port through a comprehensive approach is. The research aims to analyze the internal and external factors that influence the development of the port area in order to obtain a regional development strategy formulation using Balanced Scorecard approach. Balanced Scorecard method was first introduced by Robert S. Kaplan and David P. Norton. This method is more than just a tactical and operational measurement system and can be used as a strategic management system to manage a long-term strategy (Kaplan & Norton, 1996). Long-term strategies are formulated by taking into account the results of the implementation of development strategies and influence around the port that can affect asset management in Muara Angke Fishing Port.

Based on the theories of strategic management and asset management that are used in this study, the port's performance can be measured as an assessment of the existing internal and external aspects in order to determine the future development strategy. Internal factors can be determined by measuring the operational performance of assets, both tangible and intangible. Performance measurement of tangible asset is carried out using analysis of the utilization rate of land and port facilities, while performance measurement of intangible assets is carried out by using the Balanced Scorecard approach through four perspectives. The four perspectives are used to analyze the perception of service satisfaction and the value chain related to the trade system which took place at Muara Angke Fishing Port, in favor of the concept of industrialization of fisheries. Meanwhile, external factors can be determined by analyzing aspects of the foreland and hinterland and societal environment. Therefore, the hypotheses in this research are as follows:

1. Alleging that the main strategy in the fishing port is a strategy of growth.
2. Alleging that the implementation of strategies will improve the performance and management of the fishing port.

2 METHODS

The research is carried out using various methods. Data collection and analysis are implemented with quantitative and then qualitative techniques. A quantitative approach with

Table 1. Variable and measurement indicators of quantitative method.

No	Variable	Measurement dimension
I	Internal factors	
1	Financial perspective performance	The budget deviation, the ratio of expenditure to revenue and ratio of absorption of budget revenue
2	Costumer perspective performance	Customer satisfaction (tangibility, reliability, responsiveness, assurance, and empathy), market share, acquisition and retention, and profitability
3	Internal process perspective performance	Innovation, operation, and service towards complains
4	Growth and learning perspective performance	Employee satisfaction, employee retention, and employee productivity
5	Level of facility utilization	Percentage of level of facility utilization (pier, port land, port basin and navigation channel, fish auctions, and cold storage)
II	External factors	
1	Societal environment	Political, Economical, Socio-cultural, Technological, Ecological, and Legal (PESTEL)
2	Hinterland	Accessibility and surroundings
3	Foreland	Aquatic environments

survey method aims to determine users' perception of the port facilities and port management unit's performance through the Balanced Scorecard approach. The survey is conducted by distributing Likert scale questionnaires by random sampling onto 70 samples. Meanwhile, a qualitative approach is conducted in the form of interviews with some of the fishery business unit managers and officials, as well as focus group discussions with government and academia.

Muara Angke Port is selected using purposive method, considering that the port is one of the fishing ports managed by the Provincial Government of Jakarta, directly adjacent to the development of reclaimed G Island in the north and has the biggest fish wholesale market in Jakarta also in Indonesia. Data analysis technique in this research is descriptive statistics, using variable and indicator measurements as described in Table 1. Results of the surveys, interviews, and secondary data, such as reports, regulations, and other secondary sources, are used to analyze the internal and external factors affecting the development of Muara Angke Fishing Port using SWOT analysis (Strengths, Weaknesses, Opportunities, and Threats). Finally, TOWS matrix is applied to formulate a strategy.

3 RESULTS AND DISCUSSION

Based on the results of performance appraisal with Balanced Scorecard and the analysis of the utilization rate of port facilities, the internal factors that are identified as strengths and weaknesses that can affect the development of the port are as follows:

Strengths

1. The port has implemented E-Government.
2. The port has a unique area, which is the culinary center (Pujaseri Mas Murni).
3. The number of fishery business groups and fishermen in Muara Angke Fishing Port has increased by 39.3%.
4. Port facilities maintenance activities are performed routinely.
5. Port services are performed according to Standard Operational Services.
6. Vacant space is available, but is currently being occupied illegally.
7. The internal system of performance-based organization has been running since 2014.
8. Quality and safety assurance for fishery products is carried out regularly by the port management.
9. The market share of the amount of fish auctioned and sold without auction in Jakarta is mostly from the fishing port of Muara Angke.
10. Fishing activities take place in an integrated way and create economic agglomeration.

Weaknesses

1. Some of the port facilities still cannot accommodate the port activities.
2. Cleanliness, hygiene and environmental sanitation are in less ideal conditions.
3. Financial management and planning have not been carried out effectively and efficiently.
4. The port does not have a disaster response system.
5. The application of cold chain distribution of fish has not been running well.
6. Human resources at the port's organization unit are still lacking in quality and quantity.
7. There are seedy and poor areas.
8. Detailed Spatial Plan and Zoning Regulations Fishery Port of Muara Angke area on Local Regulation No. 1/2014 does not accommodate the current conditions of the port.
9. The facilities maintenance cost is high.

Meanwhile, external factors are derived from the analysis of societal environment and linkages of foreland and hinterland of the port area. These external factors are identified as opportunities and threats that can affect the development of the port, among others:

Opportunities

1. Strategic location of Muara Angke Angke Fishing Port.
2. The contribution of the developers on reclaimed island construction.

3. Government policy on exports restrictions on raw industrial materials.
4. The development of information technology, which aims to facilitate the work.
5. Potential economic relation between fishery industry and other sectors, including hotels, restaurants and exports.
6. Increasing interest in investing in the field of marine and fishery in Indonesia.
7. Increasing fish consumption of population of Jakarta.
8. Policies that oblige fishing industry business to have eligible certificates of quality and product safety.
9. The policy on the revitalization of the Muara Angke Fishing Port.

Threats

1. The construction plan of the G Island reclamation and Giant Sea Wall.
2. The increasing pollution in the waters of Jakarta Bay.
3. Unfeasible road infrastructure to Muara Angke Fishing Port.
4. Lack of ability of fishermen in utilizing fishery technology.
5. Decrease in number of capture fishery production worldwide by 3.5%.
6. Threat of flooding from seawater affecting port activities.
7. Low fish processing technology application.
8. Competition on fishery commodities marked caused by ASEAN Economic Community.
9. Rapid development around Muara Angke Fishing Port.

The result of the SWOT analysis can be used to identify the strengths, weaknesses, opportunities and threats, which can all be considered to establish the strategy for the Muara Angke Fishing Port and to distinguish the issues it is facing. The general strategy chosen is the growth strategy, which is consistent with the hypotheses because the Muara Angke Fishing Port plays an important role and function for the sustainability of fishery activities in Jakarta, including providing employment, fulfilling the demand for fish consumption, providing port facilities for unloading fish and running other fishery activities. The port also has various external opportunities, among others are great investment opportunities, increasing population of Jakarta, government policies that support the advancement of fisheries and strategic location. However, it also has internal weaknesses, such as excessive use of port facilities due to increasing fishing activities and the slum areas and space of land that have not been managed properly.

The selection of the growth strategy is also in-line with Regional Regulation No. 1/2014 which states that Muara Angke Fishing Port, with a total area of approximately 63.3484 hectares, is an integrated development area that functions as a fishing port, tourism spot, industrial area and warehousing, as well as residential area. Thus, Muara Angke Fishing Port is planned to be developed to become one of the coastal tourist destinations and having the wastewater treatment infrastructures.

3.1 *Development strategy of Muara Angke Fishing Port*

Based on TOWS matrix, the strategy is formulated in four types, which are (a) SO Strategies use a firm's internal strengths to take advantage of external opportunities; (b) WO Strategies aim at improving internal weaknesses by taking advantage of external opportunities; (c) WT Strategies are defensive tactics directed at reducing internal weakness and avoiding external threats; and (d) ST Strategies use a firm's strengths to avoid or reduce the impact of external threats (David, 2011). Here are the formulations of strategies that have been transformed into four perspectives of the Balanced Scorecard:

1. Customer Perspective:

    - Developing coastal tourism: culinary tourism (S2, O1, O9).
    - Improving supervision and sanction against violations of standards of quality and safety of fishery products (S7, S8, O9).
    - Increasing the availability of port facilities (W1, W5, O6).
    - Coordinating and cooperating with the developer G Block Island (W8, O2, O8).

- Empowering small-scale fishermen through diversification into aquaculture (S9, T1, T5).
- Enhancing regional promotion (S9, S10, T9).

2. Financial Perspective:

- Cooperating with private sector on asset utilization (S6, S10, O6).
- Designing a strategic asset management system (W3, T9).

3. Internal Process Perspective:

- Developing eco fishing port (S4, S5, O9).
- Diversely developing processed fish products (S9, O3).
- Streamlining the distribution marketing channel of fishery products (S9, O7, O5).
- Developing databases and integrated information systems department (S1, O4).
- Transforming the organizational structure into a Public Service Board or Regional-Owned Enterprises (W3, W9, O9).
- Improving hygiene and sanitation (W2, T2).
- Designing a disaster response system (W4, T6).
- Synchronizing the mainland and coastal spatial plan (W8, T1, T3).
- Improving processing technology of fishery products (S9, T7, T8).
- Arranging regions based on groove zoning of fishery trade system and improving the slum (S10, T9).

4. Learning and Growth Perspective:

- Improving the quality and competence of employees through training/certification (W6, O4).
- Enhancing the commitment of employees at the management unit (W7, T7).

The main objective of the implementation of development strategy, which is to achieve financial targets and customer satisfaction, is driven by a good performance from the perspective of internal processes and learning and growth. Therefore, to achieve the main goal, each strategy should be translated into performance measurements to enable agencies to measure progress towards achieving goals. According to Niven (2008), performance measurements can be considered as a standard used to evaluate and communicate performance against expected results. So, they are important in supporting the process of strategic planning and management of a non-profit organization.

Performance measurements are used to measure results and ensure accountability, including evaluating performance, tracking progress against a plan, identifying improvements and rewarding performance (Management Concepts, 2005). The performance measurements may include increased customer satisfaction, increased productivity and income, the number of monitoring and evaluation carried out, ratio of budget revenue and expenditure, the number of training followed by employees, and others.

Generally, the formulation of the development strategy of Muara Angke Fishing Port aims at increasing business for the fishermen and their communities. Another goal is to improve the environmental conditions of the region to become more organized, cleaner, and more appealing, so that it can be an alternative coastal tourism destination for city citizens. To support these objectives, port management units should be prepared through personnel competence and capability improvement.

According to Jo Santoso (2006), there are two major causes of city deterioration in Indonesia. The first is the development of policies and strategies of urban poor and the second is the basic concept of urban development planning that is not followed by proper implementation.

Similarly, the development of regional fishing ports, as part of the infrastructure of the city that determines the policies and strategies for the development of ports, should not ignore the interests of the whole community involved directly or indirectly in the activities of fisheries in the ports. The planning of the port development should also involve the community, so that the development is planned according to the needs and is long-term oriented.

Carrying out the strategy of empowering small-scale fishermen through diversification into aquaculture and slum improvement requires cooperation and rapprochement between the government and local community, as well as continuous assistances. This is because the shape of a city is the final manifestation of the accumulated increase in the number population, behavior and development policies made by the citizens (Heryanto, 2011). This means that the relationship between the government and citizens determines the development policies of the city, with residents as the main object.

3.2 *Impacts of development around the Muara Angke Fishing Port*

Based on a research conducted by Suganda (2008), the development in coastal urban areas leads to physical and social degradation because it does not integrate spatial planning land and sea areas, and the government has not been consistent in controlling the construction. Similar to what happened in the coastal area in northern Jakarta, development can be so rapid that it indirectly influences the degradation of the environment, causing the possibility of increasing intensity of tidal flooding and land subsidence. This is caused by land use that violated the spatial planning, as stated in Regional Regulation No. 1/2014. Moreover, the development of sea spatial plans, such as the reclamation and construction of the Giant Sea Wall as part from the port foreland region, does not befit the ongoing construction in the area of coastal land, as shown in Figure 1.

The disintegrated planning may affect the development and construction of the city. According to Catanese & Snyder (1988), land use planning is the key to direct the development of the city, and the initial process in land use planning is to understand what are already available. Therefore, one of strategies to synchronize mainland and coastal spatial planning is accommodating the government, society, and private sectors. This way, the proposed new spatial planning will not be a bleaching of the ongoing development process.

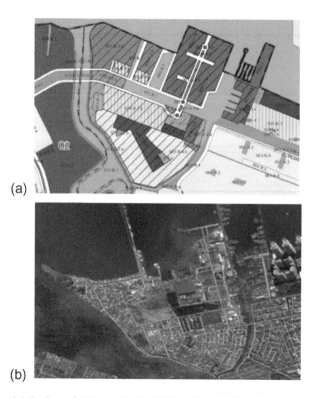

(a)

(b)

Figure 1. The Spatial Region of Muara Angke Fishing Port (a) Local regulation and (b) present condition. (Source: Perda 1/2014 and Google Maps (remodified, 2016)).

Synchronization is a form of retrenchment strategy, achieved by regrouping through cost and asset reduction to reverse declining sales and profit (David, 2011). Hence, the removal of assets and the reorganization of the port should be carried out in order to increase its competence. Adjusting to the layout of the land and sea carries its own consequences in the removal of assets belonging to the regional government, such as office buildings, workshops and overhaul of ships, breakwater and other port facilities.

3.3 *Benefits of the strategies implementation*

The application of strategic alternatives can provide social and economic benefits and advantages for businesses and communities in particular and for the economic growth of the city in general. This is because parts of the objectives of developing a region are improving socio-economic development, reducing inequalities and preserving the environment as much as possible (Blakely, 1994 in Marham & Tjokropandojo, 2015).

The social benefits are mainly intended for fishing communities, as their empowerment is expected to increase their income and improve their quality of life. Neighborhood revitalization program will improve the quality of the environment, so that the port can be used as one of the coastal tourism destinations for urban communities. The economic benefits obtained by the Jakarta Provincial Government for implementing the development strategy include the increasing local revenues from the potential of the new economy and the optimal use of the area.

4 CONCLUSION

The results obtained from the analysis of internal and external factors determine that the Muara Angke Fishing Port is ready for further development, leading to the selection of the growth strategy. However, there are constraints in executing these strategies because the spatial planning areas stipulated in Regional Regulation No. 1 of 2014 do not accommodate the current conditions of the port. An exit strategy in required to preserve the existence of the Muara Angke Fishing Port through retrenchment strategy, resulting in the formulation of four alternatives based on the Balanced Scorecard perspective.

The formulation of the development strategy of Muara Angke Fishing Port aims at increasing the fisheries business activities for the fishing community and improving the environmental conditions in the port area. Therefore, the re-planning process should include the community participation. It needs a strategic asset management system to minimize the risk of asset losses as the impact of environmental changes around the port. The implementation of the strategy can be categorized as successful and conforming to every performance measurements if it can provide benefits and improve the performance and management of Muara Angke Fishing Port in providing the best public service. Social and economic benefits that can be derived from the implementation of the strategies include improvements in the quality of life, environmental quality, and financial performance of local governments.

REFERENCES

Catanese, A.J. & Snyder, J.C. (1988) *Urban Planning Second Edition*. McGraw-Hill. Inc.
David, F.R. (2011) *Strategic Management: Concepts and Cases*. [Ebook]. New Jersey, Pearson Education.
Heryanto, B. (2011) *Roh dan Citra Kota: Peran Perancangan Kota sebagai Kebijakan Publik*. Surabaya, Brilian Internasional.
Kaplan, R.S. & Norton, D.P. (1996) *Translating Strategy into Action the Balanced Scorecard*. Harvard Business School Press. Boston Massachusetts. 1996.
Niven, P.R. (2008) *Balanced Scorecard Step By Step for Government and Nonprofit Agencies Second Edition*. New Jersey, John Wiley & Sons, Inc.
Lubis, E. (2012) *Pelabuhan Perikanan*. Bogor, IPB Press.

Management Concept. (2005) Strategic Planning and Implemenatation: Developing a Workable Plan, Course No. 4030, JD 12405. Management Concepts Incorporated. Mardiasmo, et al. Asset Management and Governance: Analysing Vehicle Fleets in Asset-intensive Organisations. *In Brown, Kerry A. and Mandell, Myrna and Furneaux, Craig W. and Beach, Sandra, Eds. Proceedings Contemporary Issues in Public Management: The Twelfth Annual Conference of the International Research Society for Public Management (IRSPM XII), pages pp. 1–20, Brisbane, Australia. 2008.*

Marham, R. & Tjokropandojo, D.S. (2015) Potensi pengembangan kawasan minapolitan di Kecamatan Berbah, Kabupaten Sleman. *Jurnal Perencanaan Wilayah dan Kota B SAPPK V4N1*, pp. 179–188.

Santoso, J. (2006) *Menyiasati Kota Tanpa Warga*. Jakarta, Centropolis & KPG.

Suganda, E. (2008) Pendekatan SWOT dan AHP pada penataan Ruang Kawasan Perkotaan Pesisir (Kasus: Pulomerak-Bojonegara). *Jurnal Tekologi*, (3), 214–222.

Triatmodjo, B. (2015) *Perencanaan Pelabuhan Cetakan Kelima*. Yogyakarta, Beta Offset Yogyakarta.

Female-headed households in Kampung Bidara Cina: Negotiation, spatial practice and reproduction of spatial structure change

I. Pratiwi & Herlily
Department of Architecture, Faculty of Engineering, Universitas Indonesia, Depok, Indonesia

ABSTRACT: The Female-Headed Households (FHH) is still not much observed in the urban planning context. In this issue, the female head and her relatives form a spatial structure to regulate the use of the dwelling environment. This study explores the understanding of spatial structure through gender perspective and its relationship with spatial practice and negotiation in the FHH. Through the method of *emic construction* and *quasi-participant,* the observation and analysis have shown that in the process of maintaining the spatial structure of daily activities, the FHH family cannot be separated from negotiation with the mother and neighbours. Through the understanding of spatial narrative and primary spatial syntax, the analysis explains that to maintain the production of spatial structure change, the negotiation is influenced by the mothers' specific perspectives in the FHH environment. This study finds that in the production of spatial structure change, its reproduction is the most important need to produce the integral spatial structure in the FHH housing, so that FHHs are able to preserve their whole activities to be continued daily in limited space, without the presence of the men.

1 INTRODUCTION

The life of people in the low-income housing of Female-Headed Households (FHH) cannot be separated from the role of a female head. They use and control over limited spaces. However, each space for its specific use is still not well-defined. One small space in their dwelling usually accommodates most of their productive and reproductive works. To struggle and maintain all works, the female head and her female relatives negotiate with their neighbours to do spatial practice together. Therefore, spatial structure is not a rigid formation. It is not

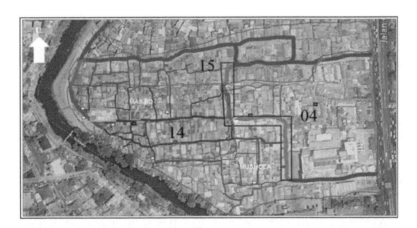

Figure 1.   Location of RW 4, RW 14, RW 15, Kampung Bidara Cina (source: Google Earth. Processed personally).

only formed by the organisation of the FHH's limited built physical space, but also through their spatial practice. Without the presence of men, the FHH members maintain their working spaces and persist in difficult situations by forming a spatial structure change within their physical dwelling space.

This study finds he production of spatial structure change is always needed by the FHH members in their routine life. Consequently, the reproduction of spatial structure change is the important thing in this process. Nevertheless, the reproduction of spatial structure change cannot occur without the presence of the mother's neighbours. This fact leads to the following research question: How does the gender perspective and social role around the FHH housing influence the spatial practice and negotiation in the reproduction of spatial structure change?

The observation methods applied are *emic construction* and *quasi-participant observant*. Meanwhile, the descriptive methods used are articles, journals, books, and digital media. The scope of this article is the subject limited to the denizen in *Rukun Warga* (RW) or citizens association 04, 14, and 15, Kelurahan Bidara Cina, Jatinegara, East Jakarta.

2 DEBATE

2.1 *Reproduction of spatial structure change through spatial practices*

Fundamentally, humans have an interest to keep surviving by doing any action. In the space difference, both socially and spatially, the individual who is considered as having the authoritative capacity—the authority of other individual spaces in this case—will undertake the spatial practices to organise the other individual spaces (Arendt, 1998; Foucault, 1997; Lefebvre, 1991). The social and spatial spaces will continue to be reproduced and the repeated relationships established—since it is formed and organised by the produced spaces (Lefebvre, 1991; Dovey, 1999; Foucault, 1997).

By means of practices, each individual is able to show that he/she is different from others. Through the social practices, the inter-individual can point out status over a particular space as a result of their differences. Through the spatial practices, each individual can actualise his/her capacity into the spatiality. This practical relationship illustrates that the spatial practices are competent to shape, arrange and restrict the spatial spaces in the social relationships (Arendt, 1998; Dovey, 1999; Foucault, 1980a; Foucault, 1980b Lefebvre, 1991). As the result, the repeated spatial chain of the connected spaces forming the spatial structure.

Structuration explains that the spatial structure is not only formed and forms in a purely spatial scope, but also the relationships between spatial space positions, social space, and its spatial practices (Giddens, 1979 in Dovey, 1999). The physical spatial spaces structure can be studied further as a place for individuals/communities to create a more complex sociospatial spatial structure, which may vary by situation as a result of the restriction of the time dimension and its social spaces (Foucalt, 1988 in Dovey, 1999; Arendt, 1998; Lefebvre, 1991; Giddens, 1986).

These theories suggest that the dwelling environment is the main medium to the complex formation of a spatial structure—various spatial objects; human actions gender and social status. The Habitus Concept explained that other formations of spatial structure can occur and are formed inside the physical spatial structure of the dwelling environment (Bourdieau, 1977 in Dovey, 1999).

'Habitus' also clarifies that the housing dwelling and its surroundings not only define the 'sense of one's place' but also the 'sense of other's place' (Dovey, 1999). It also suggests that the simple physical spatial spaces in a house will become complex as a result of its occupants, and also by the social actions given by the communities in their neighbourhood, such as negotiation. Therefore, negotiation is the important part practised by communities to maintain the production of another spatial structure inside their dwelling environment.

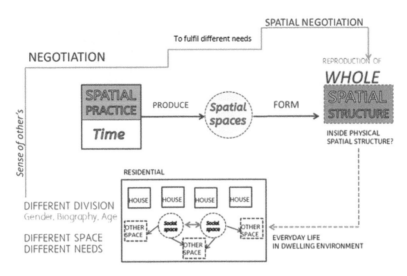

Figure 2.   Reproduction of spatial structure as spatial structure change formed by negotiation (Source: Personal data).

2.2   Gender perspective: Negotiation to reproduce spatial structure change in the female-headed household housing environment

A FHH is a family without the existence of a father, because of his death or temporary departure, and the household is organised by a mother and/or other female relationships (Chant, 1997; Fuwa, 1999).

In a *male-headed* household, a wife does not use her authority fully in undertaking the spatial practices related to domestic work and childcare because she still needs consideration from her husband (Risinda, 2013; Moser, 1993). The presence of a male was found to assist the occurring activities through negotiation with female members. However, the judgement made by men to negotiate is due more to their incomprehension towards domestic work, by making time to perform productive activities as a result of the husband's freelance job, the male physical ability which is superior compared to that of the female, as well as providing a masculine figure needed in order to gain *security needs* when negotiating with other people outside the family/community in the production works (Risinda, 2013Miraftab, 1996, 1998; Moser, 1993). According to the research byRisinda, 2013), it can be concluded that the spatial practices negotiated by the male-headed family can be restricted until inside the housing.

Different from FHH, the spatial structure change formed by the female head is negotiated and used together with the female relationships communally to form and maintain a whole spatial structure (Miraftab, 1998; Moser, 1993). Whether there are any family ties or not, women are more caring for one another. In this case, they understand each other because they play a role as a mother, so that the negotiation is based on the aspiration of motherhood which is not owned by a male head. As a result, the purpose of the negotiation tends to be cooperative and other-concerned (Bowles, 2012). In the end, they can establish the whole unitary authority in accordance with the capacity to undertake each spatial practice, so that the way they do it can be translated into the intact spatial structure formation—especially in its reproduction, to complete the form of spatial structure activities which is maintained in its environment.

2.3   Rethinking primary spatial syntax to understand spatial narrative as a process of spatial structure change

The linked cell figure created from the physical spatial structure creates a spatial narrative (Dovey, 1999). However, it is also represented from a two-way perspective. Structure containing representation appears in the whole spatial narrative, which is formed by the sequences of

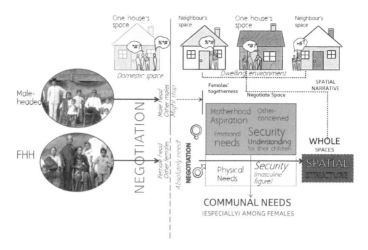

Figure 3. Gender perspective on negotiation in production & reproduction of spatial structure (Source: Personal data).

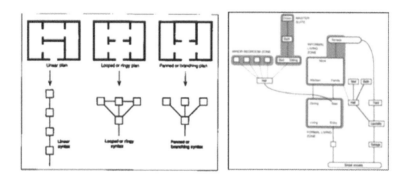

Figure 4. Primary spatial syntax by Hillier and Hanson (1984) to spatial structure change formed inside physical spatial structure (Source: Framing Place, Kim Dovey, 1999).

spaces produced in the social space, spatial space and time (Psarra, 2009; De Certeau, 1984; Lefebvre, 1991).

The whole spatial structure is not only formed by the physical spaces, but also by socio-spatial spaces with different times (Giddens, 1986). The spatial structure changes formed by a series of spatial practices compiles a complex spatial narrative as a consequence of interaction, movement, and activities of agents as well as the time difference that happens in the daily process (De Certeau, 1984; Giddens, 1986). This whole spatial structure is formed by the physical spatial structure with a spatial narrative inside it. The sequences of spatial space can also be understood as a spatial position relationship, formed by the position of the individual room units from the spatial practices accomplishments, from the beginning until the ending position in the sequences. The criterion of beginning and ending is the time standard when the process of the nonphysical spatial structure is formed (De Certeau, 1984; Dovey, 1999).

The spatial analysis in this article applies a method of primary spatial syntax which is developed by Hillier and Hanson (1984)—a spatial mapping technique into the form of cell structure (Dovey, 1999). The sequence/stages and the spaces unit proximity are substantial in establishing the activity series as the primary cluster. The relationship between the positions of primary clusters identifies the other spatial structure inside the physical space (Hillier & Hanson, 1984; in Dovey, 1999).

Figure 5. Understanding spatial narrative as production and reproduction of spatial structure change (Source: Personal data).

3 RESULTS AND DISCUSSION

3.1 *Reproduction of spatial structure change around the house*

According to the analysis outcome from the beginning observation, the analysis continues to the FHH housing, which is the highest overlapping space structure as a result of space management which has been done by the female heads and their female relationships. They, who play the role as mothers, participate to regulate the use of spaces in a relatively similar time, as a consequence of the space activities of the mothers always being bound by space activities of children, both with the birth mother or the female relationships (Chant, 1997).

In the formation of daily activities spatial structure, the mother's neighbours always play a role in producing their spatial structure. To maintain their whole activities, they are needed by the FHH to produce their spatial structure change, both inside and outside the house. Because of a difficult situation and limited space, the FHHs have to reproduce this spatial structure change to keep their activities undertaken. This spatial reproduction will not occur without the female neighbours' perspectives to the FHH member, especially the female head. In the interview of FHHs, Mrs. Fatimah and Mrs. Nur are specifically known by their neighbours. They are not only known as the original inhabitants of the region, but also because they are the active female heads working to earn livings, assisting to organise and still having the 'dependents' to be taken care of. Those factors create a 'biography' that makes neighbours feel very respectful. Moreover, the status of the female head with the single parent-daughters makes other people feel more sympathy and empathy, and makes them really understand that they are in a difficult situation.

A female head Mrs. Irma lives in RT 08 RW 04. There are three family members, who are Mrs. Fatimah (60 years old), Irma (34 years old), and her child Iki (three years old). Both Mrs. Fatimah and Irma are single parents. Their routines are busy with household chores, taking care of Iki, and trading in a kiosk which is always crowded by the buyers, as well as they being the buyers themselves.

The negotiation is undertaken in the housing between Mrs. Fatimah and Irma in forming the production of spatial structure in the kiosk and outdoor terrace, along with forming the structure of Iki's activities in the indoor terrace. There is a time when Iki and his friends in the indoor terrace are unruly. In this case, it is Mrs. Fatimah and Irma who reproduce the play spatial structure. Meanwhile, the spatial structure of the production area is created by the buyers. In this situation, the negotiation happening with the neighbours is more often offered by the buyers. They offer themselves to process their own orders and let Mrs. Fatimah and Irma to do their other activities.

On the other hand, Mrs. Nur, 57 years old, has two widowed daughters; Ipah, 33 years old and Lis, 35 years old. Ipah owns a seven-year-old child while Lis is the mother of two kids, three and four years old. Another family member is a nine-year-old girl. Mrs. Nur works as a fried food seller. Her two daughters work as employees outside Bidara China area from morning until night.

In the afternoon, Mrs. Nur creates the temporary space to trade her fried food as well as waiting for the other foods to cook. Mrs. Nur has to switch back and forth between the kitchen and terrace because she has to maintain her cooking space along with her selling space. If she has finished cooking, she also takes time to form a lying chamber. However, she is not able to constantly switch between the outdoor terrace-kitchen-balcony to ensure her fried food is bought.

Her next-door neighbour, who knows well the habits of Mrs. Nur, directly peforms the negotiation in the front area of Mrs. Nur's housing to maintain her production space position. The buyers arriving will be monitored and their movements limited by the neighbours. Meanwhile, Mrs. Nur can occasionally still see a series of fixed space still formed as it should be, both from the kitchen and the balcony.

In other situations, as a result of Mrs. Nur strolling around to sell her fried food, she has to negotiate to entrust her grandchildren to others. At the beginning, she created the structure of a children activities space. When she is trading around, she cannot ensure whether the spaces formed for kids still remain in the same position of the spatial structure she has organised before. The space can be spatially varied for children who might be bored and they can be moved to form a new activity requires space and other snacks. The neighbour who understands the habits of Mrs. Nur's grandchildren, automatically does the negotiation to directly form the meal supplying room. The position of the children activities space then returns to as it was formed by Mrs. Nur.

3.2 *Formation of spatial structure in the FHH housing environment*

In the observation process, there are situations when they are tied to different activities at the same time. The housing location of Mrs. Fatimah and Mrs. Nur that is adjacent and connected with the empty alley, makes them often meet each other when performing their respective routines around occupancy.

The situation begins from Mrs. Nur's activity forming the drying space on the neighbour's fence. The reproduction of Mrs. Nur's drying space persists as a result of negotiation that occurred in the terrace in a small alley and the terrace in a big alley. When Mrs. Nur creates her new space position in the drying area, neighbours will monitor and maintain the

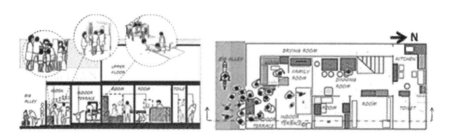

Figure 6. Section (left) and floor plan (right) of Mrs. Fatimah's house (Source: Personal sketches from field survey, September–November, 2016).

Figure 7. FHH family: Mrs. Fatimah and Mrs. Irma work in outdoor terrace and kiosk (left), They do their important activities in domestic area (centre), mother neighbours do self-ordering (Source: Personal data from field survey, September–November, 2016).

Figure 8. Diagram of producing and reproducing spatial structure change through negotiation in a difficult situation (Source: Personal data, from field survey, September-November, 2016).

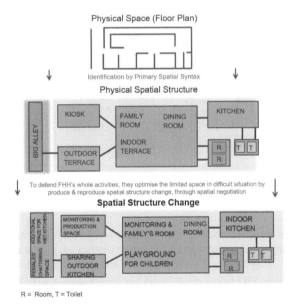

Figure 9. Diagram of Mrs. Fatimah's whole spatial structure as the result of spatial structure change (Source: Personal data).

spatial structure of Mrs. Nur's grandchildren inside or around the housing. While Mrs. Nur conducts drying, grandchildren who are in the housing areas will know that the position of the space remains constrained by neighbouring mothers who know their habits. If they are outside of any housing area, the displacement can be seen from Mrs. Nur's drying space,

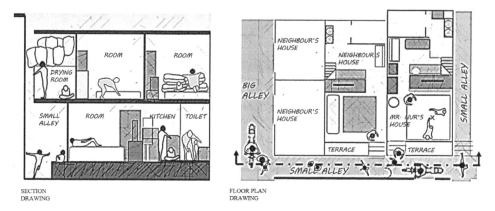

SECTION
DRAWING

FLOOR PLAN
DRAWING

Figure 10. Section (left) and floor plan (right) of Mrs. Nur's house (Source: Personal data from field survey, September-November, 2016).

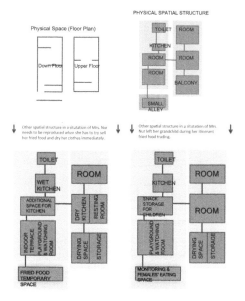

Figure 11. Diagram of Mrs. Nur's whole spatial structure as the result of spatial structure change (Source: Personal data from field survey, september-november, 2016).

because its position is directly opposite the front aisle occupancy. Therefore, the reproduction of drying space cannot be separated from the reproduction of spatial structure in the housing, which is the spatial structure of children activities. Thus, the spatial reproduction in the housing outer space becomes the space unit for the housing area spatial structure.

In such situation, the monitoring spatial structure initially formed as a result of the attachment of grandchildren space to the existence and spatial practices of their grandmother since the money is held by her. The grandchildren who need the snack cannot be far from their grandmother's drying activity and have to ask directly for a snack. On the other side, the mother's neighbours keep doing the spatial practices in the production area and serve the other buyers. The space of Mrs. Nur's grandchildren enters the position of buyer's space in the production area spatial structure and is also bound to the space around Mrs. Fatimah's housing. Consequently, in the environment scale, the spatial structure contained in each area will complete the spatial structure of other housing in the surrounding area. Thus, the whole

230

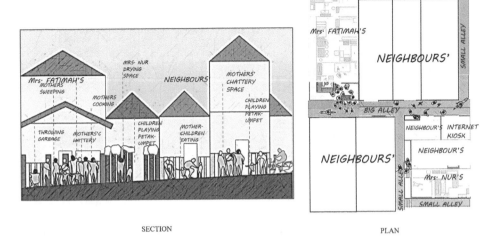

SECTION   PLAN

Figure 12. Section (left) and floor plan (right) of FHH neighbourhood with its whole activities practiced by mother and children (Source: Personal data from field survey, September-November, 2016).

Figure 13. Mother's-children's activities (left) and Mrs. Nur's drying space on neighbour's fence (Source: Personal data from field survey, september-november, 2016).

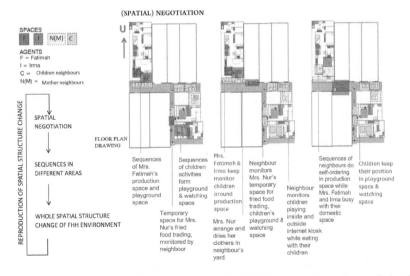

Figure 14. Reproduction of spatial structure change in FHH housing through negotiation in a difficult situation (Source: Personal data from field survey, September–November, 2016).

231

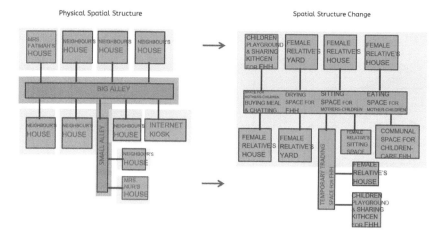

Figure 15. Whole spatial structure of FHH housing as the spatial structure change in their dwelling environment (Source: Personal data).

spatial structure still can be reproduced to retain all space for activities which persist in the FHH dwelling environment FHH.

4 CONCLUSION

In the process of the reproduction of spatial structure change inside the FHH housing area, the female head depends on the presence and position of space of spatial practices undertaken by female relationships and neighbours. To defend the FHH activities space, the reproduction of spatial structure change needs negotiation. The negotiation plays a role in regulating and occupying the maintained space in order that the spatial structure unity remains to be shaped.

According to the observation and analysis, the negotiation occurring in the housing area and the surroundings are more often undertaken by the females rather than by the males. It is influenced by the habits which everyone is already familiar with in running daily spatial practices, as well as their perspective towards the interviewee as the FHH member. Views of the mothers in doing chores and taking care of children have become the significant point in the fulfilment of the FHH mother's emotional needs. The negotiation applied by mothers will occur non-verbally, through the establishment of a spatial space unit in the reproduction process of a spatial structure change.

The point of view of the mother's neighbours, especially towards the female head that as a woman, head of family, widow, and with her life history that has been long recognised by the environment, makes it hard for her to discharge the gender and social status perspective which influences the negotiation process between her, her family, and her neighbours. On the other hand, her daughter also cannot escape from the neighbours' perspective that she is at the same time a woman, mother, and (young) widow, considered capable to struggle from a young age to live only with her mother and her child. Thus, in this context, not only a sense of kinship, but also compassion, sympathy, empathy, and mutual respect, foster a sense of togetherness and aspirations of motherhood among FHHs with the mother's neighbours. Therefore, a specific perspective formed by the mother's neighbours is really needed by the FHH mothers to directly negotiate space in reproduction of spatial structure change.

REFERENCES

Arendt, H. (1998). The human condition. In *The Human Condition*. Chicago. The University of Chicago Press.
Bourdieu, P. (1977) *Outline of a Theory of Practice,* London: Cambridge University Press.

Bowles, H.R. (2012). Psychological perspectives on gender in negotiation. In *Journal of Harvard Kennedy School, Research Working Paper Series No.46, 10(12)*. Retrieved from https://www. google.co.id/url?sa=t&rct=j&q=&esrc=s&source=web&cd=1&cad=rja&uact=8&ved=0ahUKEwj iubPWv9TQAhVHo48KHT9BCzYQFgghMAA&url=https%3A%2F%2Fresearch.hks.harvard. edu%2Fpublications%2FgetFile.aspx%3FId%3D862&usg=AFQjCNFEckyA6E6cb9b7dRvGweSxd yEzrQ& sig2=FnTeVqNVwN9H1W_GGBGFZA.

Chant, S. (1997). Introduction: Why women-headed households? In *Women-Headed Households: Diversity and Dynamics in the Developing World*. Great Britain: The Ipswich Book Company, Ltd.

De Certeau, M. (1984). *The practice of everyday life*. California: University of California Press, Ltd.

Dovey, K. (1999). Framing places: Mediating power in built form. London: Routledge.

Foucault, M. (1980a). Questions on geography. In C. Gordon (Ed.), *Power/Knowledge: Selected Interviews & Other Writings, 1972–1977* (pp. 3–77). New York: Pantheon Books.

Foucault, M. (1980b). The eye of power. In C. Gordon (Ed.), *Power/Knowledge: Selected Interviews & Other Writings 1972–1977* (pp. 146–165). New York: Pantheon Books.

Foucault, M. (1988). On Power. In L.Kritzman (Ed.), *Michel Foucault*. New York:, Routledge.

Foucault, M. (1997). Space, knowledge, and power (Interviewed conducted with Paul Rainbow). In N. Leiach (Ed.), *Rethinking Architecture: A Reader in Cultural Theory* (pp. 329–357). London: Routledge.

Fuwa, N. (1999). The poverty and heterogeneity among female headed households revisited: The case of Panama. In *World Development*, *28*(8), 1515–1542. Retrieved from https://www.researchgate.net/ publication/222663615_The_Poverty_and_Heterogeneity_Among_Female-Headed_Households_ Revisited_The_Case_of_Panama.

Giddens, A. (1979) *Central Problems in Social Theory*, London: Macmillan.

Giddens, A. (1986). Elements of the theory of structuration. In *The Constitution of Society* (pp. 1–34). Barkeley: University of California Press.

Hillier, B. and Hanson, J. (1984) *The Social Logic of Space*, New York: Cambridge University Press.

Lefebvre, H. (1991). Plan of the present work. In *The Production of Space* (pp. 1–67). Oxford: Basil Blackwell Ltd.

Mcllwaine, C. & Willis, K. (2014). *Challenges and change in Middle America: Perspectives on development in Mexico, Central America, and the Caribbean*. London: Routledge.

Miraftab, F. (1996). Space, gender, and work: Home-Based workers in Mexico. In *Homeworkers in Global Perspective: Invisible No More* (pp. 63–80). New York: Routledge.

Miraftab, F. (1998). Complexities of the margin: Housing decisions by female householders in Mexico. In *Environment and Planning D: Society and Space*, *16*, 289–310. School of Environmental Planning, Griffith University.

Moser, C. (1993). Gender roles, the family and the households. In *Gender Planning and Development: Theory, Practice and Training* (pp. 15–36). London: Routledge.

Prabowo, H. (2002). Acculturation strategies and ethnic identity: The host and guest culture in urban village of Jakarta. In *XIV International Association for Cross Cultural Psychology Congress Unity in Diversity: Enhancing a Peaceful World*. Yogyakarta. Retrieved from http://jakartapedia.bpadjakarta. net/ index.php/Perkampungan_Kota.

Psarra, S. (2009). Introduction. In *Architecture and Narrative: The Formation of Space and Cultural Meaning*. Abingdon: Routledge.

Risinda, N.F. (2013). *Women's Space of Activities in Slum Areas: Territories and Negotiation*. Unpublished Master Thesis, Universitas Indonesia, Faculty of Engineering.

Competition and Cooperation in Social and Political Sciences – Adi & Achwan (Eds)
© 2018 Taylor & Francis Group, London, ISBN 978-1-138-62676-8

Constructs of vulnerability and resilience in the coastal community of Tanjung Pasir village, Tangerang

I.B.H. Gupta, A. Rahmat, A.T. Alamsyah & D. Syahdanul
Department of Architecture, Faculty of Engineering, Universitas Indonesia, Depok, Indonesia

ABSTRACT: Response to natural hazards varies according to the perceptions of individuals and groups. In their responses, it is possible to identify spatial, social and temporal response variations in various actors within a hazard-affected area. Nevertheless, these actors share similar objectives, which are to reduce damage and protect themselves against hazard. People who are affected by flooding tend to move to higher ground when floods approach, and return when the flooding subsides. This frames a way of thinking applicable to a certain period of time and social context. Local government responds by constructing dykes that may protect the area from future flood events. The two actors named here, government and people, determine their mitigation efforts based on their exposure to risk and their resources. This research aims to identify the constructs of vulnerability and resilience in the coastal community living in Tanjung Pasir village, as well as to identify spatial, social and temporal disparities in their governance.

1 INTRODUCTION

The government of Tangerang Regency directs the development of coastal areas in relation to the businesses of aquaculture, agriculture, fishing, tourism as well as some areas designated as mineral mining zones. Some parts of the coastal area of Tangerang Regency are residential areas directly adjacent to the shoreline. However, the coastal area is a disaster-prone area in terms of both natural events and those induced by human activity. *Rencana Tata Ruang Wilayah* (RTRW) and *Rencana Zonasi Wilayah Pesisir dan Pulau-Pulau Kecil* (RZWP3 K) list natural disasters that could potentially impact coastal areas, such as flooding, downstream sedimentation, earthquakes and tsunamis. Disasters caused by human activities can include pollution of downstream and coastal areas by industrial and domestic wastes. This tension is an alarming situation for the management of the coastal area and, therefore, requires identification of the vulnerabilities and resilience of residential construction in the coastal settlement. The understanding of such vulnerabilities and resilience is greatly affected by local geographical and climatic conditions, as well as by the deep-rooted cultural traditions and patterns of interpretation of local residents, and, of course, the actual experience of disaster too (Christmann et al., 2014). Therefore, it is important to identify what the locals perceive as the vulnerabilities and make efforts to resolve them.

Vulnerability encompasses the situation and the processes that determine the exposure, sensitivity, and the reaction capacity of a system or an object in relation to dealing with dangers. Physical, socioeconomic and environmental factors all play a role (Birkmann, 2007). Birkmann (2007) describes resilience as the capacity of a system to absorb shocks and disruptions and to continue to exist with the least possible damage, and identifies three important values of resilience: first, the pure resistance of the system; second, the capacity to reestablish the original state of the system; third, the ability to learn and adapt to changing conditions.

Vulnerabilities are associated with a variety of risks that arise due to the interplay between the various processes that occur locally, as well as the human ability to cope with them (Bankoff et al. 2004). Vulnerability changes and may increase, reducing the ability to cope with it. Christmann et al. (2014) view vulnerability as the result of a social construction process in which potential threats are collectively assessed and negotiated by members of a society. Similarly, they regard resilience as a construction process that emphasises the possibilities of action and reaction within a relational network. Here, the intent is to implement more or less far-reaching changes within the relational web in order to reduce the vulnerability of a centrally placed entity by protecting its functions and ensuring its integrity.

2 METHODOLOGY

2.1 *Selection of location*

Tanjung Pasir village is a coastal settlement located on the north coast of Kabupaten Tangerang. It has a population of 2186 (*Pemkab. Tangerang*, 2011). The same data source stated that 66 per cent of inhabitants were graduates of elementary school, and only 24 per cent of the population had the opportunity to complete the standard nine-year education. Just ten per cent had completed higher education. The people in the village are predominantly fishermen and live below the poverty line. *Penyusunan Desa Pesisir Terpadu* (2012) stated that Tanjung Pasir village is affected by various natural phenomena, such as river run-off and the intrusion of seawater. Downstream sedimentation is also a major problem. The village has a settlement area of 14.64 ha. This represents two per cent of the total area of 635 ha. The majority of the village (65 per cent) is used for embankments for shrimp cultivation or as paddy fields.

2.2 *Group discussion and survey*

Group discussions were conducted on two occasions and involved a few prominent individuals as an introductory step, and group discussions with representatives of the local population, as well as discussions with the women's group in the village. These discussions aimed to collect responses in relation to living in the coastal environment, as well as about disastrous events that frequently arise. In these discussions, we also captured the perceptions of vulnerabilities as expressed by male and female groups. A survey was conducted to identify the physical conditions of the site as well as any human interventions.

2.3 *Group discussion analysis*

Analyses of the results of the focus group discussions were conducted to identify the occurrence of disastrous events as well as to further determine which of these were most disturbing to the life of the residents. Analyses were also conducted to see what, from the perspective of the men and women who live in Tanjung Pasir village, makes them feel vulnerable.

3 FINDINGS

3.1 *The hazards: Natural and man-induced*

Tanjung Pasir village is surrounded by embankments for aquaculture that belong to investors who are not residents. There is no clear information as to when these individual aquaculture entreprises appeared. However, a government-led aquaculture programme called *Tambak Inti Rakyat* started in 1980 and has been replicated by individual investors since 1995. Interestingly, one of the discussion participants mentioned that Tanjung Pasir village had never been inundated or flooded before the construction of these embankments started.

The same participant reported that the construction of the embankments was also intended to regulate the flow of seawater into the enclosed area. This is done to control the level of seawater in the pond so that it is appropriate for the needs of aquaculture. However, this arrangement has led to the residential area being positioned below the embankment walls and, therefore, sea run-off can inundate settlements before entering the embankment.

This is a typical example of the aggression of economic activity. Every human being is susceptible to economic activity or development. Disasters are not simply the product of one-off natural phenomena but can equally be the result of the action of environmentally unsustainable development projects over time (Lavell, 2014). This danger occurs when any individual modifies nature without considering the possible impact on others.

Figure 1 shows the natural and man-induced hazards in the area of Tanjung Pasir. Between December and February, seawater intrusion has reportedly reached as far as the red line. In the group discussions, infrastructure that is not well constructed and does not meet people's needs appears as one of the major issues. All of the individuals attending the discussion declared that the current existing infrastructure does not protect them from natural phenomena such as seawater intrusion and river run-off. This is an interesting remark because the interplay between the positioning of settlements and coastal characteristics is not viewed as a major cause of vulnerability.

During the group discussion, the community stated that, within the past five years, several infrastructure projects intended to further protect the settlements from natural phenomena have been carried out. In 2013, the government of Kabupaten Tangerang constructed a dyke along the river as protection against river run-off during rainy days. However, dyke construction was carried out iteratively due to budget constraints: local government only budgeted an amount sufficient to construct part of the dyke. Thus, as of September 2016, the dyke construction had not reached the downstream portion, leaving space for water to enter the settlement. The local government still needs to construct another 400 metres of dyke to completely protect the settlements.

The group discussion also reported that the government of Kabupaten Tangerang had built coastal protection in 2014. Construction of this dyke is intended to avoid the intrusion of seawater into the settlement area. However, we observed that this construction was also being conducted gradually, leaving some parts without protection.

To allow fishermen to maintain their boats, one part of the dyke is dismantled to provide entry and exit access for boats. This gap is also accessible to run-off and residential areas are

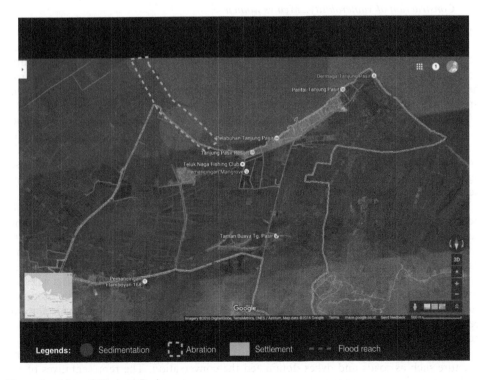

Figure 1. Map of Tanjung Pasir.

flooded as a result. This experience indicates that the dyke design does not accommodate the essential needs of the fishing community, that is, to allow them to repair their boats. In other words, technical solutions (in this case, a dyke) should also take into consideration aspects besides the main problem, which is run-off. Part of the 'opened' dyke is acting as an entry point for run-off water that inundates residential areas. Demolition of even a small part is detrimental to the investments for constructing dyke.

Fishermen rely on the wind to sail but if strong westerly winds arrive between September and January, they cannot put to sea. There is a strong reliance on nature in terms of being productive. Climate change has also brought rainy days even when the north-east monsoon wind is blowing, which adds to the existing problems for the fishermen.

The fishermen also reported sedimentation of the downstream area of the river. This sedimentation has made the downstream part shallower and makes it difficult to remove boats from their docks on the inner side of the river. A joint effort has been made to clear the sediment with the tools available. However, this effort is unsustainable because the frequency of cleaning is getting higher and available manpower is reducing. When fishermen cannot sail their boats from the dock then they will not sail.

Almost all the families in Tanjung Pasir village have shallow wells to supply clean water. The wells are typically of 50 cm depth and do not have any protective wall around them. These wells cannot be used in case of run-off because the seawater and its mud fills the pit. Although they are connected to piped water, it does not supply sufficient fresh water for their needs.

The conditions become worse when run-off is combined with rain. According to the locals, tidal flooding will quickly recede/dry. However, when combined with rain, the flooding persists for days. Residents also reported several cases of disease, such as itching experienced by children. However, this appears to be treated as though it were normal, with citizens treating their children by washing their legs. Health problems were not cited as a cause for alarm when flooding occurs.

3.2 *Construction of vulnerability: Men vs women*

The women's group discussed a better life for the fishermen. The group said that their husbands went fishing from 10 p.m. in the evening and returned home around 10 a.m. the next morning. The safety of their husband at sea was a major concern. The high capital required for fishing was also a major issue for the women's group because the monies obtained from the sale of fish were sometimes lower than the capital costs, often causing the fishermen's families to go into debt to meet daily needs.

The wives of the fishermen engage in economic activities to supplement family incomes. Some of the activities identified include selling snacks/stalls, fish trading, making shrimp paste and making salted fish. As well as helping to increase family incomes, these activities also act as pre-emptive insurance for occasions on which their husbands return from fishing without results. Even so, the revenue generated from these activities is just a single day's income. Continuity of income is not assured and is highly reliant on natural conditions.

When the fishermen are not able to sail, they switch professions and become scavengers of the plastic garbage left by the sea on the beach, such as plastic cups, plastic bags and other plastic items with economic value. The revenue from scavenging is just enough to support a fisherman's wife and children for a day. During the discussion, residents described how they scavenge plastic in the morning to cover lunch and, if they are lucky, they may be able to eat in the afternoon too. Such scavenging must be done every day because the profits will only cover a day's costs. Other activities include looking for shells and seaweed to be processed and sold later. Winterhaller, as quoted by Bankoff (2004), states that disaster can not only seen in specific events, such as the frequency or intensity of natural phenomena, but also in the sequence of events.

During the discussions with the men's group, that is, the fishermen, the provision of infrastructure such as roads and dykes dominated the conversation. The representatives of the residents walked us around the village to see the roads and areas vulnerable to run-off water.

It was also reported that young people in the village do not have any interest in becoming fishermen. They prefer to work inland, becoming factory workers or traders. Indeed, parents also encourage their children to find jobs inland with higher incomes and lower levels of risk and to avoid becoming fishermen.

Residents have no other choice but to stay where they are, even if this makes them vulnerable to a variety of natural phenomena. Besides having no other options, the residents are mostly fishermen, and therefore residing in the coastal area is preferable.

There was a fundamental difference between the men (the group of husbands) and the women (the group of wives) when they discussed vulnerability in the coastal environment. The men's group discussed many infrastructure elements, such as dykes and roads, that were deemed essential to prevent tidal or river water run-off. However, the women's group debated many non-physical elements such as sailing safety, water facilities, and maternal and child health, as well as support for the fishermen's lifestyle through maintenance of a good fish price, together with easy access to capital for boat costs. The construction of vulnerability for the two groups of actors, the husbands and the wives, is influenced by their respective roles in the household.

Although the residents of Tanjung Pasir village are exposed to hurricanes, flooding and river water run-off, they do not consider these as major vulnerabilities. They accept these conditions because this place is their home. On the other hand, efforts to reduce vulnerability to floods and associated damage need to be conducted to protect infrastructure and people's possessions from the threat of sea water or river water. The residents are highly dependent on external parties such as government or aid from non-governmental organisations to play a role in the provision of infrastructure to reduce water run-off.

The remarkable thing is the residents are aware of the time when their site is vulnerabile to natural disaster. Residents have the ability to predict the time of run-off. They also understand their local weather anomaly in which downpours of rain can occur at any time. But they only understand natural phenomena that have already happened. They do not pay attention to the natural phenomena that may appear in future periods, such as rising sea levels that could potentially drown their settlements permanently.

Overall, in relation to the three vital values in the construction of resilience against natural phenomena, namely, the resilience of the system/object, the capacity to return to its original state, and the ability to learn and adapt to changing conditions, Tanjung Pasir village has shown low resilience. People here tend to accept conditions that occur because of their limitations in protecting themselves or avoiding the negative impact of natural phenomena. Run-off and flooding have occurred since the 1980s, and various remedial efforts have been made, including presentation of the situation to local government. However, hardly any indications of adaptation to natural phenomena can be found. Thus, although the construction of the dykes can be regarded as a form of adaptation, the houses remain typical inland houses. When run-off occurs, the water can inundate the houses to heights of 50 cm. Of the three factors involved in resilience, Tanjung Pasir village suffers from inherently high susceptibility but does have the ability to return to its original state well. However, the residents of Tanjung Pasir village have not shown the ability to learn and adapt to natural phenomena. The solution of dyke construction implemented to date effectively causes villagers to live in a large basin and may potentially lead to new problems in the future.

4   CONCLUSION

This study was conducted to identify the construct of vulnerability in relation to the coastal residents of Tanjung Pasir village, as well as their adaptation as a form of resilience. Another thing that has also been identified is the degree of variety among the actors resident here in relation to their comprehension of vulnerabilities and resilience. In this context, the comprehension of vulnerability and resilience is much influenced by physical and/or geographical conditions and, most importantly, by the particular role played by the actors. This understanding is further influenced by the actors' backgrounds and their temporal structure.

On the basis of the group discussions, it can be concluded that the men's group viewed vulnerability to natural phenomena as being due to a lack of infrastructure. They reported that the lack of dykes, access roads and coastal protection, as well as construction of embankments, made their lives vulnerable. For the women's group, vulnerability revolved around their husbands' work as fishermen, as well as their dependence on high-risk fishing activities and the unreliable results of same.

REFERENCES

Bankoff, G., Frerks, G. & Hilhorst, D. (2004). *Mapping vulnerability: Disasters, development and people.* Abingdon, UK: Earthscan.

Birkmann, J. (2007). Risk and vulnerability indicators at different scales. Applicability, usefulness, and policy implications. In: Environmental Hazards 7, 20–31.

Burroughs, R. (2011). *Coastal governance.* Washington, DC: Island Press.

Christmann, G., Ibert, O., Kilper, H. & Moss, T. (2012). *Vulnerability and resilience from a socio-spatial perspective: Towards a theoretical framework* (Working paper). Erkner, Germany: Leibniz Institute for Regional Development and Structural Planning.

Christmann, G.B., Balgar, K. & Mahlkow, N. (2014). Local constructions of vulnerability and resilience in the context of climate change. A comparison of Lübeck and Rostock. *Social Sciences, 3*, 142–159. doi:10.3390/socsci3010142.

Doxiadis, C.A. (1968). *Ekistics: Introduction to the science of human settlement.* London, UK: Hutchinson.

Doxiadis, C.A. (1970). Ekistics, the science of human settlements. *Science, 170*(3956), 393–404.

Eraydin, A. & Taşan-Kok, T. (Eds.). (2013). *Resilience thinking in urban planning.* Dordrecht, The Netherlands: Springer.

Kementerian Kelautan dan Perikanan. (2012). *Penyusunan Desa Pesisir Terpadu.* KKP.

Pemerintah Kabupaten Tangerang, (2011). *Monografi Desa. Kab. Tangerang.*

Satria, A. (2015). *Pengantar Sosiologi Masyarakat Pesisir.* Jakarta, Indonesia: Yayasan Pustaka Obor.

Trading-off pattern of marine stewardship council management in Indonesia

A. Notohamijoyo
Postgraduate School, Department of Environmental Sciences, Universitas Indonesia, Jakarta, Indonesia

M. Huseini
Department of Administration, Faculty of Administrative Science, Universitas Indonesia, Depok, Indonesia

R.H. Koestoer
Coordinating Ministry of Economic Affairs, Jakarta, Indonesia

S. Fauzi
Ministry of Marine Affairs and Fisheries, Jakarta, Indonesia

ABSTRACT: In the last decade, seafood ecolabels have grown and developed as a market measurement for sustainable seafood. The most popular of these is that of the Marine Stewardship Council (MSC). Nevertheless, MSC faces immense challenges in developing countries such as Indonesia because of issues including high costs, differences in fisheries structures, challenging requirements and lack of support from stakeholders, particularly small-scale fishermen. These issues highlight the problem for MSC itself: lack of social compliance. MSC delivers on the issue of sustainability but does not address social aspects and has a tendency to be exclusively for large companies. If ecolabel certificates are to be consistent with sustainable development principles, they should be inclusive and should not ignore social issues. Some parties in Indonesia propose national seafood ecolabelling to compete with the MSC, but this faces a lack of international recognition. This research aims to find the most reliable seafood ecolabelling for Indonesia from the local stakeholder's perspective. It uses a qualitative methodology involving the Delphi method to obtain views from panels of experts. The variables employed include support from government, the private sector, fishermen and national Non-Governmental Organisations (NGOs). The result shows that a national seafood ecolabelling scheme is regarded as the best option for Indonesia from the stakeholders' perspective.

1   INTRODUCTION

Food security has become a strategic issue internationally and has been discussed in various international forums, such as ASEAN, APEC, OIC and D-8, in the last decade. It faces major threats from climate change, natural disasters, industrialisation and environmental degradation.

As one of the most important sources of nutrition, fishery commodities face a serious sustainability problem. The increase in global consumption has caused overfishing, overcapacity, and destructive Illegal, Unregulated and Unreported Fishing (IUUF).

As a fisheries producer, Indonesia is also facing the same problem. This condition is reflected in the report by the National Commission on Fish Stock Assessment, as published through Ministerial Decree No. 25/2014, concerning Fish Resources Estimation in the Fisheries Management Areas of the Republic of Indonesia. The data presented in this report are shown in Table 1.

Table 1. Fish stock status in each fisheries management area of indonesia (source: ministerial decree no. 25/2011 concerning fish resources estimation in fisheries management areas, republic of indonesia).

KELOMPOK SDI	Selat Malaka	S.Hindia (Barat Sumatera)	S.Hindia (Selatan Jawa)	Laut Cina Selatan	Laut Jawa	Selat Makassar – Laut Flores	Laut Banda	Teluk Tomini – Laut Seram	Laut Sula-wesi	S. Pasifik	L. Arafura – L.Timor	Keterangan
	WPP-571	WPP-572	WPP-573	WPP-711	WPP-712	WPP-713	WPP-714	WPP-715	WPP-716	WPP-717	WPP-718	
UDANG	O	O	O	O	O	O		O		O	F	
DEMERSAL	F	F	M	F	F	O	F	M	M	M	O(*)	(*) dampak dari pukat ikan
- Kurau	O			O								
- Manyung	O			F					M		O	
- Layur		M	M									
- Kurisi	F	F			M(1)						O	(1)Laut Jawa >40 m
- Kuniran	F	F			F						O	
- Swanggi	F	F			M(1)						O	
- Bloso	F	F			F						O	
- Gulamah	F	F									O	(2) khusus pancing
- Kakap merah	O(3)	O(3)	F (5)		O	M(2)		F	M		O	(3) khusus pancing
- Kerapu		O(4)			O	M(2)		F	M			(4) bubu beton
- Kuwe			F (5)						M			(5) pancing ulur & rawai dasar (NTT)
- Ikan lidah										F		
PELAGIS KECIL	F	O	F	O	O	O	F	F	M	M	M	
- Banyar	O	O		F	O							
- Kembung	O	O		F	O							
- Ikan terbang						O		F				
- D. kuroides			M				F	M				
- D. macarellus	F						M-F	M	M			
- D. macrosoma	F			F	O		M-F					
- D. ruseli	F			F	O							
- golok-golok	M											
- lemuru			O(6)									(6) Selat Bali
Tuna Besar :												Note : Pelagis besar non-tuna:
- Cakalang	M	M	M			M	M	M	M	M		- Tongkol
- Albakora			F									- Tenggiri
- Madidihang		F	F			O	F	F	F		O	- Setuhuk
- Mata besar		O	O		F	O	O	O	O	O		- Layaran
- SBT			O									- Lemadang
Cumi-cumi			M	M			M					

Table key: RED – Overexploited; YELLOW – Fully exploited; ORANGE – Moderately to fully exploited; GREEN – Mod-erately exploited.

With reference to Table 1, it shows that half of Indonesia's Fisheries Management Areas are overexploited, as can be seen from the level of red colouring in the table. In a report enti- tled *The Sunken Billions: The Economic Justification for Fisheries Reform*, the World Bank (2009) and the Food and Agriculture Organization (FAO) of the UN stated that, since 2006, 75% of fish resources have been depleted because of overfishing and IUUF.

This situation has encouraged some parties to introduce ecolabels as a mechanism to protect fish resources through market measurement. The first seafood ecolabel was that of the Marine Stewardship Council (MSC), which was established in 1996. The MSC certificate was initiated by the World Wildlife Fund (WWF) and Unilever. Following the MSC, various other fisheries ecolabel certificates were established, such as Friends of the Sea (FOS), Dolphin Safe, Icelandic Responsible Fisheries (IRF), and the Marine Aquarium Council (MAC).

Seafood ecolabels grew and became popular in some countries. Many big retail companies, such as Carrefour, Aligro, Manor, Wal-Mart, Aldi and McDonald's, supported and cooper- ated with the scheme. They imposed the scheme as mandatory for every fish traded in their network of stores. This particular policy is where the controversy began. Because of their strict limitations, seafood ecolabel certificates are seen as another trade barrier for producers of fish products and fishermen, particularly in developing countries. The imposition itself has, arguably, contravened the *FAO Guidelines for the Ecolabelling of Fish and Fishery Prod- ucts from Marine Capture Fisheries 2009*, which stated that seafood ecolabelling schemes should be *voluntary* and should not create unnecessary barriers to trade.

Understandably, stakeholders in developing countries have objections to seafood ecolabel- ling schemes, because of issues such as the high cost of certification, high requirements, trans- parency and credibility, among others, that are more difficult to achieve for developing country producers. A range of research has illustrated the problems that seafood ecolabel implementa- tion has presented in developing countries; for example, Bush et al. (2013), Pérez-Ramírez et al. (2012), Lay (2012), Ramírez et al. (2012), Ponte (2012), Marko et al. (2012), Oken et al. (2012), Bratt et al. (2011), Goyert et al. (2010), Gulbrandsen (2009), van Amstel et al. (2008), Raynolds et al. (2007), Rex and Baumann (2007), Jacquet and Pauly (2007), Ponte (2006), Kaiser and Edward-Jones (2006), Gale and Haward (2004), Johnston et al. (2001), and Constance and Bonanno (2000). As an island country, Indonesia faces similar problems.

Table 2. The ownership of fishing vessels in Indonesia (Source: Statistics of Capture Fisheries, MOMAF, 2013).

No.	Category	Number of vessels
1.	Total number of vessels	618,320 units
2.	Without motor	165,990 units (26.85%)
3.	Motor boat	252,590 units (40.85%)
4.	< 5 GT	137,620 units (22.26%)
5.	5–10 GT	38,740 units (6.27%)
6.	10–20 GT	11,650 units (1.88%)
7.	20–30 GT	7,620 units (1.23%)
8.	30–50 GT	920 units (0.15%)
9.	50–100 GT	670 units (0.27%)
10.	100–200 GT	1,180 units (0.19%)
11.	> 200 GT	340 units (0.05%)

Preliminary interviews with 60 respondents that are actors in the Indonesian tuna business showed that they do not accept or trust seafood ecolabels and believe that they are another way for global capitalism to control developing countries, using the excuse of protecting the environment. This is particularly problematic as fisheries in Indonesia remain precarious and underdeveloped in character.

In Indonesia, the fisheries structure is still dominated by small-scale fishermen. The definition of small-scale fishermen in Indonesia is stated by Law No. 45/2009 concerning the amendment to Law No.31/2004 regarding Fisheries, and is: "...*fishermen with the ownership of a fishing boat less than or equal to 5 Gross Tonnes (GT)*". Table 2 describes the structure of fishing vessel ownership in Indonesia.

Based on Table 2, it can be seen that 90% of Indonesian fishing vessels are under 5 GT in size. This shows how the fisheries structure in Indonesia is dominated by small-scale fishermen, typically with poor backgrounds. For them, seafood ecolabels are difficult to adopt as they are unable to afford the high costs or maintain the capacity needed for seafood ecolabelling. This makes seafood ecolabels extremely difficult to implement in Indonesia. A lack of capacity, capital and networks are the conditions typically found in small-scale fisheries. If seafood ecolabels such as MSC do not pay attention to the characteristics of these fishermen and the needs of stakeholders, implementation will be difficult.

The current research seeks to identify stakeholders' support for MSC certification in Indonesia. This paper also examines which seafood ecolabel option(s) will be best from stakeholders' perspectives in the case of Indonesia. Given that it is Indonesia's main export commodity, this research is limited to tuna fisheries.

2 PROBLEM FORMULATION

Ramírez et al. (2012) reported that research concerning MSC certification in Argentina (of anchovies, Patagonian scallops, hake and hoki) showed that the acceptance of stakeholders influences the decision to adopt or reject certification. Stakeholders can make an important contribution to the certification process and influence the opinions of other stakeholders.

Research by Christian et al. (2013) showed that the credibility of MSC certification is often questioned because MSC is viewed as inconsistent in applying its own principles. There is a duality when MSC applies its principles. Thus, a decline in species continues even while the ecolabelling process continues to operate. Hadjimichael and Hegland (2016) argued that the development of certain seafood ecolabel certifications can rapidly lead to their monopoly on sustainable fisheries management. This condition leads them to monopolise the principles only from the standpoint of the institution. It has begun to happen in the case of MSC, which is accused of imposing one-way communication and its principles regardless of the conditions in developing countries.

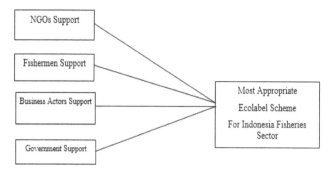

Figure 1.   Conceptual framework for the research.

Research conducted by Kvalvik et al. (2014) compares the application of MSC certification in Norway with that of Icelandic Responsible Fisheries (IRF) in Iceland. They showed that the structure of the industry and the supporting public policy of a country affect the implementation of fisheries ecolabel certificates in each country. Research by Kvalvik et al. (2014) and Ramírez et al. (2012) indicated that stakeholder support was instrumental in the successful implementation of ecolabel certificates.

Klooster (2010) stated that the success of a multi-stakeholder certification system depends on three main factors: rigour, acceptability and legitimacy. Rigour means that certification must implement strong standards compared to that applied to uncertified products. Acceptability means that the certificate must be acceptable to all parties, and not only producers and consumers. Legitimacy means that the certificate must have stakeholders' support. Bostrom (2006) stressed that inclusiveness becomes the most important factor for ecolabels in building their credibility and authority. In this sense, inclusiveness means that the certification will have to involve many participants in the process.

Thus, on the basis of this existing research, a combination of acceptability, legitimacy and inclusiveness are the requirements for the success of an ecolabelling scheme. In this paper, the researchers identified tuna stakeholders in Indonesia as consisting of business actors, fishermen, Non-Governmental Organisations (NGOs) and government. These parties thus become the independent variable in this research, such that the variables used are: support from government, private sectors, fishermen and national NGOs. The problem statement in this research focuses on the most appropriate ecolabel for Indonesia being the scheme which gets the best support from these stakeholders, as illustrated in the conceptual framework of Figure 1.

3   MATERIAL AND METHODS

This research employed a qualitative approach using the Delphi method based on purposive sampling. The number of respondents chosen for this study uses a criterion sampling formula, which is one form of purposive sampling. Criterion sampling is a sampling method that refers to specific criteria.

Because this research uses the Delphi method, the first step the researchers took was to seek a consensus for the best ecolabel option for the case of Indonesia. Based on the results of this first step, the second step was to seek consensus for the ecolabel model for Indonesia.

The calculation of the Delphi method is done using percentages. In answering the questions, respondents specify their level of agreement to a statement by choosing one of the options available. Consensus is achieved when the respondent's answer to each question exceeds 50 per cent.

The research period ran from January to June 2016. Samples were gathered in Jakarta, Bali and Cilacap. The respondents were government officials, business actors, NGOs and fishermen. The criteria applied to each of these groups is set out in the following subsections.

3.1 Government respondents

1. Have a capacity and role as a constituent of a sustainable fisheries management policy.
2. Take an active role in the preparation of ministerial decree regulation.
3. Have a good understanding of market access provisions of international marine and fisheries sector.
4. Understand sustainable fisheries management.
5. Understand the map of the world tuna market.

3.2 Business actor respondents

1. Have a Fish Processing Unit (UPI) base.
2. Have an EU Approval Number and have been active in fishery exports to Europe within the last year.
3. Have not been involved in illegal or slavery fisheries.

3.3 Fishermen respondents

1. Is the chairman of a local fishermen's cooperative.
2. Have a profession of tuna fisherman.
3. Use environmentally friendly fishing gear as permitted by regulation.

3.4 NGO respondents

1. Is the chairman of a NGO engaged in advocacy for fishermen.
2. Have an understanding of sustainable fishery management.

This research sought consensus from stakeholders about the best ecolabel model for Indonesia. Consensus is achieved if more than 50% of respondents indicate their agreement with

Figure 2. Respondents based on background (first stage) (Source: Primary data, processed by researchers).

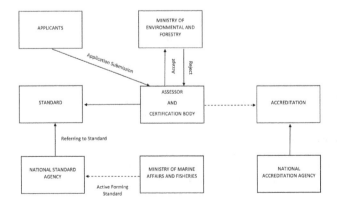

Figure 3. Existing national ecolabel model.

the statements. With purposive sampling, the number of respondents chosen was 30 (n = 30). The background of respondents was as shown in Figure 2.

The researchers described the options to the stakeholders and constructed a national seafood ecolabel model on the basis of the existing ecolabel model, as stated in Law No. 32/2014 concerning *Environmental Management and Protection*. This existing model is shown in Figure 3.

We rebuilt the existing model into a national seafood ecolabel. In this model, the Ministry of Marine Affairs and Fisheries (MOMAF) is placed at the centre. The role of the National Standards Agency (BSN) is eliminated. MOMAF also replaces the Ministry of Environment and Forestry, taking on its role too. The model thus constructed is shown in Figure 4.

The researchers then took the regional ecolabel model based on the policy paper for ASEAN Tuna Ecolabelling (ATEL) agreed by the ASEAN Tuna Working Group (ATWG) (see

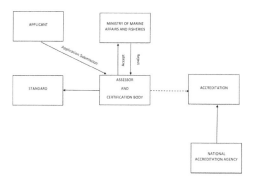

Figure 4. Proposed national seafood ecolabel model.

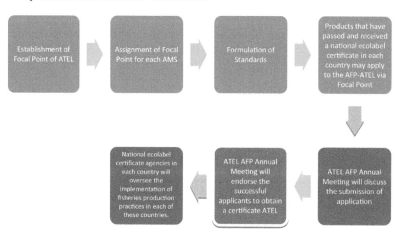

Figure 5. ATWG regional ecolabelling scheme.

Table 3. Questionnaire subjects for respondents.

No.	Description
1.	The urgency of breakthrough to handle the decline of fish stocks
2.	The urgency of ecolabelling for handling sustainability problems
3.	The effectiveness of ecolabelling in encouraging sustainable fisheries management
4.	The urgency of MSC
5.	The effectiveness of MSC
6.	The role of Government as operator of MSC
7.	The role of Government in the success of ecolabelling
8.	Urgency of national ecolabelling scheme
9.	Urgency of participation from stakeholders
10.	Urgency of a legal basis for national ecolabelling scheme

Figure 5) and showed each model to the respondents, also comparing them to the MSC model.

Having first created the questionnaire, the researchers then distributed it to the respondents. The questionnaire consisted of ten questions on the subjects shown in Table 3.

4 RESULTS

After obtaining responses to the questionnaire, the results of the first stage are shown in Table 4.

Based on these results, the necessary level of consensus was achieved on points 1, 2, 7, 8, 9 and 10. This means that the respondents *agreed* on the subjects of:

- The urgency of breakthrough to handle the decline of fish stocks.
- The urgency of ecolabelling in handling sustainability problems.
- The role of Government in the success of ecolabelling.
- The urgency of national ecolabelling scheme.
- The urgency of participation from stakeholders.
- The urgency of a legal basis for national ecolabelling scheme.

On the other hand, consensus was not achieved on points 3, 4, 5 and 6. This means that respondents *disagreed* on the subjects of:

- The effectiveness of ecolabelling in encouraging sustainable fisheries management.
- The urgency of MSC.
- The effectiveness of MSC.
- The role of Government as operator of MSC.

After the first stage of the Delphi method, the second stage sought consensus on the model of ecolabelling scheme preferred by stakeholders. For this second stage, the survey was conducted with respondents who had agreed with points 7 to 10, which essentially support national ecolabelling schemes. On the second stage, respondents were given a choice of model for a national ecolabelling scheme. This stage also gathered information from respondents about the selection of national ecolabelling schemes that had not been identified in the first stage.

A total of 22 respondents were selected to give an opinion in the second stage, comprised as shown in Figure 6. The results of this second stage are shown in Table 5.

In terms of the results, 31.82% of respondents agreed with an ecolabelling scheme from the Ministry of Environment and Forestry, as set out in question 1. This suggests that respondents did not think this scheme (the existing model) represented the interest of stakeholders. Most of the respondents wanted an active role for MOMAF in the certification process. Thus, in relation to question 2, 90.91% of respondents agreed to a national seafood ecolabelling scheme with the Ministry of Marine Affairs and Fisheries as an initiator. The same percentage

Table 4. Consensus of respondents in first stage (Source: Primary data, processed by researchers).

No.	Description	Consensus (%)
1.	The urgency of breakthrough to handle the decline of fish stocks	96.67
2.	The urgency of ecolabelling for handling sustainability problems	63.33
3.	The effectiveness of ecolabelling in encouraging sustainable fisheries management	40
4.	The urgency of MSC	46.67
5.	The effectiveness of MSC	43.33
6.	The role of Government as operator of MSC	30
7.	The role of Government in the success of ecolabelling	90
8.	Urgency of national ecolabelling scheme	63.33
9.	Urgency of participation from stakeholders	93.33
10.	Urgency of a legal basis for national ecolabelling scheme	83.33

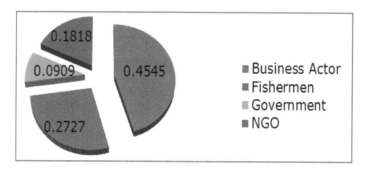

Figure 6. Respondents based on background (second stage) (Source: Primary data, processed by researchers).

Table 5. Consensus of respondents in second stage (Source: Primary data, processed by researchers).

No.	Description	Consensus (%)
1.	Consensus from respondents to the ecolabelling scheme from Ministry of Environment and Forestry	31.82
2.	Consensus from respondents to the Seafood Ecolabel with Ministry of Marine Affairs and Fisheries as an initiator	90.91
3.	Consensus from respondents for the model to be adopted as official scheme	90.91

of respondents (90.91%) agreed to adopt this model of a national seafood ecolabel as an official scheme, as set out in question 3. The result shows that the respondents supported the national seafood ecolabelling scheme as the best option for Indonesia.

The highest levels of confidence of the respondents were associated with the positive image of the Ministry of Marine Affairs and Fisheries. This was further demonstrated by the second question where respondents were asked to consent to the reconstructed model provided by the researchers. This reconstructed model replaces the role of the Ministry of Environment and Forestry with that of the Ministry of Marine Affairs and Fisheries. The researchers also eliminated the role of the National Standards Agency (BSN), again replacing it with the Ministry of Marine Affairs and Fisheries. When this is done, the level of confidence from the respondents increases to a high level, reaching 90.91%.

This consensus was repeated in relation to the third question regarding consent to the adoption of the reconstructed model as an official seafood ecolabelling scheme with the Ministry of Marine Affairs and Fisheries as an initiator, which also reached 90.91% consensus.

5 DISCUSSION

The results of these studies reflect the level of confidence of respondents in the MOMAF institution and it has, importantly, indicated that a model of seafood ecolabelling can be initiated with the support of all stakeholders. It demonstrates that government, business owners, fishermen, NGOs and regulatory support all affect the implementation of ecolabelling.

This hypothesis is consistent with the research of Ramírez et al. (2012) and Kvalvik et al. (2014), which argued that the support of stakeholders becomes the primary determinant of the success of the implementation of an ecolabelling scheme. Although the model we presented is only a reconstruction based on existing models, the results show that consensus is achieved.

Support from stakeholders is very important. As a form of best practice, ecolabelling schemes still have deficiencies in their authoritarian mechanisms and a lack of consideration for social aspects in their preparation and implementation. Research by Christian et al. (2013) showed that the MSC's credibility is often questioned because the MSC certification process confuses stakeholders in its implementation. Research by Kirby et al. (2014) and Hadjimichael and Hegland (2016) reinforced this previous research (Christian et al., 2013) by confirming that only a strong ecolabel certification and consistent application of the principles of sustainable fishery management will be accepted by stakeholders.

This suggests that, as a form of best practice, the MSC ecolabelling scheme has not yet been fully integrated with the principles of sustainable development. An ecolabelling scheme will only be successfully implemented as best practice if it has been fully integrated with these principles. This suggests that stakeholder support for ecolabelling of fisheries can affect the successful implementation of seafood ecolabelling. This condition is a strong foundation for the implementation of national seafood ecolabelling schemes.

6 CONCLUSION

The level of ecolabelling acceptance by Indonesian fisheries stakeholders is relatively low. MSC is still seen as an exclusive form of certification that does not involve participation from many stakeholders, despite this research demonstrating how support from stakeholders is an important factor in the success of seafood ecolabelling. Because stakeholders in Indonesia display a preference for initiating a national seafood ecolabel as a counterpart to MSC and other regional certificates, this opens a new avenue for future research and development. Subsequent research should thus be focused on the standards and operational tools for this particular national seafood ecolabelling scheme in Indonesia.

ACKNOWLEDGEMENTS

This research did not receive any specific grant from funding agencies in the public, commercial or not-for-profit sectors.

REFERENCES

ASEAN. (2014). Policy Paper of ASEAN Tuna Ecolabelling. Jakarta, Indonesia: Association of Southeast Asian Nations.

Bostrom, M. (2006). Regulatory credibility and authority through inclusiveness. *Organization*, *13*(3), 345–367.

Bratt, C., Hallstedt, S., Robèrt, K.H., Broman, G. & Oldmark, J. (2011). Assessment of eco-labelling criteria development from a strategic sustainability perspective. *Journal of Cleaner Production*, *19*, 1631–1638.

Bush, R. Simon, Toonen, Hilde, Oosterveer, Peter, P.J. Mol, Arthur. (2013). The 'devils triangle' of MSC certification: Balancing credibility, accessibility and continuous improvement. *Marine Policy*, *37*, 288–293.

Christian, C., Ainley, D., Bailey, M., Dayton, P., Hocevar, J., LeVine, M. & Jacquet, L.J. (2013). A review of formal objections to Marine Stewardship Council fisheries certifications. *Biological Conservation*, *161*, 10–17.

Constance, D.H. & Bonanno, A. (2000). Regulating the global fisheries: The World Wildlife Fund, Unilever and the Marine Stewardship Council. *Agriculture and Human Values*, *17*, 125–139.

Dauvergne, P. & Lister, J. (2012). Big brand sustainability: Governance prospects and environmental limits. *Global Environmental Change*, *22*, 36–45.

FAO. (2009). Guidelines for the ecolabelling of fish and fishery products from marine capture fisheries. Rome, Italy: Food and Agriculture Organization of the United Nations

Gale, F. & Haward, M. (2004). Public accountability in private regulation: Contrasting models of the Forest Stewardship Council (FSC) and Marine Stewardship Council (MSC). Referred paper presented to the *Australian Political Studies Association Conference, University of Adelaide, 29 September–1 October 2004.*

Goyert, W., Sagarin, R. & Annala, J. (2010). The promise and pitfalls of Marine Stewardship Council certification: Maine lobster as a case study. *Marine Policy*, *34*, 1103–1109.

Gulbrandsen, H.L. (2009). The emergence and effectiveness of the Marine Stewardship Council. *Marine Policy*, *33*, 654–660.

Hadjimichael, Maria, Hegland, J. Troels. (2016). Really sustainable? Inherent risks of eco-labeling in fisheries. *Fisheries Research, 174,* 129–135.

Hlaimi, D.B.B., Lucas, S., Perraudeau, Y. & Salladarré, F. (2009). Determinants of demand for green products: An application to eco-label demand for fish in Europe. *Ecological Economics, 69,* 115–125.

Jacquet, L.J. & Pauly, A. (2007). The rise of seafood awareness campaigns in an era of collapsing fisheries. *Marine Policy, 31,* 308–313.

Johnston, J.R., Wessells, R.C., Donath, H. & Asche, F. (2001). Measuring consumer preferences for ecolabeled seafood: An international comparison. *Journal of Agricultural and Resource Economics, 26*(1), 20–39.

Kaiser, J.M. & Edward-Jones, G. (2006). The role of ecolabelling in fisheries management and conservation. *Conservation Biology, 20,* 392–398.

Kirby, Seán, David, Visser, Candice, Hanich, Quentin. (2014). Assessment of eco-labelling schemes for Pacific tuna fisheries. *Marine Policy, 43,* 132–142.

Klooster, D. (2010). Standardizing sustainable development? The Forest Stewardship Council's plantation policy review process as neoliberal environmental governance. *Geoforum, 41,* 117–129.

Kvalvik, I., Noestvold, B. & Young, A.J. (2014). National or supranational fisheries sustainability certification schemes? A critical analysis of Norwegian and Icelandic responses. *Marine Policy, 46,* 137–142.

Lay, K. (2012). Seafood ecolabels: For whom and to what purpose? *Dalhousie Journal of Interdisciplinary Management, 8*(2). doi:10.5931/djim.v8i2.211.

Lopuch, M. (2008). Benefit of certification for small scale fisheries. In T. Ward & B. Phillips (Eds.), *Seafood ecolabelling: Principles and practice* (pp. 307–321). Oxford, UK: Blackwell.

Marko, P.B., Nance, H.A. & Guynn, K.D. (2012). Genetic detection of mislabeled fish from a certified sustainable fishery. *Current Biology, 21*(16), 621–622.

Oken, E., Choi, A.L., Karagas, M.R., Mariën, K., Rheinberger, C.M., Schoeny, R., Sunderland, E. & Korrick, S. (2012). Which fish should I eat? Perspectives influencing fish consumption choices. *Environmental Health Perspectives, 120,* 790–798.

Pérez-Ramírez, M., Ponce-Díaz, G. & Lluch-Cota, S. (2012). The role of MSC certification in the empowerment of fishing cooperatives in Mexico: The case of Red Rock Lobster Co-Managed Fishery. *Ocean & Coastal Management, 63,* 24–29.

Ponte, S. (2006). Ecolabels and fish trade: Marine Stewardship Council (MSC) certification and the South African hake industry. Tralac Working Paper No. 9. Retrieved from http://www.paulroos.co.za/wp-content/blogs.dir/12/files/2011/uploads/20060829_PonteMSCcertification.pdf.

Ponte, S. (2012). The Marine Stewardship Council (MSC) and the making of a market for "sustainable fish". *Journal of Agrarian Change, 12*(2–3), 300–315.

Ramírez, M.P., Phillips, B., Lluch-Belda, A. & Lluch-Cota, S. (2012). Perspectives for implementing fisheries certification in developing countries. *Marine Policy, 36,* 297–302.

Raynolds, L.T., Murray, D. & Heller, A. (2007). Regulating sustainability in the coffee sector: A comparative analysis of third-party environmental and social certification initiatives. *Agriculture and Human Values, 24,* 147–163.

Republic of Indonesia. (2014). Ministerial Decree No. 25/2014 regarding Fish Resources Estimation in Fisheries Management Area.

Republic of Indonesia. (2015). Ministerial Decree No. 107/2015 regarding The Plan of Management of Tuna, Skipjack and Tuna-like Species.

Rex, E. & Baumann, H. (2007). Beyond ecolabels: What green marketing can learn from conventional marketing. *Journal of Cleaner Production, 15,* 567–576.

Roheim, A.C., Johnston, R.J., Greer, J. & Donath, H. (2010). *Consumer preferences for ecolabeled seafood: Results of a Connecticut survey.* Storrs, CT: University of Connecticut Food Marketing Policy Center, The Rhode Island Agricultural Experiment Station, Rhode Island Sea Grant and Connecticut Sea Grant. Retrieved from http://www.fmpc.uconn.edu/research/other/Connecticut%20Final%20Ecolabel%20Report%2012%2020%2004.pdf.

Shelton, A.P. (2009). Eco-certification of sustainably managed fisheries—redundancy or synergy? *Fisheries Research, 100,* 185–190.

Sonderskov, K.M. & Daubjerg, C. (2010). Ecolabelling, the state and consumer confidence. Paper for the *60th Political Studies Association Annual Conference 29 March–1 April 2010.*

van Amstel, M., Driessen, P. & Glasbergen, P. (2008). Eco-labeling and information asymmetry: A comparison of five eco-labels in the Netherlands. *Journal of Cleaner Production, 16,* 263–276.

World Bank. (2009). *The sunken billions: The economic justification for fisheries reform.* Washington, DC: World Bank and FAO.

Women's position in *Tara Bandu* customary law: Case study on violence against women in *Suco Tibar*, Liquiçá municipality, Timor-Leste

A.D. Costa, S. Irianto & M. Siscawati
Postgraduate School, Department of Women's Studies, Universitas Indonesia, Jakarta, Indonesia

ABSTRACT: This study focuses on the experience of female victims of domestic violence in seeking justice through *Tara Bandu*, the Timor-Leste customary law. By adopting as a framework the theory of women's access to justice, feminist legal theory on the pluralism of law, and the theory of feminist multiculturalism, this research applies a qualitative approach through in-depth interview, observation and document study. The research shows that female victims have a number of legal options by which to resolve their cases. Despite having a legal option through state law, many female victims choose the *Tara Bandu* system. In so doing, they made their own decisions with, or advised by, family members. This shows that at certain points a woman is a subject of the customary law. They have the chance to choose the justice system, to negotiate and to defend themselves during the process of traditional judgement. Nevertheless, this study also finds that women have not yet had strong positions in *Tara Bandu* because of the sociocultural and power relations operating within traditional marriage in a patriarchal context. Moreover, although *Tara Bandu* is widely applied as a traditional justice system, Timor-Leste state law still does not officially recognise *Tara Bandu* as an alternative mechanism of justice in addressing violence against women.

1   INTRODUCTION

This research investigates women's position in *Tara Bandu* (the customary law of Timor-Leste), their experiences in cases of violence, and the strategy of justice seekers within a framework of legal pluralism. In using this framework of legal pluralism, it is understood that as well as customary law there is also religious law and state law. Customary law and religious law normatively have the most powerful supremacy, together with—given that Timor-Leste is a relatively new independent country—the presence of international law. This study focuses on how women victims of violence seek justice by means of *Tara Bandu*.

Women's pursuit of the process of justice cannot be separated from the situation in which they live and, in this context, women live in a patriarchal environment. In order to ensure women's rights and to prevent violence against women, the state has given a guarantee through its constitution (Republic of Timor-Leste, 2002), Article 16, Section 2: "No one shall be discriminated against on grounds of colour, race, marital status, gender, ethnical origin, language, social or economic status, political or ideological convictions, religion, education, and physical or mental condition". More specifically, in connection with the equality of rights and responsibilities between males and females, Article 17 of the same constitution states, "Women and men shall have the same rights and duties in all areas of family, political, economic, social and cultural life". The state has produced several acts to eliminate violence against women, including *Código Penal* (Republic of Timor-Leste, 2009) and the Law against Domestic Violence (SEPI, 2010). In 2002, the state also ratified the Convention on the Elimination of all forms of Discrimination Against Women (CEDAW).

Despite the above, the reality has been that many cases of violence are addressed through traditional mechanisms, or *Tara Bandu*. *Tara Bandu* consists of various regulations related to environment or nature, including rules regarding women's rights to access education and land, as well as the prevention of violence against women. Every citizen who breaks the law of *Tara Bandu* will be placed on trial on a customary basis and be punished with customary sanctions. The customary law of *Tara Bandu* was adopted by *Suco Tibar* (the village of Tibar) in September 2012. The prohibitions within *Tara Bandu* were standardised in a written form and then ratified into *Regras Kultura Tara Bandu Suco Tibar* (MSS & UNDP, 2012).

Being keen to see gender equality established in Timor-Leste, we questioned whether women's experiences, especially when the victims of violence, are appropriately reflected in *Tara Bandu*. These concerns were formulated in the research questions: How are female victims of violence positioned in the customary law of *Tara Bandu*? What are the impacts for them? What are their strategies for seeking justice?

2 THEORETICAL FRAMEWORK

This study was undertaken by utilising a feminist theory of legal pluralism, the theory of access to justice and multicultural feminist theory as tools of analysis. The application of a feminist theory of legal pluralism is intended to reflect the experiences of female victims of violence in Tibar in the context of legal pluralism. Sally Engle Merry (1998) defined legal pluralism as the existence of more than one form of law pertaining to the same social issue (Irianto, 2006). Legal pluralism involves state law, customary law, religious law, international convention, and specific rules of the economy and other areas (Irianto, 2012).

This legal pluralism is seen as the answer to the complexity of problems in society that cannot be solved by state law alone. The state law is considered inadequate to resolve many of the disputes that arise in society. The state law fails to achieve its targeted goals, and is also considered unsuccessful in explaining its non-success (Manji, 1999). The state law functions merely as a second option for females as, according to Manji, the state law does not reach all aspects of women's lives. A female is not just an individual but she also possesses different experiences to males in relation to state law (Manji, 1999).

2.1 *New legal pluralism*

In her article entitled *Pluralisme Hukum dalam Perspektif Global* (Legal Pluralism in a Global Perspective), Sulistyowati Irianto (2012) redefined legal pluralism with a new paradigm within a global perspective. In this new paradigm, she views legal pluralism as a "moving law" in the realm of globalisation (Irianto, 2012). Laws within a country or a tribe can be influenced by several types of law outside them. Thus is legal pluralism recognised as the coexistence of international law, transnational law, national law and local law, without any dichotomy among those laws in relation to particular social issues (Irianto, 2012).

2.2 *Theory of access to justice*

In this context, justice means that a victim obtains what they are entitled to in the dispute resolution process until a decision is made (Bedner & Vel, 2013). Meanwhile, according to Cappelletti and Garth (1978), as abstracted by Bedner and Vel (2013), access to justice functions to achieve the two basic main objectives of the legal system under the auspices of the state: "First, the legal system should be accessed equally by every person; second, the legal system should lead to a fair result, both for the individual and the society". Access to justice as the United Nations Development Programme (UNDP, 2005) has mentioned, as cited by Bedner and Vel (2013), defined as "the ability of a person (people) to seek and find solutions through formal or informal justice institutions, and in harmony with the values of human rights". The subject of access to justice is either an individual or a group, "especially for the poor and marginalised" (Bedner & Vel, 2013). They are marginalised on the basis of their matrimony, gender and ethnicity. These statements suggest that women have become one of the groups that tend to be marginalised in their struggle to obtain access to justice. It is therefore essential that a different approach is considered, whereby a different treatment of women is made possible, especially for female victims of injustice.

The differences between women and men, as well as among women, demand a different treatment for each woman. This difference of treatment is a recognition of the different situations of males and females. In her work entitled *Toward Gender Justice: Confronting Stratification and Unequal Power*, Stefanie Sequino (2008) describes justice (in the framework of gender equality) as demanding "to choose" and regain suitable "reward". Women and men do not have to be on the same path to achieve the same goal. Males and females have the right to be different and so they should not necessarily be treated in the same way.

At the same time, multicultural feminism has contributed by emphasising a cultural difference; that the oppression of women should be observed by the differences of culture, religion, race and class wherever they live. The women's expressionist not only because of its womanhood. In the opinion of Spelman (1988), it is a mistake if all women are considered to be one and the same (Tong, 1998). In her article entitled *Is Multiculturalism Bad for Women?*, multicultural feminist Susan Moller Okin (1999) developed her arguments based on the diversity of cultural and religious backgrounds in minority groups that survive by maintaining their cultures. Okin questioned what should be done if a conflict occurred or a contradiction of faith happened between the groups of minority religions or cultures and the norm of gender equality adopted by liberal countries. It was Okin's concern that such minority groups should receive distinctive security, because they have their own societal culture, which plays a significant role in the lives of their members (Okin, 1999). Okin inferred that these minority groups' rights are part of the solution to the question she posed. The acknowledgement of minority groups is absolutely necessary and should not be dismissed by merging a minority group into a majority culture.

3   RESEARCH METHODOLOGY

This study applies a qualitative approach with a feminist perspective. The qualitative approach is implemented here because the focus of the research is the experience of female victims of domestic violence in seeking justice through *Tara Bandu*. This research conducts in-depth observation of the female victims of violence by adopting case-tracking techniques to collect data, including in-depth interviews, observations, and document review (Poerwandari, 2013; Reinharz, 1992).

The main subjects of this study are the female victims of violence whose cases have been successfully resolved by means of *Tara Bandu*. The determination of this research subject has been undertaken by a purposive method (Poerwandari, 2013) through a case study on violence against women. The principal subjects of the research are specified by the type of case and the system of law chosen to resolve the disputes involved.

3.1   *Legal pluralism and violence against women*

Violence against women is generally realised in the form of a gender-based measure which harms women physically and/or psychologically (UN, 1993). In Timor-Leste, in the *Código Penal* (Republic of Timor-Leste, 2009) and the Law against Domestic Violence of Timor-Leste (SEPI, 2010), violence against women is understood as any action committed by the executant(s) that causes a woman mental or physical suffering, or is limiting or depriving women of their space. Ultimately, it extends to the murder of a woman. Violence against women occurs in both public and domestic arenas. Meanwhile, inside *Tara Bandu*, violence against women is defined as a verbal or physical act by others against women, leading to physical, verbal and/or economic woes, both in public and domestic spheres.

The comprehension of violence against women under *Tara Bandu* adheres to the principles contained in international and state law. This is able to happen because of the influence of a global legal movement (Irianto, 2012). At certain levels, local and international law affect each other.

The advantages and disadvantages of *Tara Bandu* for women can be divided on the basis of the contexts of culture, legal guarantee and legal subject. In these three contexts,

Tara Bandu offers a slightly positive impact at the same time as being detrimental to women. In a cultural context, *Tara Bandu* positions a woman according to her position in the *umane-mane foun* in traditional marriage relation. Females are recognised as a part of a cultural system which has its own social connections. Women gain support from *lia nain* (elders) and other family members in the customary council. This ensures that female victims are not alone in dealing with their problems. However, during the study it was apparent that the voice of women can be ignored in the *Tara Bandu* customary law process because of the power of the patriarchal culture in the *Tara Bandu* legal system. This male dominance is reflected in the view that the *Tara Bandu* customary law is the law of males (Manji, 1999). Although the 2012 version of the *Tara Bandu* customary law (MSS & UNDP, 2012) pays sufficient attention to females in theory, the actual implementation is still male-dominated.

In the context of legal guarantees, the existence of *Tara Bandu* customary law offers another legal alternative that can be selected by females. *Tara Bandu* also provides protection for women, together with an assurance of women's rights in land inheritance and participation in the field of education. However, in the case of violence against women, the legal guarantee becomes uncertain because the state has shown through the legal devices applied that it does not admit *Tara Bandu* customary law as an alternative for the solution of the case.

Moreover, in the context of legal subjects, in *Tara Bandu* customary law a female victim can be considered as a legal subject. Women have the opportunity to choose, witness, negotiate, and conduct their own self-defence in the customary council. Nevertheless, if the victim is not given the chance to participate optimally in the customary council and if the fines that must be paid by the perpetrator are not completely accepted by the victim, the position of a woman as a legal subject weakens. In our opinion, the application of custom status in the *Tara Bandu* customary law is, on the one hand, recognition of the social relations belonging to women (Manji, 1999). On the other hand, we notice that it acts as a restriction on female victims of violence to act as more than that. From the perspective of Okin (1999), the women are actually being controlled; dominated by patriarchal culture.

3.2 *Strategies of female victims in seeking justice through* Tara Bandu

In general, the females of Timor-Leste interpreted justice as a process of finding solutions to the problems they are facing (CEPAD, 2014). On this understanding, justice is something that must be fought for by way of a process. The process of finding a solution is not limited to the final decision, but can also be seen from the commencement of the process until the implementation of that decision (Bedner & Vel, 2013).

Furthermore, the process of seeking justice for women victims through *Tara Bandu* has been pursued via several strategies. These strategies include making a complaint and giving testimonies in the court. Female victims of violence could send their cases to *Conselho Suco* (Village Council), the Community Police Board and *kableha* (traditional justice institution), or directly into the police office. *Ibu* (Mrs) Bendita, a victim of domestic violence, explained that, "...*in the case of domestic violence, where my husband promised to perform violence against us, then we went to* kableha, *so that* kableha *will decide...*" (Interview, 8 March 2016).

As a result, the women's victims not always keep silence in the court, but they give the testimonies related with what happened to them. *Ibu* Ana, a victim of public violence, said, "...*in the mediation process, I retell the story as it happened. I also denied the accusation of the offender that I had incriminated him for trying to rape two kids at that time*" (Interview, 17 March 2016). Their testimonies are intended to support *kableha* and *lia nain* in making the right judgement.

Moreover, female victims also have a strategy of entering into negotiation over the sanctions imposed and demanding the presence of the third party who is considered to be the root of the problem. *Ibu* Sandra, another victim of domestic violence, explained that, "... *he must attend this so that I will be satisfied. It is because of him I am now suffering. He is the one who ruined our household…"* (Interview, 16 March 2016). Female victims must struggle to get fair treatment in the whole process, until the decision has been implemented (Bedner & Vel, 2013).

3.3   *Legal options for women*

State law is not the only law by which people seek justice (Bedner & Vel, 2013). Besides state law, there is also *Tara Bandu* customary law, which was resurrected for Tibar in 2012. The state law and the customary law presented simultaneously in the similar issue, including violence against women.

In the attempt to seek justice, female victims of violence tend to prefer the *Tara Bandu* customary law to conclude their business, either on their own or with family support. State law is the option when their cases cannot be resolved by *Tara Bandu* customary law. This preference derives from several factors, including cultural ones. The victims' choice of customary law illustrates that they are 'customary-conscious'. *Ibu* Joana, a victim of domestic violence, declared that, *"…as a person who is 'customary-conscious', we have to live according to the customary law. If we don't, our lives will be undeveloped…"* (Interview, 11 March 20016). *Tara Bandu* is a part of the culture that is not considered as ordinary law but perceived as a sacred matter. *Tara Bandu* has become an option because female victims of violence regard *Tara Bandu* as their own law (Manji, 1999).

Tara Bandu is also preferred because of the understanding the female victims have of the violence. They differentiate between serious violence and mild violence, with 'serious' violence involving bleeding having to be reported to the police, whereas 'mild' violence could be resolved by family discussion or *Tara Bandu* customary law.

Another reason why *Tara Bandu* is often the chosen option is the length of time taken to achieve a settlement. Virginia, a victim of public abuse, recounted, *"…I have been once to the police in Liquiçá and twice to the attorney in Dili; still it doesn't go on trial until now and they even asked [me] to solve the problem with customary law or family discussion"* (Interview, 31 March 2016).

Female victims have a variety of views about their choice of law in the future if the same incident reoccurred. *Ibu* Joana preferred to defer to *Conselho Suco* because, in her opinion, the perpetrator had broken the agreement made in *kableha*. By reporting to *Conselho Suco*, *Ibu* Joana wanted her case mediated by *xefe suco* (Chief of Village) and *lia nain suco* (village elders) as representatives of the government and the police. For Virginia, however, because the perpetrator had committed a crime, reporting it to the police made it more likely the person would be imprisoned, as she wanted. By contrast, *Ibu* Ana stated that, *"…now I could forgive him, but if one day he repeats what he has done, I will be ready to bring up the case anywhere I want, it could be* kableha…*"* (Interview, 17 March 2016). A similar opinion was expressed by *Ibu* Sandra, who was also prepared to take charges against the perpetrator to both *kableha* and the police. *Ibu* Bendita, meanwhile, prefers to go to *kableha* when the problem is not 'serious' enough, that is, it is without fracture or spilling of blood.

This selection of the most appropriate form of law for the women could, on the one hand, be seen as a manifestation of the ability of female victims of violence to seek justice through their chosen law system. On the other hand, we question whether women really have the ability to choose or whether they just accept the status quo. In this respect, the competence in selection represents how well women in Tibar understand the working mechanisms of *Tara Bandu* customary law. Meanwhile, state law in relation to violence against women is known only in general terms. At this stage, we think it is difficult to discern whether female victims of violence have made a decision about their choice of law consciously and properly informed.

From Manji's (1999) perspective, this can come to pass because the state has not yet actualised the justice that has been promised to the female victims of violence with any effectiveness. The state has not sufficiently socialised the laws relating to women, especially at village level. Nevertheless, through the relevant institutions of female and law policies, the state refuses to even acknowledge the existence of other legal systems in resolving cases of violence against women.

4 CONCLUSION

Overall, this research draws two important conclusions. First, that in *Tara Bandu* customary law Tibar women are positioned as legal subjects and placed appropriately for their customary status in the local cultural context. As legal subjects, female victims in Tibar have the opportunity to pursue their choice of law through law institutions created by government or through traditional institutions. As subjects, female victims look for strategies by negotiating with family and the customary law organiser in order to seek justice. In the cultural context, a Tibar woman is positioned according to her customary status in *uma lisan* or based on the status afforded by her marriage of *barlakiadu*. Second, that in this era of independence *Tara Bandu* is the product or hybrid of international, national and customary law. This is because *Tara Bandu* customary law, which is reborn within the perspective of legal pluralism, is a living law.

Furthermore, the *Tara Bandu* customary law resurrected in this era of independence incorporates the values of human rights and women's rights adopted by state and international law. From the perspective of legal pluralism, *Tara Bandu* customary law is a living law. It not only accommodates regulations based on customary law but also thrives on the spirit of other law influences, national and global. Other than that, Tara Bandu also considered and adopted as the rules of Suco by the Conselho Suco.

5 RECOMMENDATIONS

For the Timor-Leste Government, Particularly the Ministries Concerned with Law and Women

It is necessary for the government to have an open, intensive and in-depth dialogue with local leaders and *lia nain* who act as the organisers of *Tara Bandu* customary law. Government should be obliged to develop a concrete mechanism to bridge the connection between *Tara Bandu* customary law and the state law of Timor-Leste. For instance, the Ministry of Justice must ensure that the draft of the customary law includes articles which regulate the relationship between *Tara Bandu* customary law and state law. This is particularly important in relation to violence against women.

For Local Leaders and *Lia Nain*

In managing the settlement process in *Tara Bandu* customary law, local leaders and *lia nain* must place women at the centre of attention. They have to guarantee that the sanctions imposed are proportionate to the level of violence conducted by the perpetrators as well as ensuring that the victims really do receive the compensation specified.

REFERENCES

ADB. (2014). *Timor-Leste country gender assessment*. Mandaluyong City, Philippines: Asian Development Bank.
Bedner, A.W., Irianto, S., Otto, J.M. & Wirastri, T.D. (Eds.). (2012). *Kajian Socio-Legal*. Denpasar, Indonesia: Pustaka Larasan. Retrieved from http://media.leidenuniv.nl/legacy/bbrl-socio-legal-studies-final.pdf.
Belun & TAF. (2013). *Tara Bandu: Its role and use in community conflict prevention in Timor-Leste*. Dili, Timor-Leste: Belun & The Asia Foundation. Retrieved from http://asiafoundation.org/publications/pdf/1242.

CEPAD. (2014). *Women's access to land and property rights in the plural justice system of Timor-Leste.* Dili, Timor-Leste: Centre of Studies for Peace and Development. Retrieved from http://asiapacific. unwomen.org/~/media/field%20office%20eseasia/docs/publications/2015/01/cepad%20a2 j%20research% 20report_english_to%20print.pdf.

Irianto, S. (Ed.). (2006). *Perempuan & Hukum: Menuju hukum yang Berperspektif Kesetaraan dan Keadilan.* Jakarta, Indonesia: Yayasan Obor Indonesia.

Irianto, S. (Ed.). (2009). *Hukum Yang Bergerak: Tinjauan Antropologi Hukum.* Jakarta, Indonesia: Yayasan Obor Indonesia.

Irianto, S. (2012). *Perempuan di antara Berbagai Pilihan Hukum.* Jakarta, Indonesia: Yayasan Obor Indonesia.

Manji, A.S. (1999). Imagining women's 'Legal World': Toward a feminist theory of legal pluralism in Africa. *Social & Legal Studies, 8*, 435–455.

Merry, S.E. (1998). Global human rights and local social movements in a legally plural world. *Canadian Journal of Law and Society, 12*(2), 247–271.

MSS & UNDP. (2012). *Regras Kultura Tara Bandu Suco Tibar.* Dili, Timor-Leste: Ministry of Social Solidarity & United Nations Development Programme.

NSD & UNFPA. (2012). *Timor-Leste population and housing census 2010: Analytical report on gender dimensions* (Vol. 14). Dili, Timor-Leste: National Statistics Directorate & United Nations Population Fund. Retrieved from http://www.statistics.gov.tl/wp-content/uploads/2013/12/Gender_Monograph. pdf.

Okin, S.M. (1999). *Is multiculturalism bad for women?* Princeton, NJ: Princeton University Press. Retrieved from https://www.amherst.edu/media/view/88038/original/Susan2BMoller%2BOkin.pdf.

Poerwandari, E.K. (2013). *Penelitian Kualitatif untuk Penelitian Manusia.* Depok, Indonesia: LPSP3UI.

Reinharz, S. (1992). *Feminist methods in social research.* New York, NY: Oxford University Press.

Republic of Timor-Leste. (2002). *Constituicão da Republica de Democratica da Timor-Leste.* Dili, Timor-Leste: Republic of Timor-Leste.

Republic of Timor-Leste. (2009). *Código Penal.* Dili, Timor-Leste: Ministerio da Justiça.

SEPI. (2010). *Law against domestic violence of Timor-Leste.* Dili, Timor-Leste: Office of the Secretary of State for the Promotion of Gender Equality.

Sequino, S. (2008). Toward gender justice: Confronting stratification and unequal power. *Géneros: Multidisciplinary Journal of Gender Studies, 2*(1), 1–36. Retrieved from https://www.uvm.edu/~sseguino/ pdf/justice.pdf.

Tong, R.P. (1998). *Feminist thought: Pengantar Paling Komprehensif kepada Aliran Utama Pemikiran Feminis* (A.P. Prabasmoro, Trans.). Bandung, Indonesia: Jalasutra.

UN. (1993). *Declaration on the elimination of violence against women.* New York, NY: United Nations. Retrieved from http://www.un.org/documents/ga/res/48/a48r104.htm.

Competition and Cooperation in Social and Political Sciences – Adi & Achwan (Eds)
© 2018 Taylor & Francis Group, London, ISBN 978-1-138-62676-8

Preliminary study of environmental risk from endosulfan usage during the Green Revolution: Case study in central paddy area of Jombang District, East Java, Indonesia

E.S. Harsanti
Postgraduate School, Department of Environmental Sciences, Universitas Indonesia, Jakarta, Indonesia

H. Kusnoputranto
Department of Environmental Health, Faculty of Public Health, Universitas Indonesia, Depok, Indonesia

M. Suparmoko
Universitas Budi Luhur, Jakarta, Indonesia

A.N. Ardiwinata
Indonesian Agency for Agricultural Research and Development (IAARD), Agriculture Ministry, Indonesia

ABSTRACT: Endosulfan residues detected in paddy soil and rice exceed Maximum Residue Limits (MRLs). Endosulfan is bioaccumulative, persistent and semi-volatile. This study aimed to investigate the potential environmental risk of endosulfan in an agricultural environment. This research was conducted in the central paddy area of Jombang District between June and December 2015 by implementing a survey method. Farmers' agricultural behaviours, endosulfan residues in paddy fields, and endosulfan exposure in farmers' blood were all explored to establish their relationships. It was found that older farmers that used organochlorine pesticides at the time of the Green Revolution used it as recently as 2002. Younger farmers do not use it. There are several study outcomes: (1) endosulfan residues above the MRL for soil were found in 18.12% of the area, and 13.2% of the 68 samples of rice exceeded the endosulfan MRL, although only 1.4% of 71 water samples contained endosulfan at levels above acceptable ambient water quality; (2) endosulfan residues detected in farmers' blood showed that 66.7% of the 30 participating farmers had exposure above the Acceptable Daily Intake (ADI) for α-endosulfan, with the remaining 33.3% also having exposure but below this ADI; (3) endosulfan presents potential risk to the environment with a Risk Quotient (RQ) above one. These results illustrate the very significant impact of endosulfan usage during the era of the Green Revolution.

1 INTRODUCTION

The use of pesticides, including insecticides, has increased since the Green Revolution began in the 1970s (Harsanti et al., 2013). The use of organochlorine insecticides, including endosulfan, was banned in Indonesia by Agricultural Minister Regulation 434.1/KPTS/TP.270/7/2001, due to their persistent and toxic residues. Another characteristic of endosulfan is that it is bioaccumulative, meaning its concentration increases in a living organism. According to Paramita and Oginawati (2009), endosulfan residues were found in paddy soil, water and fish (0.48 to 15.54 ppb) in the Citarum watershed. In 2012, the concentration of endosulfan residues in catfish cultivated in the area around the farms of the Citarum upstream watershed was 17 times higher than that in 2010, and also higher than that of any other organochlorine residue (Rahmawati et al., 2013).

Endosulfan is potentially deposited on farmland and carried on agricultural products or accumulated in fish. It accumulates in the human body when it is consumed, resulting in potential detriment to health. Endosulfan can act as an endocrine disruptor, and exposure has also been associated with long-term immunosuppression, impaired neurodevelopment and neurodegenerative disease (Parkinson's disease) (Matsumura, 1985; ATSDR, 2013).

The use of endosulfan derives from the Green Revolution, when it was in widespread use in places such as the EU, India, Indonesia, Australia, Canada, Mexico, US (38 states), Central America, Brazil and China (Weber et al., 2010). In Indonesia, endosulfan has been found on farmland for both rice and vegetables. The maximum levels found in paddy fields were in the downstream watershed of Brantas (Jombang, East Java), where it was detected at 219.6 ppb (IAARD, 2013). On the vegetable farms of the Citarum upstream watershed, endosulfan was detected at 680 ppb (Mulyadi et al., 2011). According to Alberta Environment (2009), the Maximum Residue Limit (MRL) of endosulfan in the soil for remediation is 8.5 ppb. Therefore, endosulfan usage should not exceed the MRL.

In the Indonesian countryside, data on insecticide residues is relatively limited. In this context, a preliminary study on residue levels (soil, water, plants and farmers' blood) and the potential risks is highly necessary. Therefore, the aims of this study were: (1) to identify and evaluate endosulfan residues in soil, water and rice; (2) determine endosulfan residues in the blood of farmers; (3) evaluate the potential environmental risks of agriculture in Jombang District.

2 MATERIALS AND METHODS

2.1 Research location

The research was conducted using a quantitative approach. However, a survey method was also used to address the research questions. The research was conducted between June and December 2015 in paddy fields in Jombang District. The secondary data used are from 2013, 2014 and 2015. Blood sampling was conducted on farmers at Plandi village, Jombang subdistrict, with coordinates of 112°14'5"–112°15'15" E and 7°32'15"–7°34'30" S. Laboratory analysis was conducted at the Indonesian Agriculture Environment Research Institute (IAERI) Laboratory.

2.2 Materials and equipment

For insecticide residue analysis, the research utilised reagents including acetone, n-hexane, dichloromethane, sodium sulfate anhydrate, *Florisil*, *Celite* 545, and a 99% insecticide (endosulfan) standard (*ChemService*, Catalogue#: PS-754). Furthermore, it used Shimadzu GC-ECD (gas chromatography with electron capture detection) to detect endosulfan residue, Soxhlet extraction, and a Buchi R-114 rotary evaporator. Clean up of extract sample used chromatographic column. Other glassware was also required to support the analysis in the laboratory.

2.3 Population and sampling

The population for this research was farmers that owned land and were active as cultivators in Plandi village, Jombang subdistrict, East Java. The number of respondents was determined according to Parel et al., cited by Suparmoko (1999), as shown in Equation 1. Respondents were used to provide an interview sample and a blood sample.

$$n = \frac{NZ^2S^2}{Nd^2 + Z^2S^2} \tag{1}$$

where N: sample number; Z: area of normal curve (determined 90% = 1.645); S^2: sample variants d: error level.

Activities	Operational Diagrams	Description
A. Pre-survey and survey: Field investigation and sampling (primary data), investigate of secondary data	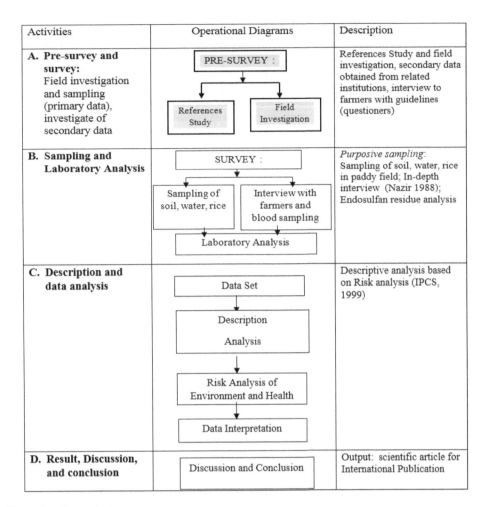	References Study and field investigation, secondary data obtained from related institutions, interview to farmers with guidelines (questioners)
B. Sampling and Laboratory Analysis		*Purposive sampling:* Sampling of soil, water, rice in paddy field; In-depth interview (Nazir 1988); Endosulfan residue analysis
C. Description and data analysis		Descriptive analysis based on Risk analysis (IPCS, 1999)
D. Result, Discussion, and conclusion		Output: scientific article for International Publication

Figure 1. Research steps.

2.4 *Research steps*

The research involved a number of steps, as illustrated in Figure 1.

3 RESULTS AND DISCUSSIONS

Previous research reported that, on the basis of foodstuffs taken from three traditional market sites, the low concentrations of Organochlorine Pesticides (OCPs) presented little risk to human health in relation to foodstuff consumption in Indonesia (Shoiful et al., 2013). The concentrations of the OCPs detected were in the following order: dichlorodiphenyltrichloroethane (DDTs) > hexachlorocyclohexanes (HCHs) > chlordane > mirex > hexachlorobenzene (HCB) > endosulfan > heptachlor. In Punjab, Pakistan, OCP-contaminated (DDTs, HCHs, chlordane, mirex, HCB, endosulfan, heptachlor) rice grains presented carcinogenic risks to health via consumption, but the Estimated Daily Intake (EDI) for non-carcinogenic effects of organochlorines was lower than those in previously published studies from China and higher than Indonesia (Mumtaz et al., 2014). Fang et al. (2016) reported endosulfan concentrations at a production site in China in the ranges of 0.01–114 mg/kg dry weight (d.w.) in soil and 4.81–289 ng/m³ in air, but in the surrounding area endosulfan concentrations were

1.37–415 ng/g d.w. in soil and 0.89–10.4 ng/m^3 in air. Inhalation and dermal risk also contributed greatly to the human health risk and could not be neglected in the soil remediation value determination. Three of these previous pieces of research (e.g. Shoiful et al., 2013; Mumtaz et al., 2014) were conducted in relation to organochlorines and were not specific for endosulfan, nor was any social aspect investigated. Although the research of Fang et al. (2016) was conducted specifically for endosulfan, it studied the impact at an endosulfan production site and its surrounding area. The current research was conducted in an agricultural area and social aspects of the impact of endosulfan use were also taken into account.

3.1 *Pesticide use in the era of the green revolution*

Food is a basic need of human beings, and this need will increase in line with human population growth. In 1979, the population of Indonesia amounted to 119.2 million people and had increased to 237.6 million in 2010 (BPS, 2013). Indonesia's population is predicted to reach 293 million by 2050 (ADB, 2011).

The Green Revolution saw the first efforts to dramatically increase food production in Indonesia with the introduction of high-yielding rice cultivars such as PB-5 and PB-8 (Abbas, 1997). These cultivars had high productivity levels but the rice was less tasty and responsive to fertiliser. Thus, the government provided a fertiliser subsidy, especially in relation to nitrogen fertilisers, and their greater use led to increasing pest attack. Thus, the use of insecticides in rice fields increased too, not least because the government also provided subsidies for pesticides. Increased agricultural inputs such as fertilisers, pesticides, and high-yielding varieties made farmers more confident in increasing rice production. These conditions changed the habits of farmers who had previously used natural resources as their inorganic inputs and farmers became dependent on the use of inorganic materials, such as fertilisers, pesticides and high-yielding varieties that were responsive to agrochemical ingredients.

Thus the changes in rice farming activities during the Green Revolution had both positive and negative impacts. The intensification of rice agriculture increased productivity, with Indonesia reaching self-sufficiency in 1984. However, the recommended use of organochlorine insecticides at the time has left its residue in the environment and endosulfan, which had been used intensively by farmers since 1970 and is persistent, bioaccumulative and highly adsorbed in soil (ATSDR, 2013), has left residues on agricultural land. Insecticide residue is still found in the soil and is also carried away in agricultural products (Ardiwinata et al., 2007), and is also carried through the food chain and water mobility. This latter can also contaminate other water and negatively impact human health (Soejitno & Ardiwinata, 1999).

3.2 *Field investigation*

The area of Jombang's rice field land is estimated at 38,801.71 hectares, which is divided into 21 subdistricts. Rice fields contaminated by endosulfan residues of more than the maximum residue limit in the soil (MRL = 0.0085 ppm) account for 7,033 hectares or about 18.1% (see Figure 2). The concentration of endosulfan residue in contaminated rice fields is 0.0032–0.2196 mg kg^{-1}. However, the maximum levels of endosulfan residue are lower than those in Jiangsu, China, where levels of 0.4150 mg kg^{-1} were found in the soil surrounding the area of endosulfan production (Fang et al., 2016), and in paddy fields at Ningde, China, where the level was 0.4539 mg kg^{-1} (Qu et al., 2014).

Figure 3 shows that endosulfan contamination in rice that is higher than the MRL (0.1 mg kg^{-1}) represents 13.2% of the total. Concentrations are in the range of <0.0017–1.0675 mg kg^{-1} with an average residue concentration of 0.0488 mg kg^{-1}. This is higher than that of β-endosulfan residue (0.00008–0.0096 mg kg^{-1}) found by Mumtaz et al. (2014) in Punjab, Pakistan. It is notable that the residues found in the rice samples taken from three traditional markets in Indonesia (Shoiful et al., 2013) were lower than that found in the fields.

According to Hamilton et al. (2003), Ambient Water Quality (AWQ) limits for endosulfan are as follows: 0.22 mg L^{-1} represents acute contamination; 0.056 mg L^{-1} represents chronic contamination. Based on water samples taken from 71 locations, it is found that

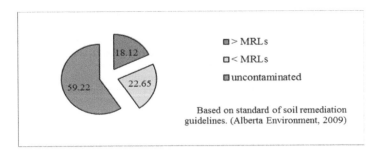

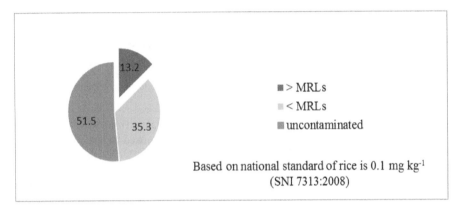

Figure 2. Percentage (%) of soil contaminated by endosulfan in Jombang District's paddy soil (Source: Computed data from IAARD, 2013).

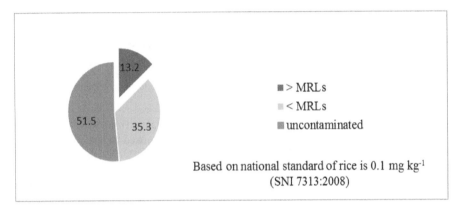

Figure 3. Percentage (%) of rice contaminated by endosulfan from 68 site samples in Jombang District's paddy soil (Source: Computed data from IAARD, 2013).

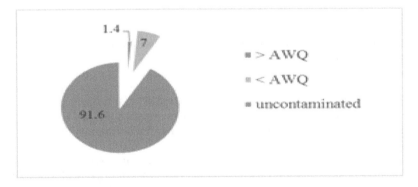

Figure 4. Percentage (%) of water contaminated by endosulfan in paddy fields in Jombang District (Source: Computed data from IAARD, 2013).

the maximum concentration of endosulfan detected is 0.50 mg L^{-1}. The lowest percentage detected is at <0.0057 mg L^{-1}; the average concentration of endosulfan residue is 0.0285 mg L^{-1}. The proportion of water contaminated by endosulfan at levels above the AWQ limit is about 1.4%; endosulfan contamination at levels below the AWQ limit represents about 7.0%, while the remaining 91.6% of water is not contaminated (see Figure 4).

Table 1. Endosulfan residues in farmers' blood in Plandi village, Jombang District.

Level of exposure	Number of farmers exposed to insecticide residue		
	α-endosulfan	β-endosulfan	Endosulfan sulfate
Farmers:			
>ADI*	20	–	7
<ADI	10	12	–
No exposure	–	18	23
Total	30	30	30
Non-farmers:			
>ADI	–	1	1
<ADI	4	4	–
No exposure	1	–	4
Total	5	5	5

Note: Samples were taken from 30 farmers and 5 non-farmers; *ADI = 0.006 mg/kg (WHO, 2012); ADI = Acceptable Daily Intake.

3.3 *Endosulfan residues in farmers' blood*

Based on the result of residue analysis of the blood samples taken from 30 selected farmers in Plandi village, it is found that as many as 66.7% have been exposed to levels of α-endosulfan in excess of the Acceptable Daily Intake (ADI). The remaining 33.3% also had exposure to α-endosulfan, albeit below the ADI level (see Table 1).

Based on its characteristics, it is possible that endosulfan residues accumulate in fat tissue because they are lipophilic. These residues have the potential to cause diseases such as endocrine disorders and reproductive disorders. Exposure to endosulfan in the blood serum causes potential risks of male reproductive disorders, such as delays to men's sperm maturity rate, thus potentially affecting their reproductive fertility (Saiyed et al., 2003; ATSDR, 2015). However, the local farmers were not aware of any such effects, possibly because they generally pay less attention to such details. According to Kiran and Varma (1988), endosulfan exposure causes hyperglycaemia and decreases the level of liver glycogen in older mice when compared with young animals. Diabetes mellitus is one of the diseases that is predominant in the region although this study cannot determine any causative relationship with endosulfan exposure; diabetes can also be caused by human diet.

The exposure to endosulfan in the farmers' blood can also be examined in terms of the farmers' behaviours in food consumption, and the use and management of insecticides. Based on the interviews with farmers in Plandi village (2015), farmers generally consume their own rice. Most farmers spray insecticide if there are pests (83.78%). Although 72.97% of the farmers have been got the Integrated Pest Management training, they select pesticide based on toxic control (72.97%).

In general (83.78% of farmers), the level of farmers success in pest control is less than 75%. Some of the insecticides commonly used by farmers are types of organochlorines which were popular before 2002 (51.35%), imidacloprid (37.84%), fenobucarb (BPMC) (37.84%), cypermethrin (84%), deltamethrin (18.92%), carbosulfan/carbofuran/carbaryl (8.11%), and organophosphates (10.81%). Farmers generally use protective clothing during spraying though with incomplete equipment (70.27%).

3.4 *Environmental risk potential*

Based on an analysis of potential environmental risk hazards, dose–response assessment, exposure assessment and risk characterisation, endosulfan contamination in the environment has potential risks to health because the Risk Quotient (RQ) is greater than 1 (IPCS, 1999).

Table 2. Environmental risk assessment in paddy soil, Jombang District.

Properties of assessment	Exposure to endosulfan		
	Rice	Soil	Water
Exposure assessment (EDI, mg/kg/day)	0.280	0.045	0.001
Reference dose (RfD; dose–response assessment)	0.0003	0.0003	0.0003
Risk characterisation (Risk quotient, RQ)	9.34	1.51	0.02

Source: Data analysis of IAARD, 2013; primary data, 2015.

Based on the exposure pathway, the most significant factor is the intake of food (rice) because it has the highest RQ value when compared with the risk of exposure through soil and water, and an EDI (0.280 mg/kg/day) that exceeds the ADI (Mumtaz et al., 2014) (see Table 2).

4 CONCLUSION

Discussion of the use of endosulfan usage has been significantly developed here in several respects, as follows:

1. It is estimated that 18.1% of Jombang's paddy field area is contaminated with endosulfan to levels that exceed the MRLs in relation to soil remediation guidelines, with levels recorded in the range 0.0032–0.2196 mg kg^{-1}. The average concentration of endosulfan residue found in the area's rice is 0.0488 mg kg^{-1} with a maximum residue of 1.0675 mg kg^{-1}. It was estimated that 13.2% of the rice exceeds the MRL (0.1 mg kg^{-1}) for endosulfan. The proportion of paddy field water contaminated by endosulfan at levels above the AWQ limit is about 1.4%.
2. Endosulfan residues were present in farmers' blood at levels above the ADI (0.006 mg kg^{-1}) for α-endosulfan in 66.7% of the total sample, with the remaining 33.3% also exposed but to a level below the ADI. The maximum level of residue detected was 0.0638 ppm. This level of exposure to endosulfan residues in farmers' blood represents a potential risk to human health.
3. Rice fields in Jombang District present a potential risk to the environment, with total value of risk quotient in excess of one. The district needs to mitigate the endosulfan contamination in these fields, based on the results of an assessment of the potential long-term environmental health risks.

ACKNOWLEDGEMENTS

We would like to thank the Indonesian Agency for Agricultural Research and Development (IAARD) for its financial support and the Indonesian Agriculture Environment Research Institute (IAERI) for its support on additional data. We also thank the technicians (Slamet Riyanto, Ariswandi, Aji M Tohir and Cahyadi).

REFERENCES

Abbas, S. (1997). *Revolusi hijau dengan swasembada beras dan jagung* [Green Revolution with Self Sufficiency of Riceand Corn]. Jakarta, Indonesia: Sekretariat Badan Pengendali BIMAS.
Alberta Environment. (2009). Soil and groundwater remediation guidelines. Alberta tier 1. Alberta Environment, Canada.
Ardiwinata, A.N., Jatmiko, S.Y. & Harsanti, E.S. (2007). Pencemaran bahan agrokimia di lahan pertanian dan teknologi penanggulangannya. In A.M. Fagi, E. Pasandaran & U. Kurnia (Eds.), *Pengelolaan lingkungan pertanian menuju mekanisme pembangunan bersih* [Management of Agricultural Environment Toward Clean Development Mechanism] (pp. 88–129). Jakenan, Indonesia: Indonesian Agriculture Environment Research Institute.

Asian Development Bank. (2011). *Asia 2050: Realizing the Asian century.* New Delhi, India: SAGE.

ATSDR. (2013). *Toxicological profile for endosulfan.* Atlanta, GA: Agency for Toxic Substances and Disease Registry.

ATSDR. (2015). *Toxicological profile for endosulfan.* Atlanta, GA: Agency for Toxic Substances and Disease Registry.

BPS-Statistics Indonesia. (2013). *Data strategis BPS.* [Strategic Data of BPS-Statistics Indonesia] Annual report. Jakarta, Indonesia: Badan Pusat Statistik.

Fang, Y., Nie, Z., Die, Q., Tian, Y., Liu, F., He, J. & Huang, Q. (2016). Spatial distribution, transport dynamics, and health risks of endosulfan at a contaminated site. *Environmental Pollution, 216,* 538–547.

Hamilton, D.J., Ambrus, Á., Dieterle, R.M., Felsot, A.S., Harris, C.A., Holland, P.T., ... Wong, S.S. (2003). Regulatory limits for pesticide residues in water (IUPAC technical report). *Pure and Applied Chemistry, 75,* 1123–1155.

Harsanti, E.S., Ardiwinata, A.N., Mulyadi & Wihardjaka, A. (2013). The role of activated charcoal in the mitigation of pesticide residues in crops of strategic commodities. *Jurnal Sumberdaya Lahan* [Indonesian Journal of Land Resources], *2,* 57–65.

IAARD. (2013). Delineation study of the residual POPs and heavy metals distribution in paddy fields (Unpublished, in Indonesian).

IPCS. (1999). *Principles for the assessment of risk to human health from exposure to chemicals (Environmental Health Criteria 210).* Geneva, Switzerland: International Programme on Chemical Safety, World Health Organization. Retrieved from http://www.inchem.org/documents/ehc/ehc/ehc210.htm.

Kiran, R. & Varma, M.N. (1988). Biochemical studies on endosulfan toxicity in different age groups of rats. *Toxicology Letters, 44,* 247–252.

Matsumura, F. (1985). *Toxicology of insecticides* (2nd ed.). New York, NY: Plenum Press.

Mulyadi, S.Y., Jatmiko, Artanti, R. & Nursyamsi, D. (2011). Identification of organochlorine in watershed area of Citarum (Unpublished).

Mumtaz, M., Qadir, A., Mahmood, A., Mehmood, A., Malik, R.N, Li, J., ... Zhang, G. (2014). Human health risk assessment, congener specific analysis and spatial distribution pattern of organochlorine pesticides (OCPs) through rice crop from selected district of Punjab Province, Pakistan. *Science of Total Environment, 511,* 354–361.

Nazir, M. (1988). *Metode Penelitian* [Research Methods]. Jakarta, Indonesia: Ghalia Indonesia.

Paramita, S.Y. & Oginawati, K. (2009). Influence of seasonal changes toward organochlorine insecticide residues in fish, water, and sediment from upper Citarum watershed segment Cisanti to Nanjung, West Java. Bandung, Indonesia: Institut Teknologi Bandung.

Qu, C., Qi, S., Yang, D., Huang, H., Zhang, J., Chen, W., ... Xing, X. (2014). Risk assessment and influence factors of organochlorine pesticides (OCPs) in agricultural soils of the hill region: A case study from Ningde, southeast China. *Journal of Geochemical Exploration, 149,* 43–51.

Rahmawati, S., Margana, G., Yoneda, M. & Oginawati, K. (2013). Organochlorine pesticide residue in Catfish (Clarias sp.) collected from local fish cultivation at Citarum watershed, West Java Province, Indonesia. *Procedia Environmental Sciences, 17,* 3–10.

Saiyed, H., Dewan, A., Bhatnagar, V., Shenoy, U., Shenoy, R., Rajmohan, H., ... Lakkad, B. (2003). Effect of endosulfan on male reproductive development. *Environmental Health Perspectives, 111*(16), 1958–1962.

Shoiful, A., Fujita, H., Watanabe, I. & Honda, K. (2013). Concentrations of organochlorine pesticides (OCPs) residues in foodstuffs collected from traditional markets in Indonesia. *Chemosphere, 90,* 1742–1750.

Soejitno, J. & Ardiwinata, A.N. (1999). Residu Pestisida pada Agroekosistem Tanaman Pangan. In S. Partohardjono, J. Soejitno & Hermanto (Eds.), *Risalah Seminar Hasil Penelitian Emisi Gas Rumah Kaca dan Peningkatan Produktivitas Padi di Lahan Sawah* [Seminar of Research Result of Greenhouse Gas Emission Research and Increasing Rice Productivity in Wetland Area] (pp. 72–90). Bogor, Indonesia: Pusat Penelitian dan Pengembangan Pertanian. [Indonesian Center for Food Crops Research and Development].

Suparmoko, M. (1999). *Metode Penelitian Praktis (Untuk Ilmu-Ilmu Sosial, Ekonomi dan Bisnis)* [Practical Research Methods (For Social Sciences, Economics and Business)]. Yogyakarta, Indonesia: BPFE.

Weber, J.C., Halsall, C., Muir, D., Teixeira, C.F., Small, J., Solomon, K.R., ... Bidleman, T. (2010). Endosulfan, a global pesticide: A review of its fate in the environment and occurrence in the Arctic. *Science of the Total Environment, 408,* 2966–2984.

WHO. (2012). Inventory of evaluations performed by the Joint Meeting on Pesticide Residues (JMPR). http://apps.who.int/pesticide-residues-jmpr-database/pesticide?name=ENDOSULFAN.

Competition and Cooperation in Social and Political Sciences – Adi & Achwan (Eds)
© 2018 Taylor & Francis Group, London, ISBN 978-1-138-62676-8

Analysis of the implementation of Sanitary and Phytosanitary (SPS) measures in the European Union (EU) on the export of Indonesian pepper and nutmeg

K. Triwibowo & T.A. Falianty
Department of Planning and Public Policy, Economic and Business Faculty, Universitas Indonesia, Depok, Indonesia

ABTRACT: The implementation of Sanitary and Phytosanitary (SPS) agreements as one of the non-tariff measures within the European Union (EU) member countries through Commission Regulation EC No. 1881/2006, setting maximum levels for certain contaminants in foodstuffs, is predicted to be influencing the export of pepper and nutmeg from Indonesia. To prove the hypothesis, this research was done using a gravity model approach with dummy SPS measures as a variable target which described regulation implemented in the EU, for analysing the impact on the export of these commodities. The analytical method used in this research is the econometric analysis model with panel data. Estimation data was made in terms of period of time from 1996 to 2015 with control variables of Gross Domestic Product (GDP), production, Real Exchange Rate (RER), dummy regulation of SPS, price, transportation cost, tariff and interaction of variables as dummy regulation of SPS and tariff. The results of this estimation show that the SPS regulations of the EU have significantly had a negative impact on the export of nutmeg and pepper from Indonesia to the EU within a 99% confidence level of estimation of each.

1 INTRODUCTION

Indonesia is one of the developing countries which still relies on agricultural products as one of its main exports. While the goods are many in variety, nutmeg and pepper are essential as spices. These kinds of foodstuffs were exported to the European Union (EU) with a degree fluctuation on their sell price, shown in Annex 1. Indonesian exports received over 500 notifications from the European Union Rapid Alert System for Food and Feed (RASFF) EU between 1 May 1985 and 30 April 2016. This concern revolves around the presence of aflatoxin in nutmeg of Indonesia as shown in Table 1. Aflatoxin itself is a toxin produced by *Aspergillus flavus*, a family of mould which grows on agricultural products such as nutmeg and pepper when the water content of this material and the humidity of the storage environment are higher than the standard (Dohlman, 2004). This kind of toxin is already regulated in the sanitary and phytosanitary (SPS) approach with the maximum aflatoxin limit set in the European Commission (EC) Regulation No. 1881/2006. Published on 19 December 2006, this regulation sets maximum levels for certain contaminants in foodstuffs (Commission of the European Communities, 2006).

Such regulation is predicted to not only influence trade, but also economy and health. RASFF data shows that several agricultural products exported by EU partners were had notifications rely on SPS requirement over the aflatoxin content, including Indonesian nutmeg, which was found to have the toxin inside (RASFF, 2016), as shown in Table 1 of Annex 2. SPS standards are not the only constraint on nutmeg exports from Indonesia, the export trend of pepper from Indonesia in general decreased (Kementerian Perdagangan, 2016) as described in Annex 1. Is there any relation between SPS regulation by the EU these export figures? While the EU was so more rigid on its limits based on food security?

These facts inspired this research to further study the impact of the SPS regulations of the EU on the export of nutmeg and pepper as commodities from Indonesia (Kementerian Perdagangan, 2015). The focus of this paper is on nutmeg (HS 090810) and pepper (HS 090411; 090412) and tries to assess the impact SPS measures have on the export of nutmeg and pepper from Indonesia to the EU. If there are any gaps, what was the reason for it? This part would be another focus as fluctuation on the trend of the graph, described in the graph in Annex 1. Another question that this research tries to answer is what can the Government of Indonesia do following this SPS barrier on the export of nutmeg and pepper to the EU?

2 METHODS

The model used in this research is a gravity model with panel data with some references including research by Nugroho (2014); Chen et al. (2008) and Ekananda (2015a; 2015b) with some adaptation on the variables. The model is as follows:

$$LGEX_{ijt} = \beta_0 + \beta_1 {}^*LGGDPCPEU_{jt} + \beta_2 {}^*LGPRODUKSI_{it} + \beta_3 {}^*D\_AFLAB1_{it} + \beta_4 {}^* \\ LG\_RER_{ijt} + \beta_5 {}^*LGTRAN\$_{ijt} + \beta_6 {}^*LG\_PRICE + \beta_7 {}^*TRF_{jt} + \beta_8 {}^* \\ D\_AFLAB1 {}^*LGTRF_{jt} + \varepsilon_{ijt}$$

where:

$LGEX_{ijt}$	:	Export of country i to country j (kg)
$LGGDPCPEU_{jt}$	:	GDP purchasing power parity per capita of importing country j (million US\$)
$LGPRODUKSI_{it}$	:	Production of pepper or nutmeg of country i, per year (kilograms)
D_AFLAB1_{jt}	:	Dummy variable of SPS
LG_RER_{ijt}	:	Variable of real exchange rate of importing country j in year t
$LGTRAN\$_{ijt}$	:	Transportation cost from country i to country j
LG_PRICE	:	Worldwide price of pepper/nutmeg in year t
$LGTRF_{jt}$	:	Tariff variable of importing country j over exporting country i
$D\_AFLAB1 {}^*LGTRF_{jt}$	:	Variable of interaction of SPS Regulation over Tariff of country j in year t
i	:	Indonesia (exporting country)
J	:	Importing country
β_o	:	constanta intercept
$\beta_1, \beta_2, \beta_3, \beta_4, \beta_5, \beta_6, \beta_7$	:	Parameters of estimation
ε_{ijt}	:	Error term

3 RESULT AND DISCUSSION

This research uses export data in kilograms of pepper (HS 090411 & HS 090412) and nutmeg (HS 090810). The pepper importing countries in the European Union include 11 countries; Belgium, Germany, Spain, France, United Kingdom, Bulgaria, Denmark, Italy, the Netherlands, Poland and Portugal. The 9 importing countries of Indonesian nutmeg are Belgium, Germany, Spain, France, United Kingdom, Italy, the Netherlands, Poland and Portugal.

Data on the export of pepper and nutmeg from Indonesia to the EU, as shown in Annex 1, generally shows decreasing exports since the year 2000 while production remains stagnant. This part predicted as the increasing growth of the Vietnamese pepper market throughout the world and over taking the Indonesian market (Kementerian Pertanian, 2014). Another reason for this is existing barriers in international trade. Between 1996 and 2015, the export of nutmeg to the Netherlands was recorded as high as 2 600 tons. This value shows some fluctuation, figures in 2015 show only 1 000 tons. In other way, these all happened was kept the Indonesian market share of nutmeg to EU over the world are still high.

Another parameter is the GDP per capita as a proxy for the average income per capita of people in the state as a mass in gravity model other than GDP. This rationale of GDP would not imply the impact on export other than the other parameters because export trends are still showing decrease since 2000. Production was also placed as one of parameters as mass other as GDP in estimation with gravity model. The production of pepper remained at the level of 52 tons per hectare in 1996 and increased to 90 tons per hectare in 2015 with traditional farming. This boost correlates with the effort of government trials by the Directorate General of Plantation of the Ministry of Agriculture, coordinated with the local government. This effort includes mobilising plantation officers to help farmers to control and handle the plantation from growth to harvest. The Indonesian Ministry of Agriculture has also published guidance on post-harvest treatment of pepper as a Rule of the Ministry of Agriculture.

The production of nutmeg has generally increased from 1996 to 2015, despite some fluctuation. This fluctuation was predicted because all farming mostly was carried out using traditional methods and needed supervising by a plantation officer. Decrease in productivity can also be caused by seasonal harvesting of nutmeg, planting using traditional methods, diseases and others. Other problems found surround agricultural practices, harvest and post-harvest because farmers have little knowledge of these processes including how to dry such grain (Patty & Kastanja, 2013).

Appreciation for this practice could reduce trade export value, while depreciation would increase export. The weakness of the exchange rate is seen as a positive sign, while depreciation of the exchange rate could increase export. In this paper, the data attached shows that the rupiah has depreciated from 1996 to 2015 and seen to support increasing export as theoretically. In fact, the export of nutmeg and pepper has been slowly decreasing since the year 2000.

Transportation is a variable that could be used in the gravity model as a trade barrier which implies that the further a place is from Indonesia there the more expensive the transportation cost. This could lead to less trading. Such length of distance caused by transportation resulted in cost getting higher. In this paper, transportation costs from Indonesia to trading partner countries fluctuate but increase year by year as fuel prices increase and fluctuate. The average cost of this parameter to Portugal increased to more than US$ 1 million in last 5 years since 2010. This has all been predicted as a reason for a reduction in exports to this country as the data shows.

Another trade barrier is import tariff (Anindita & Reed, 2008). Generally, import tariffs are used by the EU for agricultural products, including pepper and nutmeg from Indonesia. This problem decreased year by year from 7.96% in 1996 down to 6.23% in 2010 (WITS, 2015) and a further decrease is predicted in 2015 to 5.30%. Import tariff is a trade barrier which is predicted to have a negative impact on the export of Indonesian pepper to the EU. In this paper, smaller tariffs year by year have not really had any positive impact on the trend of pepper export since the year 2000. This occurrence was predicted after technical regulation as barrier and high competition on this trading was considered.

Another problem is the implementation of SPS regulations since 2006 by the EU. SPS would be closed to quality product of pepper and nutmeg from Indonesia. As shown in Annex 1, since 2006, the trade of pepper was not as fluctuate as it exports in 1996 up to 2004 that during the global crisis 1998. This trend had an impact on pepper exports from Indonesia to the EU after the depreciation of the rupiah made the price cheaper than before. Related to this case, this paper does not study the impact of the crises of 1998 and 2008 on export.

The price was an important variable in the research, especially for pepper after establishing the market through EXW Rotterdam (UNCTAD, 2016). Price is predicted to influence export while price increasing, export would follow it as supply side (Pindyck & Rubinfeld, 2013; Amoro & Shen, 2013).

4 RESULT ESTIMATION

4.1 *Impact of SPS regulations of the EU on the export of pepper and nutmeg from Indonesia*

SPS regulations in importing countries have been effective since 2006 and are particularly concerned with food security, especially with several contaminants such as aflatoxin. Based on estimation, decreasing trade flow of pepper export to 7.46% in *ceteris paribus* for other parameters with a 99% confidence level. In another case, this act influenced export of nutmeg from Indonesia to the EU as 1.77% with a 99% confidence level of estimation and assumption of other variables as *ceteris paribus* (detailed in Annex 3).

This SPS policy acts as a barrier in trade between the EU and Indonesia. The indicator places the regulation itself as a dummy variable, categorised as a non-tariff measure. SPS regulations were a problem for trade as exporting countries were unable to comply with the requirements. This fact supports the statement that all SPS regulations in the area of agriculture had negative impact on trade.

Based on this finding, SPS regulations influence pepper export from Indonesia to the EU especially to the EU-11, and the EU-9 for export of nutmeg. Indonesia should be more concerned about the quality of pepper and nutmeg for export through the implementation of standards, establishment and action of quality regulations and also address the problems in the area of plantation. Farmers need to practise Good Agricultural Practicing (GAP) for harvesting and drying. Exporters need to pay more attention to the quality of the products that would be sent, stored, fumigated and treated as it is shipped and delivered to the importing countries. These measures are required to protect the quality of goods, the trust in trade and also the sustainability of the export market.

Temporarily, the Indonesian government needs to monitor conditions on farms, in the export phase and facilities and other areas to support the export of these goods, to help maintain the quality of the export. The result of monitoring by the Ministry of Trade (2016) is the contaminants on pepper were in a good condition and does not drive the commodity as an outlier product. After testing, it is showed that the quality of pepper qualifies international methods of laboratory practice in accredited laboratory as ISO 17025:2005 and ISO 9001:2008. Otherwise, the government should continue and increase the monitoring quality and disseminate the result to all stakeholders.

Indonesia is also known as the biggest exporter of nutmeg in the world (UN Comtrade, 2016) with a market share in the EU shown in Table 4 of Annex 4. Even though Indonesia still has a great market, it should develop a good quality product and further improve on it, to maintain and sell more nutmeg in the world especially in the EU.

Data from the Ministry of Trade (2016), shows that test results of samples from monitoring displayed that aflatoxin residue on nutmeg taken in the field by an accredited sampling officer confirmed that some of nutmeg was contaminated with aflatoxin. This result was verified by a sampling method in accordance with EU Regulation EC 401/2006 on methods of sampling and analysis for the official control of the levels of mycotoxins in foodstuffs and the testing method in accordance with the international standard of AOAC official method 991.31:2012.

4.2 *Gross domestic product of importing countries*

The Gross Domestic Product (GDP) purchasing power parity per capita of importing countries theoretically has a positive impact on the export of pepper from Indonesia. As a result of estimation, this parameter showed that a GDP increase of 1% would develop export by 2.34%, *ceteris paribus* with a 95% confidence level in the analysis. This result was similar to the research of Fassarella et al. (2011), Antonucci and Manzocchi (2005), Chen et al. (2008) and Nugroho (2014), which shows that an increase in market size and purchasing power of importing countries, would support export. Gross Domestic Product (GDP) purchasing power parity per capita has a negative correlation to the export of nutmeg but is not significant in any confidence levels, *ceteris paribus*.

4.3 *Pepper and nutmeg production*

Escalating pepper production in Indonesia did support to increase rates of exports at all. This fact is empirically based on the data that shows that a growth in production of 1% would not have any impact on the export flow to the EU, in any confidence levels of analysis, *ceteris paribus*. This fact does not align with research by Nugroho (2014), that shows production as having a positive correlation to increased export. This all happened as predicted when Vietnam, as the main competitor and the biggest pepper producer (UN Comtrade, 2015), took over the market with a lower price (www.commodityonline.com, 2015; www.marketonmobile.com, 2015; Ministry of Agriculture, 2012b). Such a problem should be attended by the Indonesian Government in order to improve on production and face the competition of other producing countries.

To this end, action is required not only on GAP on farms but also on storage and delivery of the product in good way to maintain and protect the quality. Another important area for the government to act on is to give more attention to traditional farmers who plant and care for pepper fields, as some of the farms are increasingly being replaced by mining areas, especially in Bangka. This presents a bad situation as it could drive a reduction in the production of *muntok* white pepper, the EU's favourite pepper.

In another case the empirical research finds that the production of nutmeg had an impact on its export. The production was the lag from a year before. This was predicted as most of the farmers save the harvest and sells it after the price improves, or any time when the farmer needs money.

The result shows that the growth of nutmeg production by 1% would support exports to rise 0.56% on a year going, with a 95% confidence level of analysis, *ceteris paribus*. This finding is similar to research by Nugroho (2014). Following this empirical result, the government, farmer and exporter should focus more on production in order to maintain the quality of product.

Monitoring data on aflatoxin contamination (Kementerian Perdagangan, 2016) illustrates the vulnerability of Indonesian spice products to fungal contamination, which threatens production by farmers, collectors and their network. This potential could arise with sub-optimal storage. The wall and the floor have a tendency to be overgrown with the fungus *Aspergillus flavus*, a producer of aflatoxins. Good storage would reduce the potential of this, with regards to drought fumigation and storage locations as well as the water content level of the product. Moreover, the need to have conscientious farmers who pay more attention to the whole process of harvesting to storage, with due regard to all aspects of the initial planting, fertilising, spraying of pesticides through to harvesting.

4.4 *Real exchange rate*

The paper empirically found that a 1% increase in the value of the Real Exchange Rate (RER) of the rupiah against the currency of the buyer's country will lower the Indonesian pepper exports to 3.14%, with a greater than 99% confidence level, *ceteris paribus*. The result of this study is different to the approach of RER by Ekananda (2015b), which shows that the depreciation impacts Indonesian exports. This is apparently due to inflation and how the price variable is more dominant than the fall of exchange rates. The assumption is that, depreciation causes lower values indicating rupiah from Indonesian pepper products. It means that the commodity is cheaper. Then the research predicted that purchases will increase, but in these findings, depreciation actually reduces exports. Inflation and price in the market is thought of as one of the reasons why the price of pepper is higher than other countries. The high cost of shipping and transportation costs could also add to this.

Meanwhile, in the analysis of the export of nutmeg, the study also found similar trends where the empirical results of the study show that the RER of the rupiah against the currencies of the EU member states (Euro and Pound Sterling) have a significantly negative impact on Indonesian nutmeg exports to the EU. An RER increase of 1% would reduce nutmeg exports by 1.1% with a greater than 99% confidence level, *ceteris paribus*. The results of this study differ from the approach of RER by Ekananda (2015b) that shows how depreciation increases exports. This is presumably because the price of nutmeg relatively low due to the depreciation is has no impact on sell than the issue of indications of aflatoxin contamination in these products.

4.5 *Transportation costs*

Export costs were measured by the value of a country's economic distance charges. Distance economy is one of the conditions important to the gravity of economic model with the large distance seen to affect export trade flows negatively. In this study, an increase in transportation costs by 1% would lower the pepper exports to 1.76% significant with a 99% confidence level, *ceteris paribus*.

Meanwhile, different results obtained empirically for the analysis of the export of nutmeg, where, variable distance as the economic costs or transportation cost to the export destination countries is predicted to have a negative correlation to the quantity of exports of nutmeg from Indonesia, but the empirical results of this study indicate that distance costs have no effect on the export quantity of nutmeg, *ceteris paribus*. This is possible because the need for nutmeg is greater than existing costs, proven by a market share that is still sufficiently large as shown in Table 4 in Annex 4.

4.6 *Price*

The variable of price in the analysis of pepper export is found empirically positive effect on sales pepper Indonesia to the EU-11. A price increase of 1% would increase pepper exports to 1.14% significant with a 99% confidence level, *ceteris paribus*.

4.7 *Tariff*

Tariff is the cost which is falling. The reduced rate also applies to agricultural products coming into the EU. In this study, the rate increase of 1% would reduce exports of pepper at 0.44 significant with a 99% confidence level, *ceteris paribus*. Meanwhile, on the analysis of the export of nutmeg, the study empirically found that the rates in the EU member states that apply to the Indonesian export of nutmeg had significant effect with a 95% confidence level, *ceteris paribus*. An increase in tariff by 1% will decrease Indonesian nutmeg exports to the EU to 0.13%. Empirically, it suggests that a tariff reduction, will increase Indonesian pepper export to the EU. It also, however, needs to be corrected with other barriers to trade, such as SPS measures imposed against pepper and nutmeg commodities as included in the previous analysis.

4.8 *Interaction between SPS regulations and tariff*

From the analysis of the interaction between the SPS and the dummy variable tariff, the conclusion is that (see Annex 5 for detail):

a. Intercept C is decreasing so that export value of pepper and nutmeg is decreased in the period after the entry into force of the regulation of SPS while zero rates and *ceteris paribus* for other variables.
b. Coefficient rates as increasing on tariffs for every escalating on level of aflatoxin restrictions shows that the impact of it is raising after the entry into force of the SPS regulations.

4.9 *Individual effect testing*

Pepper export cross section data analysis showed that the highest individual effect was the United Kingdom while the lowest was Italy, detailed in Annex 6. This suggests that the government, exporters of pepper and all stakeholders should pay more attention to the impact of the SPS regulations that affect Indonesia's export to major countries, particularly those experiencing a decline in exports, namely Belgium, Spain, Bulgaria, Denmark, Italy and Portugal. Another consideration is that the position of the Government of Indonesia in the world market share is still relatively small, as shown in Table 5 of Annex 4. It needs to be further improved much like the quality of the sales process, in order to increase the market share in the EU-11.

Meanwhile, as detailed in Annex 6, based on the test results of individual effect on a cross section of nutmeg exports, it is showed that the effect of the highest individual is Netherlands and the lowest is Portugal with the assumption that all independent variables are disregarded. The government, exporters of nutmeg and all stakeholders therefore should be more focused on the impact of the SPS regulations that affect Indonesia's exports to major countries, particularly those experiencing a decline in exports, namely Spain, United Kingdom, Italy, Poland and Portugal.

4.10 *Government policy on quality of pepper and nutmeg from Indonesia*

The Indonesian government has responded to the policy of the European Union as one of the consequences of signing the Agreement on Sanitary and Phytosanitary Measures of the WTO and international trade. Some of the actions carried out by Indonesia are as follows: Responding to notifications on the head with a policy of product quality monitoring and guidance of nutmeg and pepper Indonesia (Kementerian Perdagangan, 2016). Monitoring directly carried out by certified sampling and then the samples were tested in a laboratory accredited to ISO 17025 and ISO 9001:2008.

5 CONCLUSIONS AND OTHERS

Based on the test results and empirical data obtained, it was concluded that; the sanitary and phytosanitary (SPS) regulations inhibit the export of pepper and nutmeg to the European Union (EU). SPS regulations inhibits export significantly by limiting the permitted levels of aflatoxin in the EU regulations of 2006 and will reduce the flow of export of pepper to 7.46%, *ceteris paribus*, with a 99% confidence level and reduce the export of nutmeg by 1.77%, *ceteris paribus* with a 99% confidence level.

Another is export of nutmeg and pepper have in common is affected by SPS barriers in the EU. There is a difference, where nutmeg has a larger market share in the EU, Indonesian pepper has a relatively small market share in the EU market.

The Government of Indonesia needs to follow up on the challenges set by the SPS barriers in a policy to support the export of Indonesian pepper and nutmeg to the EU.

6 ADVICE AND POLICY RECOMMENDATIONS

The advice and policy recommendations are:

1. The Indonesian Government should pay more attention to efforts to decrease products contaminated with aflatoxin, as follows:
   a. Improve monitoring of agricultural products, especially pepper and nutmeg, followed by dissemination to stakeholders to maintain product quality.
   b. Technically, the possible growth of the fungus aflatoxin during storage in farms, export shipments and others should be handled with fumigation and disciplinary stakeholders from farmers to exporters.
2. The Government should remain focused on the major importing countries of nutmeg and pepper, to assess which countries are likely to fall.

REFERENCES

Amoro, G. & Shen, Y. (2013). The determinants of agricultural export: Cocoa and rubber in Cote d'Ivoire International. *Journal of Economics and Finance*, 5(1), 1916–9728.

Anindita, R. & Reed, M.R. (2008). *Bisnis dan Perdagangan Internasional*. Yogyakarta, Indonesia: Penerbit Andi.

Antonucci, D. & Manzocchi, S. (2005). *Does Turkey have a special trade relation with the EU?* (LLEE Working Document no.35 - Forthcoming Economic Systems) Elsevier North Holland. Retrieved from http://www.luiss.it/ricerca/centri/llee.

Aruoma, O.I. (2006). The impact of food regulation on the food supply chain. *Toxicology*, 221, 119–127.

Badan Standardisasi Nasional. (2013a). *SNI 0004:2013 Lada Putih*. Jakarta, Indonesia: BSN.

Badan Standardisasi Nasional. (2013b). *SNI 0005:2013 Lada Hitam*. Jakarta, Indonesia: BSN.

Badan Standardisasi Nasional. (2015). *Data SNI*. Jakarta, Indonesia: BSN.

Badan Standardisasi Nasional. (2016). *SNI 0006:2015 Biji Pala*. Jakarta, Indonesia: BSN.

Bank Indonesia. (2016). *Nilai Nominal Mata Uang*. Retrieved from www.bi.go.id.

Beghin, J.C. & Bureau, J.C. (2001). Quantitative policy analysis of sanitary, phytosanitary, and technical barriers to trade. *Économie Internationale*, 87, 107–130.

Bergstrand, J.H. (1985). The gravity equation in international trade: Some microeconomic foundations and empirical evidence. *Review of Economic and Statistics*, 67, 34–78.

CCMAS. (2016) Retrieved from https://drive.google.com/folderview?id=0B85Fxk6 LWRK3Q1QzSWh CaVdINDQ&usp=sharing.

Chen, C., Jun, Y. & Findlay, C. (2008). Measuring the effect of food safety standards on China's agricultural exports. *Review of World Economics*, 144(1).

Codex. (2016). Retrieved from www.codexallimentarius.net.

Codex. (2016). Retrieved from: http://www.who.int/foodsafety/areas_work/food-standard/en/.

Commission of the European Communities. (2001). *Setting maximum levels for certain contaminants in foodstuffs*. (Commission Regulation (EC) No. 466/2001).

Commission of the European Communities. (2006). *Setting maximum levels for certain contaminants in foodstuffs* (Commission Regulation (EC) No. 1881/2006).

Dascal, D., *et al.* (2002). An analysis of EU wine trade: A gravity model approach. *Int'l Advances in Econ. Res.*, 8(2), 135–147.

Disdier, A.C., Fontagne, L. & Mimouni, M. (2008). The impact of regulations on agricultural trade: Evidence from SPS and TBT agreements. *American Journal of Agricultural Economics*, 90(2), 336–350.

Dohlman, E. (2004). Mycotoxin hazard and regulation. *International Trade and Food Safety. Economic Research Service*. USDA.

Dong, F. & Jensen, H.H. (2004). *The challenge of conforming to SPS measures for China's agricultural export* (MATRIC Working Paper 04-MWP 8). Retrieved from *www.matric.iastate.edu.*

Ekananda, M. (2015a). *Ekonometrika Dasar Untuk Penelitian Ekonomi, Sosial dan Bisnis* (1st ed.). Jakarta, Indonesia: Mitra Wacana Media.

Ekananda, M. (2015b). *Ekonomi Internasional*. Jakarta, Indonesia: Penerbit Erlangga.

Fassarella, L.M., de Souza, M.J.P. & Burnquist H.L. (2011, July). *Impact of sanitary and technical measures on Brazilian exports of poultry meat*. Paper presented at the Agricultural & Applied Economics Association's AAEA & NAREA Joint Annual Meeting, Pittsburgh, Pennsylvania.

Federal Reserve Bank of St. Louis. (1999). *Controlling for Heterogeneity in Gravity Models of Trade* (Working Paper). Cheng, I. & Wall, H.J.

Fukao, K., Okubo, T. & Stern, R.M. (2003). An econometric analysis of trade diversion under NAFTA. *North American Journal Economic Finance 2003*, 14(1), 2–24.

Fxtop.com. (n.d.) Retrieved from http://fxtop.com/en/historical-exchange-rates.php?A=1&C1=IDR&C 2=USD&YA=1&DD1=01&MM1=01&YYYY1=1989&B=1&P=&I=1&DD2=31&MM2=12&YYY Y2=2011&btnOK=Go!.

Gujarati, D.N. & Porter, D.C. (2009). *Basic Econometrics*. New York: McGraw-Hill.

Henson S.J., et.al. (1998). Impact of sanitary and phytosanitary measures on developing countries, *The University of Reading*, 1998.

Hooker, N.H. (1999). Food safety regulation and trade in food products. *Food Policy*, 24, 653–668. Retrieved from www.elsevier.com/locate/foodpol.

http://inflationdata.com/Inflation/Inflation_Rate/Historical_Oil_Prices_Table.asp.

http://www.commodityonline.com/commodities/spices/pepper.php.

http://www.marketonmobile.com/commodity/pprmlgkoc/pepper/ncdex.

International Agricultural Trade Research Consortium. (1998). *Implementation of the WTO Agreement on the Application of Sanitary and Phytosanitary Measures: The First Two Years* (Working Paper 98–4). Location: Roberts, D.

International Pepper Community (IPC). (n.d.) Retrieved from http://www.ipcnet.org/session44/index.php

International Pepper Community. (2015) Pepper Statistical Yearbook 2014. Jakarta, Indonesia: IPC.

International Pepper Community. (2016) Member of International Pepper Community. Retrieved from http://www.ipcnet.org/n/map/index.php?path=map&page=id.

Jayasinghe, S. & Sarker, R. (2004, September). *Effects of regional trade agreements on trade in Agrifood products: Evidence from Gravity Modeling Using Disaggregated Data* (Working Paper 04-WP 374). Center for Agricultural and Rural Development—Iowa State University.

Kareem, O.I. (2014). *The European Union sanitary and phytosanitary measures and Africa's exports* (EUI Working Paper RSCAS 2014/98). Robert Schuman Centre for Advanced Studies, European University Institute, San Domenico di Fiesole—Italy.

Kastner, J.J. & Pawsey, R.K. (2002). Harmonising sanitary measures and resolving trade disputes through the WTO–SPS framework. Part I: A case study of the US–EU hormone-treated beef dispute. *Food Control, 13*, 49–55.

Kementerian Perdagangan. (2014). *Laporan Rapat Persiapan Sidang CCSCH Kerala India*. Jakarta.

Kementerian Perdagangan. (2015). *10 komoditi potensial* [English translation of article title]. Retrieved from http://www.kemendag.go.id/id/economic-profile/10-main-and-potential-commodities/10-potential-commodities.

Kementerian Perdagangan. (2016a). *Pemantauan Mutu Bokor Biji Pala*. Petemuan Teknis Kementerian Perdagangan Maret 2016. Surabaya.

Kementerian Perdagangan. (2016b). *Data Ekspor Pala dan Lada*. Jakarta, Pusat Data dan Informasi Perdagangan (Pusdatin) Kementerian Perdagangan.

Kementerian Pertanian. (2012a). *Peraturan Menteri Pertanian Nomor 55/Permentan/OT.140/9/2012 Tentang Pedoman Penanganan Pascapanen Lada*. Jakarta.

Kementerian Pertanian. (2012b). *Peningkatan Produksi, Produktivitas Dan Mutu Tanaman Rempah Dan Penyegar*. Direktorat Jenderal Perkebunan Kementerian Pertanian Desember 2012.

Kementerian Pertanian. (2013, December). *Statistik Perkebunan Indonesia 2012–2014* [Tree Crop Estate Statistics of Indonesia]. Direktorat Jenderal Perkebunan Kementerian Pertanian Desember 2013.

Kementerian Pertanian. (2014). *Laporan Hasil Sidang CCSCH Kerala India*. Jakarta.

Kementerian Pertanian. (2015, December). *Statistik Perkebunan Indonesia* [Tree Crop Estate Statistics of Indonesia] *2014–2016: Lada* [Pepper]. Direktorat Jenderal Perkebunan Kementerian Pertanian Desember 2015.

Kepaptsoglou, K., Karlaftis, M.G. & Tsamboulas, D. (2010). The gravity model specification for modeling international trade flows and free trade agreement effects: A 10-year review of empirical studies. *The Open Economics Journal, 3*, 1–13.

Krugman, O. (2003). *International economics theory and policy*. Boston, MA: Addison Wesley Longman.

Kusuma, H.A. (2015). *Dampak Environmental Trade Barriers Terhadap Ekspor Kayu Olahan Primer Indonesia*. (Unpublished thesis). MPKP-FEB, UI.

Melo, O. *et al.* (2014) Do sanitary, phytosanitary, and quality-related standards affect international trade? Evidence from Chilean fruit exports. *Elsevier, World Development, 54*, 350–359.

Munich Personal RePEc Archive—TRADEAG Project. (2007). *A gravity approach to assess the effects of association agreements on EuroMediterranean trade of fruits and vegetables* (MPRA Paper 4124). Location: Coque., Garcia-Alvarez, J.M. & Marti-Selva, M.L.

Nachrowi, N.D. & Usman, H. (2002). *Pendekatan Populer dan Praktis Ekonometri untuk Analisis Ekonomi dan Keuangan*. Jakarta, Indonesia: Lembaga Penerbit FEUI.

National Bureau of Economic Research. (2001, January). *Gravity with gravitas: A solution to the border puzzle* (NBER Working paper 8079). Cambridge, MA: Anderson, J.E. & Wincoop, E.V. doi:10.3386/w8079.

National Bureau of Economic Research. (2010, December). *The gravity model* (NBER working paper 16576). Cambridge, MA: Anderson, J.E.

Nugroho, A. (2014). The impact of food safety standard on Indonesia's coffee exports. *Elsevier, Procedia Environmental Sciences, 20*, 425–433.

Nurdjanah, N. (2007). *Teknologi pengolahan pala. Balai besar penelitian dan pengembangan pascapanen pertanian*. Kementerian Pertanian.

Otsuki, T., Wilson, J.S. & Sewadeh, M. (2001). *Saving two in a billion: A case study to quantify the trade effect of European safety standard on African Export*. Washington DC: World Bank.

Otsuki, T., Wilson, J.S. & Sewadeh, M. (2001). What price precaution? European harmonisation of aflatoxin regulations and African groundnut exports. *European Review of Agricultural Economics, 28*(2), 263–283.

Patty, Z. & Kastanja, A.Y. (2013, December). Kajian budidaya tanaman pala di Kabupaten Halmahera Utara (Studi kasus di Kecamatan Galela Barat, Tobelo Selatan dan Kao Utara). *Jurnal Agroforestri, 8*(4).

Pindyck, R.S. & Rubinfeld, D.L. (2013). *Microeconomics.* 8th edition. New Jersey, USA: Pearson Education, publishing as Prentice Hall.

Pusakasari, A.S. (2014). Regresi Panel Dengan Metode *Weighted Cross-Section Sur* Pada Data Pengamatan *Gross Domestic Product* Dengan *Heterokedastisitas* dan Korelasi Antar Individu (*Cross-Section Correlation*). Jurusan Matematika F.MIPA Universitas Brawijaya.

Rapid Alert System For Food and Feed (RASFF) – European Union (EU) Rapid Alert System For Food and Feed. (2015). Retrieved from https://webgate.ec.europa.eu/rasff-window/portal/?event=searchResultList.

Salvatore, D. (1997). *International Economics.* New Jersey: Prentice Hall-Gale.

Sarfati, G. (1998). European industrial policy as a non-tariff barrier. *European Integration Online Papers (EIoP), 2*(2).

Sari, R.A. (2015). Analisis Pengaruh *Non-Tariff Measures* Terhadap Arus Perdagangan Ekspor Komoditi *Crude Palm Oil* (CPO) Indonesia Ke Negara Tujuan Ekspor Utama. Tesis IPB.

Shepherd, B. (2012). The gravity model of international trade: A user guide. ARTNeT Gravity Modeling Initiative. *United Nations Publication.*

Software: Software eviews 9, Microsoft Office Word and Excel.

Syarief, R., Ega, L. & Nurwitri, C.C. (2012). *Mikotoksin Bahan Pangan.* Bogor, IPB Press.

The World Bank Development Research Group. (2000). *Quantifying the impact of technical barriers to trade a framework for analysis* (Policy Research Working Paper 2512). Location, Country: Maskus, K.E., Wilson, J.S. & Otsuki. T.

UN Comtrade. (2015). Retrieved from: http://comtrade.un.org/data/.

UN Comtrade. (2016). Retrieved from: http://comtrade.un.org/data/www.timeanddate.com.

UNCTAD. (2013). *Non-Tariff measures to trade: Economic and policy issues for developing countries.* United Nations Conference on Trade and Development. United Nations.

UNCTAD. (2016). *Commodity Price.* Retrieved from http://unctadstat.unctad.org/wds/TableViewer/tableView.aspx?ReportId=28768.

Vancauteren, M. (2002). *The impact of technical barriers to trade on home bias: An application to EU data* (IRES Discussion Paper 2002-032). Université Catholique de Louvain, Louvain.

Wahyuni, S., Hadad E.A., Suparman & Mardiana. (2008). Keragaman Produksi Plasma Nutfah Pala (*Myristica fragrans*) di KP Cicurug. *Buletin Plasma Nutfah, 14*(2). Bogor, Balai Penelitian Tanaman Obat dan Aromatik.

Wall, H. (1999). Using the gravity model to estimate the costs of protection. *Federal Reserve Bank of St. Louis Review*, pp. 33–40.

Widayanto, S. (2011). *Fasilitasi Dan Aturan Perdagangan: Prosedur Notifikasi WTO Untuk Transparansi Kebijakan Impor Terkait Bidang Perdagangan Kewajiban Pokok Indonesia Sebagai Anggota Organisasi Perdagangan Dunia (World Trade Organization).* Jakarta, Direktorat Kerjasama Multilateral. Direktorat Jenderal Kerja Sama Perdagangan Internasional Kementerian Perdagangan Republik Indonesia.

Widiastri, M. (2014). *Dampak Implementasi ACFTA (ASEAN-China Free Trade Area) terhadap Perdagangan Produk Hortikultura antara ASEAN 5 Dengan China* (Unpublished thesis). Pasca Sarjana FEB, UI.

Winarno, W.W. (2015). *Analisis Ekonometri dan Statistika dengan Eviews Edisi 4.* Yogyakarta, UPP STIM YKPN.

World Integrated Trade Solution (WITS). (2015) World Bank. Retrieved from: http://wits.worldbank.org/#

World Trade Organization. (2015). *Non-Tariff Measures Data.* Retrieved from http://i-tip.wto.org/goods/Forms/TableView.aspx?mode=modify&action=search.

World Trade Organization. (2015). Sanitary and phytosanitary measures: Text of the agreement the WTO agreement on the application of sanitary and phytosanitary measures (SPS Agreement). Retrieved from https://www.wto.org/english/tratop_e/sps_e/spsagr_e.htm.

World Trade Report. (2012). *The Trade Effects of Non-Tariff Measures and Service Measures.* Retrieved from https://www.google.co.id/url?sa=t&rct=j&q=&esrc=s&source=web&cd=2&cad=rja&uact=8&ved=0CCEQFjAB&url=https%3 A%2F%2Fwww.wto.org%2Fenglish%2Fres_e%2Fbook sp_e%2Fanrep_e%2Fwtr12–2d_e.pdf&ei=reVHVaeYDuLOmwXJvYC4 Aw&usg=AFQjCNHxf3-SExWmpsdlclTf1HLtJ6YCcw&bvm=bv.92291466,d.dGY.

www.oilpriceinflation.com World bank. Retrieved from http://www.data.worldbank.org/.

276

ANNEX 1

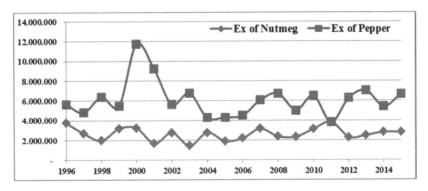

Source: Kementerian Perdagangan (2016), treated by author.

Export of Indonesian nutmeg and pepper to the European Union (EU): 1996–2015 (tons)

ANNEX 2

Table 1. RASFF notifications on nutmeg and others exported by Indonesia.

Classification	Date of case	References	Notifying country	Subject	Product category	Type	Risk decision
border rejection	03/02/2016	2016.AFI	Netherlands	aflatoxins (B1 = 170: Tot. = 210 µg/kg—ppb) in nutmeg from Indonesia	herbs and spices	food	serious
border rejection	21/09/2015	2015.BMD	Netherlands	aflatoxins (B1 = 180: Tot. = 200 µg/kg—ppb) in nutmeg from Indonesia	herbs and spices	food	serious
border rejection	06/07/2015	2015.BCQ	Netherlands	aflatoxins (B1 = 30: Tot. = 33 µg/kg—ppb) in nutmeg from Indonesia	herbs and spices	food	serious
Information for attention	07/05/2015	2015.0562	Netherlands	aflatoxins (B1 = 25 µg/kg—ppb) in aromatic ginger form India	herbs and spices	food	serious
border rejection	03/04/2015	2015.APG	Netherlands	aflatoxins (B1 = 19: Tot. = 23) in nutmeg from Indonesia	herbs and spices	food	serious
border rejection	03/02/2015	2015.AFM	Netherlands	aflatoxins (B1 = 31 µg/kg—ppb) in nutmeg from Indonesia	herbs and spices	food	serious

Source: Rapid Alert System for Food and Feed (RASFF), 2016.

Table 2.   Estimation result on panel data of analysis export of pepper.

Variable	FEM
1	2
3	4
LGGDPCPEU	2.34
	(2.05)**
LGPRODLADA	1.83
	(1.49)
D_AFLAB1	−7.46
	(−3.97)*
LG_RER	−3.14
	(−4.97)*
LGTRAN$	−1.76
	(−4.76)*
LG_PRICE	1.14
	(3.22)*
TRF	−0.44
	(−2.87)*
D_AFLAB1*TRF	1.19
	(4.16)*
C	8.59
	(1.21)
Adjusted R-squared	0.894

Notes: a. Dependent Variable = LGEXLADA (export of Indonesian pepper to the EU). b. *Significant on 1%; **Significant on 5%; ***Significant on 10%.

Then, the model equation is as follows:

$$\text{LGEXLADA} = 8.59 + 2.34*\text{LGGDPCPEU} + 1.83*\text{LGPRODLADA} - 7.46*\text{D\_AFLAB1} - 3.14* \text{LG\_RER} - 1.76*\text{LGTRAN\$} + 1.14*\text{LG\_PRICE} - 0.44*\text{TRF} + 1.19*\text{D\_AFLAB1*TRF}$$

Table 3.   Result of estimation on panel data of analysis of the export of nutmeg.

Variable	FEM
5	6
7	8
LGGDPCPEU(−1)	−1.96
	(−1.37)
LG_PRODPALA(−1)	0.56
	(2.51)**
D_AFLAB1	−1.77
	(−2.75)*
LG_RER	−1.10
	(−3.75)*
LGTRAN$	0.18
	(0.98)
TRF	−0.13
	(−2.27)**
D_AFLAB1*TRF	0.35
	(3.31)*
C	13.56
	(2.04)**
Adjusted R-squared	0.688

Note: a. Dependent Variable = LG_EXPALA (export of Indonesian nutmeg to the EU). b. *Significant on 1%; **Significant on 5%; ***Significant on 10%.

The model followed the result of estimation above is below:

$$LG\_EXPALA = 13.56 - 1.96*LGGDPCPEU(-1) + 0.56*LG\_PRODPALA(-1)$$
$$- 1.77*D\_AFLAB1 - 1.10*LG\_RER + 0.18*LGTRAN\$$$
$$- 0.13*TRF + 0.35*D\_AFLAB1*TRF$$

ANNEX 4

Table 4. Market share of the export of Indonesian nutmeg to the EU-9.

Year	Netherlands	Belgium	Germany	Spain	France	UK	Italy	Poland	Portugal
2010	47.44%	62.29%	75.07%	53.51%	62.87%	4.22%	67.34%	71.51%	0.00%
2011	41.62%	35.28%	71.71%	63.96%	60.29%	25.64%	84.61%	59.10%	10.78%
2012	70.09%	49.04%	68.84%	27.17%	60.09%	17.91%	83.97%	48.66%	32.67%
2013	51.92%	33.46%	59.73%	50.76%	60.06%	9.19%	90.86%	53.18%	0.36%
2014	50.35%	35.71%	72.15%	41.57%	58.38%	14.34%	82.60%	70.72%	0.00%

Source: UN Comtrade, 2016 (treated by author).

Table 5. Market share of the export of Indonesian pepper to the European Union (EU).

Year	Netherlands	Germany	France	UK	Poland	Belgium	Spain	Bulgaria	Denmark	Italy	Portugal
2010	10.46%	11.76%	12.10%	0.57%	8.43%	9.03%	1.02%	8.06%	1.37%	10.01%	0.00%
2011	7.56%	14.43%	10.63%	0.17%	4.63%	5.26%	0.35%	8.52%	2.89%	6.14%	1.88%
2012	16.40%	14.85%	11.24%	1.59%	10.07%	5.43%	2.58%	4.18%	0.00%	11.45%	0.45%
2013	13.52%	14.05%	12.37%	0.74%	11.94%	18.25%	0.95%	4.49%	0.00%	6.13%	0.36%
2014	13.39%	15.68%	12.20%	1.12%	4.73%	16.60%	0.33%	3.53%	0.95%	7.62%	0.00%

Source: UN Comtrade, 2016.

ANNEX 5

Interaction between SPS regulations and regulation of tariff

Interaction between SPS regulation and regulation of tariff is done in view SPS regulatory effects on rates. Referring to Ekananda (2015a), according to the results of estimation, pepper export function for the period prior to the application of SPS type restrictions on pepper Aflatoxin is as follows:

$$LGEXLADA = 2.34*LGGDPCPEU + 1.83*LGPRODLADA - 3.14*LG\_RER$$
$$- 1.76*LGTRAN\$ + 1.14*LG\_PRICE - 0.44*TRF$$

Note that the constant estimation is not significant in $\alpha = 5\%$ or $\alpha = 1\%$.
The data after the implementation of SPS regulations is:

$$LGEXLADA = 8.59 + 2.34*LGGDPCPEU + 1.83*LGPRODLADA$$
$$- 7.46*D\_AFLAB1 - 3.14*LG\_RER - 1.76*LGTRAN\$$$
$$+ 1.14*LG\_PRICE - 0.44*TRF + 1.19*D\_AFLAB1*TRF$$

$$LGEXLADA = 2.34*LGGDPCPEU + 1.83*LGPRODLADA - 7.46*1 - 3.14*LG\_RER$$
$$- 1.76*LGTRAN\$ + 1.14*LG\_PRICE - 0.44*TRF + 1.19*1*TRF$$

$$LGEXLADA = - 7.46 + 2.34*LGGDPCPEU + 1.83*LGPRODLADA - 3.14*LG\_RER$$
$$- 1.76*LGTRAN\$ + 1.14*LG\_PRICE + 0.75*TRF$$

These results with constant attention to the estimated final result is not significant in the $\alpha = 5\%$ or $\alpha = 1\%$.

Meanwhile, nutmeg export function for the period prior to the application of SPS Afla-toxin restrictions are as follows:

$$LG\_EXPALA = 13.56 - 1.96*LGGDPCPEU(-1) + 0.56*LG\_PRODPALA(-1)$$
$$- 1.10*LG\_RER + 0.18*LGTRAN\$ - 0.13*TRF$$

Note that the constants result in significant estimates $\alpha = 5\%$ or $\alpha = 1\%$. Data after the implementation of SPS regulations is as follows:

$$LG\_EXPALA = 13.56 - 1.96*LGGDPCPEU(-1) + 0.56*LG\_PRODPALA(-1)$$
$$- 1.77*D\_AFLAB1 - 1.10*LG\_RER + 0.18*LGTRAN\$$$
$$- 0.13*TRF + 0.35*D\_AFLAB1*TRF$$

$$LG\_EXPALA = 13.56 - 1.96*LGGDPCPEU(-1) + 0.56*LG\_PRODPALA(-1)$$
$$- 1.77*1 - 1.10*LG\_RER + 0.18*LGTRAN\$ - 0.13*TRF + 0.35*1*TRF$$

$$LG\_EXPALA = 13.56 - 1.77 - 1.96*LGGDPCPEU(-1) + 0.56*LG\_PRODPALA(-1)$$
$$- 1.10*LG\_RER + 0.18*LGTRAN\$ - 0.13*TRF + 0.35*1*TRF$$

$$LG\_EXPALA = 11.79 - 1.96*LGGDPCPEU(-1) + 0.56*LG\_PRODPALA(-1)$$
$$- 1.10*LG\_RER + 0.18*LGTRAN\$ + 0.22*TRF$$

With constant attention to the final result in significant estimation $\alpha = 5\%$ or $\alpha = 1\%$, then the final result obtained is positive and the rate constants are positive.

ANNEX 6

Pepper data—Individual test.

	Country	Effect
1	Netherlands	4.141147
2	Belgium	−0.827500
3	Germany	4.294359
4	Spain	−3.957831
5	France	2.227772
6	UK	4.656563
7	Bulgaria	−0.902733
8	Denmark	−0.486417
9	Italy	−6.361813
10	Poland	2.406328
11	Portugal	−5.189874

Nutmeg data—Individual test.

	Country	Effect
1	Netherlands	3.046928
2	Belgium	0.846731
3	Germany	2.303933
4	Spain	−0.597954
5	France	1.230540
6	UK	−0.043323
7	Italy	−1.223201
8	Poland	−1.213777
9	Portugal	−4.349879

The politics behind Alpalhankam: Military and politico-security factors in Indonesia's arms procurements, 2005–2015

T. Chairil
Department of International Relations, Bina Nusantara University, Jakarta, Indonesia

ABSTRACT: After the United States (US) arms embargo against Indonesia in 1999–2005, there have been efforts to meet the objective of arms autarky, or self-sufficiency, by re-establishing the Indonesia's domestic defense industry. The government has issued Law No. 16/2012 on Defense Industry and subsequent legislations to regulate the development of the domestic defense industry to meet demands of defense and security equipment (Alpalhankam). Challenges remain in Indonesia's arms procurements that can prevent the aspiration of defense industry autarky. This research aims to address military and politico-security issues in Indonesia's Alpalhankam procurements during the decade after the embargo was lifted (2005–2015) by using qualitative method. The analysis shows that Indonesia's threat perceptions during this period have increased, raising the need for major conventional weapon systems. The research also finds that the strategy of supplier diversification has prevented the dependence on a predominant arms supplier. Despite this fact, the relations between the TNI and the domestic defense industry remain challenging. The TNI as the Alpalhankam user still perceives domestic industry players less capable than foreign suppliers, whilst domestic defense industry complains about uncertainties in arms procurement, research and development difficulties due to high capital, and regulatory inconsistencies.

1 INTRODUCTION

Arms procurement is one of the most important activities in state's efforts to build its military capabilities for defense purposes. According to Matthews and Maharani (2009), states establish defense industry on strategic imperative, as well as politico-diplomatic considerations. Developing states also establish defense industry because they are subject to the threat of conflict and possible arms embargoes. Some states, like South Africa, Israel, and Sweden, built their own defense industry after they experienced military embargoes from arms suppliers.

The same goes with Indonesia. Following allegation of human rights violations in East Timor, the United States (U.S.) imposed an arms embargo on the Indonesian National Armed Forces (TNI). The U.S. embargo prevented Indonesia to purchase, use, fix, or maintain arms and military equipment from the U.S. It then severely curtailed the operational availability of Indonesia's weapon systems, of which the U.S. was its biggest supplier. It resulted in low levels of arms readiness as follow:

After the U.S. arms embargo to Indonesia and the subsequent effects on the TNI's operational capabilities, Indonesia has been seeking to diversify its suppliers away from the U.S. to prevent dependence on one supplier. There have also been efforts to meet the objective of autarky, or self-sufficiency, by re-establishing the Indonesia's domestic defense industry. The government has introduced several policies and issued several legislations to guide and regulate the re-development of the domestic defense industry to meet the demands of TNI's equipment. The domestic defense industry was included in the strategic plans (Renstra) of 15-year development for the Indonesian main forces (2010–2024).

Table 1. Indonesia's arms readiness, 2002.

Arms & equipments	Requirements	Available
SS- 1 infantry rifles	83 infantry battalions	78 infantry battalions
Cavalry and artillery rifles	100%	77% operational
Small calibre ammunitions	6x essential provisions	3x essential provisions
Large calibre ammunitions	3x essential provisions	0.6x essential provisions
Tank and armored vehicles	100%	55% operational
Army helicopeters	96	69 operational (47%)
Tactical vehicles	9658	5625 operational (40%)
2 submarines	100% condition	70% condition
6 frigates	100% condition	Decreased capability
4 corvettes	100% condition	Expired missiles
16 Parchim-class ships	100% condition	Incompatible propulsion system, unfulfilled spare
4 fast missile crafts	100% condition	Expired missiles, only 1 craft had limited availability
2 fast torpedo crafts	100% condition	100% condition
8 fast patrol boats	100% condition	100% condition
Ex-Aussie 8 fast patrol boats	100% condition	Limited availability
Support vessels	100% condition	Ex-World War II, unfit
222 flghters, transport aircrafts, helicopters	100% combat ready	93 combat ready (41.5%)
16 national air defense command radars	100% standby	11 standby (68%)
10F-16 fighter aircrafts	100% operational	4 operational (40%)
C-130 transport aircrafts	18	7

Source: Widoyoko et al. 2008, 34–36.

Rebuilding of the defense industry began with the establishment of the Defense Industry Policy Committee (KKIP), which by now has issued a number of policy products, including Law No. 16/2012 on Defense Industry. The law mandates the government to build and develop the defense industry to become more advanced, strong, independent, and competitive (Yusgiantoro, 2014). To accomplish it, the government had assigned PT Dirgantara Indonesia as the lead integrator for aerospace systems, rockets, missiles, and torpedoes; PT PAL Indonesia for ships and submarines; and PT Pindad for combat vehicles, weapons, and munitions. The Law No. 16/2012 obliges defense and security main equipment (Alpalhankam) users—TNI and Polri—to use domestic products and to conduct maintenance and repair in the country, and only when it is impossible to do so that the users are allowed to conduct foreign procurements.

The establishment of defense industry is attributed to the development of the defense force, with its main mission to meet the requirements of the domestic market for defense equipment, to be able to compete internationally, and to support national economic growth. To meet this, there are seven national programs of the defense industry development: KF-X/IF-X fighter aircrafts, submarines, propellant industry, national rockets, national missiles, national radars, and medium and heavy tanks. From 2004 to 2014, the allocated budget for state-owned defense industry production was IDR8.32 trillion, most of which (34.38%) was allocated to TNI AD. The absorbed budget coupled with allocated budget of export credit (for arms procured from abroad, but some of its components were produced by domestic defense industry) reached IDR19.6 trillion, 41.67% of which was allocated to TNI AU (Yusgiantoro, 2014).

Despite several policies and strategies on Indonesia's arms procurement and defense industry, challenges remain that can prevent the aspiration of defense industry autarky. First, there is a problem regarding political influence of TNI and TNI's relationship with the domestic defense industry. The relations between Alpalhankam users and domestic defense industry are unpleasant at times whereby users frequently complain about the defense industry's arms producing capabilities, while the defense industry often complain about being stuck in uncer-

Table 2. Indonesia's defense budget compared.

Defense budget (2015)		Per capita (2014)		Share of Govt. spending	
China	190974	Singapore	1789	Singapore	18.3%
Japan	59033	Malaysia	163	Myanmar	13.0%
India	49999	Thailand	85.3	Vietnam	8.3%
South Korea	33142	Vietnam	46	Thailand	6.6%
Taiwan	10135	Myanmar	44.2	Philippines	6.0%
Singapore	9138	Philippines	32.9	Malaysia	5.4%
Indonesia	8076	Indonesia	27.8	Indonesia	4.1%
Constant (2011) USD ($m.)		Current USD		% of general Govt. exp.	

Source: SIPRI Military Expenditure Database.

tainty of users' procurement. This challenge can be illustrated by the problem regarding the Air Force (TNI AU)'s procurement of heavy transport helicopters, including VVIP helicopters, for operational vehicles of high state officials in late 2015. TNI AU initially decided to buy British-Italian AgustaWestland AW101, although PT Dirgantara Indonesia already had the capability to produce EC725 Caracal VVIP helicopters.

The second problem concerns on the national defense budget. As defense budget increases, more is allocated to arms procurement and defense industry development. Indeed, Indonesia's defense budget has been increasing from year to year, more than tripled from US$2.174 billion in 1988 to US$8.071 billion in 2015 (SIPRI Military Expenditure Database). However, Indonesia's defense budget is still relatively low, with it always being lower than 1% of the gross domestic product (GDP) since 1999, and by comparison to other countries in East, South, and Southeast Asia as follow:

1.1 Research question

This paper aims to address the challenges in Indonesia's arms procurement as outlined above. From several challenges in Indonesia's arms procurement, this paper focuses on the military and politico-security issues in the decision-making process relating to Indonesia's Alpalhankam procurements during the one decade after the embargo was lifted (2005–2015). Hence, the research question that this paper answers is as follows: **"How do military and politico-security issues influence Indonesia's arms procurement in 2005–2015?"**

2 RESEARCH METHOD

Ravinder Pal Singh (1998, 2000) examined that there are four major themes in arms procurement decision making: (1) military and politico-security; (2) budget, financial planning and audit; (3) techno-industrial; and (4) organizational behavior and public-interest. The first theme is selected for this paper because of the prominence of military and politico-security issues in Indonesia case. This is because the Indonesian military has long history of role as a political force and immense presence in politics. Issues under the military and politico-security theme include: (1) effects of threat perceptions, security concepts and operational doctrines on force planning; (2) influence of foreign and security policies on arms procurement; and (3) the political influence of the military and of predominant arms suppliers.

The data used in this research is mainly secondary data collected from several sources, including government documents, results of studies conducted by government agencies, databases compiled by research institutes, and news articles in printed and online periodicals. The data is analyzed by a qualitative research approach, wherein instruments use more flexible style of eliciting and categorizing responses to the research questions and study design is iterative, that is, data collection and research questions are adjusted according to what is learned.

The sources are as follow:

Table 3. Data sources.

Variables	Data sources
Threat perceptions	Widjajanto and Wardhani 2008
	Indonesia Defense White Papers 2003, 2008, 2015
	Minimum Essential Force (MEF) 2010
Foreign and security policies	SIPRI Arms Transfer Database
	Indonesia's foreign policy analysis
	Indonesia's security policy analysis
The military and arms suppliers	Bappenas 2011
	Workshop on Defense Industry 2012
	Agency for Goods and Service Procurement Policy (LKPP) 2013

3 DISCUSSION

3.1 *Threat perceptions*

Historically, Indonesia's threat perceptions have been predominated by internal threats, which do not require the same high-technology defense equipment than external threats do. According to Andi Widjajanto and Artanti Wardhani (2008), out of 249 military operations launched by Indonesia in 1945–2004, 33% were directed against external threats, such as the Dutch and Malaysia, while most (67%) were deployed to fight internal threats, such as the Communist Party of Indonesia (PKI), Darul Islam (DI/TII), Free Aceh Movement (GAM), Free Papua Movement (OPM), and Fretilin. The trend continues to the last decade (2005–2015). In 2003, Indonesia published a defense white paper that stated that traditional security threats in the form of external invasion were expected to be unlikely because the role of the United Nations and the international reaction was believed to be able to prevent, or at least restrict it. Meanwhile, non-traditional threats were expected to be more probable to happen, either cross-borders or inside the state's borders.

Another defense white paper published in 2008 classified the perceived threats into military and non-military. Military threats include aggressions, territorial violations, armed insurrections, sabotages, espionages, armed terrorist acts, sea and air security threats, as well as communal conflicts; while non-military threats include ideological, political, economic, social and cultural, information technology, and public safety dimensions. The non-military threat perceptions are non-conventional issues that do not require high-technology arms; thus, barely affecting arms procurement. However, one thing to notice from the 2008 defense white paper is the rising perception of conventional threats compared to the 2003 defense white paper.

Threat perceptions can also be assessed from the 2010 Minimum Essential Force (MEF) program. Defined as "an essential and minimum standard of TNI's strength that absolutely needs to be prepared as a main and fundamental prerequisite for effective implementation of TNI's duties and functions in the face of actual threats", MEF was formulated on actual and potential threat perceptions. Actual threats are those that already existed, while potential ones are those that never happen. MEF program includes terrorisms, separatisms, violations of border areas and outer islands, natural disasters, illegal activities, horizontal conflicts, and energy scarcity as actual threats; while global warming, violations in the Indonesian archipelagic sea lanes (ALKI), environmental pollutions, pandemic, financial crisis, cybercrime, foreign military aggression, and food and water crisis as potential threats. MEF program prepared five threat scenarios of foreign military involvement, which are (1) separatist movement, (2) border conflicts, (3) securing access to energy sources, (4) securing economic routes in ALKI, and (5) war on terror. It then projected the five threat scenarios on flash point areas: Aceh, Riau and Riau Islands, East and West Kalimantan, North and Southeast Sulawesi, Lombok, East Nusa Tenggara, Maluku, Papua and West Papua, Malacca Strait, and ALKI.

The most recent defense white paper, published in 2015, explains that the forms and shapes of today's evolving threats are increasingly blurred, with the biggest potential threats taking the form of fragility of life and spirit of nationalism. The various forms of threats in the defense white paper include Asia-Pacific military modernization; border disputes; intra- and interstate conflicts; tendency of nonlinear, indirect proxy war; weapons of mass destruction; terrorism; espionage; transnational crimes; military technology innovation in conventional wars; global climate change; natural disasters; food security; water security; energy security; and epidemic. Threat perceptions in the white paper are too holistic, classified as military, non-military, or 'hybrid', and further divided into known and not yet known threats with priority given to the former.

Bob Lowry (2016) argues that by deeming a serious external threat as unlikely, and including primarily domestic threats as the known threats, the white paper provides little guidance for determining military force structures or dispositions. Reverberating that concern, Iis Gindarsah (2016) reviews that, while the white paper reflects that Indonesia's changing threat environment now includes hybrid threats or 'proxy', it does not offer strategy to anticipate a scenario of hybrid warfare. Instead, it focuses on '*Bela Negara*' (defending the state) program aimed at nurturing nationalism within the population. Keoni Indrabayu Marzuki (2016) even comments that the 'proxy war' narrative frequently aired by TNI, especially TNI AD, is an avenue to augment the military's presence in the civilian domain, also a broader effort by TNI AD to preserve its relevance under Joko Widodo government's Global Maritime Fulcrum vision.

It can be concluded that Indonesia's threat perceptions have been changing time after time, and perceptions of external threats that can only be countered by high-technology defense equipment increase after the embargo was lifted. These threats have been classified further into several subdivisions. However, the latest defense white paper assesses very holistic threats that it does little for Alpalhankam procurement decision making. Recently, even the Minister of Defense Ryamizard Ryacudu said that the government would reduce purchases of modern arms, arguing that the geopolitical and security constellation in the Southeast Asia was likely to be stable while the threats of an armed conflict between regional states were likely to decline, and ergo no need for many modern, expensive defense equipment (Putra, 2016).

3.2 *Foreign and security policies*

There are foreign and security policies that influence arms procurement decision making. In Indonesia case, it was very evident during the Sukarno's Old Order and Suharto's New Order years. During the Old Order, the relations between Sukarno's regime and the Soviet Union and the agenda of defense diplomacy against external threats from the Dutch in West Irian encouraged arms procurements dominated by imports from the Soviet Union. Later during the New Order, the relations between Suharto's regime and the U.S., coupled with the anti-communism agenda encouraged termination of arms imports from the Soviet Union, and instead the U.S. became Indonesia's main supplier. This occurred until the U.S. and its allies imposed arms embargo of defense equipment to Indonesia in 1999–2005. Because of the dominance of arms imports from the U.S. and its allies, during the embargo Indonesia struggled to barely obtain spare parts and maintenance for its defense equipment. Since then, Indonesia's arms procurements have been conducted by diversifying the suppliers to prevent defense dependence on a supplier.

The value of Indonesia's arms imports from 2005 to 2015 was dominated by imports from Russia, but with a moderate share (18.9%) not as high as the share of imports from main providers in previous eras (Soviet Union dominated arms transfer to Indonesia by 65.9% from 1950 to 1965; while the U.S. dominated by 24.5% from 1966 to 1998). During this period, new suppliers without previous transfer to Indonesia had emerged, such as China, Brazil, Brunei, Denmark, and Belgium. The U.S. had again become Indonesia's supplier after lifting the embargo in 2006. This was the result of the supplier diversification strategy. It also reflected Indonesia's foreign policy during the period of Susilo Bambang Yudhoyono (SBY)'s presidency: under SBY's 'a million friends and zero enemies' policy, Indonesia had boosted its international presence (Piccone & Yusman 2014). Rooted in nonalignment and

Table 4. Trend-indicator value of arms import to Indonesia by suppliers, 2005–2015, expressed in US$ m. at constant (1990) prices.

	2005	2006	2007	2008	2009	2010	2011	2012	2013	2014	2015	Total	%
Russia				41	165	191	59	21	351	54		880	18.9%
South Korea	2		176	5	90		120	18	182	121	3	715	15.3%
United States	4	11	17	14			0	35	85	150	364	681	14.6%
Netherlands			297	149	149					3		597	12.8%
United Kingdom										541	24	565	12.1%
France	13	30	46	21	16	33	29	21	10	16	33	267	5.7%
China		10	3	3		2	8	69	78	39	33	244	5.2%
Spain	10	10	10					24	47	35	24	161	3.5%
Germany (P)	6				1		7	1	11	77	27	130	28%
Switzerland										30	90	120	26%
Brazil								18		37	37	92	20%
Australia									18	19	37	75	1.6%
Canada	1		1	2			1	11	6	6	6	34	0.7%
Italy		16	4	4					8			32	0.7%
Brunei							24					24	0.5%
Poland			8		9							17	0.4%
Belgium									13			13	0.3%
Denmark			3		2		2				5	11	0.2%
Czech Republic				1								1	0.0%
Israel								1				1	0.0%
Total	36	61	577	239	436	225	250	218	802	1136	683	4661	100.0%

Source: SIPRI Arms Transfer Database.

pragmatism, the 'all directions foreign policy' led Indonesia to actively seeking to raise existing ties with countries in all corners of the globe to a higher level. There were concerns that this supplier diversification has led to a problem of availability and compatibility (IDSS, 2014).

3.3 The military and arms suppliers

In Indonesia, the relations between military and domestic defense industry have been influenced by the massive industrialization during the Old Order and New Order years. During the Old Order, the formation of Indonesia's defense industry was due to the military's need for defense products. The establishment of the defense industry was supported by nationalization of foreign corporations, mainly Dutch, during the West Irian dispute with the Netherlands (Hartono, 2013). Out of Dutch shipyard, the government pioneered Indonesia's modern shipbuilding. The government also set up the steel industry named Trikora Iron and Steel Project. During the New Order, defense industrialization intensified due to the role of the State Minister for Research and Technology B. J. Habibie (Karim, 2014), who also acted as chairman of the Agency for the Assessment and Application of Technology (BPPT) and the Agency for the Development of Strategic Industry (BPIS). Habibie acted as an important figure in the rise of Indonesia's defense industry, forming the Team for the Development of Defense and Security Industry (TPIH), Council for the Development of Strategic Industry (DPIS), BPIS, the Agency for the Management of SOEs, and PT BPIS and thus contributing to develop the Indonesia's defense industry (Dirwan, 2011).

During these periods, the relations of the military and domestic defense industry were considerably strong because the government's efforts to develop Indonesia's defense industry often involved the military. However, the relations deteriorated due to the 1997 financial crisis, which led to the treatment of defense industry as a regular supplier, assessed from the business as usual viewpoint of costs and economies of scale. Today after the crisis, there has been a rise in awareness of the importance of the relations between the military and the defense industry after the U.S. embargo was lifted. When Indonesia faced Maluku sectarian conflict and Aceh tsunami, the TNI had difficulty in logistics and support due to the embargo. To help disaster relief efforts, the U.S. partially lifted the embargo and reopened Foreign Military Sales (FMS) for parts of C-130 transport aircrafts (Singh & Salazar 2006). The government realized that without defense industry autarky, TNI would be constantly dependent on foreign suppliers, making them vulnerable to the threat of embargo by the supplier countries, which could jeopardize military readiness.

The government then started aligning with domestic defense industry. By issuing regulations that favored domestic defense industry, the government has sought to restore the relations between TNI and defense industry. Now, arms procurement has legal protection under the Minister of Defense's Regulation No. 34/2011 on Guidelines for TNI's Alutsista Procurement, Law No. 16/2012 on Defense Industry, and Presidential Regulation No. 54/2010 on Procurement of the Government's Goods and Services. All of these regulations were based on a spirit of arms independence and procurement from domestic defense industry. However, even now TNI often complains that the defense industry lacks managerial capability while the defense industry complains to be stuck in uncertainty of purchase and research and development (R&D) difficulties due to high capital and regulatory inconsistencies, leading to problems in the relations between TNI and domestic defense industry.

A survey on the defense industry conducted by the National Development Planning Agency (Bappenas) in 2011 found several problems related to Alutsista procurement: (1) the R&D costs charged to manufacturers increased unit production costs, making the industry uncompetitive; (2) the funding system was complex due to the use of different sources of funds to purchase the same item, or the complexity of the use of foreign debt compared to the state budget funds; (3) the KKIP had not been optimal; and (4) orders were not adapted to the production capacity. A similar topic was delivered in the workshop on defense industry conducted on July 26, 2012, in which the problems related to Alutsista procurement were classified into five categories, which then were broken down into 13 subdivisions as follow:

Table 5. Problems related to Alutsista procurement, 2012.

Problems in TNI/Polri's procurements	Poor planning
	High cost of restarting production that had been stopped
	The difficulty of domestic defense industry to enter Polr's procurement
Problems in test and evaluation	Low support from users for new technology test
	Different views between TNI/Polri and state—owned defense industry about which technology that was considered good
Problems in funding	Need for low funding and payment system
Problems in the defense industry technology development (Bangtekindhan) program	Prototypes from R&D cooperation between defense industry and the military were often not followed-up into manufacturing
	High R&D costs were not accompanied by adequate returen
	Investments made by the defense industry to support Alutsista procurement were seldom no used
Problems in ToT	Lack of purchase for products of Tot
	The decreasing value of ToT
	Need for increasing the effectiveness of ToT
	Funding premium costs in ToT

Source: Workshop on Defense Industry, 2012.

287

The government sought to resolve the problems in TNI/Polri's procurements and in test and evaluation by the obligation in Law No. 16/2012's Article 25 that Alpalhankam users propose and implement management policies that include (1) a long-term plan of Alpalhankam needs; (2) operational requirements and technical requirements of the Alpalhankam needs; and/or (3) assistance and evaluation in the production process and product development. Problems in funding were resolved by Article 41 that obliges the government to provide protection to defense industry's business expansion and increase in the production capacity by providing fiscal incentives, including exemption from import duties and taxes, guarantees, funding and/or financing of the defense industry over KKIP considerations.

Obstacles in the relations between TNI and domestic defense industry remain, as shown in the assessment of Alutsista and Almatsus procurement by the Agency for Goods and Service Procurement Policy (LKPP) in 2013. From the users' point of view, procurement was still hampered by procedures that accentuated administrative issues, suggesting that the existing legislation was still based on public procurement, not specific defense procurement that should be more concerned with the effectiveness rather than efficiency. In addition, the users also considered that the managerial capability of the defense industry was still less independent and agile. They viewed that the nature of Alpalhankam was very specific and should not be compromised; hence the fulfillment of the required specifications was essential for the effective use of the products.

On the other hand, from the defense industry's perspective, the most important issue was to open the information on users' requirement planning for access by the industry. In other words, during R&D process, the industry was stuck in uncertain purchases, which in turn not only stopped the R&D process preceded by high capital, but also prevented fulfillment of R&D to meet user requests because the corporate resources had been spent during the first R&D activity. In addition, the defense industry was also concerned about the involvement of companies that actually did not have the capability in Alpalhankam procurement if procurement was conducted through an auction system. Therefore, the defense industry saw the need for the status of preferred partner. In the macro level, the defense industry highlighted the problem of regulatory inconsistencies in the existing legislation, for example the issue of financing and taxes on the defense-related products or components.

4 CONCLUSION

The analysis shows that Indonesia's threat perceptions during the one decade after the embargo was lifted (2005–2015) has been increasing, hence increasing the need for major conventional weapon system. This paper also finds that the strategy of supplier diversification has prevented the dependence on a predominant arms supplier. However, the relations between the TNI and the domestic defense industry remain a barrier, in which the TNI as Alpalhankam user still perceives the defense industry as less capable than foreign suppliers, while the defense industry complains about the uncertainties in arms procurement, research and development difficulties due to high capital, and regulatory inconsistencies.

REFERENCES

Bappenas. (2011) *Survei Industri Pertahanan Bappenas*. Jakarta, Bappenas.
Departemen Pertahanan Republik Indonesia. (2003) *Buku Putih Pertahanan Indonesia*. Jakarta, Departemen Pertahanan Republik Indonesia.
Departemen Pertahanan Republik Indonesia. (2008) *Buku Putih Pertahanan Indonesia*. Jakarta, Departemen Pertahanan Republik Indonesia.
Direktorat Kebijakan Pengadaan Khusus dan Pertahanan Keamanan—LKPP. (2013) Kajian pengadaan alutsista dan almatsus. *Jurnal Pengadaan*, 3(3).

Dirwan, A. (2011) *Laporan Akhir Tim Pengkajian Hukum tentang Pengembangan dan Pemanfaatan Industri Strategis untuk Pertahanan*. Jakarta, Badan Pembinaan Hukum Nasional, Kementerian Hukum dan Hak Asasi Manusia Republik Indonesia.

Gindarsah, I. (2016) *Reviewing Indonesia's new defense white paper*. The Jakarta Post June 16, 2016. Retrieved from: http://www.thejakartapost.com/news/2016/06/16/reviewing-indonesia-s-new-defense-white-paper.html [Accessed on September 30, 2016].

Hartono, R. (2013) *Kronik Nasionalisasi Perusahaan Asing Tahun 1957*. Berdikari Online August 7, 2013. Retrieved from: http://www.berdikarionline.com/kronik-nasionalisasi-perusahaan-asing-tahun-1957/ [Accessed on May 9, 2016].

Hermawan, I. (2015) *Dirut PTDI: Semoga Presiden Mau Pakai Helikopter Produk Kami*. CNN Indonesia, November 26, 2015. Retrieved from: http://www.cnnindonesia.com/nasio nal/20151126101353-20-94171/dirut-ptdi-semoga-presiden-mau-pakai-helikopter-produk-kami/ [Accessed on May 25, 2016].

IDSS. (2014) *Rethinking TNI AU's Arms Procurement: A Long-run Projection*. RSIS. Retrieved from: https://www.rsis.edu.sg/wp-content/uploads/2014/09/PR_140709_Rethinking-TNI-AU.pdf.

JPNN.com. (2015) *DPR: Anggaran Pertahanan Turun, Poros Maritim Terancam*. JPNN.com September 10, 2015. Retrieved from: http://www.jpnn.com/read/2015/09/10/325699/DPR:-Anggaran-Pertahanan-Turun-Poros-Maritim-Terancam.

Karim, S. (2014) *Membangun Kemandirian Industri Pertahanan Indonesia*. Jakarta, KPG.

Kementerian Pertahanan Republik Indonesia. (2010) *Minimum Essential Force Komponen Utama*. Jakarta, Kementerian Pertahanan Republik Indonesia.

Kementerian Pertahanan Republik Indonesia. (2015) *Buku Putih Pertahanan Indonesia*. Jakarta, Kementerian Pertahanan Republik Indonesia.

Khoemaeni, S.A. (2015) *KSAU: PT DI Bikin Sayap Saja Tidak Bisa!* Okezone News November 27, 2015. Retrieved from: http://news.okezone.com/read/2015/11/27/337/1257220/ksau-pt-di-bikin-sayap-saja-tidak-bisa [Accessed on May 25, 2016].

Lowry, B. (2016) *Indonesia's 2015 Defence White Paper*. The Strategist, June 15, 2016. Retrieved from: http://www.aspistrategist.org.au/indonesias-2015-defence-white-paper/ [Accessed on September 30, 2016].

Marzuki, K.I. (2016) *Proxy Wars Narrative: TNI-AD's Quest for Relevance?* RSIS April 21, 2016. Retrieved from: http://www.rsis.edu.sg/rsis-publication/rsis/co16092-proxy-wars-narrative-tni-ads-quest-for-relevance/ [Accessed on October 6, 2016].

Matthews, R. & Maharani, C. (2009) The Defense Iron Triangle Revisited. In: Bitzinger, R.A. (ed.) *The Modern Defense Industry: Political, Economic, and Technological Issues: Political, Economic, and Technological Issues*, pp. 38–59. Santa Barbara, California, ABC-CLIO.

Parameswaran, P. (2015) *Why is Indonesia Set to Cut its Military Budget for 2016?* The Diplomat September 10, 2015. Retrieved from: http://thediplomat.com/2015/09/why-is-indonesia-set-to-cut-its-military-budget-for-2016/ [Accessed on May 26, 2016].

Piccone, T. & Yusman, B. (2014) *Indonesian Foreign Policy: 'A Million Friends and Zero Enemies'*. The Diplomat, February 14, 2014. Retrieved from: http://thediplomat.com/2014/02/indonesian-foreign-policy-a-million-friends-and-zero-enemies/ [Accessed on October 7, 2016].

Putra, W.P.A. (2016) *Menhan Akan Kurangi Pembelian Alutsista Modern*. detikNews September 21, 2016. Retrieved from: http://news.detik.com/berita/d-3303342/menhan-akan-kurangi-pembelian-alutsista-modern [Accessed on September 27, 2016].

Sagraves, R.D. (2005) *The Indirect Approach: The Role of Aviation Foreign Internal Defense in Combating Terrorism in Weak and Failing States*. Alabama, Air Command and Staff College.

Sepp, K.I. (2005) Best Practices in Counterinsurgency. *Military Review*, 8–12 (May-June 2005).

Sihaloho, M.J. (2015) *Ketua Komisi I Dukung Usulan Kenaikan Anggaran Pertahanan di 2016*. Beritasatu.com. September 25, 2015. Retrieved from: http://www.beritasatu.com/politik/309691-ketua-komisi-i-dukung-usulan-kenaikan-anggaran-pertahanan-di-2016.html.

Singh, D. & Salazar, L.C. (Eds.) (2006) *Southeast Asian Affairs 2006*. Singapore, ISEAS—Yusof Ishak Institute.

Singh, R.P. (ed.) (1998) *Arms Procurement Decision Making Volume I: China, India, Israel, Japan, South Korea and Thailand*. New York, Oxford University Press.

Singh, R.P. (ed.) (2000) *Arms Procurement Decision Making Volume II: Chile, Greece, Malaysia, Poland, South Africa and Taiwan*. New York, Oxford University Press.

SIPRI. (n.d.) *SIPRI Arms Transfers Database*. SIPRI. Retrieved from: https://www.sipri.org/databases/armstransfers.

SIPRI. (n.d.) *SIPRI Military Expenditure Database*. SIPRI. Retrieved from: https://www.sipri.org/databases/milex.

Undang-Undang Nomor 16 Tahun 2012 tentang Industri Pertahanan.

Utama, A. (2015) *KSAU Sebut PTDI Lamban Rampungkan Pesanan Helikopter*. CNN Indonesia November 27, 2015. Retrieved from: http://www.cnnindonesia.com/nasional/20151126201242-20-94378/ksau-sebut-ptdi-lamban-rampungkan-pesanan-helikopter/ [Accessed on May 25, 2016].

Widjajanto, A. & Wardhani, A. (2008) *Hubungan Intelijen—Negara 1945–2004*. Jakarta, Friedrich Ebert Stiftung, Pacivis UI.

Widoyoko, D. *et al.* (2008) *Bisnis Militer Mencari Legitimasi*. Jakarta, Indonesia Corruption Watch.

Yusgiantoro, P. (2014) *Ekonomi Pertahanan: Teori & Praktik*. Jakarta, PT Gramedia Pustaka Utama.

Pesantren education: The changing and the remaining. A case study of Bahrul Ulum Pesantren Tambak Beras in Jombang, East Java, Indonesia

M.L. Zuhdi
Program of Middle Eastern and Islamic Studies, Strategic and Global Studies, University of Indonesia, Indonesia

ABSTRACT: This study discusses the *pesantren* education system. *Pesantren* is a social and educational institution specialized in Islamic religious education. *Pesantren* as an educational institution has long historical roots. *Pesantren* is often regarded as a traditional educational institution that often resist from modernized trend and innovation. Factually, in certain times in the past, *pesantren* was identified as a symbol of backwardness and opacity. This study aims to answer the question: Is it true that *pesantren*, as a traditional boarding educational institution, is difficult to accept a change towards modernity and unable to adjust itself to social development? In the midst of globalisation, how *Pesantren* survives facing the challenges of the times? Method of this research uses historical approach, while data was collected through library research, interviews, and direct observation. The result shows that *pesantren* is a dynamic educational institution which follows the growth and the dynamics of education in Indonesia. The main guideline to perform changes in *pesantren* was based on the principle of المحافظة على القديم الصالح والاخذ بالجديد الاصلح (maintaining the existing old which is good, and ratifying the new when it is better). While *pesantren* has to develop itself into better innovation, it does not always need to replace the existing programs when are still considered beneficially important and valid.

1 INTRODUCTION

Pesantren (Traditional Islamic boarding educational centre) is a social and educational institution that teaches Islamic knowledge. *Pesantren* actually emerged in the early days of the spread of Islam in the Indonesian archipelago. In ancient times, prior to the presence of the Dutch colony into Indonesian territory, *pesantren* used to be a boarding institution that became the centre of social change through religious activities in the midst of daily living encounter among traders locally and across islands. *Pesantren* was also hold important role in the opening of new settlements. During the Dutch controlled the Indonesian archipelago, *pesantren* became the centre of resistant movement against Dutch ruler (Aboebakar, 2011; Rahardjo, 1995). Later, after Indonesia gained its independence in 1945, precisely from 1959 to 1965, *pesantren* was regarded as a 'revolutionary tool' for political changes, while in the 1970s, the New Order government considered *pesantrens* to be a 'development potency' for the country because at that time social transformation could be conducted through the role of *pesantren* (Horikoshi, 1976; Rahardjo, 1995). Beside its wide and fickle range of roles and functions as mentioned above, the characteristics of *pesantren* as an educational institution of religious knowledge have remained relatively unchanged. An early record of *pesantren* existence as an educational institution was the 1718 documents pertaining rumours of the 'schools for the devout' near Surabaya. Martin van Bruinessen, in his book titled '*KitabKuning*' ('The Yellow Book'), stated that, based on a survey held by the Dutch in 1831, is a form of Islamic schools existed in Indonesia's coastal area such as in Cirebon, Semarang, and Surabaya. The *Pesantren* have long historical roots and were often regarded as traditional educational institutions. In this context, a *pesantren* was often perceived to be a conservative institution which is reluctant to change and holds on to its traditional values. During certain times in the past, the *pesantren* was identified as a symbol of backwardness and closure. Is it true that, as an educational institution, the *pesantren* is a traditional entity that is averse to change and does not adjust to the demands of the times? If so in the midst of today's globalisation, how could *pesantren* survive facing the challenges of modernity?

2 RESEARCH METHOD

This paper discusses the patterns of *pesantren* education by observing both the changing and the constant aspects of its development. In order to observe the development of the pattern of education in *pesantren*, the author uses a case study of a *pesantren* in Jombang, East Java, namely Bahrul 'Ulum Tambak Beras *Pesantren*, which was established in 1825 AD and still persists now. The data was collected from written sources on the history of *pesantren*, as well as interviews with the caregivers of Bahrul 'Ulum *Pesantren*, Jombang.

2.1 *The education system of pesantren*

A *pesantren* has at least three elements, namely: *Kyai* who educate and teach, *santri* or students who learn, and the mosque/*mushola* (Muslim prayer house) as a place to learn Islamic knowledge or recite the Koran. The *Kyai* is the leader of the *pesantren* and the highest authority within the institution, while *santri* refers to the group of students who live in *pesantren*. Although both *santri* and *madrasa* students attend Islamic educational institutions, a *santri* is often different to a *madrasa* student. A *santri* could be a non-formal student who attend the *pesantren*, while a *madrasa* student could be both a *santri* and a formal student of a *madrasa* who attends a non-Islamic educational institution.

Pesantren teaches Islamic knowledge systematically, directly in Arabic and is based on classical *kitab* (books) readings written by great *ulamas*. The classical *kitabs*, the source of religious knowledge taught in *pesantren*, are commonly recalled as the Yellow *Kitab*. There are two ways or techniques of teaching the Yellow *Kitab*, these are *bandongan* or *wetonan* and *sorogan*. The *bandongan* or *wetonan* system is a model of teaching in which a *Kyai* reads the Yellow *Kitab* while the *santris* listen to the reading and explanation of the *Kyai*. In this system, there is no time for questions and answers. In contrast, in the *sorogan* system, a student reads the *kitab*, which has been approved by the *Kyai*, in front of the *Kyai*.

As the time flows, the *pesantren* not only teaches the Yellow *Kitab* with the teaching techniques mentioned above, but they also get to know the *madrasa* or modern school system. Even so, the focus of the *pesantren* education remains the same, which is to teach religious sciences. During the Indonesian nationalist movement in the early 1900s, pioneers in the reformation of education such as K. H. Ahmad Dahlan and Ki Hajar Dewantara began to appear. They offered new ideas in education which are oriented towards the development of science and technology. Along with his *Muhammadiyah* movement, K. H. Ahmad Dahlan established *Muhammadiyah* School. Ki Hajar Dewantara also established *Perguruan Taman Siswa* (*Taman Siswa* Institution). *Muhammadiyah* School and *Perguruan Taman Siswa* were regarded as two pioneers of Indonesian National Schools in the period after Indonesian independence. Universities and schools were once considered to be different from *pesantren* because of the different educational focus. In this context, *pesantren* educational institutions, which kept their focus on the development of religious science, were considered to be either a 'left behind' or 'traditional' educational institution in the midst of the occurring changes (Rahardjo, 1995).

The attitude of holding on to traditional values and reluctance to embrace the innovations or changes has caused *pesantren* to be regarded as a conservative institution. *Pesantren* was considered insensitive to the changes and demands of the times and society. Moreover, not all *pesantren* implement the *madrasa* system. Even today, there are still *pesantren* that apply the *bandongan* or *sorogan* system and teach classical and the Yellow *Kitab* only. This kind of *pesantren* education model is called as *salafiyah* which is considered less productive and difficult to receive the modern education system as planned by the government. As a result, *pesantren* is often seen as a close-minded and separate educational institution compared to the 'mainstream' national education system. Based on these assumptions, K. H. Wahid called *pesantren* a subculture (Rahardjo, 1995).

Based on data released by the Ministry of Religious Affairs, in 2010/2011 the number of *pesantrens* across Indonesia was 27,218 *pesantrens* (including 7,838 *salafiyah*), with 3,642,738 *santris* and 339,839 teachers (data booklet3a-1.pdf from pendis.kemenag.go.id). Although in general a *pesantren* possesses characteristics owned by all *pesantrens*, there is no common similarity in pattern shared by all *pesantrens* in Indonesia. Although located in the same area, one *pesantren* can differ from others in terms of development and education models. There are *pesantrens* that still maintain the old model of education and are not affected by the pattern of modern education. There are also *pesantrens* that continuously reform their education systems with the latest methods, and there are also *pesantrens* that still maintain old traditions and conduct development at the same time.

There are a lot of variations amongst pesantrens, each has its own uniqueness. Such variations occur due to a *pesantren*'s leadership, special features and personality which are characterised by the personal characteristics of the *Kyais*, the elements of *pesantren* leaders, and even the specific Islamic school of thought that they have embraced. Special features of the *pesantren* can be whether or not it implements Sufism and whether the education is carried out through the formal system, the informal one or a combination of the two. Some *pesantrens* or *madrasas* run their education in accordance with the curriculum created either by the Ministry of Religious Affairs or the Ministry of Education and Culture, while others, such as *Salafiya pesantrens*, do not follow the curriculum created by either of the two ministries but teach the *Yellow Kitab* only.

2.2 The history and development of Bahrul Ulum Tambak Beras Pesantren

The establishment of the Bahrul Ulum Tambak Beras *pesantren* took place in 1825 AD by a warrior *ulama* named Abdussalam. He managed to set up a *langgar* with its three small cubicle rooms and was known by the surrounding community as either *Pondok (cottage) Selawe* because the number of *santris* were twenty-five or *Pondok Telu* because the number of cubicle rooms for the *santris* were three in total (*Selawe* means 'twenty-five' and *telu* means 'three' in Javanese language). *Pondok Telu* also symbolises the three types of knowledge learned in a *pesantren*, namely *shari'ah* science, *haqeeqat* science (knowledge of truth), and *kanuragan* science (martial arts). *Pondok Selawe* or *Pondok Telu* was the forerunner of *Pondok Pesantren Bahrul Ulum Tambak Beras Jombang* (Ghofir, 2012).

In 1914, *Kyai* Hasbullah Abdul Wahab, a descendant of Abdussalam, started doing improvements in *pesantren*. The education method of Tambak Beras Pesantren, which at the time was still the *bandongan* or *wetonan* system, was then changed to the *madrasa* system by establishing the *Mubdil Fan Madrasa* in 1915. Under the leadership of Abdussalam's son, *Kyai* Abdurrahim, *Mubdil Fan Madrasa* grew rapidly because he introduced Latin inscription and algebra lessons to the *madrasa* students (Yayasan Pondok Pesantren Bahrul Ulum,).

In 1965, *Kyai* Abdurrahim's brother, *Kyai* Wahab, changed the name of *Tambak Beras pesantren* into *Bahrul 'Ulum*. Since then, the *Tambak Beras Pesantren* was known as *Bahrul 'Ulum Tambak Beras Pesantren* and rapidly progressed. In 1969, through the Ministry of Religious Affairs, the New Order government took over *Bahrul 'Ulum's Muallimin Muallimat Madrasa* and turned it into public school which was named *Madrasah Aliyah* and *Tsanawiyah Negri*. The curriculum was also adapted to the curriculum of the Ministry of Religious Affairs (DEPAG). Furthermore, The *Madrasah*, which followed the DEPAG curriculum, remained in the *Bahrul 'Ulum Pesantren* without closing down *Muallimin Muallimat Madrasa* which still applied the original curriculum of the *pesantren*. These *madrasas* still survive and remain successful in terms of contemplation, thought and their intensive interaction with the rapidly changing world (Yayasan Pondok Pesantren Bahrul Ulum, http://tambakberas. or.id/profil/bahrul-ulum/periode-pengembangan-sejarah-pondok-pesantren-bahrul-ulum/ accessed, 19 December 2016).

In December 1971, *Kyai* Wahab died and passed on the *pesantren*'s leadership to *Kyai* Fatah along with the descendants of *Kyai* Hasbullah. In 1974, *Kyai* Fatah started to establish a university named *Al Ma'had Al-Aly*. In 1977, *Kyai* Fatah died, and the *pesantren* leadership was continued by *Kyai* Najib (*Kyai* Wahab's third son). Under his leadership, *Bahrul 'Ulum Pesantren* was significantly developed, and the number of *santri's* dormitories was also increased. *Kyai* Najib, with other caregivers, turned up *Ma'had Al Aly,* who was initiated by *Kyai* Fatah, and changed the *pesantren's* name to *Sekolah Tinggi Ilmu Tarbiyah* (School of *Tarbiyah*) in 1985. *Bahrul 'Ulum Pesantren* held national recognition and a good reputation when *Kyai* Najib was appointed as the chairman of *Robithotul Ma'ahid* (Boarding Schools Association of Nahdlatul Ulama). In his capacity as the chairman of *Robithotul Ma'ahid*, *Kyai* Najib held *Usbu'ul Ma'ahid* (Javanese Pesantren Week) which consisted of a *Bahtsul Masail* event (a form of discussion on various issues) regarding Islamic law to resolve rising social issues at that time. The *Bahtsul Masail* produced a *Kompilasi Hukum Islam* (Compilation of Islamic Law) which was then used as guidelines for the *Pengadilan Agama* (Religious Courts) in Indonesia (Yayasan Pondok Pesantren Bahrul Ulum, http://tambakberas.or.id/profil/bahrul-ulum/periode-pengembangan-sejarah-pondok-pesantren-bahrul-ulum/ accessed, 19 December 2016).

When *Kyai* Najib died on 20 November 1987, the leadership of the *Bahrul Ulum Pesantren* was collectively held by *Bahr Ulum Kyais*, who were the grandsons of *Kyai* Hasbullah, together with the members of the Caregiver Council. The Caregiver Council appointed the chairman and the board of the institute whose responsibility was to manage and develop the *pesantren*. The Meeting Assembly of Caregiver and the Board Foundation Assembly Meeting Caregiver Foundation Board determined the vision of the *pesantren*, that is: to make the *Tambak Beras Pesantren* the centre of Islamic civilisation which serves as a counterweight to all aspects of human's life, so that a safe, peaceful and prosperous society can be realised. Moreover, the mission is: to create a faithful and God-devoted generation who possess a sense of responsibility to develop and spread the Islamic tenet of *Ahlussunnah Wal Jama'ah*, to create a noble generation that has a sense of social responsibility towards the benefit of the *ummah* and create a skilful, competent, independent, and scientifically capable student able to develop knowledge and skills which exist in oneself and in the surrounding environment. The purpose of *Bahrul Ulum Pesantren* is to make the *pesantren* an educational, religious and social institution as well as a catalyst centre for the development of the potential Indonesian human resources who are reliable, productive and useful, independent and consistent, strongly hold the old values as well as accommodative to the new elements.

As time passed, *Bahrul 'Ulum Pesantren* grew rapidly. Both dormitory and education units kept increasing in number. The number of students also increased to approximately 7,000 people. In 2014, in the 189th age of the *pesantren* and at the 99th age of the *madrasa*, *Bahrul 'Ulum Tambak Beras Pesantren* in Jombang had 21 boarding school dormitory units and 19 units of formal education from pre-school level to university level, which were both under the auspices of the Ministry of Religious Affairs and the Ministry of Education. *Bahrul 'Ulum Pesantren* also had a university called *Wahab Hasbullah* University (Ahdi, interview, 22 December 2013). There are also several modern educational institutions under *pesantren* adjusted by society need, such as:

1. *Sekolah Teknik Menengah Bahrul 'Ulum* (Secondary Technical School)
2. *Sekolah Menengah Kejuruan Teknik Informatika Komputer An-Najiyah Bahrul Ulum* (Vocational School of Information and Computer Science)
3. *Sekolah Tinggi Ilmu Keperawatan Bahrul Ulum* (School of Nursing)
4. *Sekolah Tinggi Ilmu Manajemen Informatika dan Komputer Bahrul 'Ulum* (School of Information Management and Computing)

In addition to the formal schools mentioned above, *Bahrul Ulum Pesantren* is required to follow the teachings of *Salaf Kitab* held in each dormitory unit. That way, the traditional *sorogan* or *bandongan* education system can still be maintained through various activities such as *takror* (Quran reading), *tadarrus* (Quran reviewing), and *tahfidz* (Quran memorising). Furthermore, to equip the students with useful skills to be able to engage with the community, each unit also conducts extra-curricular activities such as; *pesantren* magazine, a marching band, computing, sewing, electronics, tambourine art, *qasida* art (vocal art), fashion, culinary skills, martial art, scouts, youth red cross (PMR), school health unit (UKS), and a youth scientific group. Additionally, the *pesantren* also organises regular training and extra religious activities such as: foreign language training, research, leadership training, library science, organisational science, community advocacy, entrepreneurship, the rituals of Hajj, the art of reading the Quran, sermons, speeches, *Bahtsul Masail* and others (Ahdi, interview, 22 March 2016).

2.3 *Bahrul 'Ulum Pesantren: What changed and what survived*

Historian Arnold J. Toynbee argued in his book, A Study of History that a civilisation will not survive if it cannot meet the challenges of the social environment. The growth of a civilisation occurs not only when the society successfully overcomes a particular challenge, but also by being able to answer the next challenge. Civilisation emerged as an answer to a set of challenges of extreme difficulty, when a 'creative minority' reoriented the entire society. In this context, *pesantren* can be identified as a 'creative minority', that is a group of people or even individuals who have 'self-determining'. With the existence of the creative minority, a group of people will move out of archaic and primitive societies. Civilisation will thrive and survive when the creative minority find a challenge and then respond and find a way out and innovate (Rahardjo, 2002).

As mentioned in the previous description, the *pesantren* is generally regarded as a 'traditional institution'. Referring to the maintained *sorogan* and *badongan* model of education in a *pesantren* unit, *Bahrul 'Ulum Pesantren* can be considered as a traditional *pesantren*. However, this does not mean that *Bahrul Ulum Pesantren* does not follow the changing era. As an educational institution, which is labelled as 'traditional', *Bahrul Ulum Pesantren* has proven its ability to keep pace with the dynamics of education in particular and the dynamics of life in general. If we look at the history of *Bahrul Ulum Pesantren*, this *pesantren*, which at the beginning of its establishment focused on *Shari'ah* and *Sufism,* has been developed significantly. As the Islamic educational institutions in Indonesia are in a phase of modernisation, especially after the founding of the *Muhammadiyah* organisation in 1912, *Bahrul 'Ulum Pesantren* did not remain stagnant. In that era, boarding schools also reformed the education system by establishing *madrasa*s without leaving the *sorogan* or *badongan* old education model. Similarly, at the beginning of the New Order government, when the government adopted a policy to convert *madrasa* into public schools under the Ministry of Religious Affairs, *Bahrul 'Ulum Pesantren* accommodated these policies by establishing *Madrasah Aliyah Negeri* (MAN) and *Madrasah Tsanawiyah Negeri* (MTsN) without removing the existence of *Muallimin and Muallimat Madrasa* (Ahdi, interview, 22 March 2016).

A reform in educational aspects of the *pesantren* does not mean that the old educational pattern has been changed into a new one. The *pesantren* can keep running the old pattern of education, while at the same time implementing the new, better system. At the present time, the growth of a wide variety of new educational units in *Tambak Beras Bahrul 'Ulum Pesantren*, either the one under the Ministry of Religious Affairs, does not mean that *Bahrul 'Ulum Pesantren* abandons the characteristics of the traditional *pesantren* education, which are *badongan* and *sorogan*. It is exactly in line with the Arabic proverb: المحافظة على القديم الصالح والاخذ بالجديد الاصلح (Preserving the good of the old and taking the better of the new).

On the other hand, the traditional label which is attached to the education system of the *Nadhlatul Ulama* (NU) community is not only because of the education pattern that maintains the tradition. It is called 'traditional' because the lessons taught are Islamic knowledge which refers to the *Ash'arite aqidah*, based on *Syafi 'i fiqh* and adheres to Imam Al-Ghazali's teaching. Moreover, the modernist's one has made Mohammad Abduh, Jalaluddin al-Afghani as inspirator in conducting change and renewal. Modernist circles also make Ibn Taymiyyah as a reference in understanding the Quran and Al-Hadith. Therefore, the traditional *pesantren* style of education exists among the NU community while the *Muhammadiyah* community establishing formal education which become the model of modern education institution.

The rapid and up-to-date development of the *pesantren* cannot be separated from the role of the *Kyai*. The educational background and depth of knowledge of the *Kyai* will determine the development direction of the *pesantren*. A knowledgeable *Kyai* will bring the *pesantren* into a good change which follows the occurring dynamic outside the *pesantren*. Therefore, in this case, the role of the *Kyai* is not as a cultural broker as pointed out by Clifford Geertz, but as the agent of cultural change. Hiroko Horikoshi also stated that *Kyais* have a very strategic position in the social system of Islam in Indonesia. In addition to being cultural brokers, *Kyais* are the motors of change as well as being an inspiration and mediators in the community (Horikoshi, 1987).

3 CONCLUSION

Analysing the old-maintained education pattern, the *pesantren* could indeed be categorised as a traditional educational institution. However, the *pesantren* cannot be regarded as a 'left behind' entity which neither conducts any changes nor follows the changing era. In the case of *Bahrul 'Ulum Tambak Beras Pesantren* in Jombang, tradition and modernity exist altogether. *Bahrul 'Ulum Pesantren* maintains the *sorogan* and *bandongan* old teaching patterns which are held in dormitory units in the student's residency whilst also accommodating the needs of a global-facing modern society by establishing new educational units under the Ministry of Religious Affairs and the Ministry of Education and Culture. Therefore, the old education model that is considered good and useful for the *pesantren* will be maintained despite its old-fashioned or traditional label. Moreover, the modern education model will still be implemented despite its general knowledge category because it is considered as good and useful.

REFERENCES

Ahdi, Wafiyul. Caregivers of Annajiyah Bahrul Ulum Tambak Beras Jombang, Kyai Abdurrahim Hasbullah's grandson. (Interview, 22 March 2016).
Direktorat Jenderal Pendidikan Islam Kementerian Agama RI. (2011). *Data Pesantren di Indonesia.* Retrieved from http://pendis.kemenag.go.id/databooklet3a-1.pdf.
Ekayati & Toynbee, A.J. (1996). *In Ensiklopedia Nasional Indonesia (Jilid 16).* Jakarta, Indonesia: PT Cipta Adi Pustaka.
Fitriana, H.I. (2012a). *Modernisasi Sistem Pendidikan di Pondok Pesantren Bahrul Ulum Tambakberas Jombang Tahun 1915–1971.* Retrieved from http://karya-ilmiah.um.ac.id/index.php/sejarah/article/view/24869.
Fitriana, H.I. (2012b). Modernisasi sistem pendidikan di Pondok Pesantren Bahrul Ulum Tambak beras Jombang tahun 1915–1971. *Penelitian Jurusan Sejarah,* Fakultas Ilmu Sosial UM, Malang.
Ghofir, J. (2012). *Biografi Singkat Ulama Ahlussunnah Waljama'ah.* Tuban, GP Anshor.
Hadzik, M.I. (2007). *Pustaka Warisan Islam.* Jombang, Tebuireng.
Horikhoshi, H. (1987). *Kiai dan Perubahan Sosial* (D. Effendi & M. Azhari, Trans). Jakarta, Indonesia: LP3ES.
Murtaufiq, S. (2006). (Tradisi) Pesantren di mata Martin Van Bruinessen. *Jurnal Pesantren Ciganjur, 2*(1), 40–48.
Ponpes Bahrul Ulum. (2016). *Sejarah, Nama, dan Lambang Pondok Pesantren Bahrul Ulum.* Retrieved from http://tambakberas.or.id/profil/bahrul-ulum/sejarah-nama-dan-lambang-pondok-pesantren-bahrul-ulum/.
Rahardjo, D. (1995). *Pesantren dan Pembaruan.* Jakarta, Indonesia: LP3ES.
Rahardjo, S. (2002). *Peradaban Jawa; Dinamika Pranata Politik, Agama, dan Ekonomi Jawa Kuno.* Jakarta, Indonesia: Komunitas Bambu.
Ricklefs, M.C. (2013). *Mengislamkan Jawa.* Jakarta, Indonesia: Serambi.
Santoso, P. (1988). Kiprah Pesantren dalam transformasi: Catatan dari Maslakul Huda. *Jurnal Pesantren, 3*(5), 80–87.
STIKES Bahrul Ulum. (2016). *Stikes Bahrul Ulum: Perguruan Tinggi Unggul Dengan Nilai-nilai Kepesantrenan.* Retrieved from http://stikes-bu.ac.id.
STMIK Bahrul Ulum. (2011). *Sejarah STMIK Bahrul Ulum.* Retrieved from http://www.stmikbu.ac.id/index.php/2011–09–27–04–13–15/kampus/sejarah.
Van Bruinessen, M. (1999). *Kitab Kuning Pesantren dan Tarekat.* Bandung, Indonesia: Mizan.
Yayasan Pondok Pesantren Bahrul Ulum (2016). Retrieved from http://tambakberas.or.id/profil/bahrul-ulum/periode-pengembangan-sejarah-pondok-pesantren-bahrul-ulum/.

Competition and Cooperation in Social and Political Sciences – Adi & Achwan (Eds)
© 2018 Taylor & Francis Group, London, ISBN 978-1-138-62676-8

Citizen engagement: An approach to sustaining Indonesian rural water supply and sanitation?

R.Y. Kasri
Department of Environmental Sciences, Universitas Indonesia, Jakarta, Indonesia

S.S. Moersidik
Department of Civil Engineering, Faculty of Engineering, Universitas Indonesia, Depok, Indonesia

ABSTRACT: Community participation and empowerment have positively contributed to achieving development goals. Better engagement between citizens and government through an increase in ownership, participation, and reciprocal action, particularly in the rural water and sanitation sector has been implemented since 2008, which has led to greater achievements in access and service delivery in Indonesia. This paper will elaborate on the transformation in the development of policy and practice in the rural water supply and sanitation sector from Community Demand-Driven (CDD) approach to citizen engagement approach to further define citizen engagement in this sector. Taking into consideration that sustainable service delivery of water and sanitation does not only refer to quantity of access but also quality, continuity, affordability, and environmental protection, through elaborating the works of two mainstreamed government programmes namely Community-Based Water Supply and Sanitation Programme (Pamsimas and Community-Based Total Sanitation (STBM), this paper will elaborate on how citizens' engagement could lead to sustained service delivery of rural water supply and sanitation in rural Indonesia.

1 INTRODUCTION

1.1 *Sustainable Rural water supply and sanitation: Global mandates and Indonesia's progress*

Indonesia has successfully met the Millennium Development Goal's (MDGs) target to rural water supply with 79 percent of the population having access in 2015. Although slightly missing the sanitation target, the access increased sharply from 24 percent in 1990 to 47 percent in 2015 (UNICEF and WHO, 2015). In 2010, Indonesia ratified the United Nations Resolution No. 64/292 on Human Right to Water and Sanitation (UN, 2010). Indonesia also adopted the Sustainable Development Goal's (SDGs) agenda in 2015. Agenda number six is to ensure universal and equitable access to safe and affordable drinking water for all, achieve access to adequate and equitable sanitation and hygiene for all and end open defecation, improve water quality substantially increase water-use efficiency, implement integrated water resource management, protect and restore water related ecosystem, and expand international cooperation and capacity building support (UNDESA, 2015). SDG's target is beyond providing access. It aims to ensure the delivery of quantity, quality, continuity, and affordability of water and sanitation (WASH) facilities for all people as well as to protect the environment.

The Government of Indonesia (GoI) acknowledged a number of issues in increasing access to and sustaining WASH. They are low functionality infrastructures, depleting raw water, lack of synergy in the WASH development programme, lack of community awareness and empowerment, and low commitment of the local governments (Presiden RI, 2015). These issues have been known and continuously addressed. In rural WASH, they have been addressed since 2008, ultimately through shifting the development approach from subsidy-based to community-based along with the strengthening of the government's role and function in integrated water and sanitation programmes. It has had an impressive result. Statistics Indonesia reported a five-fold increase in access to rural water, from 0.54 percent annually (from 2000 to 2008) to 2.51 percent (from 2008 to 2015), while access to sanitation increased from 1.63 percent to 2.2 percent at the same period (Statistics Indonesia, 2016). However, Indonesia still needs to work four times and six times faster than the current rate should it want to reach universal access to water and sanitation in 2019, as targeted in the Indonesia Mid Term Development Plan 2015–2019, and to further sustain WASH services. Hence, a better approach to sector development is needed.

1.2 *Method and objectives*

The goal of rural WASH development, which is in line with sustainable development, is not only to increase access, but also to improve and sustain the quality of services for people and to protect the planet. The transformation of the development approach from subsidy-based to community-based along with the strengthening of the government's role and function, which is in line with the rejuvenated social concept of 'citizen engagement', has accelerated the access. However, there are still many challenges to providing access to all citizens and to sustaining services.

The citizen engagement concept is not new in social science and is considered a better way to sustain the outcomes of the development programme. This concept has been mainstreamed in the development sector since 2008. What citizen engagement in rural WASH is and how it leads to sustaining service delivery in this sector are the two main questions to be answered in this paper. Through secondary desk review utilising thematic narrative analysis focusing on project papers, regulations, evaluation reports, and academic publications of two mainstream rural WASH programmes, namely Community-Based Water Supply and Sanitation (Pamsimas) and Community-Based Total Sanitation (STBM), this paper aims to define citizen engagement in rural WASH and to assess how it could lead to the sustainability of service delivery.

2 DISCUSSION

2.1 *Shifting development approach from community-based to citizen engagement: A globally driven local policy*

The year 2008 has marked an essential transformation of development policy and practice in rural WASH. Indonesia previously implemented a geographic-based, subsidised construction approach through a number of programmes such as the Health Construction Aid Programme, known as SAMIJAGA, which started in 1974. Through this programme, the government provided free latrines and water facilities for poor villages. In 1993, Indonesia shifted its approach, from top-down to bottom-up, through the implementation of the Water Supply and Sanitation for Low Income Communities (WSSLIC) project. The WSSLIC project (1993–1999) targeted poor communities with high diarrheal diseases and no access to safe water resources in 2,209 villages. It was designed to enable community participation in the end-to-end project cycle, from design to implementation and maintenance., Although contributing substantially to better access, this programme was not successful in establishing major behavioural change in sanitation or hygiene, or in encouraging active participation of women, or delegating community control over the WASH systems. Other issues such as administrative issues and fund channelling remained partially unresolved (Bhatnagar, 2000).

Learning from WSSLIC and other rural infrastructure projects, the second WSSLIC (2000–2007) adopted a decentralised, community demand-driven (CDD) approach that channelled funds and technical assistance directly to community groups in villages, and incorporated self-monitoring by communities (World Bank, 2002). The CDD approach operates on the principle of transparency, participation, local empowerment, demand-responsiveness, greater downward accountability, and strengthening capacity of local people (World Bank,

2016). It assumes that higher participation leads to better and more sustainable outcomes (Mansuri & Rao, 2004; Briones, 2014). However, recent studies such as the meta-evaluation carried out by Mansuri and Rao (2004, 2013) of 500 World Bank CDD projects from 1993 to 2006, Rocha-Menocal and Sharma's (2008) study of 90 community-based projects in 7 countries, and Gaventa and Barret's (2012) study of 100 similar projects in 20 countries found that CDD has had mixed results. It has been successful in constructing infrastructures, increasing health quality, and encouraging behavioural change towards more transparent and accountable community planning and financing. However, it has not been effective in addressing the needs of the poor communities. It relies heavily on outside facilitators or community elites. There is no direct causality between community voices and accountability with development outcomes. It has rejuvenated social structure, such as caste and gender segregation, in order to achieve social acknowledgement or to own resources which triggers social conflict.

Implemented in 2,350 villages in 37 districts from 2000 to 2007, the WSSLIC-2 was successful in increasing access to clean water and improved sanitation. However, it was criticised for its inability to establish better behavioural change in sanitation and hygiene, promote environmental protection, improve project cycle and mechanisms to enable community participation and empowerment, ensure sustainability of service delivery, and improve inter-sectors coordination (LP3ES, 2007; Puslitsosekling Kementerian PU, 2010).

In response to the inability of the sector and hardware based programmes to encourage behaviour change and increase sanitation access, the Ministry of Health (MoH) piloted a Community-Led Total Sanitation (CLTS) approach in 2005 (MoH, 2008). CLTS is an approach that triggers demand and empowers rural communities to eliminate open defecation (OD) through collective sanitation behaviour without government subsidies for facilities (Kar, 2012; Chambers, 2009). It was introduced in Bangladesh in 1999 and according to the CLTS Knowledge Hub supported by the Institute of Development Studies, UK, the CLTS has currently been implemented in 48 countries. While CLTS focused on demand, a project named Total Sanitation and Sanitation Marketing (TSSM) was piloted in East Java in 2007 to find ways of providing supply for demand on new sanitation and hygiene behaviour such as latrines and hand washing with soap facilities. The same project was also implemented in India and Tanzania with the support from the World Bank Water and Sanitation Programme (Perez, 2011). In East Java, within the first 10 months of its implementation, 375 villages asked for CLTS intervention forwarding triggering method, 328 of which had been triggered, resulting in 262 Open-Defecation-Free (ODF) villages. The project indicated that demand and supply would be able to increase access and change behaviour only in a supportive policy environment provided by the national and local governments (Mukherjee & Wartono, 2008).

In 2008, the MoH enacted Decree No. 852 to shift the rural sanitation development approach from subsidy-based to a Community-Based Total Sanitation Programme or *Sanitasi Total Berbasis Masyarakat* (STBM). Applied nationally and led by the MoH, this programme banned subsidies for household latrines. It was implemented through three strategies: 1) creating demand through community empowerment and behavioural change, 2) increasing the supply of affordable and appropriate sanitation facilities by the private sector, and 3) enhancing availability of supportive policy environment for STBM implementation (MoH, 2008). In line with this, the GoI, led by the Ministry of Public Works (MoPW), adopted the STBM approach for sanitation component of the third WSSLIC in addition to the CDD approach in 2008. The third WSSLIC project was renamed '*Program Penyediaan Air Minum dan Sanitasi Berbasis Masyarakat*' or Pamsimas–1.

Tightened collaboration and increased ownership among national government, local government and communities have been successful in increasing access to rural sanitation, changing people's behaviour, and increasing financing for the sector. In STBM triggered communities in East Java, for example, from 2008 to 2012, the government's investment in latrine construction was IDR 20 billion while community invested IDR 139 billion (Pamsimas, 2012). The Pamsimas-1 programme was criticised for its inability to sustain infrastructures and to deploy high quality facilitators to strengthen community to provide quality service and conduct behaviour change (DitPAM, Kementerian Pekerjaan Umum, 2013). Engaging the government and community in the development might be a solution to address this criticism. Globally, there have been many studies advocating a transformation of the development approach from community-based to a collaboration between the community and the government, including Mansuri and Rao (2013), whose research found that most of the successful programmes were those implemented or led by accountable local government.

Convinced by the STBM's achievements and influenced by global evidence, in 2013 the GoI enriched the PAMSIMAS approach by complementing community-based (CDD and STBM approach) with collaborative actions from government institutions. Specifically, the sanitation component of Pamsimas-2 (2012–2015) adopted a district-wide approach to provide a demarcation of the roles and functions between government institutions and community led by local government. A recent enactment of Law No. 23/2014 on Local Government and Law No.6/2014 on Village have provided opportunity to carry out this rejuvenated collaborative approach between the government and community. To accelerate universal access to water and sanitation, targeted to be in place by 2019, the GoI reaffirmed collaboration between national government, local government and the community with specified roles, functions and responsibilities for the Pamsimas-3 programme (2016–2020). Conceptually, this new approach is known as citizen engagement. The transformation of the approach is shown in Table 1.

2.2 Mapping and defining citizen engagement in rural water supply and sanitation programmes

Citizen engagement is not a new concept. It builds on the interrelation between government and its citizens. A citizen is a person who legally belongs to a country based on bloodline, place of birth, etc. who has responsibilities and full rights as a member of that country (KBBI, 2015). Citizens give a mandate to the government to govern and fulfil their needs and, at the same time, are obliged to seek the government's accountability and ensure that they are working to meet their needs (Roberts, 2003; Mansuri & Rao, 2013; McNeil, 2016). This is different to community participation and empowerment, which is usually a one-way approach carried out by the government or external parties to empower people, citizen engagement is 'a two-way interaction between citizens and the government or private sector that gives citizens a stake in decision-making, with the objective of improving development outcomes' (McNeil, 2016). It is a system where citizens have an equal position to the government in the decision-making process to ensure that the government adheres to its responsibility and accountability to provide public goods and services (Gaventa & McGee, 2015). The main proposition of this concept is that the engagement of well-informed citizens will increase the performance of the public

Table 1.   Transformation of approach in rural water supply and sanitation development in Indonesia.

Period	Approach transformation
Local policy	
1974–1993	Geographic-based subsidy programme (SAMIJAGA)
1993–2000	Community participation (WSSLIC-1)
2000–2007	Community Demand-Driven/CDD (WSSLIC-2)
2005–2007	Piloting Community-Led Total Sanitation (CLTS)
2008	Enactment of the MoH Decree on Community-Based Total Sanitation (STBM)
2008–2012	CDD and STBM (PAMSIMAS-1)
2013–2015	CDD, STBM, and government institutions collaboration (PAMSIMAS-2)
2014- now	Citizen engagement (enactment of the MoH Regulation on STBM and PAMSIMAS-3)
Global initiatives	
Late 1990s–2008	Ineffective community participation and community demand-driven approach to meet development goal
2008- now	Rejuvenating citizen engagement approach

Source: Author.

sector and create a good governance system (Fox, 2014, MAV, 2012). It is built on a number of values: accountability, transparency, responsiveness, equality and inclusiveness, adherence to laws/regulations, effectiveness and efficiency and, most importantly, all of these values can only be implemented if both the citizens and the government participate in the decision-making process (MAV, 2012). Citizens and community are used interchangeably in this paper.

Referring to the transformation of approach as explained earlier, the 'adoption' of the citizen engagement approach has already influenced the Pamsimas and STBM programmes as reflected in:

- **Strengthened institutional framework.** A clear demarcation of the roles of the government at all levels and the roles of the citizens is enacted in the Ministry of Health Regulation No.3/2014 on STBM and in the Pamsimas' project manual 2013 and 2016. For an internal control system, the Finance and Development Supervisory Agency (BPKP) released a report on 30 December 2014, specifying roles and responsibilities of the government institutions involved in the Pamsimas programme (BPKP, 2014).
- **Accessible and open channels for engagement and monitoring.** Both the Pamsimas and STBM programmes have an open access implementation. Pamsimas launched a monitoring and information system in 2009, which can be accessed at http://pamsimas.org/., Pamsimas also provides a direct channel for grievance and redress through email and SMS. STBM launched the SMS and web based monitoring system in 2011, which can be accessed at http://stbm-indonesia.org/monev/. Updating data for STBM is done by sanitarians and government officers, while updating data for Pamsimas is done by consultants hired by the local government.
- **Empowered and heard citizens.** Both programmes are conducted based on the demand of communities following information provided by the government through a comprehensive communications strategy. This strategy includes, among others, reaching out through systematic government socialisation, public hearing, media and social media campaigns dissemination of promotional materials, etc.
- **Collaborative programme implementation.** While in Pamsimas and 2, citizens are facilitated by consultants to identify WASH problems and develop a proposal to be considered as a programme recipient. In Pamsimas-3 they prepare a proposal with the assistance from community cadres in coordination with the village government. Issues regarding the identification, planning, implementation, maintenance and sustainability of services are defined and carried out in a collaborative manner. Moreover, to ensure the sustainability of service delivery, Pamsimas establishes community-based operators (CBOs), namely *Badan Pengelola Sarana Air Minum dan Sanitasi* (BPSPAMS). Technical and organisational support to the BPSPAMS is provided by the local government, programme facilitators and the BPSPAMS' association.
- **Shared responsibility and financing.** The government contributes 70 percent of the funding for the Pamsimas programme, while 10 percent comes from the village fund and 20 percent from the community, which consists of 4 percent in cash and 16 percent in-kind (Pamsimas, 2016). With a no subsidy approach, the STBM programme requires the community to self-finance their own latrines, while the government contributes in financing software such as campaigns, capacity building such as training and learning and policy enabling. Increased ownership of the facilities is expected to sustain both the facilities and behaviour towards rural WASH.

Based on the discussion above, a citizen engagement approach requires not only an interaction but also an equal and reciprocal relationship between citizens and their government which is crystallised into a commitment to work together and share responsibility to achieve a mutual goal of better and more sustainable development outcomes.

2.3 *Citizen engagement and sustaining service delivery*

Referring to the SDG's agenda on WASH, the development of this sector is aimed at not only providing access, but also ensuring delivery of quantity, quality, continuity and affordability of services for people, and to protect planet. In Pamsimas and STBM, the transformation of the approach towards citizen engagement has increased the quantity of access; however, other aspects of the service delivery may have been left behind.

Reviewing the MoH Regulation No.3/2014 on STBM, the target for sanitation development consists of five pillars. Pillar one is to sustain open-defecation-free behaviour, to terminate faecal contamination and to provide and maintain improved latrines. Pillars two to five refer to healthy behaviour in hand washing with soap, clean and healthy water and food processing and safe disposal of solid waste and liquid waste water (MoH, 2014). Latrine quality standards are clearly defined in the regulation; however, there is no further information on requirements, methods or efforts to promote and expand the services to close the sanitation loop. Sustainable Sanitation Alliance (SuSanA) argued that sanitation is a holistic system and its sustainability should be addressed to close the loop of the existing system by considering the sustainability aspects of health and hygiene, environment, natural resources, technology, economics, social cultural and institutions (Susana, 2008). Environmental sanitation and hygiene health issues beyond household scale are pretty much left behind in the STBM although some initiatives, such as piloting a scheduled waste water disposal service, communal latrine pit emptying service, and fertiliser-based faecal production, are promoted and developed by donor agencies such as the Water and Sanitation Programme and IUWASH.

Sustainability has been continuously promoted by the Pamsimas programme; however, it is only assessed through the functionality of the infrastructures and performance of the BPSPMS, ultimately in applying a tariff (Pamsimas, 2013). Efforts to sustain water resources, protect the environment, mitigate natural disasters, social protection and gender balance are promoted by the programme, but not systematically measured and not yet seen as factors influencing service sustainability. With regards to service delivery for water supply, the GoI has already issued a service standard which has been adopted by the Pamsimas programme (Pamsimas, 2016). The standard is as follows:

Based on Law No. 23/2014 on Local Government, service delivery of WASH is the local government's mandate. However, there are unclear procedures and insufficient resources to ensure the application of this standard. For example, on water quality checking, although the MoH had 10,500 sanitarians in 2014, 30 percent of the 9,599 community health centres, which carried out the mandate for water quality's inspection, did not have sanitarians (MoH, 2014). At the service provider level, Pamsimas requires BPSPAMS to conduct water quality checking before and after facility construction (Pamsimas 2016). The BPSPAMS is also advised to conduct regular quality checking, but there is no incentive for doing or not doing it. With regards to continuity and affordability, there is no detail mechanism to ensure that this is implemented in the rural community.

Noting that there are still 47 million rural people without access to improved water and 63 million without access to sanitation, large-scale investment is needed for the facilities. The Service Delivery Assessment has estimated a capital expenditure of US$772 million to achieve universal access to rural water, which is mostly expected to come from the government, and US$442 million for rural sanitation, which will mostly come from households

Table 2. Service standard for water supply services.

Component	Standard	Regulation
Quantity	60 litre/head/day	Minister of Public Works Regulation No.1/2014
Quality	Meet physical, biological, chemical and radioactive standard	Minister of Health Regulation No. 492/2010
Continuity	24 hours/day	Minister of Home Affairs Regulation No. 23/2006
Affordability	Tariff should not exceed 4 percent of customers' income	Minister of Home Affairs Regulation No. 23/2006

Source: Author.

(WSP, 2015). According to the water and sanitation Public Expenditure Review, however, total annual investment in water and sanitation (from all sources, including private) since 2009 represents only 0.2 percent of the GDP (World Bank, 2015).

Indonesia is still a long way from sustaining rural WASH as defined by the SDGs, ultimately because of the sustainability indicators, although the standards have been defined nationally, they have not been clearly translated for implementation. There is no strong evidence that citizens have been well informed and engaged to sustain the services beyond household provision and maintenance of facilities and adherence to the water tariff.

Further, considering that a citizen engagement approach has been effective to drive collaborative actions between citizens and the government to accelerate access in both programmes, it is time to improve the access to other components of sustainability, particularly quality, continuity, affordability of services, and environmental protection. The government should inform citizens of sustainability indicators through various communication channels, forwarding messages that fit their sociocultural condition, and collaborate with citizens towards better and more sustainable outcomes, for both people and the environment. Shared responsibility, mutual commitment, as well as incentives and disincentives should be defined and monitored together. With regards to the high cost of providing facilities and sustaining the services, the government and citizens might expand their engagement to include the private sector, for example to provide cheap but standard quality drinking water, including bottled water, latrines, waste water disposal services, faecal based fertiliser or renewable energy. Moreover, considering that the critical aspect of water supply is the availability of water sources, the citizen engagement approach should be expanded beyond a village administration as implemented by the Pamsimas programme to a multi-village administration, following water sources.

3 CONCLUSIONS AND RECOMMENDATIONS

The transformation of the development approach from a community-based approach to a citizen engagement approach has contributed to accelerating access to rural water supply and sanitation. To work well, a citizen engagement could be done by strengthening institutional framework provided accessible and open channels for both government and citizens to collaborate in the program with shared responsibility and financing. An empowered citizens and responsive government is required. Transformation of the approach is taking place in the context of rural water supply and sanitation in Indonesia. However, there is still a lot to be done to sustain the service delivery in accordance with the standards for sustainable service indicators on quantity, quality, continuity and affordability of services, as well as environmental protection.

In-line with the government's target to provide universal access to safe water and improved sanitation in 2019, there is a need to improve and sustain rural water supply and sanitation services, not only in access to quantity but also the other indicators. To do that, the government should inform citizens of sustainability indicators through various communication channels, forwarding messages that fit their sociocultural condition and collaborate with them towards better and more sustainable outcomes, both for people and the environment.

Moreover, the citizen engagement concept has just been rejuvenated in development and academic works. Sharing knowledge from Indonesian experiences will be essential and fruitful to better understand the implementation of this concept and further to enrich global initiatives to meet the SDGs.

REFERENCES

Bhatnagar, D., Dewan, A., Torres, M. & Kanungo, P. (2000). *Water supply and sanitation for low income communities*. Indonesia: The World Bank.
BPKP. (2014). The PAMSIMAS II Integrated Program Controlling System (*Sistem Pengendalian Keterpaduan Program PAMSIMAS II*). Jakarta, Indonesia: Badan Pengawasan Keuangan dan Pembangunan (The Indonesia's National Government Internal Auditor/BPKP).

Briones, A.V. (2014). *Social accountability: An approach to good governance*. Manila, Philippines: Affiliated Network for Social Accountability in East Asia and the Pacific.

Chambers, R. (2009). *Going to scale with community-led total sanitation: Reflections on experience, issues and ways Forward* (IDS Practice Paper 2009 (1)). Brighton, UK: Institute of Development Studies.

DitPAM. (2013). Progress on Rural Water Supply System and Pamsimas (*Capaian Program Pengembangan SPAM Perdesaan dan PAMSIMAS*). Jakarta, Indonesia: Virtual Information Centre on Water Supply and Sanitation, Directorate General of Human Settlements, Ministry of Public Works (Pusat Informasi Virtual Air Minum dan Penyehatan Lingkungan, Direktorat Jenderal Cipta Karya, Kementerian Pekerjaan Umum). Retrieved from http://www.slideshare.net/OswarMungkasa/dir-pam-capaian-spamdesaampl20ags13–1.

Fox, J. (2014, September). *Social accountability: What does the evidence really say?* (GPSA Working Paper No. 1). Washington DC, USA: Global Partnership for Social Accountability.

Gaventa, J. & Barret, G. (2012). Mapping the outcomes of citizen engagement. *World Development*, *40*(12), 2399–2410. doi: 10.1016/j.worlddev.2012.05.014.

Gaventa, J. & McGee, R. (2015). *The impact of transparency and accountability initiatives*. London, UK: DFID.

Kar, K. (2012). Why not basics for all? Scopes and challenges of community-led total sanitation. *IDS Bulletin, 43*(2), 93–96. Oxford, UK: Blackwell Publishing Ltd.

KBBI. (2015). *Kamus Besar Bahasa Indonesia Daring. Online Indonesian Dictionary*. [Online] Pusat Bahasa Kemendiknas: Retrieved from http://kbbi.web.id.

LP3ES. (2007). *Independent evaluation WSLIC-2 and PAMSIMAS Program*. Jakarta: Lembaga Penelitian, Pendidikan dan Penerapan Ekonomi dan Sosial. Retrieved from http://siteresources.worldbank.org/INTINDONESIA/Resources/Projects/288973–1224059746389/INDEPENDENT.EVALUATION.WSLIC-2.AND.PAMSIMAS.PROGRAMS_en.pdf.

Mansuri, G. & Rao, V. (2004). Community-Based and driven development: A critical review. *The World Bank Research Observer*, *19*(1), 1–39. 10.1093/wbro/lkh012.

Mansuri, G. & Rao, V. (2013). *Localizing Development: Does Participation Work*? Washington D.C.: The World Bank.

Mc Neil, M. (2016). *A game changer for development? Citizen engagement: History and definitions* [Motion Picture]. The World Bank. Retrieved from https://olc.worldbank.org/content/engaging-citizens-game-changer-development-1.

Menocal, A.R. & Sharma. (2008). *Joint evaluation of citizens, voice & accountability*. London, UK: ODI, DFID.

Ministry of Health Indonesia. (2008). *National Strategy on Community Based Total Sanitation* (Minister of Health Decree No. 852/2008). Jakarta, Indonesia: Kementerian Kesehatan RI.

Ministry of Health Indonesia. (2014). *Community Based Total Sanitation* (Minister of Health Regulation No.3/2014). Jakarta, Indonesia: Kementerian Kesehatan RI.

Ministry of Health. (2010). *Requirements for Drinking Water Quality* (Minister of Health Regulation No. 492/Menkes/per/IV/2010). Jakarta, Indonesia: Kementerian Kesehatan.

Ministry of Health. (2014). *Bank Data*. Retrieved from http://www.bankdata.depkes.go.id/.

Ministry of Home Affairs. (2006). *Technical Guideline and Procedures to Regulate Water Tariff* (Minister of Home Affairs Regulation No. 23/2006). Jakarta, Indonesia: Kementerian Dalam Negeri.

Ministry of Public Works. (2014). *Minimum Service Standard for Public Works and Spatial* (Minister of Public Works Regulation No.01/PRT/M/2014). Jakarta, Indonesia: Kementerian Pekerjaan Umum.

Mukherjee, N., Wartono, D.. (2008, November 19). Indonesia: Indigenous strategies for scaling up CLTS. *IDS Global Knowledge*. Retrieved from www.ids.ac.uk.

Municipal Association Victoria. (2012). *Good Governance*. The Municipal Association of Victoria (MAV), Victorian Local Governance Association (VLGA), Local Government Victoria (LGV) and Local Government Professionals (LGPro). Retrieved from http://goodgovernance.org.au/about-good-governance/what-is-good-governance/.

Pamsimas. (2012). Handbook for Pamsimas II: Health Component, District-Wide STBM Approach (*Buku Saku PAMSIMAS II: Komponen Kesehatan, Pendekatan STBM Skala Kabupaten (District-Wide)*. Jakarta, Indonesia: Sekretariat CPMU PAMSIMAS.

Pamsimas. (2013). *Glossary Logbook*. Jakarta, Indonesia: Sekretariat PAMSIMAS.

Pamsimas. (2016). Pamsimas Program General Guidance (*Pedoman Umum Program Pamsimas*). Jakarta, Indonesia: Sekretariat PAMSIMAS.

Perez, E. (2011). *Sustainable rural sanitation at scale: Lessons and results from India, Indonesia, Ethiopia and Tanzania*. Water and Sanitation Program, the World Bank.

Presiden RI. (2015). The Second Attachment to the Presidential Regulation No 2/2015 on National Mid Term Development Plan 2015–2019 (Buku Kedua Lampiran Peraturan Presiden No. 2/2015 tentang RPJMN 2015–2019). Jakarta, Indonesia: Sekretariat Negara.

Puslitsosekling Kementerian PU. (2010). The Final Report of the Pamsimasn Social Economic Assessment (*Laporan Akhir Kajian Sosial Ekonomi Pengelolaan PAMSIMAS*). Jakarta, Indonesia: Centre for Social Economic and Environment Research, Ministry of Public Works.

Republic of Indonesia. (2014a). Law No 23/2014 on Local Government.

Republic of Indonesia. (2014b). Law No. 6/2014 on Village.

Roberts, N. (2003). Direct Citizen Participation: Building a Theory. *7th National Public Management Reseach Conference, October 9–11.* Washington D.C.: Georgetown University.

Rocha Menocal, A dan Sharma, B. (2008). *Joint Evaluation of Citizens' Voice and Accountability, Synthesis Report.* London: DFID.

Statistics Indonesia. (2016.) *Statistik Indonesia 2016.* Jakarta, Indonesia: Badan Pusat Statistik.

Susana. (2008). *Sustainable sanitation—the definition.* Retrieved from Sustainable Sanitation Alliance (SuSanA). www.susana.org.

UNDESA. (2015). *Transforming our world: the 2030 agenda for sustainable development.* Retrieved from: https://sustainabledevelopment.un.org/post2015/transformingourworld.

UNICEF & WHO. (2015). *25 Years Progress on Sanitation and Drinking Water, 2015 Update and MDG Assessment.* Retrieved from: http://www.who.int/water_sanitation_health/monitoring/jmp-2015-update/en/.

United Nations (UN). (2010) *The Human Right to Water and Sanitation* (Resolution No. 64/292). United Nations General Assembly.

World Bank. (2002, April 15–19). *WSSLIC 2 project, Indonesia. Gender and socially inclusive form of community organization and process in infrastructure development.* Workshop: Strengthening Operational Skills in Community Driven Development, Washington, D.C. Retrieved from the World Bank http://siteresources.worldbank.org/INTCDDTRAINING/1335815–1119647414398/20558633/WSSLIC%20paper%20CDD%20 session.pdf.

World Bank. (2015). More and better spending: connecting people to improved water supply and sanitation in Indonesia. *Water Supply and Sanitation Public Expenditure Review* (WSS-PER). Jakarta, Indonesia: The World Bank.

World Bank. (2016, March). Community Driven Development. The World Bank. Retrieved from http://www.worldbank.org/en/topic/communitydrivendevelopment/overview.

WSP. (2015). *Water Supply and Sanitation in Indonesia, Turning Finance into Service for the Future.* Jakarta, Indonesia: World Bank Group.

Competition and Cooperation in Social and Political Sciences – Adi & Achwan (Eds)
© *2018 Taylor & Francis Group, London, ISBN 978-1-138-62676-8*

Indonesian multi-track diplomacy for Palestine: Indonesian Red Crescent's (Bulan Sabit Merah Indonesia) support for education of Palestinians

L. Taqwa & M.L. Zuhdi
School of Postgraduate, Department of Politics and International Relations in Middle East, Universitas Indonesia, Jakarta, Indonesia

ABSTRACT: This study aims to examine the effectiveness of multi-track diplomacy that has been built by non-state actors, in order to improve non-political cooperation between Indonesian and Palestinian society. Additionally, this research also seeks to answer how cooperation in education can promote and strengthen the issue of Palestinian education. The approach used in this research is multi-track diplomacy that is carried out by non-state actors in a non-formal relationship with the state or its citizens, who are the intended target of the cooperation and diplomacy. This paper reveals that multi-track diplomacy strengthens the diplomatic efforts of non-state actors for colonized Palestine, especially in education. Furthermore, increasing the capacity of education is very important for the sustainable development of Palestine and, even further, for the preparation of its independence.

1 INTRODUCTION

Since the beginning of their conflicts with Israel, Arab countries and Palestine are yet to recover from Nationalism turmoil. Until 1948, the area that both groups claimed was known internationally as Palestine. But following the of 1948–1949, this land was divided into three parts: The state of Israel, The West Bank (of The Jordan River), and the Gaza Strip (Beinin & Hajjar, 2014) after The Israel actors declared their modern state. The Arab countries then responded with attacks and military aggression at territories occupied by Israel, which culminated in 1967. However, Israel won the war, and Arab countries were repelled. The failure to win the war led to the growth of movements that support sovereign Palestine, carried out by the Palestine Liberation Organization (PLO) (The United Nation Booklet, 2008). The conflict continues to roll out until the late 1960s and early 1970s, when some Palestinian military groups launched various waves of attacks against Israelis all over the world. Since then, the conflicts between Israel and Palestine have become more pointed, and there has been no solution for both parties. There are several factors that have been hampering the peace and reconciliation process between both sides, among other things (Sihbudi, 2007).

First, peace agreements between the two sides have so far failed to produce solutions that are sustainable, especially in the implementation. Many results of treaties and peace conferences, from the Oslo Accords to the Annapolis Conference, could not be carried out because of the deadlock of approaches between the two parties in seeking peace. Conflicts and cross-fire between civilians and the military on both sides were often reported. Yet, differences between the sides over core issues, such as borders, security, settlements, the status of Jerusalem, refugees, and water rights, have not been overcome, despite the third-party involvement of various international actors—the United States, in particular (Zanotti, 2010).

Second, there has been no conflict mediator who genuinely seeks to resolve the conflict entirely. The United States that has several times acted as the mediator can be considered less serious in carrying out this role.

Third, Likud political party is reigning in Israeli government. The conflict resolution has become more difficult after the Likud won the election, as they are well known as an opposition to the establishment of a sovereign Palestine.

One of the consequences of the prolonged conflict is the poor quality of education for the Palestinians. According to UNESCO, from 2003 to 2005, more than 180 attacks were launched by the Israeli army at Palestinian schools, causing the death of 180 students and teachers. During that period, more than 1,500 schools were destroyed. This fact is in inverse proportion to education in Israel, where the government spent an average of ILS 2,300 for education development of each Jewish student (www.seamac.org).

Education discrepancies between the two countries further exacerbate the education for the Palestinians. Out of billions of dollars of aids given by the World Bank in 2012 (www. birzeit.edu), only a few were earmarked to build education infrastructure and increase education quality as most of the fund, if not obstructed by Israel's boycott, were depleted on humanitarian cost and construction of other infrastructures.

In the meantime, Indonesia as a Muslim country has officially contributed to and supported Palestine. The relations between Islam in Indonesia and their implications for the country's foreign policy have attracted considerable attention in recent years especially for Palestine (Anwar, 2010). But the aids are not only limited to only formal diplomatic channels. Non-formal approach, as a part of multi-way diplomacy driven by non-governmental components, provides greater opportunities in supporting Palestine. Humanitarian agencies in Indonesia have showed continuous commitments in helping the Palestinians by preparing "capacity building" program, including the improvement of the quality of education.

One of the non-governmental organizations in Indonesia that share concerns on education for the Palestinians is the Indonesian Red Crescent (BSMI). It is the only public institution in Indonesia that prioritizes education quality for the Palestinians.

Palestine has become a concern for BSMI in its efforts to build humanitarian missions that cross borders of states and nations, while carrying the Indonesian flag and the Red Crescent emblem. Since 2002, BSMI has been sending groups of volunteers to the Middle East to provide medical aid and ambulance assistance to Palestine, Iraq, and other countries impacted by war and civil resistance (Direktorat Timur Tengah Kemenlu RI, 2014). The aid programs are not limited to only physical development, but are also closely related to human resource capacity building through education and scholarships for Palestinians.

2 METHODOLOGY AND LITERATURE STUDY

2.1 *Research methodology*

This study uses a qualitative approach and content analysis method. This method attends to the content or contextual meaning of the texts to analyze the primary data, derived from books, articles, websites that are directly related to the Indonesian Red Crescent (BSMI), and secondary data, derived from books, journals, news and articles on the internet, as well as interviews.

2.2 *Literature study*

International mediation is not conducted solely by official actors such as states or international organizations (Bohmelt, 2010). Non-formal parties, such as individuals, community groups and non-governmental organizations, also play important roles in diplomacy, as mediators in conflict resolution and in the development of relations between countries. The role of non-state actors, also include the Indonesian public that now expects Indonesia to have an influence in far-off corners of the globe, according to Indonesian foreign affairs officials. Twenty per cent of Indonesians nominated Indonesia as one of the ten most influential countries in the world (McRae, 2014).

Besides, the important of the non-state actor, according to Hujgh research, have profound implications for Indonesia's engagement with civil societies at home and abroad. Indonesia in turn is also influenced by its own civil society to build the country's regional and international image and relations, its credibility and efficacy (Huijgh, 2016). As Rizal Sukma, Deputy Executive at the Centre for Strategic and International Studies (CSIS), argues: "Without sufficient public support (—including for the new president, who has fallen in popularity recently—) all of Indonesia's efforts to build its international reputation and relations can quickly evaporate into thin air (Huijgh, 2016)." So, Indonesia's government is compelled to conduct public diplomacy both in the domectic and international areas (Sukma, 2011).

Conceptually, this research is using multi-track diplomacy as an approach, which is based on the original distinction made by Joseph Montville in 1981 between official, governmental actions to resolve conflicts (Track one) and unofficial efforts by nongovernmental professionals to resolve conflicts within and between states (Track two) (Notter & Diamond, 1996). This concept was later developed after John McDonald expanded the study of diplomacy by dividing it into several disciplines –including media, conflict resolution professionals and business– that reconstructed phases of diverse diplomatic efforts into multi-path, hence the birth of Multi-Track (Mcdonald, 2014). Multi-track diplomacy becomes a subject that centers at least nine diplomatic approaches, including education, that were carried out by non-state actors during peace negotiation process and the development of human resources of a country. Multi-track diplomacy was founded because of awareness that formal and official interactions, including those between government representatives, had not always been effective in achieving international cooperation to resolve conflicts and create a mutual relation.

In other words, Multi-track diplomacy is a conceptual way to view the process of international peacekeeping as a living system. It looks at the web of interconnected activities, individuals, institutions, and communities that operate together for a common goal: a world at peace, as an expansion of the "Track one" and "Track two" diplomacy. The development of several tracks (multi-track) of diplomacy demonstrates that conflict is a complicated phenomenon, requiring a multi-dimensional approach if it is to be effectively managed and resolved. Since each track of diplomacy has its own strengths and weaknesses, it is important to involve the multi-track diplomacy concept to improve the chances of resolving conflicts without the loss of lives and material (Mapendere, 2005). Multi-track diplomacy represents one key set of foreign policy strategies that can be deployed to achieve development assistance for the people (Dudouet & Dressler, 2016) by non-state actors.

Education, as the focus of this study, is a part of multi-track diplomacy that is carried out by non-state actors in a non-formal relationship with the state or its citizens who are the intended target of the cooperation and diplomacy. Education is beneficial in improving the quality of productive generation of Palestinians by providing framework of world peace, including reducing (resolving) conflicts, tensions and misunderstandings between groups or nations in preparing for its independence.

3 DISCUSSION

3.1 *Profile of the Indonesian red crescent*

Historically, the Red Crescent organization was founded and first used by Turkey on June 11, 1868, as a symbol during the armed conflict between the Ottoman Empire and Russia (1877–1878), (www.bsmi.or.id). The symbol of the Red Crescent was later officially recognized in an international convention in 1929, when the Geneva Conventions was amended. It is currently being used by 33 countries that have majority Moslem population.

After the collapse of the Ottoman Empire, the Red Crescent emblem was then inherited by the Turks and later the Egypt. The Red Crescent later became the symbol of almost every country with majority Moslem population. Pakistan (1974), Malaysia (1975) and Bangladesh (1989), for example, have officially changed their humanitarian organizations' names and symbols, from the Red Cross to the Red Crescent. In Indonesia, although it was initially objected because it was considered part of the Red Cross, the Indonesian Red Crescent was founded as a humanitarian institution on June 8, 2002 in Jakarta and was officiated by the Decree of the Minister of Justice and Human Rights of the Republic of Indonesia No. AHU-72.AH.01.06. Year 2008 (www.bsmi.or.id).

According to Ambari (2016), the establishment of the Indonesian Red Crescent was initiated during the social conflicts in Ambon. At that time, several people with different backgrounds gathered in Jakarta to strive for social and humanitarian assistance for victims of the conflict, with primary focus in health, logistics and education. The group was formalized by KH Amidhan from the Indonesian Council of Ulema (MUI) with the vision of being a national humanitarian agency in Indonesia and working together with other humanitarian agencies at the national, regional and international level. The missions of the organization are: carrying out humanitarian and peacekeeping efforts, protecting lives during conflicts and other situations, preventing sufferings by improving and strengthening humanitarian laws and universal humanitarian principles, providing the best service for humanity and peace and collaborating with humanitarian agencies and other institutions at the national, regional, and international level in achieving its goals.

Today, other than the special program for the Palestine, BSMI is managing a number of significant programs all over Indonesia in the following fields: Disaster Response, Emergency, Health Care, Maternal and Child Health, Red Crescent Youth, Ambulance Service, Health Education, Refugees, and Community Empowerment.

3.2 BSMI's education scholarship for Palestinian

According to "Save the Children Alliance, West Bank and Gaza" report (Save the Children Alliance, 2001), there were around 20,000 Palestinian children who could not continue their education in 2000 due to closure of schools and the fear of violence caused by Israeli attacks. This situation was worsened as between 29 September and 31 December 2000, Israeli residents and military massacred 94 children under 18 years old, 47 of which were under 15 years old. Additionally, more than 2,100 children were injured, 80% of which were shot with rubber bullets, while 70% were killed by a shot in their upper body.

Since 2000, with the commencement of the second Intifada movement, the Palestinian educational institutions have been experiencing tremendous destruction from the Israeli military aggression. Approximately 803 children have been killed, following the destruction of more than 300 schools. At the university level, the Bir Zeit University was taken over by the military from 2001 to 2004, while the Hebron University and the Palestine Polytechnic University were closed for 3 to 6 months in 2003.

The situation of Palestinian education has been increasingly erratic, inviting sympathy from Indonesia, as we have seen at the Post-summit of the 50th anniversary of the Asian-African Conference (KAA) in 2005 (Hermawan, 2014). Indonesia, the host, raised the issue of Palestine as the main agenda of the New Asian-African Strategic Partnership (NAASP). As a follow up to the agreement, Indonesia held a ministerial meeting in 2008, which led to an agreement on the provision of technical assistance for the Palestinians (www.sameaf.mfa.go.th). The aid program, known as the "Capacity building" program, was intended to support development in the Palestine. The training focuses on the improvement of human resources capability in managing socio-political development, should they gain independence. Through these efforts, Indonesia officially confirms its consistency in supporting Palestinian human capacity building and further expects its independence.

Corresponding to the programs launched by the government as a result of the KAA, BSMI started to observe the Middle East as part of its humanitarian missions abroad. In late 2002, BSMI sent a team of aid workers to Iraq and later carried out a wide range of humanitarian assistance, including becoming a mediator when Indonesian journalists were taken hostage. The humanitarian aid to Palestine began when a war broke out between Palestine and Israel in 2008, causing civilians to take refuge (Ambari, 2016).

After carrying out various relief efforts, BSMI as a non-governmental organization believes that there is one thing that would be useful for future Palestinian productive generation: improving the capacity of human resources in education. Moreover, BSMI believes that capacity building is playing more influential role than physical development, as Israeli's blockades at the Gaza Strip have hampered the distribution of construction materials. Thus, within BSMI's framework, developing and prioritizing education mission to Palestine have become very essential for several considerations, such as (Ambari, 2016):

1. It improves the reputation and the dignity of Indonesia as a nation that is committed to its Constitution, that colonialism should be abolished.
2. Precedence must be given to aid in conflict areas.
3. The Palestinians have recognized and appreciated the role of Indonesian diplomacy, including those carried out by Indonesian citizens, more than those of its neighboring Arab countries.
4. Education in Palestine has been poor in quality and infrastructure since the occupation.
5. Palestinian youth are derailed from continuing their education to a higher level, such as universities.

Those considerations are reflecting the reality of education for Palestinian youth, who have their schools and universities closed periodically during the military invasion (www.stopthewall.org). To respond to the situation, BSMI collects and utilizes fund from the local level to international region. BSMI initiated "Education for Palestine" mission to trigger the improvement of human resources capacity. This program has been BSMI's priority for Palestine since 2009. At that time, two students from Palestine received scholarships to further their education in medicine at Universitas Indonesia and Universitas Gadjah Mada.

As the only non-governmental organization that manages capacity building program for Gazan Palestinians, BSMI has finally garnered supports from educational institutions in Indonesia, such as Universitas Indonesia, Universitas Islam Negeri Jakarta, Universitas Gadjah Mada, Universitas Brawijaya, Universitas Solo, and Universitas Airlangga. These partnerships are carried out in sharing system, in which the Ministry of Education and Culture of the Republic of Indonesia covers the tuition fee, while BSMI, with the support of Indonesians, bear the living cost for the students.

3.3 *BSMI's achievement in improving education for Palestinians*

The BSMI scholarships are not limited to medical school only. Engineering and political science are also available for students from Palestine who are studying under the scholarships for the next six to seven years. All the alumni are obliged to return to Palestine after completing their studies. Therefore, according to Abdel Rahman, who has been receiving scholarships for medical studies at Universitas Indonesia, BSMI scholarship should be highly applauded. Mueen Zayed Elshurafa, a medical school graduate from Universitas Gadjah Mada who earned his graduate degree with cumlaude, also praised the institution's initiative.

Within 2010 to 2016, six students from Palestine have been continuing their education at universities in Indonesia. However, this opportunity is still limited to men only as female Gazan Palestinian students are having difficulties in adjusting themselves to Indonesian culture.

BSMI's programs to improve Palestinian productive generation capacity are not limited to scholarships only. Scholarships awardees are also given Indonesian language training program as a requirement a year before they enroll at the universities. Furthermore, as complementary, soft skills development trainings are also given at BSMI headquarters and other places.

Thus, we have to admit that educational diplomacy, as initiated by BSMI, is essential in improving human development capability that is needed to enhance human security to compete in global competition (Sen, 2002). Education will create new paths and opportunities to enhance a nation's ability, to win the competition at global level, which may all lead to the welfare and prosperity of the society (Soesilowati, 2015).

Moreover, educational diplomacy by non-state actors is necessary to improve strong psychological people-to-people relationship, without being hampered by lengthy and bureaucratic procedures. Educational diplomacy, as a multi-track and cultural diplomacy, is believed to have enhanced mutual understanding, mutual interest, mutual respect and mutual trust

(Asia Pulse, 2011) between the Palestinians and Indonesians. This approach is an effective strategy in an effort to foster mutual concern and common interest. One of the reasons is that cultural and educational diplomacy will have a positive effect in embracing the "heart and mind" of a nation (Nye, 2004).

Additionally, this approach will foster a sense of togetherness between the citizens of Indonesia and Palestine and has also become a strategic concept in soft power diplomacy that wield official influence for the government, such as: (1) increasing the "attractiveness" of Indonesian government, which will increase the interest and bond between the people and the government of Indonesia because, in many cases, there are discrepancies between the government's foreign policy and the people's expectation; (2) creating opportunities for Indonesian government for better bargaining position in expanding its influence, particularly within the Islamic society and the Middle East, in terms of developing education for colonized countries; (3) receiving helps and supports from the Indonesian citizens to improve the quality of education in other countries, in this case the Palestine, which will raise Indonesian prestige in foreign relations and diplomacy.

4 CONCLUSION

The explanation of multi-track diplomacy above further elaborates the significance of non-state actors in diplomatic efforts, especially in education for the colonized Palestinians Inclusion of the society in diplomacy, by giving as many opportunities as possible, is very important. Diplomacy is carried out by not only elite officials from the Ministry of Foreign Affairs or the Ministry of Education and Culture, but also the public or stakeholders actively, such as the NGOs, professionals and also university students, as they can especially focus on issues that are more related to the needs of public and to non-political diplomacy approach.

Diplomacy therefore has to be able to accommodate non-state actors and promote human resources development as the main issue in international relation. In multi-track diplomacy, achieving mutual interests and benefits among the actors is a must. The ability to understand the interests and priorities of citizens, especially Palestinians as the partner in this diplomatic process, is very significant because this is essentially a trade-off of interests between the actors. Thus, it takes "a real closeness" with appreciation and similarity of ideas, history and values, a form that is influenced by the proximity of "heart and mind". Increasing the capacity of education has become a sustainable and very important issue for the development of a country.

REFERENCES

Ambari, Djazuli. The Director of Indonesian Red Crescent, Lemhanas Building, Jakarta. (Interview, September 16, 2016).

Asia Pulse. (2011) *Indonesia to Use Education as Cultural Diplomacy with US, 5 April.*

Beinin, J. & Hajjar, L. (2014) *Palestine, Israel and the Arab-Israeli Conflict; A Primer.* The Middle East Research & Information Project.

Bohmelt, T. (2010) The effectiveness of tracks of diplomacy strategies in third-party interventions. *Journal of Peace Research; Peace Research Institute Oslo,* pp. 167–178. Available on: DOI: 10.1177/0022343309356488.

Direktorat Timur Tengah Kemenlu RI. (2014) *Bantuan Kemanusiaan Indonesia Untuk Palestina.* Jakarta, CV. Hilda.

Dudouet, V. & Dressler, M. (2016) *From Power Mediation to Dialogue Facilitation: Assessing the European Union's Approach to Multi-Track Diplomacy.* Germany, Berghof Foundation. Retrieved from: www.woscap.eu

Fortuna, A.D. (2010) Foreign Policy, Islam and Democracy in Indonesia. *Journal of Indonesian Social Sciences and Humanities,* 3, 37–54. Retrieved from: http://www.kitlv-journals.nl/index.php/jissh/index.

Hermawan, Y.P. (2014) *Indonesia in international institutions: Living up to Ideals. National* Security College Issue Brief No. 8. ISBN 978-1-925084-12-2; ISSN 2203-5842 [online].

http://bsmi.or.id/id_ID/capacity-building-palestine-prog-beasiswa-bsmi/, retrieved on September 16, 2016.

http://bsmi.or.id/id_ID/profile/history/, retrieved on September 08, 2016.

http://bsmi.or.id/id_ID/profile/history/, retrieved on September 19, 2016.

http://bsmi.or.id/id_ID/profile/history/, retrieved on September 08, 2016.

http://imtd.org/multi-track-diplomacy, retrieved on September 09, 2016.

http://sameaf.mfa.go.th/en/organization/detail.php?ID=902, retrieved on September 21, 2016.

http://www.birzeit.edu/en/blogs/palestinian-universities-under-occupation, retrieved on September 11, 2016.

http://www.seamac.org/equalrights.htm, Retrieved on September 11, 2016.

Huijgh, E. (2016) *The Public Diplomacy of Emerging Powers Part 2: The Case of Indonesia.* Los Angeles, Figueroa Press.

Mapendere, J. (2005) Track One and a Half Diplomacy and The Complementarity of Track. *COPOJ – Culture of Peace Online Journal,* 2(1), 66–81. ISSN 1715-538X. Retrieved from: www.copoj.ca.

McDonald, J.W. (2014) Multi-Track Diplomacy: A Positive Approach to Peace. Washington DC, Institute for Multi-Track Diplomacy.

McRae, D. (2014) *More Talk than Walk: Indonesia as a Foreign Policy Actor.* Australia, Lowy Institute for International Policy.

Notter, J. & Diamond, L. (1996) *Building Peace and Transforming Conflict: Multi-Track Diplomacy in Practice.* Occasional Paper Number 7: The Institute for Multi-Track Diplomacy.

Nye, J.S. Jr., (2004) "Soft Power and American Foreign Policy", *Political Science Quarterly,* 119 (2); *The Academy of Political Science,* pp. 255–270.

Palestinian grassroots Anti- Apartheid Wall Campaign. Students in Palestine; Education under Occupation and Apartheid. Retrieved from: www.stopthewall.org. [Accessed on August 18, 2016].

Save the Children Alliance; West Bank and Gaza. (2001) *Palestine, The education of children at risk.*

Sen, A. (2002) "Basic Education and Human Security" Background paper for the workshop on Basic Education and Human Security. *Jointly organized by the Commission on Human Security, UNICEF, the Harvard University, Kolkata, 2–4 January 2002.*

Sihbudi, R. (2007) *Menyandera Timur Tengah.* Jakarta, Mizan.

Soesilowati, S. (2015) Diplomasi soft power Indonesia melalui Atase Pendidikan dan Kebudayaan. *Jurnal Global & Strategis,* 9(2), 293–308. Universitas Airlangga. ISSN 1907–9729. Retrieved from: http://journal.unair.ac.id/diplomasi-soft-power-indonesia-article-10289-media-23-category-8.html.

Sukma, R. (2011) Soft Power and Public Diplomacy; The Case of Indonesia. In: Lee, S.J. & Melissen, J. (Eds.) *Public Diplomacy and Soft Power in East Asia.* Palgrave Macmillan Series in Global Public Diplomacy.

The United Nation. (2008) *The Question of the Palestine and the United Nation.* New York, The United Nation.

Zanotti, J. (2010) *Israel and the Palestinians: Prospects for a Two-State Solution.* Congressional Research Service.

Competition and Cooperation in Social and Political Sciences – Adi & Achwan (Eds)
© 2018 Taylor & Francis Group, London, ISBN 978-1-138-62676-8

The role of the Middle Eastern first lady in the public sphere: A case study of Queen Rania of Jordan

R.N. Fitria & Apipudin
School of Postgraduate, Department of Islamic and Middle Eastern Studies,
Universitas Indonesia, Jakarta, Indonesia

ABSTRACT: This research investigates the importance of the Middle Eastern first lady expand her role not only in private but also in the public sphere. So far, the significance of the Middle Eastern first lady's role has not been widely studied. Most Middle Eastern women, including the first lady, are normally considered to have minimal roles in society. In contrast to that statement, this study suggests that a region, where the patriarchal culture is particularly strong, the first ladies show significant contributions to the wider community. Queen Rania of Jordan shows that being the spouse of the king in the Middle East does not preclude her potential to contribute. In this paper, the roles of Queen Rania are analysed using a qualitative case study approach. This study proves that a number of Middle Eastern first ladies play their role in society according to opportunities and challenges that occur in their region. In contrast to other Middle Eastern first ladies, Queen Rania has her own characteristic of contributing and delivering her ideas in the public sphere.

1 INTRODUCTION

This article supports the argument that the first ladies of the Middle East have considerable impact on the community. Paul Kennedy, a British historian, places women's roles as one of three main elements that help society face the 21st century, besides economy and education. A woman determines the quality of the people around her and the next generation (Kennedy, 1993). Women in this century are expected to become self-sufficient figures who balance between responsibilities in the family and in the community (Agrawal & Joglekar, 2013). Women are not always positioned as the main leader who is able to play a strategic role. The first lady, for example, is usually portrayed as a behind-the-scenes figure. However, they may have important roles in various aspects in a country even though there are not any responsibilities explicitly written in the constitution.

In international scientific literature, the significance of the Middle East first ladies' role has not been explored and discussed excessively. Nevertheless, a long study of the first ladies of other regions is, in fact, not a new concept. In the United States, for example, a paper by Lewis L. Gould explains the roles of the first ladies in the country from the past until now (Gould, 1985). Another article, written by Karen O'Connor, Bernadette Nye and Laura Van Assendelft, discusses the influence of the United States' first ladies in politics and public policies. The writers start off the article by giving an introduction and description of the first ladies' political activities through comparative-quantitative data on 38 women who have married presidents. Furthermore, they explain the variety of roles played by the first ladies and evaluate the influence of their roles on politics and public policies (O'Connor, et al., 1996). The public role of the first lady in this country has been well studied for decades, for example, by Betty Houchin Winfield. Winfield (1988) states that the public relations aspect of this White House position is now all too important. The American first lady is a public national woman. If she leaves a legacy in this status, she plays a large part in making the first lady a visible, modern, feminine ideal (Winfield, 1988). In Europe, the concept of the first lady was written about by Ina Woodcock in her dissertation. Woodcock (1998), who learned about the Empress Livia case study, states that the first lady does not have the access to be part of the official government structure of a country. However, the first ladies are allowed to bring massive influence from all resources available in the private sphere. The private sphere does not hinder nor restrict them to use all their potential to contribute.

The Middle Eastern region is often portrayed as a monolithic, traditional, and patriarchal culture (Schwedler, 2006). According to the research of Global Gender Gap Index 2015, Middle Eastern women have the lowest percentage of awareness regarding equality. Only 59% of women have full awareness of the matter (Bekhouche, 2013. However, the assumption regarding the lack of female roles in the Middle East, or in this context, the first lady's roles, needs to be reanalysed. Even though it is not often studied by researchers, the region that is famous for its strong patriarchal culture has turned out to have a large number of first ladies who contribute to the wider community. Among the Middle Eastern first ladies who have roles in the public sphere, Queen Rania Al-Abdullah stands out as an interesting subject to be studied Queen Rania, who is of Palestinian origin and became Queen of Jordan at the age of 28, has more than just youth, a pretty face and an elegant sense of fashion. Queen Rania has an intensive commitment and contribution to the public sphere. She has contributed both in national and international arenas. Moreover, she never leaves her critical roles as a wife and a mother of four children.

Information regarding the public role of the Middle Eastern first ladies, Queen Rania and her contributions in the national and international realm can be obtained through various references and other studies about the role of the first lady. This information will be analysed to find Queen Rania's main roles in the public sphere, using a qualitative case study approach proposed by John W. Creswell (2009). This study is expected to support other studies about first ladies in the Middle East, especially in Jordan. Specifically, this study can be used as a reference for governments in regulating policies regarding the contribution of the first lady of a country.

2 THE MIDDLE EASTERN FIRST LADIES' ROLES AND THEIR CONTRIBUTIONS IN THE PUBLIC SPHERE

The concept of a role cannot be separated from the concept of status. Neil J. Smelser (1981), in his book titled Sociology, states that every person occupies a certain position in society. Each social position with rights and obligations is called status. In accordance with Smelser's explanation, Linton (as cited in Lewis A. Coser, 1976) stated that 'a status, as distinct from the individual who may occupy it, is simply a collection of rights and duties.' Adding to this explanation, Horton and Hunt (1984), define status as a rank or position of an individual in a group. They also add that status is a rank or position of a group in relation to the other groups. Meanwhile, role is the behaviour which is expected from an individual who occupies a certain status. In some cases status and role are two aspects of the same phenomenon. Status is a set of rights and duties while a role is the actualisation of that set of rights and duties. Every individual has roles, and it is called a role set. The term role set is used to show that status not only has a single role, but also has several correlated and matching roles (Merton, as cited in Horton & Hunt, 1984).

The concept of a series of roles can be explored further to discuss the role of the first lady. In 1961, First Lady Jacqueline Kennedy (as cited in Westfall, 2016), stated that the main role of a woman who occupied first lady status was 'to take care of the president so that he can best serve the people. And not to fail her family, her husband, and children'. However, the role of the first lady does not end there because each individual holds several roles and positions in society. Gould (1985) adds that the first lady is expected to take on the role of a woman who is able to represent and contribute to herself, her family and also her country.

In the Middle East, a number of first ladies contribute to their countries in the public realm, such as Iffat AlThunayan from Kingdom of Saudi Arabia, Emine Erdoğan from Turkey, Susan Mubarak from Egypt, as well as Noor and Rania Al-Abdullah from Jordan.

Iffat Al-Thunayan, King Faisal bin 'Abdul 'Aziz Al Sa'ud's wife during her husband's reign, brought many progressive ideas into the Kingdom of Saudi Arabia such as improving education and health care in the country for both men and women (Kechichian, 2014). At the end of Faisal's rule, Iffat was referred to as queen: this was a token of special respect, since the wives of Saudi kings were never usually regarded as queens (Vassiliev, 1998). In Turkey, Emine Erdoğan plays a significant role in political and social life (WISE Summit, 2014). According to a web page sponsored by the Presidency of The Republic of Turkey, Emine Erdoğan has practiced philanthropy and supported important projects concerned with the education of women and children. Besides that, Mrs. Erdoğan has also encouraged women all over the country to take part in business as well as in politics (Turkey, n.d.). In Egypt, Suzanne Mubarak was 'a founder and chairperson of the Integrated Care Society established in 1977, a non-profit organisation with the main objective of providing social, cultural and health care to school children. (FAO, 2009).' This organisation carries out various activities such as a vigorous campaign on 'Reading for all' that has stimulated a national movement focused on young and adult readers. She also spread her wings internationally by participating in conferences on women and children held by the United Nations (FAO, 2009).

In Jordan, one of the influential first ladies is Queen Noor. Nationally, she has been involved in education, conservation, sustainable development and human rights issues. Her contributions were mainly international relations. She was interested in people worldwide who wanted to better understand the Arab world. Her areas of interest include the relationship between the western and Muslim worlds and conflict management (King Hussein Foundation, 2008). After her husband's death in 1999, she wrote her memoir of her life when she accompanied the King. The book is expected to contribute to greater awareness, especially in the West, of events that have shaped the modern Middle East, and encourage a deeper understanding of contemporary challenges facing the Arab world, as well as an appreciation for the true values of Islam (Queen Noor, 2003).

Since 1999, the first lady position in Jordan is held by Queen Rania. To fulfil her role as the first lady, Queen Rania always attempts to understand both the opportunities and the challenges of the nation. As stated by Leuenberger (2006), the previous king, King Hussein's reign 'brought advancements for women, including a rise in the literacy rate from 33% to over 85%.' Queen Rania is planning to continue working towards the advancement of women and to open up the country to modern influences. She is expected to be able to continue shaping the role of women in Jordanian societies and balancing her country's traditions with newer influences (Leuenberger, 2006).

Socially, there remain challenges within schools in Jordan, although access to education is relatively high. The Government of Jordan (GOJ) recognises that the quality of Jordanian education is variable and is not as competitive as that in other countries with international standards, both in urban and rural areas. This issue is exacerbated by unqualified learning environment system and school performance. The pressure of population growth and the increasing population of Syrian refugees on school facilities requires a prominent system, particularly in disadvantaged areas. More than 50% of students in Jordan are learning in overcrowded classrooms. Not only that, the learning objectives do not thoroughly equip students with the ability to handle real challenges in society such as understanding gender inequalities in jobs and how they can survive in this rapidly changing world (USAID, 2015). In response to this challenge, The Jordanian Ministry of Education stated, as cited by Jarrar and Shawareb (2013), that 'the national education strategy stresses the importance of involving students, teachers, directors, and the local community in the development and sustaining of an effective, safe, supportive, and healthy learning environment.'

As cited on her official website, Queen Rania's passion is education. She has helped the Ministry of Education a lot with her initiatives. Rania's contribution to education is not limited to Jordanian or Arab nationals, but it spreads out across the world. Queen Rania is UNICEF's Top Global Advocate for Children and an Honourable Chief in the Global Campaign for Education Act Week. She also supported Education for All, the UN Girl's Education Initiative, In My Name Campaign, and Class of 2015 to achieve the targets set in the Millennium Development Goals. Queen Rania also launched a program called Co-Founder and is a Global CO-Chair for the 1GOAL program that promotes global education in partnership with FIFA, South Africa Cup 2010, and Global Campaign for Education. In addition, Queen Rania gave a speech in Le Web, the biggest technology conference in Europe, with the world's giant online websites, such as Facebook, Twitter, MySpace, Microsoft, and PayPal for 1GOAL on the list. In order to acknowledge her creative way of promoting education through soccer, Queen Rania received the FIFA Presidential Award. She is also a Member of the UN United Nation foundation's council and a supporter of Girl Up (UN Foundation, 2012). This suggests that Queen Rania, as Jordan's first lady, focuses on many aspects as the United Nations Foundation stated: '1) improving the quality of Jordanian family life; 2) promoting quality, access, excellence and innovation in education in Jordan; 3) advocating access to quality global education; 4) championing cross-cultural and inter-faith dialogue; 5) encouraging sustainability throughout the public and private sector; 6) tackling issues affecting youth' (UN Foundation, 2012).

Queen Rania's activities, which have been elaborated above, are the antithesis of the negative stigma surrounding Middle Eastern women. Her role as a mother raising four children doesn't stop her from becoming the mother of the country. Like her own children, she opens opportunities for society to develop through micro-finance, start-ups and new technologies which she believes are very vulnerable in the business world. To enable these opportunities, empowering education and women became her agenda. She is well-known, not only in her country, but also globally. This recognition represents a long process of bridging different cultures and religions, and challenging Arab and Muslims stereotypes ('Her Majesty', 2016).

In carrying out those activities, several Middle Eastern first ladies show that being a king's wife in the Middle East does not hinder nor lessen her potential to contribute. Their roles are parallel to the role of modern women stated by Valentine M. Moghadam (2003):

> …women are not simply passive recipients of the effects of social change. They are agents, too; women as well as men are makers of history and builders of movements and societies. This holds equally true for the Middle East and North Africa. Women are actively involved in movements for social change—revolution, national liberation, human rights, women's rights, and democratization.

Queen Rania and the New Public Sphere: Queen Rania thinks that the role of the Jordanian Queen changes with time. She argued, "Ten years ago, for example, the need for cross-cultural dialogue was not as pertinent as it is now. Today it is an integral part of what I do. Having said that, many aspects of the role remain un-changed—primarily to listen, to care, and to serve." (Time Magazine, 2007).

As the first lady of Jordan, her interesting background leads her to give more attention to problems dealing with cross-cultural and inter-faith dialogue as well as improving social welfare. In the California Governor and First Lady's Women's Conference (2007), Queen Rania explained about her childhood experience which has shaped her character and her mind. 'And while I cannot say I learned everything that I really need to know in kindergarten, I did—thanks to this cross-cultural exposure.' ('California Governor and First Lady's Women's Conference', 2007).

Queen Rania was born in Kuwait on 31 August 1970 to a notable Jordanian family of Palestinian origin who recently came under fire for leaving the country when Israel was stepping up military operations in the West Bank and Gaza ('The Hashemites, n.d.' and 'BBC News', 2001). In 1991, she moved from Kuwait to Amman, where her parents settled after fleeing from the Gulf war that liberated Kuwait from Iraq (Halaby, 2015). Abdullah II married Rania El Yassin from a non-royal family in 1993. Abdullah II ascended the throne when King Hussein died in 1999. Six weeks later, he officially proclaimed his 28-year-old wife as the Queen of Jordan (Halaby, 2015). Rania became the world's youngest queen ('Modern Monarch', 2010). Although she has always lived in the Middle East region, Queen Rania obtained her degree from western educational institutions. She attended the New English School in Kuwait City (Halaby, 2015) and received her Bachelor's Degree in Business from the American University in Cairo (Leuenberger, 2006).

Compared to the other first ladies in the Middle East, Queen Rania has adequate mass media and social media to facilitate her activities and views to be accessed by public. The term 'public' is usually used in the context of events or occasions that are open to everyone—in the same way as we talk about public places or buildings. Indeed, it is in contrast to the meaning of closed or exclusive. However, Habermas (as cited in Iqbal, 2016) describes the public sphere as a sphere of mediation between society and the state. In further discussion, Habermas stated, 'depending on the circumstances, either the organs of the state or the media, like the press, which provide communication among members of the public, may be counted as "public organs"' (Habermas, 1991).

Another perspective regarding the public sphere appears in the digital era. In this era, internet and social media concurrently appear as both public and private spaces. Through home computers or smart phones, internet users can express their thoughts without being hampered by distance and time (Papacharissi as cited in Iqbal, 2016). Reading rooms, libraries, cafes, and places mentioned by Habermas as facilitators of intellectual discussions have transformed into something called the virtual space. Online media is one realisation of the public sphere in this internet era (Poster, 1995).

Queen Rania is a royal figure who considers herself a person who loves to interact and share ideas with many people ('Queen Rania Talks', 2009). Queen Rania airs her views through mass media. For instance, she wrote fourteen blog posts for The Huffington Post and The Washington Post in 2009–2016 (The Huffington Post, 2016 and Al-Abdullah, 2016b). She even delivers her thoughts to children through her story books for children entitled Eternal Beauty, Maha of the Mountains, The Sandwich Swap, and The King's Gift ('Publications', 2016).

She is also a first lady of the 21st century who stays connected to the people and fans from all over the world through the internet. New media is her tool to spread her ideas. She has Facebook, Twitter, Instagram and YouTube accounts. Each of her official social media accounts has millions of followers. In an interview by France 24, Queen Rania answered a question about her interest in using new media. In the dynamic world, Queen Rania has found that social media has helped her to develop a social network. It is a tool that allows her to demystify herself and her actions to people around the world. Compared to the real world, which is full of protocols, social media is more convenient for people to give their opinions. It helps her to get closer to people and enables her to raise awareness of certain issues and to rally people behind them. According to Queen Rania, social networking improves her sense of humanity, especially if it leads to something positive in the real world. ('Queen Rania Talks', 2009).

The Kuwait-born monarch has been known for her passionate work related to education, health, community empowerment, youth, cross-cultural dialogue, and micro-finance in recent years, and she uses her social media pages to help spread her messages of awareness (Binding, 2016). One of the social media tools that Queen Rania uses is YouTube. Since its launch in 2008 until the beginning of December 2016, Queen Rania's official YouTube account, has published 104 videos. According to Playlist (2016), here is the list of videos published in Queen Rania's official YouTube account:

– 9 tourism videos
– 22 videos about fighting extremism and promoting peace
– 5 videos about children
– 7 videos about dialogue and diversity
– 5 videos about community empowerment
– 12 videos about women
– 4 videos about technology and social media
– 40 videos about education

Queen Rania also has an official website at http://www.queenrania.jo/. Through the website, everyone can access various information regarding Queen Rania and her works. It ranges from information about her personality, initiatives, global advocacy (including international affiliations, global education, and UN advocacy), and media centre (including press releases, interviews, speeches, and social media). Halegoua and Aslinger (2016) state that while she is still constructing herself as royalty, this website presents her in a more pluralistic and approachable manner. The significance of this more pluralistic representation enables Queen Rania to be constructed as not only a citizen of Jordan and its Queen Consort, but also a humanitarian and global citizen.

Queen Rania believes that new media can produce a social change to improve the quality of society. As an example, in December 2009, Queen Rania carried out a campaign called 1GOAL, which aimed at encouraging 75 million dropout children to come back to school. The campaign, which was carried out prior to the World Cup 2010, utilised the media to gain as many supporters as possible. Queen Rania explained 'We want to channel the excitement of football behind education, and we're using the media to reach out, social media to reach out to people'. As a result, a total of 30 million signatures were achieved through this campaign. Those signatures were submitted to world leaders as evidence that the program was worthwhile. 'So, this is really what we're looking to see how the internet can finally make this leap from being just online interest which can sometimes be fleeting to something that can lead to real change in our real world', she said (Queen Rania on 'Queen Rania Talks', 2009).

As Tiffany D. Reed (2011) states, a fresh image of Middle Eastern women has risen in the 21st century, Queen Rania also carries out activities to increase the prosperity of her society. Her position as the first lady allows many opportunities to be maximised in order to achieve their goals. Thus, the public roles she carries out in the new public sphere have the potential to create a social change in society.

3   CONCLUSION

Middle Eastern first ladies have an important position in their families. They have a significant role to take care of the leader of the country so that he can best serve the people. They also have the full responsibility of raising their children and preparing them to be well-equipped generation. However, their role as the first lady is not limited to their status as a wife and a mother. They are also encouraged to make contributions from all the available resources they have in their position. Their position as first ladies allows them to support positive social changes in the Middle East. Queen Rania, the first lady of Jordan, fulfils her role through various activities in the public sphere and utilises media publication to achieve her goals. In the national scope, Queen Rania has been trying to understand the challenges and the potential of her country to improve the quality of life of society. Meanwhile, as part of her international scope, Queen Rania has been focusing on bridging the cross-cultural relations between the East and the West. The activities carried out by Queen Rania can provide a new perspective on the position of women in the Middle East, particularly on the role of the first lady in the region. Contributions given by Queen Rania illustrate that the first lady is expected to understand the challenges faced by the nation. By maximising opportunities, the first lady can do something for the country and society in addition to her important role as a wife and a mother in the family. Therefore, it is very important for the first ladies of the Middle East to maximise their roles not only in the private sphere, but also in the public domain.

REFERENCES

Agrawal, A. & Joglekar, A. (2013). *Role of women in The 21st century. Research Journal of Family, Community, and Consumer Sciences*, *1*(2), 14–17. Retrieved from http://www.isca.in/FAMILY_SCI/Archive/v1/i2/4.ISCA-RJFCCS-2013-009.pdf.
Al Abdullah, R. (2016a). Queen Rania: The Syrian refugees I met are experiencing something worse than death. *Washington Post*. Retrieved from https://www.washingtonpost.com/opinions/queen-rania-the-

syrians-refugees-i-met-are-experiencing-something-worse-than-death/2016/05/05/d09f41f4-1156-11e6-93ae-50921721165d_story.html?utm_term=e5dedeaa0b6e.

Al Abdullah, R. (2016b). Retrieved from http://www.queenrania.jo/en/rania.

Al-Abdullah, R. (n.d.). Retrieved from http://www.kinghussein.gov.jo/queen_rania.html.

BBC News. (2001). *Profile: Jordan's Queen Rania*. Retrieved. http://news.bbc.co.uk/2/hi/middle_east/1632614.stm.

Bekhouche, Y. (2013). Top 5 countries for gender equality in the Middle East. *Website/journal name*. Retrieved from https://www.weforum.org/agenda/2013/05/top-5-countries-for-gender-equality-in-the-middle-east/

Binding, L. (2016). *Queen Rania of Jordan: World's most influential online royal turns 46. Website/journal name*. Retrieved from http://www.ibtimes.co.uk/queen-rania-jordan-worlds-most-influential-online-royal-turns-46-her-birthday-will-rule-1578948.

California Governor and First Lady's Women's Conference. (2007). Retrieved from http://www.queenrania.jo/en/media/speeches/california-governor-and-first-ladys-women%E2%80%99s-conference.

Coser, L.A. & Rosenbergh, B. (1976). *Sociological Theory: A book of Readings. Fourth Edition*. New York, NY: Macmillan Publishing. Creswell, J.W. (2009). *Research Design: Qualitative, Quantitative, and Mixed Methods Approaches*. New York: SAGE Publications.

Food and Agriculture Organization (FAO) of the United Nations. (2009). *H.E. Suzanne Mubarak*. Retrieved from: http://www.fao.org/fileadmin/user_upload/Get_Involved/Ambassadors/Mubarak_cv.pdf.

Gould, L.L. (1985). Modern first ladies in historical perspective. *Presidential Studies Quarterly*, 15(3).

Habermas, J. (1991). *The structural transformation of the public sphere*. Massachusetts: MIT Press.

Halaby, J.J. (2015). Queen Rania of Jordan: A mum and wife with a really cool day job. Retrieved from http://www.thearabweekly.com/Opinion/1744/Queen-Rania-of-Jordan:-%E2%80%98a-mum-and-wife-with-a-really-cool-day-job%E2%80%99.

Halegoua, G.R. & Aslinger, B. (2016). *Locating Emerging Media*. New York, NY: Routledge.

Her Majesty Queen Rania Al Abdullah. (2016). Retrieved from http://www.queenrania.jo/en/rania/publications.

Horton, P.B. & Hunt, C.L. (1984). *Sociology*. New York, NY: McGraw-Hill.

Iqbal, M. (2016). *Ridwan Kamil for mayor: A study of political figures on twitter*. (Unpublished master's thesis). Stockholm University, Sweden.

Jarrar, A. & Shawareb, A. (2013). Factors affecting teachers' excellence from the perspective of Queen Rania award-winning teachers: (A Jordanian case). *Journal of Education and Practice*, 4(8), 71–82. Retrieved from www.iiste.org.

Kechichian, J.A. (2014). *'Iffat Al Thunayan: An Arabian Queen.'* Sussex, UK: Sussex Academic Press.

Kennedy, P. (1993). *Preparing for the twenty-first century*. New York, NY: Knopf Doubleday Publishing Group.

King Hussein Foundation. (2008). *Her Majesty Queen Noor*. Retrieved from http://www.kinghussein-foundation.org/?pager=end&task=view&type=content&pageid=61.

Leuenberger, D.Z. (2006). Building leaders for the future: Women in the Middle East. *Bridgewater Review*, 25(2), 3–6. Retrieved from http://vc.bridgew.edu/br_rev/vol25/iss2/5.

Modern Monarch: Queen Rania of Jordan. (2010) Retrieved from http://www.oprah.com/oprahshow/queen-rania-of-jordan.

Moghadam, V.M. (2003). *Modernizing women: Gender and social change in the Middle East*. Boulder: Lynne Rienner Publisher.

More Questions with Queen Rania. (2007). *Website/journal name*. Retrieved from http://content.time.com/time/world/article/0,8599,1619826,00.html.

O'Connor, K., Nye, B. & Van Assendelft, L. (1996). Wives in the White House: The political influence of first ladies. *Presidential Studies Quarterly*, 26(3), 835–853.

Playlist. (2016). Title. Retrieved from https://www.youtube.com/user/QueenRania/playlists?sort=dd&shelf_id=6&view=1.

Poster, M. (1995). *Cyber democracy: Internet and the public sphere*. Retrieved from http://faculty.humanities.uci.edu/poster/writings/democ.html.

Publications. (2016). Retrieved from http://www.queenrania.jo/en/rania/publications.

Queen Noor. (2003) *Leap of Faith: Memoirs of an Unexpected Life*. New York, NY: Hyperion.

Queen Rania Talks to France 24 about Social Media's Power to Bring Positive Social Change. (2009). Retrieved from http://www.queenrania.jo/en/media/interviews/interview-france-24.

Reed, T.D. (2011). *Modern Middle Eastern Women and their Rising Impact on Society*. Monticello: Running Head, University of Arkansas.

Schwedler, J. (2006). The third gender: Western female researchers in the Middle East. *PS: Political Science & Politics*, 39(3), 425–428. Retrieved from //doi.org/10.1017/S104909650606077X.

Smelser, N.J. (1981). *Sociology*. New Jersey: Prentice-Hall.

Time Magazine. 11 Mei 2007. *More Questions with Queen Rania*. Diakses pada 19 November 2016. http://content.time.com/time/world/article/0,8599,1619826,00.html.

The Huffington Post. (2016). *Queen Rania al Abdullah: Hashemite Kingdom of Jordan*. Retrieved from http://www.huffingtonpost.com/author/rania-al-abdullah

Turkey. Presidency of The Republic of Turkey. (n.d.) *Biography*. Retrieved from https://www.tccb.gov.tr/en/emineerdogan/biography/

United Nation Foundation (UN Foundation). (2012). *Who we are: Her Majesty Queen Rania Al-Abdullah (Jordan)*. Retrieved from http://www.unfoundation.org/who-we-are/board/al-abdullah.html?referrer=https://www.google.co.id/

United States Agency International Development (USAID). (2015). *Jordan Country Development Cooperation Strategy 2013–2017*. Retrieved from https://www.usaid.gov/sites/default/files/documents/1883/Amended-Jordan-Country-Development-Strategy-March-015.pdf

Vassiliev, A. (1998). *The History of Saudi Arabia*. London, UK: Al-Saqi.

Westfall, S.S. (2016). Realm of the first lady: Headquartered in The East Wing, The President's spouse plays many roles: Partner, Parent, Hostess, and Crusader. In *Timeinside the White House: The history, secrets, and style of the world's famous home*. New York, NY: Time.

Winfield, B.H. (1988). Anna Eleanor Roosevelt's White House legacy: the public first lady. *Presidential Studies Quarterly, 18*(2), 331–345.

WISE (World Innovation Summit for Education) Summit. (2014) *H.E. Mrs. Emine Erdoğan*. Retrieved from http://www.wise-qatar.org/emine-erdogan

Woodcock, I. (1998). *Inventing the first lady role: The Empress Livia and the public sphere*. (Unpublished doctoral dissertation). The University of Queensland, Australia.

Study of recreational function quality of public green open space around Jakarta's East Flood Canal

H. Septa & J. Sumabrata
Postgraduate School, Department of Urban Development Studies, Universitas Indonesia,
Jakarta, Indonesia

ABSTRACT: Public Green Open Spaces (GOS) play a significant role in urban sustainability. These places should provide high-quality recreational experiences for urban residents. However, they are often overused. The area around the East Flood Canal (*Banjir Kanal Timur*) in eastern Jakarta, Indonesia, was transformed into a natural recreation area. An on-site survey among public GOS visitors reveals that most of them consider the area to be overcrowded on Sundays or public holidays. They also report a perceived increase in visitor numbers over recent years. All users of the public GOS report perceptions of higher crowding. A significant proportion try to avoid these crowds, relying on behavioural coping strategies such as inter-area displacement. While urban renewal has provided an attractive recreation area, urban densification around the green space appears to have reduced its recreational quality. Monitoring recreational quality indicators, such as crowding perceptions, seems to be useful for sustainable management of urban green space and city planning. This study uses a qualitative approach to seek information and public opinion about the existence of a GOS that was developed as a public open space, as well as information from stakeholders and other relevant agencies.

1 INTRODUCTION

Green open spaces are very important for sustainable development (Haq, 2011). This is because green areas are the most important open spaces for making life in cities more comfortable and of a better quality (Jurkovič, 2014). Urban green open spaces, like an oasis in the city, render great benefits to urban sustainable development from the perspectives of ecological, economic and social equity (Zhou & Parves Rana, 2011).

According to Law of the Republic of Indonesia No. 26 Year 2007 on Spatial Planning, it is anticipated that local government should provide 30% of city space as green open space and about 20% as public green open space, because Green Open Space (GOS) has ecological, social and cultural functions, as well as aesthetic functions, vital in the support of urban sustainability (Ministerial Regulation of Public Works No. 05/PRT/M/2008).

Daerah Khusus Ibukota (DKI) Jakarta is one of the cities in Indonesia that has difficulties in providing 30% of its total area for GOS (Sammy, 2016). Therefore, the government of DKI Jakarta and the Ministry of Public Works and Public Housing developed public GOS located on the green belt of river banks, such as the East Flood Canal (EFC) that extends along *Jalan Kolonel Soegiono* through *Jalan Jenderal R.S. Soekamto, Duren Sawit* and *Jalan Basuki Rahmat*, Cipinang Besar Urban Village, East Jakarta, as seen in Figure 1.

Since the launch of the public GOS of EFC by the government of DKI Jakarta at the end of 2012, the area has been visited by many people, giving rise to the emergence of various activities in that area. This phenomenon is worthy of further investigation and analysis. Thus, the aims of this research are as follows:

1. Identifying the characteristics—in terms of age, occupation/education, and location of domicile—of the people using the GOS of EFC along *Jalan Kolonel Soegiono* and *Jalan Jenderal R.S. Soekamto* through *Jalan Basuki Rahmat*;

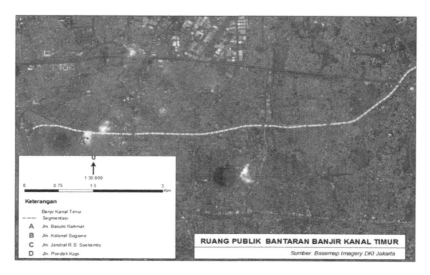

Figure 1. The green open space research area.

2. Identifying the activities undertaken by citizens in the public GOS of EFC;
3. Analysing the developing activities in the public GOS of EFC and the reasons that cause people to be interested in visiting the area;
4. Identifying the perceptions of public GOS users towards the expanded recreational functions in the public GOS of EFC and the quality of those functions;
5. Uncovering the adaptation strategies employed by public GOS users to remain comfortable and enjoy the expected recreational activities while in the public GOS of EFC.

2 THEORY REVIEW

2.1 *Public open space and green open space*

Open space is space designed to act as a place for meeting, gathering and other communal activities conducted by people in the open air (Budihardjo & Sujarto, 2005). The open space that thrives in the city is a formative element of a city's appearance (Lynch, 1960). According to Shirvani (1985), public open space should, as a minimum, be able to support public activities such as: (1) pedestrian activity triggered by the presence of pathways and/or plazas; (2) ensuring circulation through connections between points of activity; (3) the provision of food services, entertainment and physical objects or scenery that can become a focus of appeal; (4) widening of the sidewalk through the availability of off-street parking, as well as giving attention to the canopy, paving and landscaping of pedestrian amenities, to support entertainment and retail trading activities.

Green Open Space (GOS) and Public Open Space (POS) have almost identical definitions. GOS is an area that can be shaped as a pathway or be clustered or open, usually supplemented by trees or green vegetation that are either naturally present or deliberately planted (Ministerial Instruction of Home Affairs No. 14 Year 1998, Ministerial Regulation of Public Works No. 05/PRT/M/2008; Trancik, 1986; van Rooden, within Grove & Gresswell, 1983, in Malawat, 2010).

2.2 *Definition of recreation*

Recreation can be defined as the use of leisure time for pleasurable activities, which may be performed indoors or outdoors. Recreational activities are generally planned not only to provide various forms of fun, but also for the enrichment, expansion, and development of the ability of a person in relation to a new or more satisfying subject. Recreation can take the form of physical recreation (sports, travelling) and psychological recreation, which involves thoughts, feelings and comfort (Torkildsen, 2003; Gold, 1980).

According to McLean and Hurd (2012), there are three recreational categories: recreation in the form of activities undertaken in certain conditions or with particular motivations; recreation as a process that occurs within oneself; recreation as a social institution or in a professional field. Based on these three categories, the motivation behind the recreation will affect the behaviour and activities carried out in the place designated as a recreational space. Several elements implicit in this definition of recreation are: (1) recreation in general is considered as an activity, either physical, mental or social; (2) recreation can take the form of activities with a very broad scope, such as sports, games, hobbies and social activities, which can be performed individually or in groups; (3) recreation is usually based on self-motivation and the desire to attain self-satisfaction; (4) the purpose of recreation is ordinarily to seek happiness, but it also serves to support the fulfilment of intellectual, physical and social needs.

2.3 *Public perception and recreational function quality of public green open space*

Quality assessment in relation to the recreational function of public GOS in an urban area is really influenced by user satisfaction levels, namely, the satisfaction of the citizen in this context. According to Nasution et al. (2011), as quoted in Imansari and Khadiyanta (2015), aspects that can affect the satisfaction levels of citizens in relation to public open space in an urban area include its distance from their residence, its accessibility, the area of the public space, the facilities available, informal retail trading (e.g. street vendors), the vegetation, sanitation, aesthetics, recreational functions, social interactional functions, and the activities undertaken in the public space.

With the current levels of population growth within cities, urban green corridors are the most in-demand green spaces, acting both as recreational areas and as alternative paths for bicycle commuters or pedestrians. Such population growth also poses new management challenges for the development of adequate public use of these sites (Alonso-Monasterio et al., 2015).

According to Carr et al. (1992), the qualification of public space can be derived from the various activities occurring in the space, for example, people interacting, playing, greeting, or simply passive contact, such as sitting to look at the surroundings and listen to the crowd. A public space should possess three main characteristics to become a great public space, namely, (1) the public space should be responsive, (2) the public space should be democratic, and (3) the public space should have meaning.

3 DISCUSSION AND CRITICISM

3.1 *Public GOS visitor characteristics in the East Flood Canal (EFC) area*

The visitors to the public GOS of EFC are very diverse. This review of visitor characteristics is based on groupings according to three categories, namely, age, occupation and domicile (as seen in Table 1). The sources of this data were questionnaires that were distributed to 160 visitors.

Based on this survey of the characteristics of the users of the public GOS of EFC, it can be concluded that they consist of people of all ages, from toddlers to the elderly, people from different occupational backgrounds, and people not only from Jakarta but also from outside Jakarta. All of them can enjoy and take benefit from the public GOS of EFC. It shows that the GOS of EFC is quite justifiably referred to as a public space due to its accessibility to, and utilisation by, all citizens.

Table 1. Characteristics of visitors to GOS of EFC.

| Category | Visitor characteristics based on survey (160 respondents) | |
	Group	Percentage (%)
Age	<20 years old	50
	20–30 years old	35
	31–40 years old	8.25
	>40 years old	6.75
Occupation	Student	58
	Employee (private or public sector)	32
	Housewife	10
Domicile/residence	Jakarta Timur area:	73
	Kelurahan* Duren Sawit (30%)	
	Kelurahan Pondok Kelapa (6.88%)	
	Kelurahan Pondok Bambu (8.13%)	
	Kelurahan Pondok Kopi (7.5%)	
	Kelurahan Malaka Jaya (12.5%)	
	Kelurahan Klender (1.25%)	
	Kelurahan Cipinang Muara (3.13%)	
	Kelurahan Jatinegara (3.75%)	
	Outside Jakarta Timur area:	27
	Bekasi (1.88%)	
	Bandung (0.63%)	
	Bogor (0.63%)	
	Jakarta Pusat (2.5%)	
	Jakarta Selatan (Pejaten, Pasar Minggu, Tebet, Permata Hijau) (13.75%)	
	Depok (5.63%)	
	Bintaro (1.88%)	

Source: Survey, 2016.
*Kelurahan – sub-district.

3.2 *Activities in GOS of EFC*

The activities of the public GOS are determined by the motivation and purpose of users that come to the public GOS, and vice versa. The users of public GOS does not just refer to society in general, but also to more specific representatives of those who are citizens of the city, such as businessmen, civil servants, street vendors, youths and housewives (Kurniawati, 2011).

The type of activities that predominantly thrive in the public GOS of EFC are recreational ones, and are as follows: (1) sports, such as jogging, leisurely strolling, and cycling; (2) leisurely sitting; (3) playing in the interactive park, which has children's games; (4) informal retail trading. Table 2 illustrates the types of activities that have developed in the public GOS of EFC.

3.3 *Recreational function quality of public GOS of EFC*

Within this study, the recreational function quality of the public GOS of EFC is reviewed from the perspective of several indicators, including (1) public attitudes toward the existence of the public GOS of EFC, which is supported by an assessment of the completeness of its facilities and design, and (2) an assessment of non-physical factors, such as amenities, security and accessibility of the public GOS of EFC. Based on a survey of 160 users, opinions about the need for the existence of the public GOS of EFC as a recreational area for the public are described in Table 3.

After reviewing the public need for the GOS of EFC, the survey shows that the majority of users (88%) agree that the existence of the public GOS of EFC is necessary for the public.

At the same time, the survey results also illustrate the opinion of GOS users as to whether or not the public GOS of EFC can serve as one of the recreational places for urban people. That 90% agree that it can indicates that the quality of the public GOS of EFC as a recreational area is good for the community.

Table 2. Types of activities taking place in the public green open space of East Flood Canal.

No.	Activity type	Subject	Analysis result	Picture
1	Sports: jogging/ leisurely strolling, cycling, gymnastics, yoga	People of all ages. Jogging/leisurely strolling most often undertaken by groups of teens and parents. Cycling is undertaken by children, teenagers and adults.	The most dominant activity undertaken by the public in the public GOS of EFC.	
2	Leisurely sitting	People of all ages.	The public GOS of EFC provides places to sit that are built into the maintenance roads and in interactive parks.	
3	Playing activities	Most widely undertaken by children, from toddlers up to the age of 10 years.	The interactive parks in the GOS of EFC are equipped with children's rides, like swings and slides, as well as sandpits, which serve to engage the children.	

(Continued)

Table 2. (*Continued*)

No.	Activity type	Subject	Analysis result	Picture
4	Retail trading (informal sector)	Street vendors.	By design, the GOS of EFC does not provide space for informal retail activity. In addition, government regulation states that trading activities are forbidden in the entire EFC area. Street vendors usually sell food and drinks in the jogging and cycling areas of the maintenance roads but, since January 2016, these tracks have been cleared of street vendors and patrolled by the Municipal Police (Satpol PP). The street vendors have been given space to sell near the Flood Gate.	

Source: Survey, 2016.

Table 3. Users' perceptions of the existence of the public GOS of EFC as a public recreation area.

Statement	User perception (persons)				
	SA	A	NS	D	SD
The existence of the public GOS of EFC is essential to the public	61	81	11	6	1
The public GOS is suitable for consideration as a recreational area	57	88	10	3	2

Source: Survey, 2016.
Note: SA – Strongly Agree; A – Agree; NS – Not Sure; D – Disagree; SD – Strongly Disagree.

Another indicator which shows that the quality of the public GOS of EFC is useful for citizens can be seen in how busy the place is with people who utilise it as a recreational area for sports (jogging, cycling), leisurely sitting with children and family, playing, and community gatherings. The design of the public GOS of EFC, which is planted with vegetation, makes the atmosphere of the EFC area fresh and cool. This, together with the support of available facilities such as the interactive parks, raises the public's interest in visiting the public GOS of EFC. These conditions benefit people in terms of their physical and mental health. Therefore, according to users who were interviewed, the functional quality of the GOS of EFC as a recreational area is very good: 36% stated that this was because of its freshness, because there are trees and green borders; 32% because there is a jogging track facility; 14% because of the existence of informal retail trading (street vendors).

Table 4. Users' perceptions of the amenity and safety of the public GOS of EFC.

Statement	User perception (persons)				
	SA	A	NS	D	SD
I feel comfortable and secure while doing activities in the public GOS of EFC	39	98	16	5	2
The condition of the public GOS of EFC is always clean and fresh	41	67	37	14	1

Source: Survey, 2016.
Note: SA – Strongly Agree; A – Agree; NS – Not Sure; D – Disagree; SD – Strongly Disagree.

Table 5. User strategies for adaptation to conditions in the public GOS of EFC.

No.	Form of adaptation	Percentage (%) (160 respondents)	Analysis results
1	Come earlier	37	Includes people who desire recreation such as early-morning jogging, cycling, and other sports. This group comes earlier because the air is still fresh and the public GOS of EFC is not yet crowded.
2	Avoid the crowded and congested areas	32	This group usually uses the public GOS of EFC as a recreational area to achieve tranquillity and enjoy a relaxed atmosphere.
3	Undisturbed by the crowds and busyness of the public GOS of EFC	20	This group consists of families with children who treat the public GOS of EFC as a recreational area by utilising the facilities of the interactive parks that incorporate children's games.
4	Enjoy the crowds and busyness within the public GOS of EFC	11	This group generally has no specific intentions for recreational activity so that any condition of the public GOS of EFC is enjoyed and activities are undertaken as desired at the time of their visit.

Source: Survey, 2016.

The assessment of recreational function quality of the public GOS of EFC based on users' perceptions is also affected by non-physical factors, such as the amenity and safety while carrying out activities in the public GOS of EFC. Table 4 shows users' perceptions of safety and amenity while they are undertaking activities in the public GOS of EFC.

Based on Table 4, it can be seen that most respondents (85%) feel comfortable and secure while carrying out activities in the public GOS of EFC. Comfort in performing activities increased after the removal of the street vendors who usually sold along the maintenance road. Since the beginning of 2016, such street vendors have been allocated a special location for their trade, which is near the water gate and along the part of the maintenance road that is close to housing areas.

The availability of interactive parks and the condition of trees and greenery are considered good by the public GOS users. Through the sub-agency of the Landscape Gardening and Cemetery of Jakarta Timur, the government of DKI Jakarta tries to keep plants and trees attractive and to keep the area clean by establishing a special unit that works to monitor and maintain the public GOS of EFC.

The analysis of user adaptations referred to in this research concerns the adjustments made by visitors to the public GOS of EFC in order to undertake recreational activities according to their needs. The public GOS of EFC is a space which can be visited by citizens without restriction and is free of charge. Thus, all citizens have the same right to recreation and enjoyment of the atmosphere of the public GOS of EFC, which often makes the area very crowded, with a variety of people pursuing a variety of activities, especially during weekends and public holidays. For some users, these conditions influence their level of comfort and sense of freedom in using the public GOS of EFC and, as a result, users often need to make adaptations to such conditions.

Some of the most common adaptation patterns adopted by users, as revealed by our survey, can be seen in Table 5.

4 CONCLUSIONS AND RECOMMENDATIONS

Based on our survey of visitors to the public GOS EFC, a number of conclusions can be drawn from study of its recreational function quality, including:

1. The characteristics of the primary users of the public GOS in terms of age and occupation, that is, children, students, and college students. Most of the visitors are those who live in areas within Jakarta Timur, such as *Kelurahans* Pondok Bambu, Duren Sawit, Cipinang Muara and Pondok Kopi, as well as from other sub-districts around the East Flood Canal area.
2. Recreational activities such as sports (e.g. jogging, cycling) are the most widely performed activities, supplemented by other activities such as lounging and playing. The public GOS of EFC is also utilised by some users as a means of social interaction, such as gatherings with friends and community groups. One of the activities that has come about as a result of the development of the public GOS of EFC as a recreation area is the rise of the informal sector, which includes street vendors who sell foods, clothes, children's toys, bicycle parts, and *odong odong* owners who offer rides to children.
3. In terms of the quality of the recreational function of the public GOS of EFC, the survey results demonstrate that it is rated as excellent in terms of public perception, and its existence is considered important by citizens. Users also feel that the amenity and safety of the public GOS of EFC as a place to undertake recreational activities is good.
4. The crowded condition of the public GOS of EFC, especially at weekends, means that users have to be able to adapt in order to remain comfortable undertaking the recreational activities anticipated. Visitor adaptation strategies include coming earlier, avoiding the most crowded areas, and simply taking pleasure in the crowds.

The recommendations for the government and for citizens in relation to the GOS are:

1. The overall level of public GOS needs an increase in both quality and quantity, because the function of the public GOS represents essential city development. Moreover, in addition to the ecological need to ensure the sustainability of the urban environment, public GOS also has sociocultural functions, including a recreational function, the knowledge development and providing an interaction space for urban communities.
2. The management of public GOS and efforts to maintain its sustainability are not only the responsibility of government. Public awareness and concern is also required, especially from GOS visitors. There is an urgent need, therefore, to increase public awareness of, and concern for, public GOS through the dissemination of information about the importance of GOS and how the public can help to maintain the cleanliness, comfort and sustainability of the GOS and all the facilities that exist there.

REFERENCES

Alonso-Monasterio, M., Alonso-Monasterio, P. & Viñals, M.J. (2015). Natusers' motivations and attitudes in urban green corridors: Challenges and opportunities. Case study of the Parc Fluvial del Túria (Spain). *Boletin de la Asociacion de Geografos Espanoles, 68*, 369–383.

Budihardjo, E. & Sujarto, D. (2005). *Kota Berkelanjutan*. Bandung, Indonesia: Alumni.

Carr, S., Francis, M., Rivlin, L.G. & Stone, A.M. (1992). *Public space*. New York, NY: Cambridge University Press.

Gold, S.M. (1980). *Recreational planning and design*. New York, NY: McGraw Hill.

Haq, S.M.A. (2011). Urban green spaces and an integrative approach to sustainable environment. *Journal of Environmental Protection, 2*, 601–608.

Imansari, N. & Khadiyanta, P. (2015). Penyediaan hutan kota dan taman kota sebagai ruang terbuka hijau (RTH) publik menurut preferensi masyarakat di kawasan pusat Kota Tangerang. *Jurnal Ruang, 1*(3), 101–110.

Jurkovič, N.B. (2014). Perception, experience and the use of public urban spaces by residents of urban neighbourhoods. *Urbani Izziv, 25*, 107–125.

Kementerian Dalam Negeri. Instruksi Mendagri No 14/1988 tentang. Penataan Ruang Terbuka Hijau Di Wilayah Perkotaan.

Kementerian Pekerjaan Umum & Perumahan. (2008). Permen PU No 05/PRT/M/2008 tentang Pedoman Penyediaan dan Pemanfaatan Ruang Terbuka Hijau di Kawasan Perkotaan.

Kurniawati, W. (2011). Public space for marginal people. *Procedia—Social and Behavioral Sciences, 36*, 476–484.

Lynch, K. (1960). *The image of the city*. Cambridge, MA: MIT Press.

Malawat, N.F. (2010). Bab 2: Tinjauan Teoritis Bentuk Insentif Dan Disinsentif untuk Melestarikan Ruang Terbuka Hijau Kota. Retrieved from https://www.scribd.com/doc/56170125/TEORI-RTH.

McLean, D. and Hurd, A. (2012), Kraus' Recreation and Leisure in Modern Society, Jones & Bartlett Publishers, Ninth Edition, USA.

Sammy. (2016). Hanya Berjumlah 9,98 persen Jakarta Minim Jumlah RTH. Retrieved from http://megapolitan.harianterbit.com/megapol/2016/02/09/55897/18/18/Hanya-Berjumlah-998-Persen-Jakarta-Minim-Jumlah-RTH.

Shirvani, H. (1985). *The urban design process*. New York, NY: Van Nostrand Reinhold.

Torkildsen, G. (2003). *Leisure and recreation management*. New York, NY: Routledge.

Trancik, R. (1986). *Finding lost spaces: Theories of urban design*. New York, NY: John Wiley and Sons.

Undang-Undang Republik Indonesia No 26 Tahun 2007 Tentang Penataan Ruang.

Wardhani, S.T., Hanurani, D., Nurhijrah & Ridwan. (2015). Identifikasi Kualitas Penggunaan Ruang Terbuka Publik pada Perumahan di Kota Bandung. Retrieved from http://temuilmiah.iplbi.or.id/wp-content/uploads/2015/11/TI2015-B-021-026-Identifikasi-Kualitas-Penggunaan-Ruang-Terbuka-Publik-pada-Perumahan-di-Kota-Bandung.pdf.

Zhou, X. & Parves Rana, M. (2011). Social benefits of urban green space: A conceptual framework of valuation and accessibility measurements. *Management of Environmental Quality, 23*, 173–189.

Competition and Cooperation in Social and Political Sciences – Adi & Achwan (Eds)
© 2018 Taylor & Francis Group, London, ISBN 978-1-138-62676-8

The optimisation of society as an environment conservation actor: Study of Citanduy upstream watershed conservation

H. Herdiansyah, A. Brotosusilo, H. Agustina & W. Berkademi
Postgraduate School, Department of Environmental Science, Universitas Indonesia, Jakarta, Indonesia

ABSTRACT: The understanding and participation of society in maintaining the good condition of natural resources are needed to preserve the environment. Many instances demonstrate that individual societies have unique and different perspectives when it comes to preserving their natural resources. One example of such a society is the community within the Citanduy upstream watershed area that keeps the water supply and river ecosystem clean. So far, the government has made various attempts to overcome problems in the watershed. These have ranged from reforestation to criminal proceedings against those destroying nature. However, these actions are not yet optimal, because there is insufficient involvement of society as the main actor. This study describes the management of the Citanduy upstream watershed on the basis of local society empowerment. The study method used was a year-long participatory observation. The result shows the need for more social intervention, in the form of a culture of environmental law enforcement and a basis of local wisdom, to handle various problems that threaten watershed sustainability. It also reveals the use of environmental law enforcement in the form of social punishment for the person who breaks the law and declines to participate in watershed conservation.

1 INTRODUCTION

Water is a natural resource that can be renewed, but environmental changes within the watershed of such water resources can lead to disturbances of the *natural water cycle* that can cause the water to become a non-renewable resource. Kementerian Lingkungan Hidup (2009) describe current conditions of water quality in Indonesia are judged to be very critical over an area of about 460 acres (0.67%), critical for an area of about 14,793 acres (4.05%), and potentially critical for an area of about 117,730 acres (32.27%).

Water quality changes have been caused by environmental damage in the upstream area. This environmental damage will increase if the upstream activities of people increase. The associated problems in the upstream area are changes in land use, land preparation activities, and increasing levels of development and human activity. These have an impact in the form of downstream changes in the amounts of water discharge and sediment load, as well as the material it contains (Sastrowihardjo, 2015). Therefore, watershed management is required to regulate land use patterns and optimise the rational use of land for various purposes.

Watershed damage in upstream areas could potentially affect midstream and downstream areas. Flooding and sedimentation in the watershed is an example of the problems that stem from the upstream use of land in the watershed and processing methods that are not appropriate. This can create impacts on critical land, that is, land in a condition such that it can no longer function properly in accordance with its designation as a medium of water production and management (Sastrowihardjo, 2015).

Because of this dynamic, it is essential that society initiatives, through community involvement, are included within the arrangements for upstream river management. This community involvement should be based on local knowledge, with local ecological knowledge providing the basis for natural resource management. This is necessary for the establishment of an understanding of the relationship between local ecological knowledge and local community practices in managing natural resources. Public participation should also be enhanced through the appreciation of local knowledge, as described by Reddy (2005), where local knowledge of a body of water can be a major element in the management of natural resources. This can provide the integration between the social and environmental elements in the management of water bodies.

2 THEORETICAL BACKGROUND

2.1 *Watershed management*

A watershed is generally defined as the area or region restricted by limiting topographical overlays (e.g. hills) that receives and collects rain water, sediments and nutrients and then discharges it through streams to a larger river and then to the sea or a significant lake. Linsley et al. (1980) describe a watershed as the drainage basin for the entire area drained by a river. So, a watershed is the flow system that connects to the river so that water that originated in a region is discharged through a single outlet. From the definition above, it can be argued that the watershed is an ecosystem in which the biological, biophysical and chemical elements of the environment interact dynamically, including the balance of the inflow and outflow of materials and energy.

As a system, watershed management should be integrated as an aspect of natural resource management to reduce damage and control the distribution of water flow. The watershed ecosystem can be classified into upstream, midstream, and downstream components. Typically, upstream watershed areas are intended for conservation and downstream watershed areas for fertilisation. The upstream watershed is of great importance, especially in terms of the protection of water function. Thus, every instance of activities in upstream areas will have an impact in the downstream areas in the form of changes in fluctuation and transport of sediment and dissolved material in the water flow system. Therefore, protecting the function of the water ecosystem is very important. So upstream watershed management often become the focus of attention in a watershed. The upstream and downstream areas have biophysical linkages through the hydrological cycle.

In terms of their different functions, upstream watershed management can be based on a conservation function to protect the environmental conditions from degradation. This can be guided by the nature of the land cover, land watersheds, water quality, the ability to store water (discharge), and rainfall. Then, midstream watershed management can be based on utilisation, with the function successfully providing benefits for local social and economic interests. This can be reflected in the water quantity and quality, the ability to deliver water, and the height of groundwater levels, as well as in the stable maintenance of water infrastructure, such as the management of rivers, reservoirs and lakes. Finally, downstream watershed management can also be based on a utilisation function, again successfully providing benefits for social and economic interests, as indicated by the quantity and quality of water, the ability to deliver water, the management of rainfall events, and related to the needs of agriculture, clean water provision, and waste water management.

2.2 *Participation*

According to Verhagen (1980), participation is defined as the active involvement of the target group in the planning, implementation and control of programmes or projects. In participation, people are actively involved in carrying out the tasks that have been identified. The principle of participation is to ensure the interests of the beneficiary's own account, and should prove effective in meeting the needs of the development and that of the participants if they perceive a fair share of all the benefits, and not just reduced cost. Another voice in support of enrolment is Moulik (1977), who describes participation as part of the process of encouraging individual states to take the initiative and mobilise people to work for the development of society.

Other literature, by Cohen and Uphoff (1980), describes the general participation that shows that the involvement of people in a situation or action increases in accordance with the aim of improving their welfare. According to Banki (1981), participation is a process of group dynamics such that all group members contribute to the achievement of the objectives of the group, share the benefits of the group activities, exchange information and experiences of common interest, and follow the rules, regulations and other decisions made by the group. The equitable distribution of the benefits of development represents the main benefit of participation. Community involvement in development should involve knowledge improvement to provide consistency, and ensure that there is no clash between development goals and the values of the people in the related sociocultural, ecological, economic and engineering aspects. Community involvement should be conducted in such a way that the community feels a sense of ownership of the programme. The involvement of local communities, especially in rural areas, can be seen from their commitment to the implementation of a programme.

3 METHODS

The method used in this research is participatory observation. The study is based on the society empowerment programme in the Citanduy upstream watershed in Tasikmalaya. The study was conducted through observation and involvement in community activities in this area. The observations were made through in-depth interviews and structured discussions with stakeholders. The study lasted a year.

4 RESULTS AND DISCUSSION

This study is based on community empowerment programmes conducted in the upstream watershed of Citanduy. One key element of this programme is a watershed development project created for the local people in which they contribute to the workforce and invest in the development of soil and water conservation structures in their fields and community land. They also participate in the maintenance aspect, which is necessary because without the protection and care of the local upland communities the programme cannot succeed. Local involvement in programme evaluation is also needed, to identify the things to be considered for improvement in planning future programmes. There are three factors that determine the establishment of natural resource management among the public in the Citanduy upstream area. These factors affect either the stance taken on conservation or the consideration given to sustainability in the course of exploitative actions, and they are: (1) the conditions created as a basis by the rules and values generated by local culture, religion, and morality; (2) the state of the economy; (3) policy intervention, either in the form of material resources or knowledge support.

In the context of the Citanduy watershed, success in the implementation of the concept of sustainable development within this programme:

1. Provides the community with understanding of the interrelationship between natural resources (objects) and man (the subject of development). There is an intertwined relationship between man and river. Humans need the river to support their goals and activities. Conversely, the river can also be affected by human activities in both its utilisation and maintenance, not only in relation to the use of the water in the rivers and streams, but also in integrating the management of the river. It also depends on the characteristics of the river and the social and cultural conditions (Darmanto & Sudarmadji, 2013).
2. Establishes community understanding of the principles of hydrology or ecology of watersheds and the implementation of these principles in the form of sustainable watershed management (Asdak, 2004).

The sources of Citanduy villagers' local ecological knowledge form an integrated resource that comes from internal and external communities. The acquisition of knowledge alone is not enough to drive a person or group of people to make decisions or act. Economic factors are among those that affect the realisation of an action or actions in natural resource management. Both measures a conservative attitude and exploitative actions that would be considered sustainable in the upstream of Citanduy to manage their natural resources.

A new approach, a 'bottom up' approach, is taken in which the local community and/or local landowners become the subject rather than the object in overcoming the destruction of forests, and land forms the conceptual basis for this participation. Most people know the origin of the land in their use and how to cultivate and care for it on a long-term basis in the traditional (indigenous) hereditary manner (Sastrowihardjo, 2015). To meet the goal of community-based watershed management, people gather in the village whereby the unity of society is legally based on wisdom, showing respect to the origins and ways in which local customs are recognised and respected in the Indonesian governance system. But this legality not popular. Community development in the upstream Citanduy watershed allows people to survive and develop themselves to achieve the goal of a civilised society in increasing the value and dignity that cannot escape from poverty, ignorance, unhealthiness and backwardness. The active participation of local communities is a successful strategy for biodiversity conservation. A matter that should be of concern is how to ensure people's interest (Liambi et al., 2005), as this interest is very dependent on the perceived benefits, either direct or indirect, their knowledge of the ecosystem, their value systems and their relationship with nature.

It is appropriate to describe a watershed mega-system complex as being built on physical, biological and human systems in which every system and subsystem interacts. As such a complex, Citanduy watershed will maintain its presence and function as a single entity through the interaction between its components. The output of an ecosystem is largely determined by the quality of the interaction between the components, so that in this process the role of each component and the relationship between the components determine the overall quality of the watershed ecosystem (Kartodihardjo et al., 2004).

According to Brotosusilo et al. (2016), society in general has proven that indigenous knowledge can preserve natural resources. Community restrictions have been used to preserve biodiversity and utilise natural resources in a sustainable way. Judging from the striking similarities in all three areas, namely the upstream, midstream and downstream Citanduy watersheds, it can be seen how every society demonstrates that indigenous knowledge is important in preserving natural resources.

On the other hand, the results of these observations suggest environmental law enforcement in the form of social punishment for those who violate or decline to participate in watershed conservation. Positive law is applied on the principle of *ultimum remedium* when the enforcement of environmental law proves ineffective, with the enforcement of criminal law being the last option.

5 CONCLUSION

Upper watershed areas must be managed and maintained to conserve both upstream and midstream areas. This must be supported by infrastructure and central facilities to enable further support for the functions and benefits of basins in the lower reaches, benefitting agriculture, forestry and the need for clean water in the community as a whole. With an extensive area of watershed both in terms of administration and geography, management coordination is necessary in managing watersheds for various stakeholders, including cross-sectoral and cross-regional stakeholders. The social intervention requirements, in the form of local wisdom and a culture of environmental law enforcement, serve as a solid basis for dealing with problems that threaten the sustainability of the watershed. Environmental law enforcement in the form of social punishment is more effective than criminal law enforcement.

ACKNOWLEDGEMENT

Thanks are due to the Directorate of Research and Community Service (DRPM), Universitas Indonesia, which has supported and facilitated this Community Engagement Programme, 2016.

REFERENCES

Asdak, C. (2004). *Hidrologi dan Pengelolaan Daerah Aliran Sungai*. Yogyakarta, Indonesia: Gajah Mada University Press.

Bagdi, G.L. (2002). *People's participation in soil and water conservation for sustainable agricultural production in the Antisar watershed of Gujarat* (Doctoral thesis, Department of Extension and Communication, Faculty of Home Science, The Maharaja Sayajirao University of Baroda, Vadodara, India).

Banki, E.S. (1981). *Dictionary of administration and management*. Los Angeles, CA: Systems Research Institute.

Brotosusilo, A., Dyah, U. & Afrizal, A. (2016). Sustainability of water resources in the upstream watershed- based community engagement and multistakeholder cooperation. *IOP Conference Series: Earth and Environmental Science, 30*, Conference 1.

Cohen, J.M. & Uphoff, N.T. (1980). Participation's place in rural development: Seeking clarity through specificity. *World Development, 8*(3), 213–235.

Darmanto, D. & Sudarmadji. (2013). Pengelolaan Sungai Berbasis Masyarakat Lokal di Daerah Lereng Selatan Gunung Api Merapi. *Jurnal Manusia dan Lingkungan, 20*(2), 229–239.

Kartodihardjo, H., Murtilaksono & Sudadi, K.U. (2004). *Institusi Pengelolaan Daerah Aliran Sungai: Konsep dan Pengantar Analisis Kebijakan*. Fakultas Kehutanan, Institut Pertanian Bogor.

Kementerian Lingkungan Hidup. (2009). *Status Lingkungan Hidup 2007* [*State of Environment Report (SoER) of Indonesia 2007*]. Jakarta Timur, Indonesia: Kementerian Lingkungan Hidup.

Liambi, D.L., Smith, K.J., Pereira, N., Pereira, A.C., Valero, F., Monasterio, M. & Davila, V. (2005). Participatory planning for biodiversity conservation in the high tropical Andes: Are farmers interested? *Mountain Research and Development, 25*(3), 200–205.

Linsley, R.K., Kohler, M.A. & Paulhus, J.L.H. (1980). *Applied hydrology*. New Delhi, India: Tata McGraw Hill.

Moulik, T.K. (1977). *Techniques of mobilizing rural people to support rural development programme*. IIMA Working Paper WP1977–08–01_00247. Ahmedabad, India: Indian Institute of Management Ahmedabad (IIMA).

Reddy, A.R. (Ed.). (2005). *Watershed management for sustainable development*. New Delhi, India: Mittal Publications.

Sastrowihardjo, S. (2015). *Panduan Pengelolaan Daerah Aliran Sungai Mikro Berbasis Masyarakat*. Ditjen Bina Pengelolaan DAS dan Perhutanan Sosial.

Verhagen, K. (1980). How to promote people's participation in rural development through local organizations. *Review of International Cooperation, 73*(1), 11–28.

Indonesia's support for Palestine in international forums

N.S. Azani & M.L. Zuhdi
Postgraduate School, Department of Middle Eastern and Islamic Studies, Universitas Indonesia, Jakarta, Indonesia

ABSTRACT: This research discusses the role of the Indonesian government and its commitment to support the Palestinians in international forums such as the Organisation of Islamic Cooperation, the United Nations, the Asian–African Conference and the Non-Aligned Movement. The theory employed in this research is the foreign policy approach of James N. Rosenau, which focuses on the roles of leadership, internal conditions and external conditions in determining policy. The method used is a qualitative approach, while the data was obtained from primary and secondary sources. The conclusion is that Indonesia's support of Palestine from Sukarno to the era of Joko Widodo has demonstrated different interests and styles but has been consistently guided by national interests and the many internal communities that support the Indonesian government in continuing to voice support for the Palestinians. Indonesia's support of Palestine in international forums has opened the door to both opportunities and challenges for Indonesia.

1 INTRODUCTION

The conflict between Israel and Palestine has been debated since the beginning of the displacement of Jews from around the world, which gave rise to the establishment of Israel in 1948. This event forced many Palestinians to leave their homeland and increasingly led to the rise of Palestinian resistance to Israeli occupation. Despite various efforts at conflict resolution mediated by other parties, the Israeli–Palestinian problems continue to this day. The important thing to note is that the Israeli and Palestinian conflict is not distinguished simply as a religious conflict. The political interests involved are far more significant, although some issues have often been peppered with religious issues.

As the country with the largest Muslim population in the world, it is inevitable that Indonesia would have attitudes and opinions related to the Israeli–Palestinian conflict. Since its independence in 1945, Indonesia's position on Israeli–Palestinian issues has tended to be consistent.

However, in viewing these issues, the Indonesian government did not adopt a position driven by religious solidarity, but rather by humanitarian solidarity. Indonesian's firm attitude was reflected in a series of foreign policies not only in relation to those on both sides of the conflict, but also in international forums.

This study poses two research questions: How consistent has the Indonesian government's support of the Palestinians been between the era of Sukarno (1945–1967) and that of Joko Widodo (2014 to present), especially in the international arena? What challenges face Indonesia and what steps should be taken? On the basis of these questions, this study aims to understand the extent of Indonesia's support for the Palestinians and what kind of attitude must be adopted by Indonesia in relation to the Israeli–Palestinian conflict. This study is important because the Palestinian issue has become a subject of Indonesian society's attention and is thus expected to be given due consideration by the government.

2   RESEARCH METHOD

This research employed a qualitative approach and the analysis was derived through descriptive analysis. Data collection techniques involved reviewing the sources of both primary and secondary data obtained from books and website articles published by the Ministry of Foreign Affairs of the Republic of Indonesia. Secondary sources included books, journals, theses, news and articles on the internet, and also interviews.

3   LITERATURE REVIEW

Foreign policy is determined by four things: the conduct of foreign policymakers, the military and economic conditions of a country, the role of internal demands, and the international situation (Coplin, 1992). Thus, foreign policy takes the form of a decision that is preceded by a process of considering the internal demands of the state and its economic and military capabilities. These factors then influence policymakers in the process of foreign policy formulation in response to the international situation (Coplin, 1992). The research described here employed three variables of the foreign policy approach formulated by the American political scientist, James N. Rosenau. The first of these is the *idiosyncratic* variable, which concerns the individual factors defined as unique and special in the nature of a leader or decision maker that influence the determination and implementation of foreign policy (Hara, 2011). *National* variables are associated with the internal conditions of a country and their effect on international issues, and include the dominant value orientation in society, the level of national unity, the level of political development, industrialisation, and the state of the economy, all of which contribute to the content of a country's aspirations and its foreign policy (Hara, 2011). The *systemic* variables deal with aspects of the external environment of a state or the actions that occur abroad and influence the choices of decision makers (Hara, 2011).

There is a growing amount of literature on the role of Indonesia in the problems of the Middle East, especially the Israeli–Palestinian conflict. Sihbudi (2011) focused on the relationship between Indonesia and the Middle East during the Arab Spring phenomenon and found that Indonesia adopted a cautious attitude in order to maintain good relations with both the Middle East and the West. In contrast, Indonesia preferred to demonstrate a more obvious position of opposing the West with regard to the Israeli–Palestinian conflict. Aslamiah (2015) emphasised Indonesian multilateral diplomacy in supporting Palestine to become a UN "observer state" in 2012. The research focused on the consistency of view between the Indonesian people and their government, but only described the policies that have been formulated to date and there was not any further review of directions and challenges of Indonesia's diplomacy toward the Palestinians for the future. Moreover, Wulansari (2013) showed that Indonesian foreign policy could not be separated from the role of the Indonesian president. The pattern of international communication through which foreign policy was conducted determined the development of bilateral relations between Indonesia and Palestine. The research assessed the close relations successfully established by President Susilo Bambang Yudhoyono (2004–2014).

Taken together, this research suggests that Indonesia has established a firm stance in the Israeli–Palestinian conflict through continued support of the Palestinians, although it has not been successful in lobbying Israel to relinquish the colonised lands. It is important to analyse the further approaches that could be implemented by the Indonesian government in an attempt to improve Indonesia's diplomatic influence at international level.

4   DISCUSSION

4.1   *Indonesian foreign policy*

The independence of Indonesia in 1945 established the spirit of fighting against colonialism, which had strongly influenced the Indonesian struggle. The 1945 constitution of the Republic of Indonesia summarised the obligation that must be borne by Indonesia to "participate in the establishment of world order based on freedom, lasting peace and social justice" (Sabir, 1987). The orientation of Indonesian foreign policy is built on the basic principles of independence and activity. Mohammad Hatta noted that "independent" meant that Indonesia's position was not to align itself to either of the two dominant global power blocs but to adopt its own approach to solving international problems (Hatta, 1953, p. 444). Being "active" meant attempting to work harder in order to maintain peace and ease tensions between the two blocs (Hatta, 1953, 1976).

The purpose of an "independent and active" foreign policy is to maintain the role of Indonesia in acting against imperialism and colonialism in all their forms and manifestations (Sabir, 1987). It aims to respect Indonesia's struggle for national independence and is dedicated to the struggle for national independence of all nations in the world, as well as serving in the struggle to defend world peace (Sabir, 1987).

There have been evolutions in this "independent and active" foreign policy orientation. In the reign of President Sukarno, it was intended not to show partiality to any bloc, an orientation that shifted slightly when the Indonesian government tended to get closer to the Communist bloc (Widjaja, 1986). According to Suryadinata (1998), in the period of *Demokrasi Terpimpin* (Guided Democracy) (1959–1965) there were three main political actors: Sukarno, the Indonesian Communist Party (PKI), and the military. Unofficially, Indonesia "allied" with Communist and socialist countries, although some political observers took the view that Sukarno's policy was not just black and white, with Sukarno also building good relations with the US. Under President Suharto, Indonesian foreign policy prioritised the development of economic and domestic political stability as a national interest. Suryadinata (1998) argues that unlike Sukarno, who was often confrontational with the West, Suharto emphasised good relations with the West and opened Indonesia up to Western foreign investment for purposes of national development. In the Reformation era, an "independent and active" foreign policy was translated into a "one thousand friends, zero enemies" policy by President Susilo Bambang Yudhoyono. The change in the international political constellation that was no longer dominated by competition between two major blocs guided Indonesia's policy orientation in opening up in all directions. Indonesia had greater space in the regional and world stage, and pursued independent, active and transformative diplomacy to realise the national interest, being increasingly active in international organisations (http://www.presidenri.go.id/2010/08/16).

Thus, it is understood that the basis of Indonesian foreign policy is to be independent and active, although the concept of an independent and active foreign policy tends to be interpreted differently according to the interests of the government of each era and its elaboration of Indonesia's national interests. As Wulansari (2013) has stated, this independent and active foreign policy cannot be separated from the role of the president in each era. It can be equated to politics that can move in any direction, as long as it is in accordance with the national interest.

4.2 *Indonesian attitude to the Palestinian struggle*

During the era of President Sukarno, Indonesia refused to recognise the sovereignty of Israel over the Palestinians and refused to establish diplomatic relations with Israel (Aulia in Lesmana, 2014). Indonesia considered Israel to be an ally of imperialist groups and an enemy of the Third World. During the 4th Asian Games, held in Jakarta in 1962, Sukarno prohibited the Israeli delegation from entering Indonesia (Leifer, 1986).

In the era of President Suharto, support for the Palestinian struggle against Israeli occupation continued. In 1984, the leader of the Palestine Liberation Organization (PLO), Yasser Arafat, attended a meeting with Suharto at Merdeka Palace (Suryadinata, 1998). After the PLO set up a government in exile in 1987 and Yasser Arafat proclaimed the independence of Palestine in Algeria in 1988, Indonesia, under Suharto's government, acknowledged the Palestinian struggle and, in 1989, Indonesia and a representative of the PLO signed an agreement for the commencement of diplomatic relations (Lesmana, 2014).

After the fall of Suharto, Indonesia continued to support Palestinian efforts to achieve their absolute independence. At the same time, there was intensification of Israeli efforts to establish diplomatic relations with Indonesia. The situation was exacerbated in the era of President Abdurrahman Wahid, who urged cooperation with Israel, especially in the trade sector (Suhartono, 2001). For Wahid, this was consistent with Indonesia's interests for recovery after the economic crisis. Israeli trade relations with Indonesia would also strengthen Indonesian lobbying on the international stage and offer new potential for Indonesia's export activities (Suhartono, 2001, p. 184). However, this was met with harsh reactions from many interest groups in Indonesia, such as the political party *Partai Keadilan Sejahtera* (PKS), and Middle Eastern countries that denounced the plans (Suhartono, 2001; Akbar, 2013).

Since Abdurrahman Wahid until the reign of Joko Widodo today, support for the Palestinians has continued. When Yasser Arafat passed away, President Susilo Bambang Yudhoyono attended a memorial service held in Cairo. When Mahmoud Abbas was elected in 2005 to replace Yasser Arafat, Indonesia issued a press release through the Ministry of Foreign Affairs (numbered 002/PR/ 1/2005 on 10 January 2005) that affirmed support for Mahmoud Abbas and hoped for the establishment of a sovereign and democratic Palestine (Basyar, 2006). Through the statement of former Foreign Minister Marty Natalegawa, Indonesia expressed support for Palestine's achievement in gaining "observer state" status in the UN in 2012. (http://www.tempo.co/pbb-akui-palestina).

In the era of President Joko Widodo, the implementation of Indonesia's support for Palestine has been reflected through the opening of the Honorary Consulate of Indonesia in Ramallah, in the Palestinian Territories (Kementerian Luar Negeri Republik Indonesia, 2016). Joko Widodo issued a Presidential Decree (No. 172/M 2015 on 21 December 2015) regarding the appointment of Mrs Maha Abu-Shusheh as the Honorary Consul of Indonesia in Ramallah. Although the opening was conducted and facilitated by the Indonesian Embassy in Amman, Jordan (Kedutaan Besar Republik Indonesia di Amman, 2016), it was an indication of a significant advance in Indonesian foreign policy towards Palestine. It sent a firm signal to the international community regarding Indonesia's intent to continue defending the Palestinians, in accordance with its constitutional mandate.

In addition to political support, the Indonesian government also provided assistance and foreign aid to the Palestinians. In 2007, Indonesia provided one million US dollars in aid to the Palestinians at the Paris Donors Conference, establishing the Indonesian Cardiac Cente at Al-Shifa Hospital in Gaza. It delivered over one million US dollars in assistance to the recovery process at the time of the Cairo International Conference on Palestine following the 2014 attack on Gaza. The aid included public facilities such as solar cells and ambulances (http://www.kemlu.go.id/indonesia-commitment-palestine-israel). Between 2008 and 2013, Indonesia initiated 128 technical assistance programmes and trained 1,338 Palestinians in the framework of cooperation under NAASP (New Asian–African Strategic Partnership for Palestinian Capacity Building Programme) and CEAPAD (Conference on Cooperation among East Asian Countries for Palestinian Development), which has been extended until 2019. In 2016, further cooperation in the capacity building programme was earmarked for the Palestinian police and civil service (http://www.kemlu.go.id/The-Palestinian-Issue).

Indonesian society has also become an important factor in determining Indonesia's attitude towards Palestine. This has been characterised by the emergence of non-governmental organisations that focus on humanitarian issues and help promote the efforts of aid delivery to the Palestinian people, such as *Aksi Cepat Tanggap* (ACT), *Sahabat Al-Aqsha, Mer-C, Bulan Merah Sabit Indonesia*, and *Komite Nasional Untuk Rakyat Palestina* (KNRP). These organisations often provide inputs to the government in terms of aid to the Palestinians; KNRP cooperates with the government in terms of humanitarian aid delivery and generates feedback and demands related to the delivery of social assistance (A. Suhaemi, Chairman of the Bureau of Research and Information of KNRP, personal communication, August 25, 2016; M. Zakaria, Chief of Public Relations Bureau of KNRP, personal communication, August 25, 2016).

In terms of the relationship between Indonesia and Palestine described above, it is clear that every President of Indonesia agreed to give encouragement to the Palestinians by supporting the Palestinian struggle, although this took on varying shades. Sukarno's foreign policy was to seek international sympathy and gain recognition in the international political arena and to campaign on a firm stance against any form of colonisation. Thus, there was Indonesian support for the Palestinians based on efforts to halt the Israeli occupation. Suharto's foreign policy tended to be more opportunistic, positioning Indonesia as a major player in international organisations, as well as improving relations with the West for the benefit of Indonesia's own economic development. Besides supporting the Palestinians, this was also an attempt to attract the sympathy of Middle Eastern countries in order to create opportunities for Indonesian cooperation with them. Similarly, in the Reform era, as competition in the constellation of international politics grew increasingly strong, there was a continuing challenge to seek such opportunities. However, although there were attempts to build diplomatic relations with Israel, Indonesia still adhered to the primacy of support for Palestinian independence. This cannot be separated from the internal conditions, or the national variable, of Indonesia as the country with the largest Muslim population in the world. Nevertheless, Indonesian people demanded continuing support for the Palestinians not only out of solidarity based on religion, but also on humanitarian grounds. Israeli colonisation not only affected the Muslims of Palestine, but also afflicted people of other faiths. In addition, the concern of Indonesian communities for the Palestinians is often based on feelings of repulsion for the practice of colonisation and this too had an impact on the animosity with which Israel is viewed. The Indonesian government's foreign policy towards the Israeli–Palestinian issue cannot be separated from public sentiment. To open greater diplomatic links with Israel or reduce support to the Palestinians could lead to tensions that could have a negative impact on domestic political stability.

4.3 *Support for Palestine in international forums*

During the reign of President Sukarno, Palestinian issues were raised at the first Asian–African Conference (KAA), which was held in 1955 in Bandung, Indonesia, and produced the *Dasa Sila Bandung* (Ten Principles of Bandung) (Widjaja, 1986). The KAA summit inspired the establishment of the Non-Aligned Movement and, at the first summit of the Non-Aligned Movement, held in Beograd, Yugoslavia in 1961 (Widjaja, 1986), Indonesian support for Palestine was actively echoed by the other member countries.

Furthermore, when the Non-Aligned Movement summit was held in Jakarta on 10 September 1992, it created the "Jakarta Message", which articulated a firm position against Israeli occupation and demanded the withdrawal of Israeli forces from all occupied Arab territories (Sihbudi, 1997). In the UN forums, when Indonesia became a non-permanent member of the UN Security Council for 2007–2008, Indonesia consistently continued to support the rights of the Palestinians, especially the right to self-determination and also the right to establish an independent state. Indonesia urged the UN Security Council to adopt resolutions regarding the issue of Palestine, including the Presidential Statement (PRST) and resolution (http://www.kemlu.go.id/The-Palestinian-Issue). As a result of Indonesia's insistence, the UN Security Council passed resolution No. 1850 on 16 December 2008 regarding the Middle East peace processes, especially the Israeli–Palestinian one. Indonesia also became one of the principal initiators of the Special Session of the UN General Assembly in Palestine from 15 to 16 January 2009 and called on the international community to continue to assist with the humanitarian crisis in Gaza (http://www.kemlu.go.id/The-Palestinian-Issue).

As a member of the Human Rights Council, Indonesia also pushed for a Special Session of the Council on 9–12 January 2009. The outcome was a draft resolution on "The grave violation of human rights in the occupied Palestinian territory, particularly due to the recent Israeli military attacks in the occupied Gaza Strip" (http://www.kemlu.go.id/The-Palestinian-Issue). At the Asian–African Conference in 2015, the leaders, including Indonesia, approved a Declaration on Palestine as an important achievement of the conference. The declaration reflects support from Asian and African countries in relation to the Palestinian struggle for independence and the achievement of a two-state solution alongside Israel. The conference also gave birth to a declaration strengthening the New Asia–Africa Strategic Partnership, which extended support and assistance to the Palestinians until 2019. In December 2015, Indonesia also hosted the International Conference on the Question of Jerusalem in Jakarta. The conference was held in collaboration with the Organisation of Islamic Cooperation (OIC) and the United Nations Committee on the Inalienable Rights of the Palestinian People (http://www.kemlu.go.id/The-Palestinian-Issue).

On 6–7 March 2016, the 5th Extraordinary Summit of the OIC was held in Jakarta with the theme of "United for a Just Solution". The result was the ratification of a resolution which reaffirmed the principles and commitment of the OIC to Palestine and Al-Quds Al-Sharif, and the Jakarta Declaration, an initiative of Indonesia, which included a concrete plan for OIC leaders to resolve the issue of Palestine and Al-Quds Al-Sharif (http://presidenri.go.id/mengakselerasi-kemerdekaan-palestina). This plan consisted of: the strengthening of political support to revive the peace process; reconsideration of the 'Quartet' (the Israeli–Palestinian peace mediators, comprising the UN, US, EU and Russia) and the possibility of adding members; strengthening pressure on Israel, including a boycott of Israeli products produced in the occupied territories; increasing pressure on the UN Security Council to provide international protection for the Palestinians; an expiration date for the termination of the Israeli occupation (http://presidenri.go.id/mengakselerasi-kemerdekaan-palestina).

Indonesia's particularly active role in supporting Palestine in international forums should be maintained for the sake of fulfilling its constitutional mandate. Such a role would certainly be of positive value in international views and Indonesia would be perceived as a country consistently and actively involved in international issues. Despite the support to the Palestinians that Indonesia has been providing for decades, it has still not succeeded in pressing for the withdrawal of the Israeli colonisation of Palestine, but the contribution of Indonesia has been recognised by the Palestinian people (F. Mehdawi, Palestinian Ambassador to Indonesia, personal communication, April 6, 2016). It is a challenge for Indonesia to continue maintaining the trust of the Palestinian people. Nevertheless, Indonesia could become an example for the Palestinians as well as other conflicted countries as a nation that has been built on the differences among its people.

Another challenge for Indonesia is not only to be an initiator for development programmes in Palestine but also a mediator for the Israeli–Palestinian conflict. This could be difficult given the absence of diplomatic relations with Israel. Alternatively, Indonesia could become a mediator in the rivalry between Hamas and Fatah in Palestinian internal politics. Although Indonesia has officially established relations with Fatah, it should also be able to maintain good relations with Hamas.

5 CONCLUSION

Indonesia's support of Palestine from the era of Sukarno to that of Joko Widodo is an implementation of the mandate of the 1945 Constitution that requires Indonesia to participate actively in opposing all acts of colonisation and in preserving the peace. Although there have been diverse shades of interest on the part of each leader of Indonesia in relation to support for the Palestinians, the consistency in maintaining a firm stance in defence of Palestine should be appreciated. Policies that are based on Indonesia's interests and internal conditions could lead to both opportunities and challenges at the international level.

REFERENCES

Akbar, A.A. (2013). *Menguak Hubungan Dagang Indonesia-Israel*. Serpong, Indonesia: MarjinKiri.
Aslamiah, S. (2015). Diplomasi Indonesia dalam mendukung Palestina menjadi negara peninjau di PBB tahun 2012. *Jom FISIP*, 2(2).
Basyar, H. (2006). Agresi Israel 2006 dan Masa Depan Perdamaian di Timur Tengah. In *Ambiguitas Perdamaian: Pusat Penelitian Politik, Year Book 2006*. Jakarta, Indonesia: Yayasan Obor.
Coplin, W.D. (1992). *Pengantar Politik Internasional*. Bandung, Indonesia: Sinar Baru.

Hara, A.E. (2011). *Pengantar Analisis Politik Luar Negeri dari Realisme sampai Konstruktivisme*. Bandung, Indonesia: Nuansa.

Hatta, M. (1953). Indonesian Foreign Policy. *Foreign Affairs*, 31(3).

Hatta, M. (1976). *Mendayung Antara Dua Karang*, Jakarta, Indonesia: Bulan Bintang.

Kedutaan Besar Republik Indonesia di Amman. (2016, March 13). Menlu RI Lantik Konsul Kehormatan RI untuk Palestina [Indonesian Minister of Foreign Affairs inaugurates Honorary RI Consul for Palestine]. Amman, Jordan: Kedutaan Besar Republik Indonesia di Amman. Retrieved from http://www.kemlu.go.id/amman/id/berita-agenda/berita-perwakilan/Pages/MENLU-RI-MELAN-TIK-KONSUL-KEHORMATAN-RI-UNTUK-PALESTINA-DI-AMMAN—YORDANIA.aspx.

Kementerian Luar Negeri Republik Indonesia. (2016, March 10). Pembukaan Konsul Kehormatan RI di Ramallah [Opening of Honorary Consul of RI in Ramallah]. Jakarta, Indonesia: Kementerian Luar Negeri Republik Indonesia. Retrieved from http://www.kemlu.go.id/id/berita/siaran-pers/Pages/Pembukaan-Konsul-Kehormatan-RI-di-Ramallah.aspx.

Leifer, M. (1986). *Politik Luar Negeri Indonesia*. Jakarta, Indonesia: Gramedia.

Lesmana, D. (2014). *Dukungan Indonesia untuk Palestina*. Jakarta, Indonesia: Direktorat Timur Tengah Kementerian Luar Negeri RI.

Ministry of Foreign Affairs Republic of Indonesia. (2016, January 20). The Palestinian Issue. Retrieved from http://www.kemlu.go.id/en/kebijakan/isu-khusus/Pages/The-Palestinian-Issue.aspx.

Ministry of Foreign Affairs Republic of Indonesia. (2016, June 4). Indonesia Affirms Commitment to Push Palestine-Israel Peace Process Forward. Retrieved from http://www.kemlu.go.id/en/berita/Pages/indonesia-commitment-palestine-israel-peace.aspx.

Presiden RI. (2016, March 8). Mengakselerasi Kemerdekaan Palestina [Accelerating Palestinian independence]. Retrieved from http://presidenri.go.id/internasional/mengakselerasi-kemerdekaan-palestina.html.

Presiden Susilo Bambang Yudhoyono. (2010, August 16). Speech on the 65th Anniversary Celebration of the Independence Day of Indonesia. Retrieved from http://www.presidenri.go.id/index.php/pidato/2010/08/16/1457.html.

Sabir, M. (1987). *Politik Bebas Aktif*. Jakarta, Indonesia: Haji Masagung.

Sihbudi, R. (1997). *Indonesia-Timur Tengah: Masalah dan Prospek*. Jakarta, Indonesia: Gema Insani Press.

Sihbudi, R. (2011). Indonesia dan dinamika politik Timur Tengah (Januari–November 2011). *Jurnal LIPI: Membaca Arah Politik Luar Negeri Indonesia*, 8(2).

Suhartono. (2001). Tim Peneliti, *Analisis Kebijakan Luar Negeri Pemerintahan Abdurrahman Wahid (1999–2000)*, Tim Peneliti Hubungan Internasional Pusat Pengkajian dan Pelayanan Informasi Dewan Perwakilan Rakyat Republik Indonesia.

Suryadinata, L. (1998). *Politik Luar Negeri Indonesia di Bawah Soeharto*. Jakarta, Indonesia: LP3ES.

Tempo. (2012, November 30). PBB Akui Palestina. *TEMPO*. Retrieved from http://www.tempo.co/read/fokus/2012/11/30/2672/pbb-akui-palestina.

Widjaja, A.W. (1986). *Indonesia, Asia-Afrika, Non-Blok, Politik Bebas Aktif*. Jakarta, Indonesia: Bina Aksara.

Wulansari, I. (2013). Komunikasi Internasional Indonesia Dalam Upaya Mendukung Palestina Sebagai Negara Yang Berdaulat. In H. Budianto, L.M. Ganiem & D.S. Tanti (Eds.), *Identitas Indonesia Dalam Televisi, Film Dan Musik* (pp. 64–71). Jakarta, Indonesia: Pusat Studi Komunikasi dan Bisnis Program Pasca Sarjana Universitas Mercu Buana Jakarta.

Competition and Cooperation in Social and Political Sciences – Adi & Achwan (Eds)
© 2018 Taylor & Francis Group, London, ISBN 978-1-138-62676-8

Negotiating identities: National identity vs global branding and cooperation in the case of Real Madrid CF

V. Syamsi
Department of English Studies, Faculty of Humanities, Universitas Indonesia, Depok, Indonesia

ABSTRACT: Sport has taken on a more important role in the political realm. In the past, countries such as the Soviet Union and the German Democratic Republic linked the health of the state to its ability to perform successfully in the international sporting arena (Galeano, 2013; Kuper, 2006). In Spain, General Franco, military dictator from 1939 to 1975, used sport as a part of the state's machinery. He used football to muster support, a sense of national identity, and nationalism from the Spaniards. While repressing the regional identities of the Basque Country and Catalonia (represented by the football clubs Athletic Bilbao and FC Barcelona, respectively), he used Real Madrid CF as a symbol of Spain. Hence, matches between Real Madrid CF and FC Barcelona have never been a mere game of football; they represent a game between "oppressor" and "oppressed" political groups, a match with the highest audience in the world, dubbed *El Clásico*. In 2012, Real Madrid CF cooperated with the Bank of Abu Dhabi and the royal family of the United Arab Emirates and, in the process, removed the 'Christian' cross from its club crest because the project targeted a Muslim region. This paper analyses the conflicting interests involved and how identities compete and are negotiated.

1 INTRODUCTION

Sport has been a huge magnet, attracting millions of spectators. The 2008 Beijing Olympics attracted an audience of 4.7 billion over the 17 days that the event was staged (Kennedy & Hills, 2009), equivalent to 70 per cent of the world population, making it the "most watched Games ever", according to Nielsen Media Research (2008). To emphasise how popular sport is, World Cup USA 1994, Inc. released a booklet stating that the TV audience for the Italian World Cup was 25.6 billion (five times the world population), and that 31 billion are expected to watch the American World Cup (Kuper, 2006). Of all the different kinds of sports, football has been regarded as the most popular sport (Giulianotti & Robertson, 2004), with an estimate of 250 million people playing the game, and a further 1.4 million people attending matches in stadiums, while another 3.4 billion people become spectators through the variety of television channels that air the sport.

Having a large number of fans can be a big advantage, and is in line with Tuñón (2012), who stated that: "in modern society, sports become important for the identification of individuals with the communities to which they belong". Through their identification with a sports team, people can express their identification with the city it represents or perhaps with a particular sub-group within it, such as a class or ethnic group (Dunning, 1999). Sport clubs, therefore, offer a sanctuary in which their fans can feel at home, a place where they belong.

Tuñón (2012) adds that, from the very beginning, football has been a useful tool in stimulating symbolic integration in order to build state ("imagined communities") identities. The development of football has not only served as a vehicle for a nation-state to internally organise its communities; it also served to demonstrate its superiority to other nations in the field of symbolic disputes. Moreover, football's ability to break through various social class barriers can create a feeling of common membership among people. Hobsbawm (in Tuñón, 2012) has argued that "after both World Wars, football turned into a mass spectacle where opposing teams symbolised nation-states". Hence, football can be used as a medium by which to build and show patriotism.

Spain has a lot of football clubs, and each of them is representative of a city, region or even a nationality. This makes the domestic football championship a symbolic field for a competition of identity. Football matches update historical rivalries and serve as vehicles for attitudes of both supremacy and revenge (Tuñón, 2012). Tuñón explains further that, even today, football in Spain still maintains the ethno-territorial roots it showed during the twentieth century. This is what has subsequently been embodied as the symbolic importance of FC Barcelona, Athletic Bilbao and Real Madrid CF. However, of these three clubs, this paper will focus on the competition between the two largest in the Spanish league, *La Liga*: Real Madrid, its triumphs leading it to be considered as the perfect embodiment of the collective identity of the Spanish people, and Barcelona, a longstanding symbol of resistance toward Madrid for the local Catalan population.

2 EL CLÁSICO: THE EVERLASTING DOMESTIC COMPETITION AND FIERCE RIVALRY

FC Barcelona has been used as a resistance identity by the Catalans; an identity generated by those actors in positions or conditions that are devalued and/or stigmatised by the logic of domination, thus creating trenches of resistance and survival on the basis of principles different from, or opposed to, those permeating the institutions of society (Castells, 1999). Thus was FC Barcelona used by the Catalans as a symbol of a resistance toward Madrid and Franco's government of the 1940s and 1950s. Barcelo et al. (2014) state that "perhaps the most internationally famous as symbolising the substate national Catalan identity in Spain is FC Barcelona". Founded in 1899, the club quickly became associated with Catalan identity, underlining Catalonia's struggle against Spain; *més que un club* (something more than a club). For many Barcelona fans, the football club was a proxy where they could exhibit their suppressed identity. Indeed, FC Barcelona is more than just a club. It is a uniting factor for the Catalans that feel oppressed, a shrine for the Catalans to pray in and to realise their dream of independence.

As a consequence, the competition between the two biggest clubs in Spain has always served as a battlefield for the identity of Spain. The match between the two clubs, which is known as *El Clásico*, is without doubt the most awaited match on the planet. This game is exaggeratedly described as the game that can stir emotions, and divide families, friends and even nations. It attracts the largest TV audience in the world in terms of football, and thus the biggest income from TV stations. Referring to its history throughout the previous century, *El Clásico* was frequently presented as a version of David vs Goliath, with Barcelona relishing its status as the David-like underdog and Real Madrid more than happy to play the role of the big guy (Balague, cited in Fitzpatrick, 2012).

3 REAL MADRID AND WORLD CONQUEST

Florentino Pérez, the president of Real Madrid CF, revolutionised the business model of running a football club (Carlin, 2012). His big idea from the start of his presidency in the year 2000 was to buy the very best players, *Los Galácticos*. He believed that if the club bought the very best players, the club would always win in the end because the players paid their own way. It is the same logic that is followed by Hollywood producers when they elect to pay vast sums of money to the top box-office actors to appear in their films. "We're content providers, just like a film studio," explained Jose Angel Sanchez, Real Madrid's director of marketing, "Having a team with Zidane in it is like having a movie with Tom Cruise". That is why Pérez broke the world transfer record to bring Figo to Real Madrid in 2000, Zidane in 2001, Ronaldo in 2002, Cristiano Ronaldo in 2009 (for €94 million, or $131.5 million) (Hughes, 2009), and Gareth Bale in 2012. Yet, each year the club's profits grew. It turned out that the recruitment of *Los Galácticos* has proved to increase the number of fans, and eventually the sales of replica shirts.

As president of Real Madrid, Florentino Pérez was successful in transforming the club into a modern sport and media company; it has become a good model of a corporation possessing a long-term vision that has been way ahead of its time. One of the fundamental pillars of this model has involved designing and implementing a new marketing strategy aimed at strengthening the value of the club's brand (Callejo & Forcadell, 2006). By adopting this new model, Real Madrid increased its income from marketing, making it, from a marketing and business perspective, the leader in the football world. This new management model and strategy has drawn a lot of attention from other sports clubs, the economic press, and even the academic world in the field of business management.

One of the strategies pursued by Pérez is the principle of *espectáculo* (spectacle) (Carlin, 2012). Pérez crushed longstanding orthodoxies, changing the whole conception of the game of football. He wanted to buy "*Los Mejores*", "*Quiero a los mejores*"; "The best, I want the best; let the other teams get centre-halves and defensive midfielders: against us they are going to need them!" Rather than buying players that might better suit the club's needs, Pérez bought the most expensive players because football is mainly about presenting *espectáculo*. With Real Madrid, football is beyond sport; it is business and entertainment for the entire football fan base.

Landor, an American firm considered to be a leader in brand and design consultancy, was hired by a company that was interested in becoming a sponsor of Real Madrid to conduct a survey and study of the club. Landor's analysis reached a surprisingly subversive conclusion about what Pérez and Sanchez had achieved at Real Madrid (Carlin, 2012); subversive in the sense that it violated the game's dominant orthodoxy, especially dominant in Italy, namely, that winning was paramount. "A good show", or *espectáculo*, was extremely important in terms of a football club's global appeal, and Landor found that for Real Madrid it might even have been more important than victory. Real Madrid fully embraces the idea that the charisma of the showmen on display is an integral part of the show. The greater their charisma, the more they were talked about, the bigger the reach of their own name and that of the club they played for.

4   REAL MADRID AND GLOBAL EXPANSION

In transforming the business model, Pérez created project identity at the same time, the third concept of identity formation of Castells (1999), which describes how "social actors, on the basis of whichever cultural materials are available to them, build a new identity that redefines their positions in the society and, by doing so, seek the transformation of overall society structure". In this case, Pérez has successfully changed the football model. To further strengthen the club's brand and value, Pérez made several business deals with the royal family of the United Arab Emirates. There are at least three business initiatives with the Emirates: a Real Madrid theme park in Abu Dhabi (Timms, 2014); cooperation with the Bank of Abu Dhabi in the issue of a credit card bearing the (modified) crest of Real Madrid CF (Jiménez, 2014); a massive deal with Abu Dhabi's International Petroleum Investment Company (IPIC; 100% owned by the royal family of Abu Dhabi), which has reportedly secured the naming rights to Real Madrid's Santiago Bernabéu stadium (Marca, 2014).

The deals bore a vast amount of money, or as the Madrid-based online sports newspaper *Marca* put it: "The figures are jaw-dropping: the club is to rake in between €450 and 500 million over 20 years. ... In short, the Santiago Bernabéu would be re-baptised in exchange for an astronomical sum of money" (Marca, 2014).

To gain acceptance in the Emirates, which is a predominantly Muslim-populated country, Real Madrid was willing to adjust its crest, removing a small cross, as part of the credit card deal with the National Bank of Abu Dhabi, the leading bank in the United Arab Emirates (Sharkov, 2014). *Marca* suggested that the reason the Christian symbol had been removed was "to avoid causing offence or discomfort among Muslim customers" (Jiménez, 2014).

This deal was referred to by Pérez as a "strategic alliance with one of the most prestigious institutions in the world" (Jiménez, 2014). This is not a trivial consideration, given that the crown in the crest from which the cross was removed was a symbol of the Spanish Royal Kingdom, added in 1920 when King Alfonso XIII endowed the club with the title "Real", meaning royal (Sharkov, 2014). Removal of this cross illustrates the extent to which the club was willing to compromise its historical identity (Opelka, 2012).

The fact was that Real Madrid had already led the ranking of clubs with the highest revenue for eleven consecutive years (Real Madrid, 2016), demonstrating that the club was not in dire need of cash. The same resource reported that the club's 2015 revenue increased by 28 million euros compared to the previous year, 2014. According to the 19th edition of the report *Football Money League*, published by the consulting firm Deloitte following analysis of financial data corresponding to the 2014/15 season, Real Madrid's revenue totalled 577 million euros, and they led the rankings of football clubs with the highest revenue in the world for the eleventh consecutive year (Real Madrid, 2016). Nevertheless, the club needed to increase its brand value and keep strengthening its image and reputation in the world.

In terms of the concept of project identity (Castells, 1999), it could be said that the new business model and strategy developed by Pérez have resulted in a successful transformation of the football club model. He has built a new identity for a sports club that has resulted in a new way of managing a club. He transformed a mere game of sport into a showbiz entertainment, football players into world-famous 'actors', and a football club into a giant corporation that has become a role model for many sports clubs around the world.

The case of Poland serves to illustrate how a project identity can be formed, in this instance to represent a nation. Poland (Aronczyk, cited in Calhoun & Sennett, 2007) also made an effort to create a new national persona when its Ministry of Foreign Affairs hired an advertising agency to create a new image—including logo, slogan and symbol—for the nation: a red-and-white toy kite emerged as Poland's metonymic mark of identification. Poland's flag has been in existence for centuries; it accompanied King Wladyslaw II Jagiello during the Battle of Grunwald in 1410. The branding expert hired to do the redesign, Wally Olins, explained that "flags have nationalist, military or political connotations, while the kite is 'post-political'". The kite represents "a break from the past and at the same time ... it is joyful, modern" (Boxer, 2002). The process of adapting the Polish King's banner and flag that were transformed into the new country's logo was analysed by Aronczyk, cited in Calhoun & Sennett (2007).

Real Madrid's venture in creating its "new" identity or brand could be compared to what the Polish government did, because there are similarities between the two, by reference to what Aronczyk (cited in Calhoun & Sennett, 2007) says:

> *The shift from political to "post-political" representations of national identity appears to signal a change in the way we think about the ideas of a nation. It is now possible to form allegiances with regard not to shared traditions and rituals, kinship and ethnicity, language or geographic proximity, but to the profit-based marketing strategies of private enterprise.*

> *The interpenetration of corporate and state interests in creating and conveying national identity through a specific branding and marketing process is a growing trend, one that has been adopted in countries with emerging market economies and with established capitalist economies alike.*

By compromising its crest, Real Madrid is indicating to the world that it may no longer be closely associated with government, which has been associated with dictatorship and oppression in the past. Instead, due to its history, it can be a representative of a country that can work and embrace people from different backgrounds. Real Madrid is still a football club, and it can still be a representative of Spain that brings glory to the nation but, at the same time, it can collect more and more financial capital to itself to build an even stronger club.

REFERENCES

Ball, P. (2002). *White storm: 100 years of Real Madrid*. Edinburgh, UK: Mainstream Publishing.

Barcelo, J., Clinton, P. & Celo, C.S. (2014). National identity, social institutions and political values. The case of FC Barcelona and Catalonia from intergenerational comparison. *Soccer & Society, 16*(4), 469–481.

Benjamin, C. (2012). Visca el Barça! Ideology, nationalism, and the FIFA World Cup. *Kroeber Anthropological Society, 101*(1), 66–78.

Boxer, Sarah. (2002). The Way We Live Now: 12-01-02: Process; A New Poland, No Joke. Retrieved from http://www.nytimes.com/2002/12/01/magazine/the-way-we-live-now-12-01-02-process-a-new-poland-no-joke.html?mcubz=1.

Burns, J. (2012). *La Roja: How soccer conquered Spain and how Spanish soccer conquered the world*. New York, NY: Nation Books.

Calhoun, C. & Sennett, R. (Eds). (2007). *Practicing culture*. Abingdon, UK: Routledge.

Callejo, M.B. & Forcadell, F.J. (2006). Real Madrid football club: A new model of business organization for sport club in Spain. *Global Business and Organizational Excellence, 26*(1), 51–64.

Carlin, John. (2004). *White Angels. Beckham, Real Madrid & the New Football*. London: Bloomsbury Publishing.

Castells, M. (1999). *The power of identity (The information age: Economy, society and culture Vol. II)*. Oxford, UK: Blackwell.

Dunning, Eric. (1999). Sport Matters. Sociological studies of sport, violence and civilization. London: Routledge.

Fitzpatrick, R. (2012). *El Clasico: Barcelona vs Real Madrid: Football's greatest rivalry*. London, UK, Bloomsbury.

Galeano, E. (2013). *Soccer in sun and shadow*. New York, NY: Nation Books.

Giulianotti, R. & Robertson, R. (2004). The globalization of football: A study in the glocalization of the 'serious life'. *The British Journal of Sociology, 55*(4), 545–568.

Goldblatt, D. (2008). *The ball is round: A global history of soccer*. New York, NY: Riverhead Books.

Hobsbawm, E.J. (2000). *Nationalism in the late twentieth century*. Cambridge, UK: Cambridge University Press.

Hughes, R. (2009, June 11). Ronaldo to join Real Madrid for record price. *The New York Times*. Retrieved from http://www.nytimes.com/2009/06/12/sports/soccer/12iht-RONALDO.html?_r=0.

Jiménez, R. (2014, November 25). Real remove cross from badge for Abu Dhabi Bank. *MARCA*. Retrieved from http://www.marca.com/en/2014/11/25/en/football/real_madrid/1416936669.html.

Kennedy, E. & Hills, L. (2009). *Sport, media and society*. Oxford, UK: Berg.

Kuper, S. (2006). *Soccer against the enemy: How the world's most popular sport starts and fuels revolutions and keeps dictators in power*. New York, NY: Nation Books.

Le Miere, J. (2014, October 24). Real Madrid vs Barcelona 2014: Rivalry Preview, Kickoff Time, TV Channel; Duopoly Good for La Liga? *International Business Times*. Retrieved from http://www.ibtimes.com/real-madrid-vs-barcelona-2014-rivalry-preview-kickoff-time-tv-channel-duopoly-good-la-liga-1712663.

Ligaligaeropa. (2010, November 23). Sejarah La Liga Spanyol. Retrieved from https://ligaligaeropa.wordpress.com/2010/11/23/liga-spanyol-primera-divisionla-liga/.

Lowe, S. (2013). *Fear and loathing in La Liga: Barcelona vs Real Madrid*. London, UK: Yellow Jersey Press.

Marca. (2014, September 21). The Abu Dhabi Santiago Bernabéu. *MARCA.COM*. Retrieved from http://www.marca.com/en/2014/09/21/en/football/real_madrid/1411295611.html.

Marca. (2015, September 30). Camp Nou crowd aim catcalls at UEFA. *MARCA.COM*. Retrieved from http://www.marca.com/en/2015/09/30/en/football/barcelona/1443568466.html.

Nielsen Press Release (2008). Retrieved from http://www.nielsen.com/content/dam/corporate/us/en/newswire/uploads/2008/09/press_release3.pdf.

Opelka, M. (2012, April 2). Why did soccer's Real Madrid remove the cross from its logo? *The Blaze*. Retrieved from http://www.theblaze.com/news/2012/04/02/why-did-soccers-real-madrid-remove-the-cross-from-its-logo/.

Payne, S. (2008). *Spain and the great powers in the twentieth century*. Abingdon, UK: Routledge.

Peck, B. (2012, March 30). Real Madrid make small but deliberate change to their crest for UAE resort island partnership. *Yahoo!* Retrieved from https://ca.sports.yahoo.com/blogs/soccer-dirty-tackle/real-madrid-small-distinct-change-crest-uae-resort-202900831.html.

Quiroga, A. (2013). *Football and national identities in Spain: The strange death of Don Quixote*. Basingstoke, UK: Palgrave Macmillan.

Real Madrid. (2016, January 21) Real Madrid lead ranking of clubs with highest revenue for the eleventh consecutive year. Retrieved from http://www.realmadrid.com/en/news/2016/01/real-madrid-lead-ranking-of-clubs-with-highest-revenue-for-the-eleventh-consecutive-year.

Reuters. (2015, July 23). Barcelona fined by Uefa for pro-Catalan banners at Champions League final. *The Guardian*. Retrieved from http://www.theguardian.com/football/2015/jul/23/barcelona-fined-uefa-pro-catalan-banners-champions-league.

Salvatore, D. (2015, February 25). Barcelona vs Real Madrid: Spanish soccer clubs by far the most popular sport teams on social media in the world today. *Latin Post*. Retrieved from http://www.latin-post.com/articles/39647/20150225/barcelona-vs-real-madrid-spanish-soccer-clubs-by-far-the-most-popular-sports-teams-on-social-media-in-the-world-today.htm.

Sarmento, H., Pereira, A., Matos, N., Campanico, J., Anguera, M.T., & Leitao, J. (2013). English Premier League, Spain's La Liga and Italy's Serie A—What's different? *International Journal of Performance Analysis in Sport, 13*, 773–789.

Sharkov, D. (2014, November 27). Real Madrid Changes Logo in Deal with Abu Dhabi Bank. *Newsweek*. Retrieved from http://www.newsweek.com/real-madrid-changes-logo-deal-abu-dhabi-bank-287664.

Singer, J. (2014, September 22). Real Madrid close to renaming Bernabeu for £400 m as Abu Dhabi royal family offer deal. *MailOnline*. Retrieved from http://www.dailymail.co.uk/sport/football/article-2765112/Real-Madrid-close-renaming-Santiago-Bernabeu-400million-Abu-Dhabi-royal-family-offer-huge-deal.html.

Solís, F.L. (2003). *Negotiating Spain and Catalonia: Competing narratives of national identity*. Bristol, UK: Intellect Books.

Sporteology. (2014, December 3). Top ten richest football clubs in the world. Retrieved from http://sporteology.com/top-10-richest-football-clubs-in-the-world/.

Sporteology. (2015). Top 10 most popular sports in the world. Retrieved from http://sporteology.com/top-10-popular-sports-world/.

Timms, E. (Ed.). (2014, September 24). A Real Madrid theme park is in the works in Abu Dhabi. *Buro 24/7*. Retrieved from http://www.buro247.me/lifestyle/news/abu-dhabi-real-madrid-theme-park.html.

Townson, N. (2007). *Spain transformed. The late Franco dictatorship, 1959–75*. New York, NY: Palgrave Macmillan.

Tuñón, Jorge & Elisa Brey. (2012). Sports and Politics in Spain—Football and Nationalist Attitudes within the Basque Country and Catalonia. *European Journal for Sport and Society* 2012, 9 (1+2), 7–32.

Wilson, D. (2009, June 8). Why Real Madrid is so rich. *Bleacher Report*. Retrieved from http://bleacher-report.com/articles/194885-why-real-madrid-is-so-rich.

Competition and Cooperation in Social and Political Sciences – Adi & Achwan (Eds)
© *2018 Taylor & Francis Group, London, ISBN 978-1-138-62676-8*

Spiritual ecofeminism of indigenous women in Indonesia: A celebration of women's strength, power and virtue

G. Arivia

Department of Philosophy, Faculty of Humanities, Universitas Indonesia, Depok, Indonesia

ABSTRACT: Women in East Nusa Tenggara and the mothers of Mt. Kendeng in Northern Central Java have been in the public spotlight for fighting against environmental injustice. Nine mothers of Mt. Kendeng cemented their feet in concrete blocks to protest against the construction of a cement factory in 2016, while women in East Nusa Tenggara received death threats and were badly beaten for their stand against mining companies in 2006. These women were not politically driven in their protests against big companies, but instead stressed earth-based spirituality: the woman–nature link, an interconnecting and expanded view of self, and caring or promoting a compassionate lifestyle. This study examines their indigenous cultural identities, which include naturism, and how these shape their gender structures. The findings of this study indicate that although these women are "saving the earth", they restrict themselves to their local beliefs and do not question other forms of oppression such as sexism.

1 INTRODUCTION

Aleta Baun, Wilfrida and Sukinah are ordinary women living in their village. They farm, take care of the family, and take serious care of their environment. Their lives are close to nature because it guarantees their livelihood. Mothers, living in the village, work all day in the farm to grow the daily food, collect firewood for cooking or sale; they spend the money that they earn on their children's education and to honour the forest, which is part of their religious practices. They believe in the prowess of nature and its connection with women and not necessarily with feminism.

The paper describes the cases of three women leaders from different indigenous communities who fought for their land against mining companies. Using ecofeminist perspectives,[1] it uses qualitative interviews to investigate the three women's experiences. The research focuses on the following questions: What is the meaning of nature and its connection with women? How do they experience discrimination against women in their environment? These questions are pertinent to understanding whether these women are ecofeminists or if they are merely practising their indigenous cultural belief without any connection to women's rights.

The paper further explores the term ecofeminism and how it fits, particularly in the indigenous context. Do their framework fall into essentialism? How do they regard gender bias in a patriarchal community? Is there any interconnection with domination over nature and domination over women? And, how do women deal with gender bias in their community?

---

1. The ecofeminist perspective that I employ is social and political that rejects essentialist and universalizing assumptions about women and the importance of understanding oppression, subordination and domination of women (emphasizing feminism). See Karen J Warren work on ecofeminist philosophy.

2 LITERATURE

Several Indonesian-perspective studies on ecofeminism, which relates to the indigenous community and sees it from the spiritual side, are published in *Ecofeminism III: Mining, Climate Change & Womb Memory* (Candraningrum & Hunga, 2015). They describe case studies in Indonesia, such as the one written by Asriani and Mangililo (2015) on Mollo women from the secluded area in the South Central Timor Regency, East Nusa Tenggara, and Saturi's (2015) article that discusses indigenous women's problems and communal rights in an agrarian conflict, drawing examples from the case of Dayak Maanyan in Central Kalimantan. Those writings and others compiled in the book are very useful because they provide a general description of the cases found in Indonesia and the position of women there. However, even though the book does mention the area of ecofeminism, it does not provide an in-depth debate about women's oppression and domination the essentials of ecofeminist theory, nor does it address gender issues in the indigenous community.

Another study that is also useful in seeing the problem of indigenous communities and spiritual ecofeminism is the perspective of Mies and Shiva (2014). Mies and Shiva are ecofeminist thinkers who were well-known in the 1970s and whose ideas contributed significantly to the development of ecofeminism. From reflection on their studies, one can see how ecofeminism could integrate with almost all domains, such as economic theory, women's empowerment, women's reproduction rights, patriarchal culture, development theory, philosophy, and modern technology. Mies and Shiva also underline the problem of violence against women in many forms, including sexual aggression and economic exclusion. In addition, I also refer to the work of Karen J. Warren, *Eco-Feminist Philosophy*, published in 2000. The book highlights the domination and oppression of women, including social oppression. Warren emphasises the importance of a multicultural ethical understanding, but she also specifies transcultural values such as justice and freedom. In this paper, I also draw on Starhawk's (1982) book to understand spiritual ecofeminism, which relates particularly well to the Mt. Kendeng mothers.

3 WOMEN, INDIGENOUS COMMUNITY AND ENVIRONMENT

Throughout the history of women in Indonesia, women from indigenous communities have experienced some form of discrimination. In the pre-colonial era, the experience of indigenous women was possibly better even though they suffered exclusion as a result of male-led customs. During the Deutch colonial times (Indonesia was colonialized 350 years by the Deutch)), there were many stories about women brought as slaves to Batavia from the villages to serve in the houses of the Dutchmen who had mistresses. The story, dated 25 April 1775, of Tjindrais is an example. She told her doctor, David Beijlon from Kampung Baru, who was examining her, that she was beaten by her employer's mistress with firewood and rattan (Jones, 2010). Hundreds of other women like Tjindrais, living in the eighteenth century, became slaves and suffered from torture and violence in their workplace. During the VOC[2] era, the Dutch companies employees were permitted to marry or to own a mistress. The arrival of the Dutch in Indonesia had classified women into a hierarchy: the whites, who had the highest status, followed by the Chinese and Arabs, the mixed-bloods, and the brown-skinned (Indonesian natives) who were at the lowest level. The control of the social system in Southeast Asia was in the hands of the men. They had mistresses whose social status was escalated. Violence against slaves happened in the domestic space.

Aleta Baun, a figure from the indigenous community from East Nusa Tenggara and an environmental activist and fighter, believes that the problem of colonialism is one of the many problem causing women from the indigenous community to experience violence and

---

2. VOC (The Deutch East Indoa Company) founded in 1602 was a company of combination of commercial organisations in Holland traded between Asia and Europe.

to live in poverty. She cautions that women's position is weakened because of colonialism: "When a woman becomes someone's wife, she must work at home. Indigenous people do not recognise women staying home but they work in the field. They are working women." (Baun, Interview 9 July 2016)

The role of Indonesian indigenous women became prominent in 2012 after a meeting was held in Tobelo, Maluku, gathering 200 women from indigenous communities to form a group. They were committed to upholding the rights of women and to protecting nature (McVeigh, 2012). At the meeting, it was noted that women from indigenous communities were at the frontline in discussing environmental issues because they were knowledgeable about the science of nature and were skilful in arranging and monitoring their land. They were also well connected with their community so they could be the agent of change.

One famous figure is Aleta Baun from the Mollo community, East Nusa Tenggara. She won a Goldman Environmental Prize in 2013 for her action in organising hundreds of people in her village to fight against the mining company that would damage the sacred land on Mutis Mountain. The Mollo community believes that they are united with nature and that this cannot be disturbed. In her struggle, Baun was pressured and intimidated, but she kept on fighting because she believed in a saying from her region: *Pohon bagai rambut dan kulit, air bagai darah, batu bagai tulang, tanah bagai tubuh* (Tree as hair and skin, water as blood, stone as bone, land as the body). Aleta Baun describes the relationship between women and nature as follows:

> *There is no difference between women and environment because both depend highly on each other. When we speak about women, we speak about food. When we speak about food, we speak about women. Women are responsible for the seed and for the food put on the table. When we talk about nature, we talk about women.* (Baun, Interview 9 July 2016)

The presence of Aleta Baun as a figure is fascinating in the ecofeminism discourse because it exemplifies the women's movement as initiated by an ordinary group of women. For example, in India, Chipko was a well-known women's movement in 1986, led by an ordinary woman, Chamundeyi from Nahi-Kala village. When she was looking for firewood, she heard the sound of big trucks climbing up to the mountain. As she was not in favour of the mining company's plans to conduct activity in her region, she, together with women from her village, put up tents to blockade the mining company. When she heard the trucks entering the area surrounding her mountain, she stood in front of the trucks and stopped them. On one occasion she was dragged by a truck before it stopped (Shiva, 2014).

Similar to the figure of Chamundeyi, Wilfrida Lalian was an environmental fighter from Oekopa, North Central Timor (TTU), and East Nusa Tenggara. She was a modest figure from the Usatnesi Sonaf K'bat indigenous group, which is the majority group in Oekopa. She dared to fight against a large mining company, PT Gema Energi Indonesia, in 2012. One day, she saw large trucks coming and being driven up her mountain, and then she decided to take action.

She was shocked as she realised that she had been cheated. Initially, the company said that they would come to do research. However, they blockaded, drilled and exploited the land. On one occasion, they gathered thousands of people to intimidate the local community. Wilfrida was so upset that she faced the mob alone. The following is an excerpt from the interview with Wilfrida on 8 July 2016:

> *Their number was thousands and I was alone. My mother told me not to go. But I went to challenge them. The police were in front of my nose. I was emotional and my hands were on my hips, challenging them. I took a deep breath, "Good afternoon. What are you coming for?" The Chief of Police Resort studied me from top to toe; I wasn't afraid. The Village Chief had run away, my husband was hiding at home. So I challenged them; I said, "If you dare, I will trample you down, tie you up and throw you to the river". They said, "Arrest her!" But I said, "I am the host: even before the government existed, the land belonged to the indigenous people, to the community".*

They deployed thugs and police personnel and they threatened me. I was courageous because I am not wrong. They are thieves, and they came as guests, so I need to challenge them. I provoked them. I said if you want to bring me (incarcerate me), the first party responsible should have been the village government. I never take their money, not even a penny. Never. The land belongs to the community.

What was unique about Wilfrida is that she fought alone because her community was too frightened, and she could not force the mothers to join, especially if their families did not allow it. She tells the story that, sometimes, she drove away company and local officials when patrolling.

At night the company cut trees, I came with my motorbike, they ran away. I scolded them. Now they know, if there is the sound of my motorbike, they run. (Wilfrida, Interview 8 July 2016).

Wilfrida attests that her strengths came from nature. She is very devoted in practising the rituals in her community, such as giving pigs and chickens as offerings based on the customary practice, so nature will protect her. Wilfrida reflects the spiritual dimension in the life of the indigenous community.

The customary ritual (adat) ceremony is supported by the clergy, nuns and the indigenous community. We walk and bring offerings such as pigs and chickens, so nature will help us. It is a habit. If we ask something and give offerings, the wish will come true. Wusam (primary ritual place) is our place where we ask for rain, food and anything. We believe in customary ritual. (Wilfrida, Interview 8 July 2016).

The ceremony is called "na hake neu paham ma nifu" (that the land and water should be treated well for the children and grandchildren) in Dawan, the local language. Similar to Wilfrida's "adat" (customary practice), the mothers in Mt. Kendeng also practice customary ritual to worship and protect nature. The Kendeng mothers are part of the Sedulur Sikep community, which is the indigenous community living among the Mt.Kendeng from Blora and Pati in Central Java, up to Bojonegoro in East Java. They are the followers of Surosentiko Samin's[3] teachings, dubbed as *Raden Kohar* due to his noble Majapahit descent. He believed in "manunggaling kawula gusti" (God is within me), thus the Samin people do not practice the Muslim faith. Saminists believe in the "Faith of Adam". (Kartopradja, 1990). The Samin people regarded natural forests as open to all. This practical teaching attempts to convey a message on how to live well through traditional Javanese songs.

One prominent figure from the Kendeng mothers was Sukinah, who delivered her messages via a song entitled *ibu bumi wis maringi, ibu bumi dilarani* (Mother Earth has given, Mother Earth is hurt, Mother Earth will judge). Sukinah believes that everything in the world comes from Mother Earth, who gives us life. The responsibility of women in protecting nature included a struggle against the cement factory in her region, PT Semen Indonesia. Her protest against PT Semen Indonesia started in 2014.

We are women; Mother Earth is likened as myself. If women give birth to human beings, then Mother Earth gives birth to water, plant and edible sources for humans. If all gone, automatically, mothers will suffer from the impact. So women have to care for nature and are responsible for it. (Sukinah, Interview 20 June 2016).

My interview with these three women figures shows that women have a strong connection with nature. For them, woman is nature. The connectedness of women with nature links to spirituality, as ecofeminist Starhawk (1989) emphasises the relationship between women and nature. She argues that the experience of women in menstruation, pregnancy and breastfeeding

---

3. Surontiko Samin was the leader of a non violent movement in the late 19th and early 20th century which rejected the Dutch colonial forced taxes on the people of Indonesia. The Dutch also monopolized public forest lands which contains teak forests for trade. See Harry J. Benda and Lance Castles on their work on the Samin people.

is an earth-based spirituality. According to her, earth-based spirituality involves three concepts. The first one is immanence. Immanence is a spirituality that is not addictive (unlike the Semitic religion), but it is a spirituality that grows and has energy to take action. The second is the existence of interconnection and self-expansion. It is not only our body that is natural, but also our way of thinking. Starhawk emphasises: "Our human capacities of loyalty and love, rage and humour, lust, intuition, intellect, and compassion are as much a part of nature as the lizards and the redwood forests" (Starhawk, cited in Tong, 1998, p. 262). The third and the most important is the existence of a compassionate lifestyle often possessed by women. Starhawk reminds us of the necessity of "reweaving the world" or "healing the wounds". Aleta Baun shows the interconnection between nature and women by referring to the weave made by women:

> Through the weave, the earth is symbolised as a mother that needs a cover. When the earth is destroyed, it means that we skin the human. The hair is cut off, the skin is reddish, and the earth disapproves that. The motive of the weave symbolises respect for nature. (Baun, Interview 8 July 2016).

As for Starhawk, there is an intervention of the meaning of women who weave with the women healing process against the nature.

4   ECOFEMINISM, A MOVEMENT AGAINST OPPRESSION OF WOMEN

In ecofeminist discourse, there are two contrasting views between those who stress the interconnection between women and nature only, and those who see the struggle of ecofeminism as a struggle of anti-domination over women. Explicitly, Warren (2000) emphasises that the struggle to end naturism exists because the objective of feminism is to end sexism. Although feminism is a movement to end sexism, sexism is conceptually linked with naturism through an oppressive conceptual framework characterised by the logic of domination. Thus, feminism is (also) a movement to end naturism (Tong, 1999).

Warren's statement shows that ecofeminism still aims to fight for women's rights. Several opinions consider that spiritual ecofeminism can also indirectly end men's domination. However, Warren asserts that the goal of ecofeminism must come from an understanding of feminism, and that the concepts of ecofeminism are based on the following: (1) there is an important connection between the oppression of women and the oppression of nature; (2) understanding the nature of these connections is necessary to any adequate understanding of the oppression of women and the oppression of nature, and (3) feminist theory and practice must include an ecological perspective as well as solutions to ecological problems (Warren, 1987). Further, she emphasises the three things that make feminism important in ecofeminism. First, there is the scholarly issue of accurately representing the historical and empirical realities of the interconnections among the domination of women, "other", and nature. Second, the domination of women, "other", and the domination of nature are "justified" by the oppressive and logical domination framework. Third, the use of the prefix "feminist" helps to clarify just how the domination of nature is linked to patriarchy, the domination of women, and other human subordinates (Warren, 2000). Warren's criticises the spiritual position in ecofeminism which does not emphasise the prefix "feminist". Based on the case study, Aleta Baun is the only figure who remains true to the struggle for women's equality and questions domination over women.

> The sign near my house (is) Stop Violence against Women. I once joined Yayasan Sanggar Perempuan (Women's Community Foundation). For me, violence against women should be discussed with the indigenous community. We should take the approach of storytelling, because most indigenous communities don't read. (Baun, Interview 8 July 2016).

As a member of the Provincial Council Representatives (locally known as DPRD), Baun is now thinking about how to decrease the amount of violence against women in East Nusa Tenggara, as it is a problem in the area. However, she insists that in the philosophy of the indigenous community, women's status is equal to men. She draws an example from the pillars of the indigenous house which clearly depict women and men as equal. Baun is very articulate in talking about women's rights and connecting them with environmental issues, so she has a gender-perspective analysis.

Meanwhile, Wilfrida talks about the high maternal mortality rate in her region, but she does not connect it specifically with the environmental issues. What I find interesting about Wilfrida is her autonomy, as she protested against the mining company because of her own decision, so she is an independent agent. Autonomy is significant in the feminism perspective. On the other hand, Sukinah's narrative differs from Wilfrida's. She was "selected" by Gunretno, the man who is the coordinator of the Mt. Kendeng Care Community Network (JMPKK). He was the one who contacted Sukinah and invited her to do a rally and to stage a protest in Jakarta. He saw Sukinah as someone who possessed good verbal skills, so he asked her to become his "Senopati" (Commander) of Rembang. Gunretno also composed the songs and organised the Kendeng Mothers protest (he named it "My Action"). He went as far as determining what the mothers should do and should wear (specific clothes, namely white cloth) as they marched to the State Administrative High Court (Dhewy, 2016).

The Kendeng mothers show strong feminine values, but not feminist values. Their arguments are not coming from the active voice (first person), but from the passive voice (third person) instead. They tend to give their voice in the name of what Mother Earth needs and how Mother Earth regulates all. This passive voice shows that the environmental movement is not something that is political or "socially constructed", but it is fixed (because it is based on their belief not knowledge).

5 CONCLUSION

All three subjects interviewed for this research claimed that women were essentially or biologically closer to nature. There was a strong view among the three subjects, particularly with the Mt. Kendeng mothers, that "biology is destiny" or that the role of women was that of nurturing mothers. This approach attributes to body-based, and the belief that women are connected to nature because of their nurturing role. This view is problematic in two ways. First, if we believe that women are connected with nature and are the primary caretaker of nature (because men are not equipped biologically), then women are simply fulfilling their traditional role as nurturing mothers and not as emancipated women. Second, connecting women with nature because of their feminine values (essentialism) does not question the patriarchal culture and the domination and exploitation of women. Actually, the claim that women are biologically closer to nature reinforces patriarchal ideology.

The movement of indigenous women against the exploitation of the environment is important, and it should be supported and celebrated. However, the movement should not rely solely (or overly identify) with women's bodily experiences, and should not be uncritical of gender injustice and its various forms. The reason for this is because, ultimately, the goal of feminism is to end sexism.

REFERENCES

Aleta Baun, in Kota SOE, Timur Tengah Selatan, NTT (Interview 9 July, 2016).
Asriani, D. (2015). Perempuan Mollo Merawat Tubuh & Alam: Aleta Baun, Paham Nifu & Pegunungan Mutis. In Candraningrum & Hunga (Eds.), *Ekofeminisme III: Tambang, Perubahan Iklim & Memori Rahim*. Yogyakarta, Indonesia: Jalasutra.
Dhewy, A. (2016). Gunretno: Ibu Bumi Wis Maringi, Ibu Bumi Dilarani, Ibu Bumi Kang Ngadili. *Jurnal Perempuan 91, Pedagogi Feminis, 21*(3), 217–230.
Jones, E. (2010). *Wives, slaves and concubines: A history of the female underclass in Dutch Asia*. DeKalb, IL: Northern Illinois University Press.

Kartopradja, K. (1990). *Aliran Kebatinan dan Kepercayaan di Indonesia*. Jakarta, Indonesia: CV Mas Agung.

Mathuki, J. (2006). Challenging structures: Wangari Maathai and the Green Belt Movement in Kenya. *Agenda: Empowering women for gender equity, 20*(69), 82–91.

McVeigh, C. (2012). *The indigenous women perspective: A personal meeting space*. Retrieved from http://www.downtoearth-indonesia.org/story/indigenous-women-s-meeting-personal-perspective.

Mies, M. & Shiva, V. (2014). *Ecofeminism*. London, UK: Zed Books.

Mowforth, M. (2014). *The violence of development*. London, UK: Pluto Press.

Ruggie, J. (2013). *Just business: Multinational corporations and human rights*. New York, NY: Norton & Company.

Saturi, S. (2015). Perempuan adat dan hak ulayat dalam Konflik Agraria: Kajian Ekofeminisme. In Candraningrum & Hunga (Eds.), *Ekofeminisme III: Tambang, Perubahan Iklim & Memori Rahim*. Yogyakarta, Indonesia: Jalasutra.

Starhawk (Ed). (1982). *The politics of women's spirituality*. Garden City, NY: Anchor.

Starhawk. (1989). Feminist earth-based spirituality and ecofeminism. In J. Plant (Ed.), *Healing the wounds: The promise of ecofeminism*. Philadelphia, PA: New Society Publishers.

Sukinah, in Jakarta (Interview, 20 June, 2016).

Tong, R. (1998). *Feminist thought*. Australia: Westview Press.

Warren, K. (1987). Feminism and ecology: Making connections. *Environmental Ethics, 9*, 3–20.

Warren, K. (2000). *Ecofeminist philosophy*. Lanham, MD: Rowman & Littlefield Publishers.

Wilfrida, in Desa Oenbit, Kecamatan Insana, Kabupaten Timor, Tengah Utara, NTT (8 July, 2016).

Competition and Cooperation in Social and Political Sciences – Adi & Achwan (Eds)
© 2018 Taylor & Francis Group, London, ISBN 978-1-138-62676-8

The dynamics of Japan–ASEAN diplomacy: The significance of the 1991 Nakayama Proposal upon the establishment of the ASEAN Regional Forum in 1994

D. Afiatanti & I.K. Surajaya
Japanese Studies, Faculty of Humanities, Universitas Indonesia, Depok, Indonesia

ABSTRACT: The Nakayama Proposal is an initiative of multilateral security dialogue submitted by Japan's Former Foreign Minister Tarō Nakayama to the Association of Southeast Asian Nations (ASEAN) on 22 July 1991. It was a significant initial phase that was related to the establishment of the ASEAN Regional Forum (ARF) in 1994. The ARF is regarded as a considerable instrument of community security because it was the first forum where high-ranking representatives from the majority of states in the Asia-Pacific region gathered specifically to discuss political and security cooperation issues following the end of the Cold War in the 1990s. Various scholars such as Midford and Nishihara claimed that the Nakayama Proposal, despite ASEAN's rejection of it, had contributed positively to the establishment process of the ARF by accelerating and formalising ASEAN's collaborative capacities for collective actions. This paper authenticates this by analysing ASEAN's policy characteristics in the process of establishing this regional security forum before and after the submission of the proposal. The paper concludes that the Nakayama Proposal triggered and catalysed the founding process of the ARF.

1 INTRODUCTION

Japan's foreign policies towards the Association of Southeast Asian Nations (ASEAN) from 1973 to the late 1980s focused mainly on commercial affairs such as economics or cultural exchanges and tended to avoid security and multilateral issues. Most of Japan's prime ministers' policies emphasised Japan's demilitarisation. Prime Minister Takeo Fukuda, in his political speech, known as the Fukuda Doctrine, in Manila, on 18 August 1977, stated that Japan would not play any role in international military affairs (Sudo, 1995). Further, commercial sectors such as economics, trade and cultural exchange dominated the first to the twelfth ASEAN–Japan Forum's agendas (ASEAN Secretariat, 1988, 1989a, 1991, 1994). However, the end of the Cold War in the 1990s led to dramatic upheavals in Europe and overshadowed the profound changes occurring in Asia. The collapse of communism and the Soviet Union did not necessarily mean that there would be more peace and stability, particularly in South Asia and Southeast Asia, since the term of military pact with the U.S. is no longer relevant. In Asia-Pacific, several proposals regarding a multilateral regional security forum were proposed to ASEAN, mostly by Australia, Canada, and Japan.

Japan, in particular, proposed its idea of a future security community through its Foreign Minister, Tarō Nakayama. The Nakayama Proposal was an initiative for dealing with security issues in the Asia-Pacific region which was submitted during the ASEAN Post-Ministerial Conference (PMC) held in Kuala Lumpur in 1991. The ASEAN-PMC are series of annual meetings between ASEAN foreign ministers and their counterparts known as 'dialogue partners'. The primary agendas of PMCs focused on economic issues, while political issues formed part of its agenda in the 1990s. The ASEAN-PMC was regularly held following the ASEAN Ministerial Meeting (AMM) every year. These meetings had been initiated gradually since the mid-1970s. However, from 1994 onwards, the PMC was replaced by the

ASEAN Regional Forum (ARF), one of the security community instruments in Asia-Pacific (Acharya, 2001).

Japan participated in the PMCs from 1978 to 1993. In 1991, the ASEAN-PMC was held in Kuala Lumpur on 22–24 July. On the first day of the conference, Japanese Foreign Minister Nakayama delivered a major political speech at the general session (Ogasawara & Oobashi, 1991). Yuzawa (2007) showed that Nakayama not only promoted a regional multilateral security dialogue, but also proposed a direct mechanism for executing the dialogue, namely the Senior Officials' Meeting (SOM). The proposal for establishing an SOM under the ASEAN-PMC was the most important point in the Nakayama Proposal.

The submission of the Nakayama Proposal, particularly the SOM mechanism, led to various disparate responses from countries attending this ASEAN-PMC. Australia and Canada agreed with the idea, yet, dismissive responses came from the ASEAN countries. As reported by several Japanese local newspapers, the ASEAN reaction to the Nakayama Proposal was rather careful. There were no ASEAN countries which expressed clear agreement with the proposal (Aso, 1991). In the plenary session of the PMC, ASEAN clearly stated that it would consider the detail of the meeting's mechanism with or without Nakayama's idea. This statement was also reinforced by Indonesian Foreign Minister Alatas and Malaysian Foreign Minister Abdullah Badawi (Ikeuchi & Kandia, 1991; Midford, 2000).

Despite ASEAN's rejection of Nakayama's idea, some scholars expressed their support and argued that the proposal had contributed positively to the foundation of ASEAN's first security community instrument, namely the ASEAN Regional Forum (ARF), by promoting the idea of a multilateral security dialogue connected with ASEAN. Thus, Midford (2000) and Nishihara (2003) both agree that the proposal implicitly accelerated the foundation of the ARF.

Based on the aforementioned arguments, this paper authenticates the Nakayama Proposal's contribution in accelerating the 1994 ASEAN Regional Forum by analysing ASEAN's policy characteristics in the decision-making process of the regional security forum before and after the submission of the Nakayama Proposal in 1991, using Japan and ASEAN's official documents as the main corpus. In addition, several interviews with officials, researchers and journalists were conducted to strengthen the argument. By doing so, this paper seeks to respond to discourses regarding the success or failure of the 1991 Nakayama Proposal, and prove the impact of the 1991 Nakayama Proposal on ASEAN's policy alteration in the regional security forum, which indirectly led to the foundation of the ARF, and offer new interpretations of the issues surrounding the 1991 Nakayama Proposal.

2 THE ABSENCE OF REGIONAL SECURITY FORUM POLICY PRIOR TO THE SUBMISSION OF THE NAKAYAMA PROPOSAL (JANUARY 1989–JULY 1991)

The following section identifies patterns in ASEAN's policy prior to the submission of the Nakayama Proposal, mainly in the regional security forum. In order to apprehend ASEAN's policy direction, each agenda of ASEAN's official meetings and conferences from January 1989 to June 1991 was explored in chronological order.

The first policy instance is the ASEAN Ministerial Meeting (AMM) and Post-Ministerial Conference (PMC) 1989, which were held on 3–8 July in Bandar Seri Begawan. The forum was dedicated to economic consultation and cooperation among the Pacific Rim countries (ASEAN Secretariat, 1989b). In cooperation with the dialogue countries and other organisations, such as United Nations, ASEAN's foreign ministers noted considerable progress and satisfaction over the results of its cooperation with organisational partners and forums, for example, the United Nations Development Programme (UNDP), the ASEAN–Australia Forum, Uruguay Round, ASEAN–Canada Economic Cooperation, and the ASEAN–Japan Forum (Bangkok Post, 1989a).

The second policy instance is Singapore offers of US military base facilities in August 1989. Approximately 40,000 soldiers, civilian employees and military dependents of the US were stationed at the six installations in the Philippines' Clark Air Base and Subic Bay

Naval Base. However, the land lease for the bases was due to expire in 1991. Singapore's support for the continued American military presence in this region had been known since 1989. The offers led to various responses from ASEAN members. The Thai Foreign Minister and National Security Councils said that they understood and would not object to Singapore's move. However, this idea was openly opposed by Malaysia. Prime Minister Mahathir Mohamed said that the establishment of a US military base in Singapore would lead to arms escalation and threaten efforts to bring peace and stability to the region. Singapore's offer would be detrimental to ASEAN's goal of establishing a Zone Of Peace, Freedom And Neutrality (ZOPFAN). Meanwhile, Indonesian Foreign Minister Ali Alatas said that Singapore's offer to provide facilities for US forces had not created disunity within ASEAN. Alatas said that no commitments had been made on the issue, while a statement by Singapore Foreign Minister Wong Kan Seng declared that the small republic had the right to decide its future without interference (Bangkok Post, 1989c).

The next ASEAN policy instance is ASEAN Ministerial Meeting (AMM) and Post-Ministerial Meeting (PMC) 1990, which were held on 24–29 July in Jakarta. Cambodian issues still dominated this conference as well as the meeting in the previous year (Bangkok Post, 1990a). Moreover, in this meeting, Australia proposed the creation of a new Asia-Pacific forum for addressing and resolving regional security problems modelled on the Conference on Security and Cooperation in Europe (CSCE). ASEAN rejected the proposal because it was not compatible with ASEAN's principles, the "ASEAN way".

It was a month before the regular meeting of the AMM and PMC in 1991 that the ASEAN member states, particularly the Philippines, had actually begun the process of reviving their dialogue on regional security issues and of widening their exchanges of views with their dialogue partners. A seminar on "ASEAN and the Asia-Pacific Region: Prospects for Security Cooperation in the 1990s" held on 5–7 June, 1991 was a vivid example of this initiative. The seminar was sponsored by the Philippines and Thailand governments. The participants were mostly researchers from ASEAN-Institutes of Strategic and International Studies (ASEAN-ISIS) Malaysia and professors from various universities in Asia-Pacific. According to *ASEAN and the Asia-Pacific Region: Prospects for Security Cooperation in the 1990s* (Department of Foreign Affairs Philippines & Ministry of Foreign Affairs Thailand, 1990), two Japanese speakers, Director-General Yukio Satō of the Information Analysis Bureau Ministry of Foreign Affairs and Professor Masashi Nishihara, a researcher of the Institute for Peace and Security, were invited to illuminate participants regarding Japan's security interests in Southeast Asia in the third session of the seminar. At the conference, Sato's vision for regional security multilateralism, as articulated in the concept of a multifaceted approach, successfully drew support from ASEAN's participants (Yuzawa, 2007). By this time, (the ministries of foreign affairs) and intellectuals in ASEAN-ISIS had also reached a consensus on the need to establish a region-wide forum for multilateral security dialogue (Kompas, 1991).

3 THE ALTERATION OF ASEAN'S POLICY ON REGIONAL SECURITY FORUM AFTER THE SUBMISSION OF THE NAKAYAMA PROPOSAL (JULY 1991–JULY 1994)

This section explores ASEAN's policy on a regional forum after the submission of the Nakayama Proposal on 22 July, 1991 until July 1994 when the first ARF working session was held. Following the submission of the Nakayama Proposal, ASEAN's ministerial leaders held a special closed assembly. The meeting focused on the Japan issue because ASEAN had a collective opinion regarding the country, both as a partner and a threat. ASEAN will afford a regional security-community, through confidence-building activities by each country (Kompas, 1991). The Malaysian Foreign Minister, Abdullah Ahmad Badawi, stated that Japan did not ask ASEAN to decide then but requested that they thought about it. He believed that the dialogue partners could still raise the regional security issue at the PMC without turning it into an exclusive regional security dialogue (The Star Malaysia, 1991).

The fourth ASEAN summit, commonly known as the Singapore Summit, was held on 27–28 January 1992. According to an article published on 20 January by The Star Malaysia (1992), this summit was a significant summit for the ARF because this was the first time the concept of a regional security forum mechanism was discussed. Prior to the summit, ASEAN-ISIS had actually made a policy recommendation called *A Time for Initiative* (ASEAN-ISIS, 1991). The recommendation urged the Singapore Summit meeting of ASEAN to address the issue of advancing towards a new regional order in Southeast Asia as well as enriching and strengthening the ASEAN-PMC process so that the dialogue partnerships could be further developed in new directions. The above development eventually persuaded ASEAN's leaders, more or less, to decide its policy direction towards a new order in Asia-Pacific.

In contrast to the AMM of 1991, the member states agreed to address security matters through the ASEAN-PMC. The Singapore Summit generated the Singapore Declaration, which clearly mentions that ASEAN could use established forums to promote external dialogues on enhancing security in the region, as well as intra-ASEAN dialogues on ASEAN security cooperation, by taking full cognizance of the Declaration of ASEAN Concord. To enhance this effort, Emmers (2001) argued that ASEAN should intensify its external dialogues in political and security matters by using the ASEAN-PMC. Therefore, on 21 July 1992, ASEAN's foreign ministers opened their annual meeting with a call for more discussions on regional security as it was requested by the dialogue partners although the talks regarding political-security had long been a taboo among the six grouping members (The Star Malaysia, 1992).

A first ever ASEAN-PMC Senior Officials' Meeting (SOM) was organised in Singapore on 22 May, 1993 and brought together the heads of the foreign ministries of the ASEAN countries and their dialogue partners. The political and security context in the Asia-Pacific region following the Cold War was discussed at this meeting. They agreed to form a multilateral process of cooperative security in order to promote cooperation in the region. Significantly, the ASEAN-PMC SOM registered balance of power considerations in signalling the forthcoming establishment of the ARF (Emmers, 2001).

The AMM and ASEAN-PMC 1993 were held in Singapore on 23–24 and 26 July, At the meetings, ASEAN stressed the importance of developing a relationship among equals between ASEAN and its dialogue partners for peace, stability and economic growth in the region and the larger Asia-Pacific region (Thaitawat, 1993). This was the first time that the major players in the Asia-Pacific region had the opportunity to discuss regional security issues collectively (Chew, 1993). Further, Japan's Foreign Minister Kabun Mutō emphasised that Japan intended to actively take part in the political and security dialogues at the ASEAN-PMC and its SOM. He added that Japan would play a positive role by strengthening its cooperative relations with the ASEAN countries and others in the region (Thaitawat, 1993).

Finally, the year 1994 was a period of consolidation for ASEAN. Two years after the summit in 1992, member countries had to transform specific cooperation initiatives into concrete plans of action in various fields. Regional political and security cooperation had been enhanced on the intra-ASEAN and Asia-Pacific levels. The 27th AMM was held in Bangkok on 22–23 July while the PMC and the ASEAN Regional Forum (ARF) were scheduled for 25–28 July 1994 (Bangkok Post, 1994). Thus, the first meeting of the ARF was held in accordance with the 1992 Singapore Declaration of the 4th ASEAN Summit, where the ASEAN heads of state and governments proclaimed their intention to intensify ASEAN's external dialogues in political and security matters as a means of building cooperative ties with states in the Asia-Pacific region (ASEAN Secretariat, 1994).

4 ANALYSIS AND DISCUSSION OF ASEAN'S POLICY SHIFT: FROM INDIFFERENCE TO CAUTIOUS DIRECTION

By observing each policy of ASEAN before the submission of the Nakayama Proposal on 22 July 1991, we can observe that ASEAN initiated no idea of a regional security forum. ASEAN policies and cooperation were dominated by economic fields such as trade and industry as well as the issues of the Cambodian conflict. However, it is important to highlight the following points.

First, a couple of individualdecisions preoccupied ASEAN's members throughout 1989–1990. In August 1989, Singapore decided to make some offers to accept some American military facilities. In the end, the US stated that it could not accept the offers because it did not believe that Singapore was large enough to replace US bases in Southeast Asia (Bangkok Post, 1989).

Second, ASEAN leaders, taking into consideration the security conditions in Southeast Asia, had begun to realise the need for a regional security forum to be held as soon as possible during the AMM 1990 talks. Unfortunately, this idea was left untouched because there was no further action related to the mechanism of such a forum until January 1992, when the Singapore Summit was held. Rather, on 7 June 1991, Southeast Asian nations agreed to consider forming a security consultative body to deal with such diverse issues as international terrorism and pollution (Bangkok Post, 1991).

Third, the idea of a new regional security approach had not originated from ASEAN's leaders. Instead, the initial recommendation came from ASEAN-ISIS at the Manila Conference in June 1991. Despite the sponsorship by the Philippines and Thailand governments of the Manila Conference in June 1991, it was very unfortunate that the only ASEAN member-state representatives who attended the meeting were the Philippines's Foreign Minister Raul Manglapaus and President Corazon Aquino. This shows the low level of ASEAN's seriousness in dealing with the issues of a new regional order in the context of a security forum. In addition, several days before the ASEAN-PMC 1991, a CSIS researcher, Jusuf Wanandi, had actually suggested a similar concept for a regional forum as did Foreign Minister Nakayama (Wanandi, 1991). As a matter of fact, as also emphasized by Yukio Satō in the interview with the author, ASEAN showed little interest in the suggestion from ASEAN-ISIS. ASEAN acted cautiously in approaching its diplomatic partners while planning and debating a specific regional dialogue mechanism. We can therefore infer one feasible explanation for ASEAN's choice of policy in the regional security forum before the Nakayama Proposal: ASEAN had not been ready to expand its role by outlining any mechanism for a regional security forum. It is assumed that ASEAN acted cautiously due to ASEAN's internal process of *adjustment to diverse members' interests* and the *normative adherence to the ASEAN way* in deciding any policy.

ASEAN's consciousness of the importance of a regional security forum was presumably awakened by Japan's bold proposal of the SOM concept incorporated in the Nakayama Proposal, which was also presumed by Masashi Nishihara during the interview with the author. Although ASEAN denied the role of Japan and other dialogue partners' proposals, we obviously cannot ignore the fact that a number of tangible policies regarding a regional security forum began to appear in most of ASEAN's meetings and discussions in the period after the submission of the Nakayama Proposal. Never before had ASEAN held a closed assembly after the annual PMC, except for the one held after the submission of the Nakayama Proposal. The Jakarta Post (1991) article on 25 July noted that it was the Japanese initiative, taken by Foreign Minister Taro Nakayama, that on this occasion caused the issue to acquire a special significance for many observers of ASEAN developments. Alatas even admitted to James Baker, Secretary of States of the U.S., that they would study Japan's proposal first, which in fact he had not done when Australia proposed a CSCE-like forum two years earlier.

The shift in ASEAN's policy characteristics and direction before and after the submission of the 1991 Nakayama Proposal, as demonstrated in the two previous sections, leads us to one inference: that the proposal was indeed affecting ASEAN's leaders' paradigms in rather positive ways, as a trigger and as a catalyst. In addition to Midford's statement, it can further be asserted that the proposal was also acting as a trigger for ASEAN's further actions. The Oxford Dictionary defines the word 'trigger' as "an event that is the cause of a particular action, process or situation". Exactly what action and process were triggered by the proposal? The Nakayama Proposal triggered a shift in ASEAN's policy in terms of influencing and deflecting its leaders' ways of acting. This can be seen in the difference in attitude as they started to cautiously take real steps of action after Nakayama proposed the idea of a SOM mechanism, bolstering Midford's claim that the proposal was a catalyst for regional security multilateralism. According to the Oxford Dictionary, the word '*catalyst*' is defined as "a person or thing that precipitates an event". The 1991 Nakayama Proposal was a catalyst in terms of precipitating the first ever discussion of regional security forum issues during the Singapore Summit in January 1992, although the meeting was only supposed to be dealing with economic issues. This discussion then led to other important meetings, boosting the founding process of the ARF.

5 CONCLUSION AND FURTHER RESEARCH

The Nakayama Proposal, composed between 1990 and 1991, was Japan's impactful foreign policy laying the foundation for multilateral cooperation, following the end of the Cold War. The Nakayama Proposal was met with rejection from most of ASEAN's leaders. However, many scholars expressed positivity towards the impact of the Nakayama Proposal.

From the document analysis of ASEAN's meetings and policies on a regional security forum from 1989 to 1994, there are obvious changes in ASEAN's characteristics in policy decision-making. Before the submission of the Nakayama Proposal, ASEAN's policy-making process was dominated by economy-enhancing policies and collective decisions regarding security matters among members, and a lack of further action in outlining regional security forum mechanisms despite perpetual suggestions from ASEAN's researchers. However, after the submission of the Nakayama Proposal in July 1991, a change in ASEAN's policy-making process occurred as ASEAN began to study a mechanism for a regional policy forum so that a number of noticeable discussions related to the issues started to emerge, including the Singapore Declaration in January 1992 and the first working session of the ARF in July 1994.

Japan in general, and the 1991 Nakayama Proposal in particular, had actively played important roles in initiating security multilateralism in the region. In accordance with its proponent's intention, the 1991 Nakayama Proposal clearly served its purpose as a general message for ASEAN's leaders as it re-inspired them by triggering and accelerating the working process of the 1994 ARF.

This study's approach has relied on various parallel studies in the printed literatures of both Japan and ASEAN, as well as verbal confirmations from Yukio Satō, Masashi Nishihara and Tarō Nakayama themselves. Yet, in contrast to the number of Japanese interviewees, the study could only identify one official source from ASEAN, namely, Rene Pattiradjawane, the official journalist during the ASEAN-PMC. The argument would be more valid if there were interview statements officially confirmed by ASEAN's side, in particular the drafters of the 1992 Singapore Declaration, ASEAN-ISIS researchers active during the early 1990s, and all actors involved in the 1991 ASEAN-PMC and the establishment of the ARF itself.

REFERENCES

Acharya, A. (2001). *Constructing a security community in Southeast Asia: ASEAN and the problem of regional order*. New York, NY: Routledge.
Anon. (1991). PMC must not be solely a security forum: Abdullah. *The Star Malaysia*, p. 4.
Anon. (1991). RI reacts to Japan's plan on security. *The Jakarta Post*, p. 4.
Anon. (1991). *Threats in Asia-Pacific from Interaction of the Four Great Countries*. (Ancaman di Asia-Pasifik dari Interaksi Empat Negara Besar). *Kompas*. pp. 1 & 11.
Anon. (1992). Security tops ASEAN agenda. *The Star Malaysia*, July 24, 1992, p. 4.
Anon. (1992). Talk of security plan crops up for the first time. *The Star Malaysia*, p. 2.
ASEAN-ISIS. (1991). *A time for initiative: Proposals for the consideration of the fourth ASEAN summit*. Manila, Philippines: ASEAN Institutes of Strategic and International Studies.
ASEAN Kakudai Gaishou Kaigi (PMC) no Gaiyou. (n.d.). Retrieved from http://www.mofa.go.jp/mofaj/area/asean/asean_7.html.
ASEAN Secretariat. (1988). *ASEAN Document Series 1967–1988* (3rd ed.). Jakarta, Indonesia: ASEAN Secretariat.

ASEAN Secretariat. (1989a). *ASEAN Document Series 1988–1989* (Suppl. ed.). Jakarta, Indonesia: ASEAN Secretariat.

ASEAN Secretariat. (1989b). *22nd ASEAN ministerial meeting and post-ministerial conferences with the dialogue partners.* Bandar Seri Begawan, Indonesia: ASEAN Secretariat.

ASEAN Secretariat. (1990). *The first ASEAN Regional Forum.* Bangkok, Thailand: ASEAN Secretariat.

ASEAN Secretariat. (1991). *ASEAN Document Series 1989–1991* (Suppl. ed.). Jakarta, Indonesia: ASEAN Secretariat.

ASEAN Secretariat. (1992). *Singapore declaration of 1992 in meeting of the ASEAN heads government.* Singapore: ASEAN Secretariat.

ASEAN Secretariat. (1994). *ASEAN Document Series 1992–1994* (Suppl. ed.). Jakarta, Indonesia: ASEAN Secretariat.

Bangkok Post. (CSEAS Kyoto University: July 5, 1989a). *ASEAN calls for settlement.* Microfilms, p. 8.

Bangkok Post. (CSEAS Kyoto University: August 23, 1989b). *RP seeks ASEAN views on bases.* Microfilms, p. 1.

Bangkok Post. (CSEAS Kyoto University: August 5, 1989c). *S'pore offers to accept US bases.* Microfilms, p. 1.

Bangkok Post. (CSEAS Kyoto University: August 16, 1989d). *Mahathir: US base will trigger arms race.* Microfilms, p. 6.

Bangkok Post. (CSEAS Kyoto University: July 25, 1990a). *ASEAN calls for realism in bids to restore peace.* Microfilms, p. 6.

Bangkok Post. (CSEAS Kyoto University: July 28, 1990b). *Australia seeks Asian security forum.* Microfilms, p. 6.

Bangkok Post. (CSEAS Kyoto University: July 29, 1990c). *ASEAN mulls over security stance.* Microfilms, p. 5.

Bangkok Post. (CSEAS Kyoto University: July 22, 1994). *Forum seeks trust, security.* Microfilms, p. 4.

Chew, L.K. (1993). ASEAN launches new initiative on security. *The Straits Times,* p. 1.

Chua, R. & Kwang, M. (1992). Ministers focus on regional security. *The Straits Times.*

Department of Foreign Affairs Philippines & Ministry of Foreign Affairs Thailand. (1990). ASEAN and the Asia-Pacific region: Prospects for security cooperation in the 1990s (pp.73–83).

Emmers, R. (2001). The influence of the balance of power factor within the ASEAN Regional Forum. *Contemporary Southeast Asia, 23*(2), 275–291.

Jimbo, Ken (Interview, March 26, 2015). Assistant Professor of Faculty of Policy Management of Keio University in Canon Institute of Global Studies, Shin-Marunoichi Bld.,Tokyo.

Mainichi Shinbun Tokyou Choukan 1991/07/22, 1991/07/23 edition (Aso).

Midford, P. (2000). Japan's leadership role in East Asian security multilateralism: The Nakayama proposal and the logic of reassurance. *The Pacific Review, 13*(3), 367–397.

Nakayama, Tarou (Interview, November 29, 2013). Former Foreign Minister of Japan 1990–1991, PM Kaifu's Cabinet in Tarou Nakayama Nishi-Ku Osaka Office.

Nihon Keizai Shinbun 1990/01/05 edition (Ishikawa), 1990/06/29, 07/17, 07/29, 07/30 edition (Oobashi), 1990/09/15 edition.

Nihon Keizai Shinbun 1991/07/23 edition (Ogasawara and Oobashi), 1990/06/29, 07/17, 07/29, 07/30 edition (Oobashi), 1991/07/22, 1991/07/24 edition (Ikeuchi), 1991/07/28, 1991/07/24 edition (Ikeuchi), 1991/07/23, 1991/07/28 edition (Kandia), 1990/01/05, edition (Ishikawa), 1990/09/15, 1991/06/15, 1991/07/22, 1991/07/28, 1991/07/28, 1991/03/23, 1991/07/15 edition.

Nishihara, M. (2003). ASEAN's political and security relations with Japan. In P.J. Noda (Ed.), *ASEAN–Japan cooperation: A foundation for East Asian community.* Tokyo, Japan: Japan Center for International Exchange (JCIE).

Nishihara, Masashi (Interview, October 30, 2014). President of Research Institute for Peace and Security, Minato-Ku, Tokyo.

Nusara Thaitawat. (CSEAS Kyoto University: July 27, 1993). *Bangkok Post. ASEAN calls for equality at annual dialogue.* Microfilms, p. 6.

Pattiradjawane, Rene (E-mail correspondence May 11, 2014 ~ November 5, 2014). Senior Journalist of *Kompas.*

Salamm, A. (1991). *ASEAN and Spartly Dispute* Settlement. (ASEAN dan Penyelesaian Konflik Spartly). *Kompas.* p. 4.

Sato, Yukio (Interview, December 9, 2014). Initiator of the Nakayama Proposal, Former Director of the Foreign Ministry's Information and Analysis Division 1991 in The Japan Institute of International Affairs, Chiyoda-Ku, Tokyo.

Sudo, S. (1995). *Idea and the External Decision-making Process Theory: Factory Policy Process on Fukuda Doctrine* ("Idea" to Taigai Seisaku Ketteiron: Fukuda Dokutorin wo meguru Nihon no Seisaku Kettei Katei). Journal *Buki Idou no Kenkyu*, 108. Nihon Kokusai Seijigaku Kaihen Publisher.

Wanandi, J. (1991). Peace and security in Southeast Asia. *The Jakarta Post*, p. 5.

Yuzawa, T. (2007). *Japan's security policy and the ASEAN Regional Forum: The search for multilateral security in Asia-Pacific*. London, UK: Routledge.

Yuzawa, Takeshi (Interview, May 28, 2015). Professor of Faculty of Global and Interdisciplinary Studies in Ichigaya Campus, Hosei University, Tokyo.

Social transformation of Kuwaiti women and their contribution to Kuwait's economic development

A.T. Hidayati & Apipudin
Department of Middle East and Islamic Studies, Universitas Indonesia, Jakarta, Indonesia

ABSTRACT: This research was conducted to explore the social transformation of Kuwaiti women. In the past, Kuwaiti women only had roles in the private sphere of the family and home. Kuwaiti women were also not allowed to obtain higher education qualifications. Women in Kuwait were in a marginal position compared to men. However, there has recently been a social transformation in the role of Kuwaiti women. Kuwaiti women have started gaining higher education qualifications and participating in the workforce. Gender equality in Kuwait is increasing as seen in the report 'Global Gender Gap Index 2015' (World Economic Forum, 2015), which announced that Kuwait has the highest level of gender equality among the Gulf States. This change certainly contributes to economic development in Kuwait. By using a qualitative method and descriptive analysis approach (Creswell, 2010), this study aimed to analyse the factors which led to the social transformation of Kuwaiti women and their contribution to the economic development of the country.

1 INTRODUCTION

In the life of Kuwaiti society, gender construction provide considerable effect to the status of Kuwaiti women. In the past men were always placed a level higher than women. This patriarchal concept is very common in Kuwaiti society and other Middle Eastern countries. This has placed certain limitations on women according to the social norms of the country. Women were placed in a marginal position which means that they were not able to do things beyond their role as mothers or their domestic life at home. However, men were given the freedom of self-determination and to run their lives as financial controllers who also made major decisions in the family.

Kuwaiti men possess important roles in the economy, politics, and education institutions because men are expected to be the decision makers in Kuwaiti society. They attend school, go to work, and also participate in politics. On the hand, women in Kuwait were confined to household activities where they accepted a limited role in the domestic sphere. As a result, Kuwaiti men occupy the patriarchal positions in the society while women only occupy domestic positions with limited freedom and opportunity. However, this situation has been changing. The initial point of this social change in Kuwait's social order was the discovery of oil in 1938 (Alsuwailan, 2006). This discovery encouraged Kuwait to carry out oil exploration which increased mass development in Kuwait. It encouraged social changes in the life of Kuwaiti society including that of Kuwaiti women. Women began to make efforts to gain self-determination and obtain equal rights to men. Kuwaiti women fought for their rights to work and to participate in politics and they also argued that Kuwaiti society should acknowledge modernity. During this period, Kuwait experienced many changes to its social order, which account for much of the development up until the present day.

Nowadays, Kuwait is one of the countries which possess high gender equality levels in the Middle East region. To be specific, Kuwait is the country with the highest gender equality, which means that Kuwait is the country with the smallest gender gap compared to the other Gulf States. The 2015 Gender Gap Index (World Economic Forum, 2015) ranked Kuwait in position 117 out of 145 countries, with an index of 0.646 (out of 1). On this basis, it can be seen that Kuwait remains in a considerably low position from a global perspective. However, it remains an achievement for Kuwait, remembering that the Middle East region still possesses larger gender gaps.

Because Kuwait possesses the smallest gender gap compared to the other Gulf States, it raises many questions and is academically interesting to explore deeper in order to investigate the factors which caused these social transformations in Kuwait. The social transformation of women's roles in Kuwait also contributed to the development of the economy in Kuwait. Therefore, it is important to discuss the factors which led to the social transformation of Kuwaiti women and what their contribution is to Kuwait's economic development after the social transformation occurred.

2 RESEARCH METHODS

In doing this research, the writers use qualitative methods with a descriptive analysis approach. Creswell (2010) states that a qualitative method is a method to explore and comprehend the meaning of the actions of some individuals or a group of people which are expected to change social or human problems. According to Creswell, the process of qualitative research has several stages. The stages are proposing questions and procedures, collecting data, analysing the data inductively from the special theme to general theme, and interpreting the data (Creswell, 2010. In analysing a problem, a reference in a form of theory or concept is needed to help the researchers in finding empirical answers. Here we use one theory that is correlated with this research: the social change theory of Charles L. Harper (1989). This research focuses on two questions: What are the factors which led to the social transformation of Kuwaiti women? What are the contributions of Kuwaiti women to the economic development of the country?

3 ANALYSIS AND DISCUSSION

3.1 *Social change theory*

According to Indonesia's National Encyclopaedia (Cipta Adi Pustaka, 1991), the term 'social transformation' is described as a total change in the form, shape, characteristics or nature of mutual relationships, both in terms of individuals and as a group. Social transformation cannot occur without any cause, but it is always influenced by several factors. The factors which influence social transformation are connections with other cultures, heterogeneous society, social disorder, and the social change itself. The process of social transformation will involve the society, technology, cultural values, and social movement. In Indonesia's National Encyclopaedia (Cipta Adi Pustaka, 1991), it is often stated that the term 'social transformation' is interpreted as meaning the same as social change.

According to Sztompka (1994), social change is defined as a change which occurs in the social system. According to Harper (1989), social change is defined as a significant change of social structure over a certain period of time. There are several types of structural change. The type of structural change considered in this research is the personal change which is related to roles and individual changes in the existence of structure in human life (Harper, 1989). This type of change has gradual characteristic and there are not many new elements or missing elements.

The changes of this type can be seen in the change of women's roles and functions in society. Women were positioned as subjects who carry and functioned in the domestic space, but now women in modern society participate in the public space which had been dominated by men. This change gives impact to the other sectors. In this case, as Harper has stated, social transformation or social change in the role of Kuwaiti women is considered as a social structure change related to a role change in a social group.

3.2 Kuwaiti women in the pre-oil-discovery period

Some phenomena which have occurred in Kuwait, including the way that the Kuwaiti society think, behave, and communicate with others, were correlated with the discovery of oil and its development. To observe the social transformation of women's roles in Kuwait, it is necessary to divide the period concerned to show the initial point of the transformation. The transformation period can be divided into two: a pre-oil-discovery period and a post-oil-discovery period (Alghaith, 2016). Before the discovery of oil, Kuwaiti men played an important role in the family economy. Kuwaiti men were the decision makers in the family and they also had higher authority than women, especially because men were responsible for providing financial support for the family, while women's working role was limited. Women's work was confined to dealing with domestic jobs, such as, taking care of family members and. Because of such things, women were considered as playing inferior roles to men.

Before the discovery of oil, nomadic life was one of the characteristics of Kuwaiti society. At that time, Kuwait was one of the poorest countries in the world which had limited natural resources. Furthermore, the per capita income of Kuwait was no more than US$100. During this period, there was economic diversity, which included commerce, fishing, pearling and diamond exploration (Shelash, 1985). In addition, the development of Kuwait was strongly related to its identity as a British protectorate. The first oil exploration was carried out in 1936; the oil discovery came later in 1938. Because of the oil discovery, the Middle East states, including Kuwait, became highly strategic and competent countries competing on an international level, especially in 1946 when Kuwait became an oil exporting country (Shelash, 1985).

Education during the pre-oil-discovery period was only provided to male children. At that time, study consisted only of Quranic studies and basic arithmetic. In addition, the male children were taught in home schools or classes. In 1936, the first education institution, The Council of Education, was established in Kuwait. However, education for female children was not officially established until 1957. In the 1950s, universal education was established for both male and female Kuwaiti children. At the outset, during the years of 1961 and 1962, 60.4% of the pupils were male. However, by 1983–1984 the percentage of male and female students was almost balanced, with 51.8% male and 48.1% female.

3.3 The role of Kuwaiti women in the post-oil-discovery era

Oil discovery in Kuwait began in 1913 when Sheikh Mubarak signed an agreement with a British company to prospect for oil in Kuwait. Twenty-five years later, in 1938, oil was first found in Bugran, Kuwait (Alsuwailan, 2006). At the beginning of the 1950s, a significant transformation of Kuwait's development occurred when Sheikh Abdallah al-Salim governed the country, turning Kuwait from a poor British colony into an independent, modern and rich country. During this period, the country's infrastructure, such as public roads, hospitals and office buildings, was developed and established.

Kuwait also repaired its public service systems, such as water supply, electricity, healthcare, education and housing. This mass development made Kuwait a more urban, cosmopolitan country. In due course, it forced Kuwait to take in many foreign workers in order to expand its oil industry and other infrastructure, because Kuwait had a small citizen population, which was not able to support the need for expert workers. Between 1957 and 1975, the foreign workers who came to Kuwait increased the population of the country by 557% (Alghaith, 2016).

This drastic increase in the population of Kuwait resulted in social and cultural changes in Kuwaiti society. The mixture of social and cultural values between local citizens and immigrants caused Kuwaiti women to become more aware of the outside world. Kuwaiti women started to demand gender equality along with socio-political rights. In the post-oil-discovery era, the Kuwaiti government gave their attention to public services and education. In education, the government built schools, employed new teachers, and supported the development of education. In 1949, the government passed a regulation which gave women and men the same education curriculum. This was done due to social changes in Kuwait and to prepare women to enter the workforce. Since then, female student numbers have increased dramatically from 1% in 1950 to 80% in 1987 (Alsuwailan, 2006).

The social transformation of women's role in Kuwait was influenced by the discovery of oil in 1938. The transformation introduced new concepts in terms of the social activities of women, including participation in employment, exercising their political rights, and supporting Kuwaiti women in accepting modernisation. This oil discovery is commonly accepted as an influence on the dynamics of Middle Eastern countries, especially in the Gulf States such as Kuwait, Saudi Arabia, United Arab Emirates, Qatar, Bahrain and Iraq. This oil discovery is one of a few indicators on social transformation in the roles of Kuwaiti women. These transformations gradually supported Kuwaiti women in increasing their social status in society. Previously, female workers had fewer rights than males; through education, Kuwaiti women were enabled to work on a more equal basis.

Furthermore, Kuwait became the country with the highest percentage of female workers compared to other countries in the Middle Eastern region. After the oil discovery, Kuwait experienced a mass transformation in many sectors, including the social and economic sectors. Kuwaiti women started to gain access to education, demand gender equality, and to get jobs, expanding the scope of Kuwaiti women's activities beyond the domestic sphere.

3.4   *The social transformation of Kuwaiti women*

The transformation which happened in Kuwait was a result of the oil discovery, development and arrangement. This oil discovery altered the social order in Kuwait. Kuwait was initially in domestic level, but now Kuwait is an independent country in international level. Population movement during this era of transformation was high. In 1947, Kuwait's population was estimated at 120,000 citizens, consisting mostly of indigenous Kuwaitis. Kuwait's first census in 1957 recorded that the population had increased to 206,478 citizens (Al-Sabah, 1980). In that year, 113,622 citizens were indigenous Kuwaiti citizens while 92,851 were non-Kuwaiti citizens. Thus, the huge increase in Kuwait's population was caused by the immigrants who came to Kuwait. The number of immigrants kept increasing, eventually exceeding the population of indigenous Kuwaitis. According to the statistical data of 1984, the total population leapt to 1,910,856 citizens. Specifically, there were 682,507 indigenous Kuwaiti citizens, while non-Kuwaitis accounted for almost twice this number at 1,228,349 (Shelash, 1985).

The 1984 statistics on the number of Kuwaiti citizens show that the populations of indigenous male and female Kuwaitis were nearly equal at 335,111 and 347,396, respectively (Shelash, 1985). For non-Kuwaiti citizens, the number of women was 412,437, while the number of men was twice this figure at 815,912.

Statistics for workers in 1965 recorded that there were 41,926 Kuwaiti men in employment and only 1,092 Kuwaiti women. For non-Kuwaitis, there were 133,603 men and 7,676 women (Shalesh, 1985). This data suggests that at that time men occupied more roles in the field of work than women. In the course of Kuwait's transformation to a modern, industrial country, the 1984statistics show that the number of women workers dramatically increased, to a total of 14,172 Kuwaiti women and 49,105 non-Kuwaiti women. Although men were still dominant in terms of numbers, this change showed that there was a transformation of women's roles in Kuwait because of the increasing employment of women. In terms of percentages, the increase in women's participation in employment changed dramatically, from 1.8% in 1965 to 10.3% in 1980 (Shalesh, 1985).

One of the most significant changes was the discovery of oil, followed by the rapid increase in the education of Kuwaiti women, which spread throughout the country. Kuwaiti and non-Kuwaiti citizens both had access to free state-funded education. In the 1965–1985 period, the percentage of educated Kuwaiti women increased from 28% to 63% (Shah, 1990). At the same time, the percentage of educated non-Kuwaiti women also increased, from 58% to 81% (Shah, 1990). Today, women in Kuwait generally have a high educational status. This has led to a change in women's role, which was initially focused on the domestic field, to a level where they are able to obtain equal status with men with regards to working opportunities. This sociodemographic change has increased the status of women's role, especially economically.

Development is a process synonymous with social transformation. So, in this case, the social transformation of the roles of Kuwaiti women also contributes to the development of Kuwait economy. Previously, women could not contribute to activities other than performing domestic duties. Because of the transformation of the roles of Kuwaiti women, their contribution to the economy's development increased. Nowadays, Kuwaiti women are free to express themselves and to participate in every field, including the economic development of Kuwait.

Kuwaiti women are very active in many global economic development activities. Kuwaiti women have participated in four of the UN's biggest conferences, hosted in several countries. One of these conferences was the 5th World Conference of Women, held in Beijing in 1995. The resulting Beijing Declaration and its associated programme have promoted gender and information issues, communication and technology for women, and upgrading women's skills, knowledge, and access to and use of information. Kuwait is one of many countries that were involved in this discourse on the development of women (UN Women, 2005). In 1999, Kuwait Gas ratified CEDAW which aimed to integrate all of the women in the process of the country development.

As one of 189 countries that signed the Millennium Declaration in support of the Millennium Development Goals (MDGs), Kuwait is committed to implementing the onjectives of the MDGs. Goal 2 specifically relates to the role of women, while Goal 3 is concerned with education and the gender gap in education. Kuwait has established an equal education system for both men and women since 1999. Thus, Goal 3 is a continuation of the equal education system for men and women, such that they can both have equal working opportunities too.

Kuwait has the highest women workers' participation among the Gulf States. Moreover, the increase in women workers' participation can support development prospects in the future. The women workers' participation is 53%, above the average for other countries in the world, which is 51% (Alghaith, 2016). In addition, it also exceeds the average percentage in other Middle Eastern and North African countries, which is just 21%.

In recent years, Kuwaiti women have been given support not only to become workers but also to start their own businesses. This support came not just from family but also from society, social institutions, and the Kuwaiti government. There were programmes which were designed to support women in contributing more to the Kuwaiti economy's development. One of the programmes created by the Kuwait Economy Society (KES) is entitled "Women's Leadership and Entrepreneurship" (Algaith, 2016). This project aims to support Kuwaiti women in expanding their businesses, in training their employees, in broadening their links, and in gaining experience. Many Kuwaiti women choose to develop businesses because they want to raise their social status and make their own decisions (Alghaith, 2016).

In 2006, according to the Kuwait News Agency (KUNA), the population of Kuwaiti women was more than 50% of the total Kuwaiti population, with 24.5% of them designated as businesswomen. These businesswomen contribute to the creation of more than 40% of Kuwait's employment (Alghaith, 2016). This report shows that Kuwaiti women have made a considerable contribution to the development of the economy in Kuwait.

Many Kuwaiti women are now entering higher education in order to be more involved in economic development in Kuwait. Kawther Al-Joan, Seham Al-Rezouki, Dr. Rasha Al-Sabah, Dr. Fayza Al-Kherafi, Dr. Masoma Al-Mubarak, Noreya Alsabeh, Dr. Modai Al-Hamoud, Sara Al-Duwaisan, Nabeela Alanjery, Wafa Aljassem and Sana Jumah are all examples of Kuwaiti women who hold important roles in their respective work (Al-Suwaihel, 2010). Kuwaiti women can now work in many fields, as teachers, doctors, engineers, businesswomen, ambassadors, lawyers, managers, administrators, and even as government ministers. Kuwaiti women have successfully raised their social status by entering many fields and taking senior positions that have not been occupied by women before.

4 CONCLUSION

Oil exploration, has caused the national development of many sectors in Kuwait, especially the economic sector. To support Kuwait's development, the government took many initiatives, for example, focusing on the development of the education sector in order better educate Kuwaiti society so that it could participate in Kuwait's development, which needed many workers.

These changes in Kuwait directly enabled changes to women's roles and empowered them to participate in Kuwait's development. Kuwaiti women are now allowed to gain higher education qualifications and hold high positions in organisations and institutions. Furthermore, Kuwaiti women also actively participate in Kuwait's economic development by participating in a variety of economic efforts, such as creating employment for Kuwaiti society as a whole and building women's organisations focused on the empowerment of women in Kuwait.

REFERENCES

Al-Suwaihel, O.E. (2010). Kuwaiti female leaders' perspectives: The influence of culture on their leadership. *Contemporary Issues in Education Research, 3*(3), 29–39.

Alghaith, S. (2016). Understanding Kuwaiti women entrepreneurs and their adoption of social media: A study of gender, diffusion, and culture in the Middle East (Doctoral dissertation, Department of Journalism and Media Communication, Colorado State University, Fort Collins, CO). Retrieved from https://dspace.library.colostate.edu/bitstream/handle/10217/173380/Alghaith_colostate_0053 A_13525.pdf.

Alhamli, S.A. (2013). Impact of the level of women participation in the workforce on economic growth in Kuwait (Doctoral thesis, Department of Middle East & Mediterranean Studies, King's College London, UK). Retrieved from https://kclpure.kcl.ac.uk/portal/files/12371970/Studentthesis-Sahar_ Alhamli_2013.pdf.

Alsuwailan, Z.F.M.M. (2006). The impact of societal values on Kuwaiti women and the role of education (Doctoral dissertation, University of Tennessee, Knoxville, TN). Retrieved from http://trace. tennessee.edu/cgi/viewcontent.cgi?article=3141&context=utk_graddiss.

Cipta Adi Pustaka. (1991). *Ensiklopedi Nasional Indonesia* [National Encyclopaedia of Indonesia]. Jakarta, Indonesia: Cipta Adi Pustaka.

Creswell, J.W. (2010). *Research design: Pendekatan Kualitatif, Kuantitatif, dan Mixed.* Yogyakarta, Indonesia: Pustaka Pelajar.

Dianu, A.H. (2003). Contemporary short stories by Kuwaiti women: A study of their social context and characteristics. *MELA Notes, 75/76,* 70–84.

Harper, C.L. (1989). *Exploring social change.* London, UK: Prentice Hall.

Martono, N. (2016). *Sosiologi Perubahan Sosial.* Jakarta, Indonesia: Rajawali Press.

Meleis, A.I., El-Sanabary, N. & Beeson, D. (1979). Women, modernization, and education in Kuwait. *Comparative Education Review, 23*(1), 115–124.

Sanad, J.A. & Tessle, M.A. (1988). The economic orientations of Kuwaiti women: Their nature, determinants, and consequences. *International Journal of Middle East Studies, 20*(4), 443–468.

Shelash, Mesad F. (1985). Change in The Perception of the Role of Women in Kuwait. Ohio State University.

Sztompka Piotr. (1994). Agency and Structure: Reorienting Social Theory. Amsterdam: Gordon and Breach.

Tetreault, M.A. & Al-Mughni, H. (1995). Gender, citizenship and nationalism in Kuwait. *British Journal of Middle Eastern Studies, 22*(1–2), 64–80.

Tétreault, M.A. & Al-Mughni, H. (1995). Modernization and its discontents: State and gender in Kuwait. *Middle East Journal, 49*(3), 403–417.

UN Women. (2005). Kuwait: Brief overview of Kuwait's experience in the implementation of the Beijing Platform for Action (Beijing + 10). New York, NY: UN Women. Retrieved from http://www. un.org/womenwatch/daw/Review/responses/KUWAIT-English.pdf.

World Economic Forum. (2015) *Insight report: The global gender gap report 2015.* Geneva, Switzerland: World Economic Forum.

A strategy of inter-state institutional cooperation for conflict resolution and maritime security in Indonesia

A. Brotosusilo & I.W.A. Apriana
Faculty of Law, Universitas Indonesia, Depok, Indonesia

ABSTRACT: The new Indonesian government regime under President Joko Widodo introduced a new goal of establishing Indonesia as the world's maritime axis. However, the maritime situation is an uneasy issue because Indonesia shares sea borders with neighbouring countries that sometimes involve conflicts of interest. In terms of strengthening maritime security, Indonesia should also give more attention to its seas becoming a peaceful traffic route for international shipping. Prevention of conflict potential and attention to maritime security are important to study as a country's reference to preparing and confronting government policies. This study uses quantitative methods for Internal Environment analysis with quantification of data relating to internal "strength" and "weakness" out. It also for External Environment analysis with data of external "opportunities" and "threats" in. The analysis shows that the weighting results of "strength" and "weakness" are in the range of (–1, 227), while the weighting results of "opportunity" and "threat" are (0, 282). These results indicate that a feasible strategy of inter-state institutional cooperation for conflict resolution and maritime security in Indonesia is a strategy of "consolidation". It requires efforts to utilise and optimise the existing opportunities in order to minimise inherent weaknesses, for example reforming regulations to handle conflict resolution and maritime security.

1 INTRODUCTION

Maritime conflict is basically related to issues of space. As we know, in geopolitics interactions happen between space and human beings, which makes space consciousness directly or indirectly related with security interests and people's welfare. This spatial consciousness begins at the individual and family levels which then, ultimately, progress to a wider level. In the context of the modern state, the concept of spatial consciousness is realised by a sovereignty claim that is established by state boundaries, with a set of laws and administrators to ensure security and sovereignty.

Some geopolitical experts have organised their concepts in terms of the relationships and interactions between spaces and human beings, such Friedrich Ratzel and Karl Haushofer. According to Ratzel (1901), life is an inevitable struggle that occurs even in the case of a nation-state for a struggle for space. Nations-state must have a spatial concept of territorial initiative. Ratzel considers that the state, as a human groups to the spatial units where they develop. Borders change dynamically as a reflection of the behaviours of aggressive, expansionist countries. Therefore, according to Ratzel, if there is any degradation of a country's spatial concept, the state and nation can face collapse. This is known as the theory of Lebensraum (living space).

The theory of Lebensraum was further developed by Karl Haushofer. According to Haushofer (1934), space (raum) is a platform of political and military dynamics, hence, the taking over of a space or sphere of influence is itself a spatial phenomenon. If a given space is enlarged in this way, it will be beneficial for one side and disadvantageous for the other. This seizure of space will have tremendous implications for politics, economics and regional security because it is directly related to energy and maritime security, two issues that have recently been the subject of international attention and will continue to be in the future.

The threat of conflict with neighbouring countries over sea territories can become real. Indonesia has had the experience of losing the islands of Sipadan and Ligitan to Malaysia. This happened because of the weakness of Indonesia's attention to the management of border areas and failure to deal with conflict among nations. Indonesia's territorial sea border is still a problem because there is still no agreement between neighboring countries.

There are several seas that have become a source of potential conflict in Southeast Asia, such as the Andaman Sea, South China Sea, Thailand Gulf, Tonkin Gulf, Malacca Strait, Celebes Sea, Sulu Sea, Arafura Sea, Timor Sea and Torres Strait. These sources of conflict are related to territorial seas, the continental shelf, and exclusive economic zones. Nine of ten members of the Association of Southeast Asian Nations (ASEAN) are claiming these seas as territorial. This is a dispute and border conflict that represents a definite threat for maritime security in Southeast Asia.

By looking at the configuration of Indonesian geography, it is clear that Indonesia understands that the sea can act as a uniting medium among islands and even between countries. By ratifying the 1982 United Nations Convention on the Law of the Sea (UNCLOS), Indonesia has confirmed that it is an archipelagic country. In this context, all aspects of life and state administration should consider issues of geostrategies, geopolitics, geo-economics, and geo-social culture from the perspective of an archipelagic country. The mindset, attitudes and behaviours of the nation should be concerned with maritime spatial consciousness, and that is why a clear maritime vision is a necessity for Indonesia (Abdulrazaq et al., 2013).

A consequence of Indonesia's strategic geographical position is its importance to global parties who need to make use of the sea. It means that if Indonesia can take advantage of these opportunities and challenges, it will increase Indonesian welfare. Historically, Indonesia has recorded evidence that the ancestors of Indonesia had legal authority over the Nusantara oceans, and had been able to sail across the oceans as far as the Madagascan coast of South Africa. This demonstrates that the ancestors of Indonesia have long had a maritime spirit in creating relationships with other nations of the world. In addition, the Indonesian ancestors understood the meaning and benefit of the sea as a medium to secure the interests of nations, such as trade and communication. Therefore, it can be understood that the traditional use of the sea is for transport media and most of the world's trade continues to increase until now (Pujayanti, 2011).

The Preamble of the 1945 Constitution of the Republic of Indonesia mandates that the State of Indonesia shall protect the whole people of Indonesia and the entire homeland of Indonesia, as well as advance general prosperity. The implementation of safeguard measures, to include the maritime territories, should therefore be conducted properly, in a coordinated and integrated manner with a security- and welfare-based approach.

The Unitary State of the Republic of Indonesia (NKRI) is the biggest archipelagic state in the world, with 17,499 islands. The area of the sea is 5.8 million km^2, the coastline length is 81,000 km, and the continental shelf is relatively wide. The Indonesian sea therefore has great potential in terms of biological and non-biological marine resources. In terms of aesthetics, the Indonesian sea also has a high value for marine tourism, and in terms of economy and industry, the Indonesian sea is very beneficial as a national and local transport route as well as being a valuable fishing and marine resource (Keliat, 2009).

The Indonesian geographical position between the intersection of two continents and two oceans results in a globally strategic position. The Indonesian sea territory contains very important sea lanes for national and international shipping traffic. This position has provided Indonesia with an important position and role in the international community as the maritime focal point of Asia-Pacific. The logical consequence of this geographical factor is that the Indonesian sea has a political value and an important security role, not only for Indonesia but also for other countries, especially those in the Asia-Pacific region (Abdulrazaq et al., 2013).

The maritime security challenge which arises from the possibilities of conflict among countries is an issue that should be managed correctly. This conflict is due to the competition among countries in obtaining natural resources and claiming national and territorial boundaries. Currently, Indonesia still has several sea border disputes with Singapore (the

Strait of Malacca/the Phillips Channel), Vietnam (northern part of Natuna Islands), and Timor-Leste post-separation from Indonesia (sea border problems around Timor islands) (Perwita, 2004). Exclusive Economic Zone Mapping (EEZ) with neighboring countries also needs to be clarified because of potential conflicts related to strategic issues and national interests such as the military and the economy for Indonesia. This is because the Indonesian sea territory and its surroundings have considerable marine resources, including crude oil and natural gas. The area of the South China Sea, for instance, is estimated to contain nearly 18 trillion tons of natural gas. It is therefore not surprising that this area has become a source of conflict for neighbouring countries that can pose a threat for the maritime security of Indonesia (Perwita, 2004).

The relationship between potential conflict issues with neighboring countries relates to Indonesia's geographical position as an archipelagic country, as well as maritime security issues where the maritime sector is currently being developed as a national policy objective, an interesting theme for study. This research is trying to solve this problem while the output suggests a system of cooperation to prevent and resolve conflict potential and maritime security. A multidisciplinary approach is necessary here because the complexity of the research problem means that it cannot be resolved by following a basic legal approach alone; it will also need to involve considerations of political economy, international relations, sociology and institutions.

2 METHODS

A SWOT (Strengths, Weaknesses, Opportunities and Threats) analysis is used as a tool to formulate a strategy, by identifying the factors that influence civil–military relationship. Internal factors such as strengths and weaknesses, together with external factors such as opportunities and threats, are considered.

The process of strategic planning is done through three stages of analysis, which are data collection, analysis and decision-making. In the data collection stage, internal and external factors are evaluated. The analysis stage involves conducting a measurement of influential elements, designating a rating point, calculating the external and internal factors on a SWOT matrix, and arranging a strategy matrix. The decision-making stage entails determining a grand strategy matrix and an alternative strategy.

The SWOT analysis of the data is executed through the following steps:

1. The researcher collects the primary and secondary data, and then processes this initial data.
2. The processed data is classified into influencing external and internal factors, and each factor is limited to no more than ten elements.
3. Identification of the external factors that are categorised as an opportunity or threat. The internal factors are then categorised as a strength or weakness of the civil–military relationship in Indonesia. This identification is done by the researchers and teams using the Consensus Decision-Making Group (CDMG) method.
4. Deciding the value and weighting of each element including factors of opportunity, threat, strength, and weakness by CDMG method. The value and weight are multiplied and the product becomes a score of the value to arrange the External Factor Evaluation (EFE) and Internal Factor Evaluation (IFE) matrix.
5. According to the calculation of the external and internal factors score values in the EFE and IFE matrix, a strategy priority can then be determined, whether in quadrants I/SO (expansive strategy), quadrant II/SO (diversification strategy), quadrant III/SO (survival strategy), and quadrant IV/SO (consolidation strategy).
6. After determining the priority strategy, the research then elvove an alternative strategy which is suitable for the conditions of the civil–military relationship in Indonesia.

The next stage involves conducting an analysis to determine the order of the alternative operational strategies, using a Quantitative Strategic Planning Matrix (QSPM). The value of each element in the matrix of EFE and IFE is multiplied by an Attractiveness Score (AS) which is determined first. The multiplication product is called a Total Attractiveness Score (TAS), which is used to determine the sequence of strategies that will be conducted.

3 RESULTS AND DISCUSSION

In order to formulate the right strategy for the realisation of a trans-state institutional cooperation model for conflict resolution and maritime security to achieve the goal of Indonesia as a maritime axis, the researcher conducted a SWOT analysis. This analysis is done by identifying the strengths and weaknesses caused by internal influences, as well as the opportunities and threats that come from the external environment. The benefits of this analysis provide a reference point from which to strengthen the strengths and exploit the opportunities, as well as minimising the weaknesses and counteracting the threats.

The first step of analysis was to identify the various internal and external factors that are important to the objective of establishing Indonesia as a maritime axis, then calculating the value of various strategy factors in accordance with the SWOT analysis stages.

3.1 *Internal environmental analysis*

Data of Internal Environmental Analysis related with strength and weakness that comes from the inside out. Related with Internal Environmental Analysis which according to its urgency is as follows:

a. A strength to overcome conflict and maritime security in realising Indonesia's position as a maritime axis
 i. The change of paradigm and national development focus with maritime insight;
 ii. The strong political will from the current government with its Nawacita (term for nine programs of Joko Widodo);
 iii. The step and strong action from Ministry of Marine Affairs and Fisheries (*Kementerian Kelautan dan Perikanan*/KKP) is highly important;
 iv. The forming of the Maritime Security Board (Bakamla) as an institution that has authority as a coordinator of maritime security and law enforcement in the Indonesian seas;
 v. Indonesia's foreign policy of independent and freely-active becomes a strength for Indonesia to enhance its role.

b. Weaknesses in overcoming conflict and ensuring maritime security in order to realise Indonesia's position as a maritime axis
c. Economic interests utilized by foreign and domestic "players" to steal fish and other Indonesian marine resources;
 i. The weakness of government diplomacy becomes one factor of weakness;
 ii. The misunderstanding in marine management between central and local government as well as mis-coordination among related administrators;
 iii. Many overlapping regulations about marine natural resource management and maritime security;
 iv. The lack of socialisation for the coastal and local communities about government's policy of maritime.

3.2 *External environment analysis*

Data of related with opportunity and threat that comes from the outside in. Related with External Environment Analysis which according to its urgency is as follows:

a. Opportunities for Indonesia to strengthen its role in order to overcome conflicts and ensure maritime security in order to realise Indonesia's position as a maritime axis

i. In 2015 marked the beginning of the ASEAN Economic Community (AEC) and this was an opportunity;

ii. The conflicts in the South China Sea and the Strait of Malacca are becoming an opportunity;

iii. The joint security cooperation between Indonesia, Malaysia and Singapore in the Strait of Malacca and South China Sea has become an inter-state cooperation that should be developed in the effort for conflict resolution and maritime security enhancement in the process of realising Indonesia's position as a maritime axis;

iv. The ASEAN Regional Forum (ARF) can be optimised by ASEAN countries for common goals in the effort of conflict resolution and maritime security improvement in the process of realising Indonesia's position as a maritime axis.

b. Challenges/Threats in conflict resolution and maritime security in realising Indonesia's position as a maritime axis

i. The high frequency of illegal fishing by foreign vessels in the Indonesian seas is a challenge;

ii. The regional political tension and China's plan to build a military base in an area (the Spratly Islands) disputed over by several countries could disturb the effort of conflict resolution and stability in Southeast Asia;

iii. Many treaties about state boundaries, especially the unclear territorial sea border, can disturb the effort of Indonesia;

iv. The different perspectives of each country when it comes to securing its own national interests could halt efforts of conflict resolution and maritime security.

According to the results of the SWOT analysis shown in Figure 1, the best possible strategy in comprehensive action of trans-state institutional model for conflict resolution and maritime security is the strategy of "consolidation" in quadrant IV. This strategy requires an effort to utilise and optimise the opportunities that could minimise the weaknesses Indonesia have.

According to the 1982 UN Convention on the Law of the Sea (UNCLOS), there are three Indonesian archipelagic sea lanes (*aluv laut kepulauan Indonesia, ALKI*) that function as shipping lanes and routes for ships or international flights (UNCLOS, 1982). These three lanes see 45% of total world trade value pass through them, or approximately 1,500 US dollars.

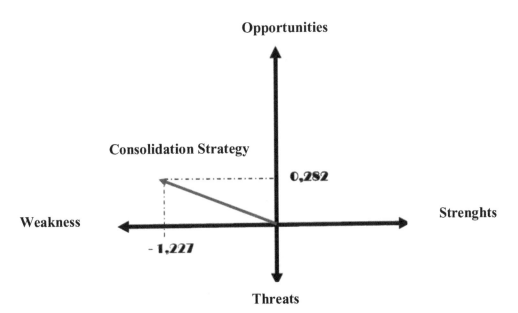

Figure 1. Graph of positioning of trans-state institutional model for conflict resolution and maritime security based on SWOT analysis.

The recognition of Indonesia as an archipelagic country is a significant achievement, following a long process of struggle, because of the many interests of other countries, especially those major maritime countries with a strong desire to maintain their hegemony. UNCLOS 1982 was ratified by the Indonesian government into Law No. 17 in 1985.

Even though the government of Indonesia has a long history of carrying the concept of an "archipelagic country", it still faces complex problems over sea territories that are caused by different visions in developing the state as an archipelagic country. This is evident in the conflicts of interest between state institutions in taking their "domestics" separately to develop marines' affairs. This also happens because every institution has authority and legitimacy emanating from various overlapping laws (Wahyudin & Mahifal, 2013, p. 3).

4 CONCLUSION

A maritime axis can be understood as three things. First, a maritime axis can be seen as a vision or ideal of Indonesia that it wants to build. In this context, the idea of a maritime axis is an opportunity for Indonesia to regain its national identity as an archipelagic country which has unity, maritime power, prosperity and dignity. Second, a maritime axis is seen as a doctrine which gives guidance for a sense of a common purpose. As a doctrine, Joko Widodo invited all Indonesian elements to see maritime axis as "the strength between two oceans". Third, the idea of a maritime axis is also an important part of the national development agenda (Speech of President Joko Widodo in Beijing, 2014).

Even though the government still faces complex problems at sea territories that are caused by different visions in developing this state as an archipelagic country. In addition, the conflict also happened with the neighbouring countries, which have sea border areas with Indonesia, and also have their own interests. Hence, based on the analysis of this research, the best possible strategy in comprehensive action of trans-state institutional model for conflict resolution and maritime security is the strategy of "consolidation" on quadrant IV. That strategy requires an effort to utilise and optimise the opportunities that could minimise the weaknesses Indonesia have.

ACKNOWLEDGEMENT

Thanks are due to Directorate of Research and Community Engagement (*Direktorat Riset dan Pengabdian Masyarakat*/DRPM) Universitas Indonesia, which has supported this 2015 multidisciplinary research.

REFERENCES

Abdulrazaq, Abdulkadir, Sharifah Zubaidah Syed Abdul Kader. 2013. Port Security Vs Economic Gain: An Exposition of the Malaysian Port and Maritime Security Practice. IIUM Law Journal. 21(2).

Brotosusilo, A. (1986). *Masyarakat dan Kekuasaan (Society and Power)*. Jakarta, Indonesia: Rajawali Press.

Brotosusilo, A. . (1997). . Paper for Seminar on Prevention of Social-Unrest, Javanesse Christian Chruch, Depok*Hubungan Antar Kelompok Sosial di Indonesia: Antisipasi Masa Mendatang (Inter-Groups Relation in Indonesia: the Future)*. Depok, DRPM UI (prosiding internasional).

Crouch, H. (1986). *Militer dan Politik di Indonesia (Military and Politics in Indonesia)*. Jakarta, Indonesia,: Sinar Harapan.

Hasenclever, A., Mayer P. & Rittberger V. (2000). Integrating theories of international regimes. *Review of International Studies, 26* (1), 3–33.

Haushofer, K. (1934). Atemweite, Lebensraum und Gleichberechtigung auf Erden. *Zeitschrift für Geopolitik, 11*(1), 1–14.

Huntington, S.P. (1957). *The Soldier soldier and the Statestate: The Theory theory and Politics politics of Civilcivil-Military military Rrelations*. New York, NY,: Free University Press.

Joko Widodo Speech in APEC. (2014). https://www.google.co.id/url?sa=t&rct=j&q=&esrc=s&source=video&cd=1&cad=rja&uact=8&ved=0ahUKEwiwx-zOpMjWAhXDKo8KHbg8CuEQtwIIKjAA&url=https%3 A%2F%2Fwww.youtube.com%2Fwatch%3Fv%3DLo2 jx_IFAoU&usg=AFQjCNEiq66KV_VVAcOkvBXsy3c8q_KURQ.

Keliat, M. (2009). Keamanan Maritim dan Implikasi Kebijakan bagi Indonesia (Maritime Security and Policy Implications for Indonesia). *Jurnal Ilmu Sosial Ilmu Politik*, *13* (1), 111–129.

Kusumaatmadja, M. & Agoes, E.R. (2003). *Pengantar Hukum Internasional (Introduction to International Law)*. Jakarta, Indonesia: Alumni.

Perwita, A.A.B. (2004). Sekuriti Isu Maritim: Koordinasi Nasional dan Kerangka Kerja Sama Maritim Regional di Asia Tenggara (Security of Maritime Issues: National Coordination and Framework for Regional Maritime Cooperation in Southeast Asia). *Jurnal Global*, *7* (1), 35–47.

Pujayanti, A. (2011). *Budaya Maritim Geopolitik dan Tantangan Keamanan Indonesia (Maritime Culture of Geopolitics and Indonesia's Security Challenge)*. Pusat Pengkajian Pengolahan Data dan Informasi Sekretariat Jenderal DPR RI.

Ratzel, F. (1901). *Der lebensraum: Eine biogeographische studie*. H. Laupp.

Sundhaussen, U. (1986). *Politik Militer Indonesia, 1945-1967: Menuju Dwi Fungsi ABRI (Indonesian Military Politics, 1945-1967: Towards a dual function of ABRI)*. Jakarta, Indonesia,: LP3ES.

Victor, M.S. (2014). Indonesia Menuju Poros Maritim Dunia (*Indonesia Towards World Maritime Axis*). *Jurnal Hubungan Internasional.* Vol. VI, No. 21/I/P3DI/November/2014. Pusat Pengkajian Pengolahan Data dan Informasi Sekretariat Jenderal DPR RI.

Wahyudin, Yudi and Mahifal. (2013). Strategi Pembangunan Negara Kepulauan (*Strategic Development for Archipelago State*). *Wawasan Tridharma: Majalah Ilmiah Kopertis Wilayah IV, No.6 Tahun XXV Januari 2013*.

Competition and Cooperation in Social and Political Sciences – Adi & Achwan (Eds)
© 2018 Taylor & Francis Group, London, ISBN 978-1-138-62676-8

Waste management and waste minimization study in manufacturing (analysis of human involvement in the implementation of waste minimization in the auto component industry)

L. Handayani
School of Environmental Science, Universias Indonesia, Depok, Indonesia

S.S. Moersidik
Environmental Engineering Program, Department of Civil Engineering, Faculty of Engineering, Universitas Indonesia, Depok, Indonesia

ABSTRACT: To achieve sustainable manufacturing, industries would have to prevent waste generation, reduce material, energy, and pollution through waste minimization hierarchy. The implementation of waste minimization without considering the social aspect tends to be ineffective, as the successful environmental performance of industry was influenced by the role of workers. Auto component industries have conducted waste management but a holistic review of environment impact is needed to meet sustainable manufacturing requirements. The objectives of this research are i) identification of waste management performance in auto component industry iii) identification of waste minimization implementation ii) understanding the implementation waste minimization through human resource involvement. This study suggests that in its production process, auto component industry generates wastewater, hazardous waste and air pollution. Waste minimization implementation has been conducted with the approach in raw material management, testing process, technology modification, and good housekeeping. Here, the role of human resources in implementation of waste minimization is essential for success especially good knowledge of employee and motivation in cost reduction.

1 INTRODUCTION

One of the Sustainable Development Goals is to promote inclusive and sustainable industrialization. The industry which has an enormous influence on the environment is the automotive industries. Industries develop an environmental management strategy to manage its environment needs proactively (Bangwal & Tiwari, 2015). Environmental management in industry requires interaction between human resources management and environmental management (Paille *et al.*, 2013).

To meet sustainable industry requirements, waste minimization implementation is required as an important element in sustainable development. According to the United Nations Environmental Programme (2014) waste minimization is a strategy that is aimed to prevent waste at its source through upstream intervention. Prevention and waste reduction at the source is implemented through efficiency of raw materials usage, energy and other natural resources consumption related to production process.

Automotive industry manages their waste by approaching end of process stage known as "end of pipe." This approach tends to ignore various kinds of environmental problems generated from the operational process of the overall industry and was not effectively implemented (Hasibuan, 2000). On the other hand, some industries are more focused on the dimensions of delivery, quality, and costs in business aspects (Sutherland *et al.*, 2004). To implement waste minimization in industry, holistic assessment is needed. Waste management without involving social aspects tend to be ineffective, making it necessary to study waste minimization with consideration of human resources. Sustainable manufacturing practice in Malaysian automotive industry, for example, is well developed due to the involvement of social responsibility in its business aspect (Habidin *et al.*, 2015).

It has been suggested that engaging employees in addressing environmental concerns is one of the most significant challenges faced by organizations today and in the future (Frank *et al.*, 2004). The attitudes and behaviors of individuals involved in this industry influence its growth and performance (Begum *et al.*, 2009). Geffen & Rothenberg (2000) states that the role of industrial workers is an important part in achieving improved environmental performance. There are many factors that require good environmental governance in industry such as hierarchy, harmonization by fulfilled legal requirement, employee participation, accountability in preventing environmental pollution, and effectiveness of legal regulation (Nistor, 2006). Some drivers in waste management are human drivers, economic drivers, environmental drivers, and institutional drivers (Agamuthu *et al.*, 2007).

Research on waste minimization has been conducted by Zakarya (2004) in the pharmaceutical industry which leads to the conclusion that waste minimization efforts can be made directly to the reduction of waste at source, waste recovery, modification of raw materials, process modifications, and water savings. Minimization efforts indirectly through counseling, education and training of all human resources involved in the industry. Waste minimization in the paint industry is done by good housekeeping, reuse of used washing water, as well as separating contaminated liquid waste from uncontaminated waste (Hernadewita *et al.*, 2007). Waste reduction from sources is closely related with material and energy efficiency, which will therefore benefit from data for qualitative evaluation. Calculation of waste minimization is a process that is complex and difficult to do (Zorpas & Lasaridi, 2013).

The concept of waste minimization looks easy, but in practice it is relatively difficult due to factors which are not directly related, such as government regulation, quality human resources from various disciplines and strong commitment from top management (Panggabean, 2000). Lee & Paik (2001) state that the availability of technology and facility for waste minimization can affect the social behavior as well as pro-environment attitudes. Nevertheless, waste management behavior might have been motivated to avoid the cost for waste disposal, not due to the pro-environment attitude. Assem & Karima (2011) reveal that indirect causes of waste were found to be the lack of legal and contractual incentives. Saunders *et al.* (2002) note that the ability of operative contribution to a waste reduction in industry as well as an offered management support contribute in waste minimization in the industry. In industrial activities, the implementation of waste minimization activities has not been effective due to some barriers. The absence of appropriate policy, lack of awareness and information, inadequate guidelines, limited time, lack of cooperation and application of old technologies such as end of pipe approaches are the common obstructive factors towards the practice of efficient waste minimization by industries (Doniec, 1995; Mallak *et al.*, 2014).

Research on waste minimization in Indonesia has been done by Zakarya (2004) in the pharmaceutical industry which lead to the conclusion that waste minimization efforts can be made directly to the reduction of waste at source, waste recovery, modification of raw materials, process modifications, and water saving. Indirect waste minimization efforts include counseling, education and training of all human resources involved in the industry. In addition to this, waste minimization program in the industry are also dependent on regulatory, behavioral and industrial strategy of awareness of the benefits of the waste minimization program. Khuriyati *et al.* (2015) conducted a research in small scale industry state that implement good housekeeping by improving the worker thoroughness in the production process and cleaning the scrap on the production equipment before washing to improve the environmental performance in terms of biological oxygen demand (BOD) and chemical oxygen demand (COD).

Objectives of this research are i) identification waste management performance in automotive component industry iii) identification waste minimization implementation, and ii) understanding implementation waste minimization through human resource involvement.

2 METHODOLOGY OR EXPERIMENT

Research was carried out at an automotive component industry located in Cibitung, Bekasi, West Java Province Indonesia. The study was conducted from July 2015 to October 2015. Identification of waste management and waste minimization implementation was conducted through observation in auto component industry process.

A structured questionnaire was used to collect the data from industry workers. This was used to collect the information of their knowledge in waste minimization, what industry do in relation to waste minimization, implementation and motivation of industry in waste minimization. A total of 39 responses were used for data analysis. On the other hand, qualitative methods were used to obtain information from 6 interviewees in management level. Semi in-depth interview was developed to find out current environmental management practices as well as their background of attitude and motivation.

The results of the questionnaire calculation with certain criteria will be presented in graphs spider webs. For the assessment of the overall analysis of knowledge workers, the attitude of the management, and motivation in the implementation of waste minimization will be rated a score from 0 percent (poor) to 100 percent (good). The reference of scoring refers to research conducted by Jafari (2015) in Qualitative analysis with strategic approach by spider web technique.

3 RESULT AND DISCUSSION

3.1 Waste management performance

3.1.1 Wastewater
Waste generated is the liquid waste that comes from the process of painting. The results showed that the management of liquid waste produces peak discharge of 37.6 m³ per month. Wastewater discharge was not influenced by the total production per unit product due to water consumption for wet spray booth (to catch fume from painting sprays). It was not controlled based on necessity.

There was a parameter exceeding quality standard such as BOD at the level 286 mg/L and COD at 853 mg/L which exceeds quality standard for BOD of 150 mg/L and COD of 300 mg/L. High value of BOD and COD in the painting process was caused by accumulation of chemical contents such as solvent and paint in wet spray booth where cleaning activity was done regularly.

3.1.2 Solid waste
Solid waste in auto component industry consists of general waste (including packaging waste), steel scrap from material, and hazardous waste. Waste management for general waste and scrap from material is conducted in offsite industry area. Auto component industry generates hazardous waste from production process then hands it over to waste management industry. Hazardous waste is converted as raw material alternative in cement industry. Types of hazardous waste in auto component industry are scale quenching, shot peen ash, painting sludge, saw contaminated, and wastewater treatment sludge. Total of hazardous waste volume a year in a row are 69,840; 10,800 kg; 44.950 kg; 475 kg; 12.800 kg. The characteristic of hazardous waste that meet raw material standard in cement industry can be seen in Table 1.

3.1.3 Air emission
Emission in auto component industry originated from chimneys or furnaces in production process. Air emission measurement was conducted during the period of July 2014-June 2015 where the measurement frequency was conducted every six months. The result of air measurement parameter that was generated from production process can be seen in Table 2.

Table 1. Characteristic of hazardous waste in auto component industry.

Parameter	Unit	Scale quenching	Shot peen ash	Painting sludge	Saw contaminated	WWTP sludge	Standard*
Chromium (Cr)	ppm	2743,58	4384,67	341,79	782,87	1205	1500
Arsen (As)	ppm	0,073	1,35	0,073	0,073	48,3	200
Cadmium (Cd)	ppm	37,24	57,36	0,002	9,86	5,2	70
Mercury (Hg)	ppm	0,0001	0,0001	0,0001	0,62	0,0001	5
Lead (Pb)	ppm	0,73	28,64	57,08	191,96	98,7	1000
Nickel (Ni)	ppm	70,75	56,71	196,03	96,33	172	1000
Cobalt (Co)	ppm	53,16	91,46	193,51	30,9	33	200
Copper (Cu)	ppm	0,013	431,94	11,89	61,22	59	1000
Zinc (Zn)	ppm	53,72	236,14	397	2854,7	162	5000

* Standard refers to Decision of Ministry of Environment and Forestry of Republic Indonesia No. SK 478 / 2015 about Hazardous Waste Management and Recycle as Raw Material in Cement Industry, PT. Holcim. Source: Hazardous Waste Characteristic Data, 2015.

Table 2. Characteristic of Air Emission in Auto Component Industry.

Parameter	Discharge (kg/year)
Particle	5.012,465
SO_2	2.196,992
NO_2	479,439

Source: Auto component Industry Air Emission Discharge Data in 2015

Table 3. Waste minimization implementation in auto component industry.

Waste type	Management method				
	Raw material management	Setting process condition	Technology modification	Good housekeeping	Offsite treatment
General Waste including packaging materials	o	o	o	o	√
Steel scrap, material scrap	√	o	√	o	√
Saw Contaminated	o	o	o	√	√
Hazardous waste painting sludge	o	√	o	o	√
Scale Quenching	o	√	o	o	√
Shot peen Ash	o	√	o	o	√
WWTP Sludge	o	o	o	o	√

Source: Research Analysis, 2015

3.2 Waste minimization implementation

Waste reduction from source has been implemented on production process with various approaches; waste minimization implementation in auto component industry can be seen in Table 3.

Based on Table 3, auto component industry has implemented waste minimization with the approach of raw material management, setting process condition, technology modification, good housekeeping, and offsite treatment.

3.2.1 *Raw material management*

Material inventory is conducted by tracking the quantity of raw material that flows into production line. Material inventory is constantly carried out every day to check out of the raw material, then do the recording. To see availability of stock, physical check is conducted regularly.

3.2.2 *Setting of process condition*

Operation process has to avoid leaking of oil quenching and chemical contained such as chemical contained painting and shotstell to minimize shotpeen ash. Monitoring of production equipment, especially for machines and furnace, is carried out periodically in order to control the optimization of process.

3.2.3 *Technology modification*

Efforts in terms of technology for waste reduction consist of modification molding in the material cutting machine, and other machines (punch holes, roll eyes, taper roll) to improve efficiency and reduce scrap.

3.2.4 *Good housekeeping*

Housekeeping is one of the efforts in production process to maintain the cleanliness of the environment in production area by preventing spills of raw materials, oil, and chemical content.

3.2.5 *Off-site treatment*

General waste management includes packaging material managed by local communities around industrial area. For scrap metal, off-site treatment is set by collecting on-site wastes to a waste contractor who then transport them off-site for off-site recycling. Hazardous waste such as contaminated saw, painting sludge, scale quenching, shot peen ash, and WWTP sludge after collected on-site and stored in a segregated area based on their characteristic. This then get transported off-site by a licensed regulated waste transporter, to a licensed facility for treatment.

3.3 *Human resources involvement in waste minimization*

This study analyses of industrial behavior in waste minimization programs by reviewing some elements that was conducted. To implement good waste minimization governance in auto component industry some elements are analyzed such as employee knowledge, auto component industry policy, industry attitude and motivation in waste minimization implementation. Data analysis was conducted by using descriptive statistics techniques.

Assessment of this element based on Agumuthu *et al.* 2007stated that some drivers in waste management are human drivers, economic drivers, environmental drivers, and institutional drivers. Human drivers related with this topic include employee knowledge in reducing waste from its source, reuse, and recycle. Economic drivers here refer to effort of industry in implementation of waste minimization program that is accordance to availability of budget for this program. Environmental driver's implementation here are based on motivation of industry to run waste minimization activity. While institutional drivers here are closely related to policies of industry with regards to the efficiency of its resource uses and energy consumption through waste minimization. Overview of all of these elements can be seen in Figure 1.

3.3.1 *Knowledge*

Data on knowledge of employee regarding waste minimization implementation are obtained from questionnaires. In the questionnaire, three questions were asked related to efforts to reduce waste of resources and align workers conduct with the standardized approach, the approach of the production process and production equipment as well as the approach of the two questions related to the utilization of solid waste and wastewater management. Knowledge workers have the highest scoring of 98.3%.

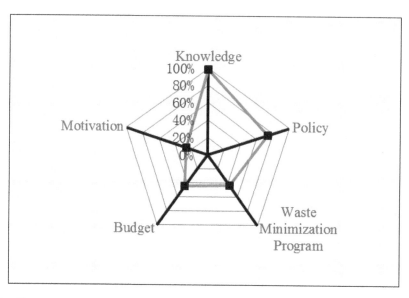

Figure 1.   Human resources involvement in waste minimization (Source: Research Analysis, 2015).

Workers are suggested to have a good knowledge on waste minimization. Although profile of workers, education background and length of work vary, it did not affect the associated knowledge of waste minimization. This is due to the activity of waste reduction that has been applied in everyday production activities. This situation is contrary to the statement of Hakim *et al.* (2014) whereby it was claimed that unpleasant work experience greatly affects knowledge of workers on waste management.

3.3.2 *Attitude of auto component industry in waste minimization*
Assessment of attitude of industry in the application of waste minimization consist of several elements, such as company's policy, waste minimization program, and the allocation of costs in implementing waste minimization program. Based on the answers of respondents, it was obtained that scoring for the company's policy was 74%, the implementation of waste minimization 44%, and the allocation of costs by 45%.

The result suggests that auto component industry has some policy to protect environment and prevent pollution due to production process but it does not specifically state any distinct policy in waste minimization implementation. On the other hand, waste minimization program has not been effective due to non-optimal attention from management. Most of the industry's attention is more focused on development of projects for new products based on consumer demands. Through interviews with the management, the data that was obtained shows how problems were found in the implementation of waste minimization activities were related to cost, time, labor constraints the effort to conduct improvement ideas, following lack of support from the top management. This result is in line with Zorpas & Lasaridi's (2013) statement that calculation of waste minimization is a complex process due to the needs of quality and multidiscipline human resources and a requirement for strong commitment from top management.

This is consistent with studies that have been done in the implementation of waste minimization in various sectors of the industry that found a similar problem such as lack of awareness (awareness), not knowing how and information, there's no time, there are no available technology, and economic factors (Doniec, 1995; Mallak *et al.,* 2014).

3.3.3 *Motivation*
Data for motivation of companies in implementing waste minimization in this study was obtained through the perception of workers. Based on respondents' answers, it can be seen

that the motivation of waste reduction in the protection of the environment is in the final sequence with a percentage of 26.7%. Most of the motivation of Auto Component Industry in waste reduction lies in cost savings. Waste management efforts based on cost savings are also in line with the idea of Lee & Paik (2011) in which the waste management activities have the background of the motivation to avoid the cost of elimination (disposal) of waste. In addition, research conducted by Assem & Karima (2011) on some industrial contractors in the United Arab Emirates shows that the main benefits obtained by the industry in the application of waste minimization is to increase profits while benefiting environmental protection.

4 CONCLUSION

Based on study result, it can be concluded that in the fulfillment of a sustainable industry with the approach of waste minimization, auto component Industry has yet to develop a sustainable industry. There is some reasoning that support this statement:

First, waste minimization implementation through some approaches was ineffective, most of the approach involve types of off-site waste treatment and there were no estimated data volumes of each waste type during operational industry. This process was complex and difficult to do in the auto component industry. Second, waste minimization implementation was not accompanied by top management commitment and because of it met some barriers such as cost and time.

REFERENCES

Agamuthu, P., Fauziah, S.H., Khidzir, K.M. & Aiza, A.N. (2007) Sustainable Waste Management-Asian Perspective. *Proceeding of the International Conference on Sustainable Solid Waste Management, Chennai, India. pp. 15–26.*

Assem, A.H. & Karima, H. (2011) *Material Waste in the UAE Construction Industry Main Causes and Minimisation Practices.* Research Gateway. Heriot-Watt University. doi: 10.1080/17452007.2011.594576.

Bangwal, D. & Tiwari, P. (2015) Green HRM- A way to greening the environment. *IOSR Journal of Business and Management (IOSR-JBM).* 17 (12), 45–53. doi: 10.9790/487X-171214553.

Begum, R.A., Chamhuri, S. & Pereira, J.J. (2009). Attitude and behavior factors in waste management in the construction industry of Malaysia. *Journal of Resources, Conservation, and Recycling.* 53 (6), 321–328. doi: 10.1016/j.resconrec.2009.01.005.

Doniec, A. (1995) *Obstacles in Application of Cleaner Production in the Polish Industry.* Faculty of Process and Environment Engineering. Technical University of Lodz. Poland.

Frank, F.D., Finnegan, R.P., Taylor, C.R. & Keepers, T. (2004) The Race for Talent: Retaining and Engaging Workers in the 21st Century. Human Resource Planning. 27.3, 12–25. Human Resource Planning Society.

Geffen, C.A. & Ronthenberg, S. (2000) Suppliers and Environmental Innovation: The Automotive Paint Process. *International Journal of Operations and Production Management. Proceedings of the Twelfth Annual Conference of the Production and Operations Management Society, POM-2001, March 30 April 2, 2001, Orlando Fl.* 20 (20), pp. 166–186.

Habidin, N.F., Zubir, A.F.M., Fuzi, N.M., Latip, N.A.M, & Azman, N.M.A. (2015) Sustainable manufacturing practices in Malaysian automotive industry: confirmation factor analysis. *Journal of Global Entepreneurship Research.* 5, 14.

Hakim, S.A., Mohsen, A. & Bakr, I. (2014) Knowledge, attitudes and practices of health-care personnel towards wase disposal management at Ain Shams University Hospital, Cairo. *Eastern Mediteraanean Health Journal.* 20 (5), 347–354.

Hasibuan, S. (2000). Karakteristik dukungan industri terhadap upaya implementasi produksi bersih. *Jurnal Teknologi Lingkungan.* 1 (1), 54–62.

Hernadewita, Rahman, M.N.A. & Deros, B.M. (2007). Penanganan limbah industri cat ditinjau dari sisi *clean technology* dalam manajemen industri. *Jurnal Teknik Mesin.* 4 (2), 108–113.

Jafari, M. (2015) Qualitative analysis of the status of twenty-one religious, cultural, and social criteria with strategic approach by spider web technique. *International Journal of Academic Resarch in Business and Social Sciences.* 5 (7), 140–155. doi: 10. 6007/IJARBSS/v5-i6/1705.

Khuriyati, N., Wagiman, & Kumalasari, D. (2015). Cleaner Production Strategy for Improving Environmental Performance of Small Scale Cracker Industry. *Agriculture and Agricultural Science Procedia.* 3, 102–107. doi: 10.1016/j.aaspro.2015.01.021.

Lee, S. & Paik, H.S. (2011). Korean household waste management and recycling behavior. *Building and Environment.* 46, 1159–1166. doi: 10.1016/j.buildenv.2010.12.005.

Mallak, SK., Ishak,MB., & Mohamed, AF. (2014). Waste Minimization Benefit and Obstacles for Solid Industrial Waste in Malaysia. *IOSR Journal of Environmental Science, Toxicoogy and Food Technology (IOSR-JESTFT).* 8 (2), 43–52.

Nistor, L. (2006) How an environmental good governance should look like? The impact of EU on Romania's environmental transition. Faculty of Sociology and Social Assistance. Babes-Bolyai University. Romania.

Paille, P., Chen, Y., Boiral, O., & Jin, J. (2003) The impact of human resources management performance: An employee-level study. *Springer Science and Business Media.* 121 (3), 451–466 doi: 10.1007/s10551-031-1732-0.

Panggabean, S.M. (2000) Minimisasi Limbah pada Pusat Pengembangan Pengelolaan Limbah Radioaktif. *Buletin Limbah.* 5 (1), 1–6.

Saunder, J., Construction, H. & Wynn, P. (2002) *Attitudes towards waste minimization amongs labour only sub-contractor.* Department of Built Environment, Anglia Polytechnic University, Chelmsford, UK.

Sutherland, J., Gunter, K., Allen, D., Bauer, D., Bras, B., Gutowski, T., Murphy, C., Piwonka, T., Sheng, P., Thurston, D., & Wolff, E. (2004) A global perspective on the environmental challenges facing the automotive industry: state of the art and direction for the future. *International Journal Vehicle Design.* 35 (1–2), 86–110.

United Nation Environmental Programme. (2014) *Waste Minimization.* Divison of Technology, Industry and Economics.

Zakarya, E.N. (2004) Minimasi Limbah Pada Industri Farmasi. Program Studi Ilmu Lingkungan. Program Pascasarjana. Universitas Indonesia. Jakarta.

Zorpas, A.A. & Lasaridi, K. (2013) Measuring waste preventive. *Journal Waste Management.* 33, 1047–1056. http://dx.doi.org/10.1016/j.wasman.2012.12.017.

Author index